Metalloporphyrins in Catalytic Oxidations

Metalloporphyrins in Catalytic Oxidations

edited by
Roger A. Sheldon
Delft University of Technology
Delft, The Netherlands

Marcel Dekker, Inc. New York • Basel • Hong Kong

Library of Congress Cataloging-in-Publication Data

Metalloporphyrins in catalytic oxidations / edited by Roger A. Sheldon.
p. cm.
Includes bibliographical references and index.
ISBN 0-8247-9228-9 (acid-free paper)
1. Porphyrin and porphyrin compounds. 2. Catalysis. 3. Oxidation. I. Sheldon, Roger A.
QD401.M578 1994
547'.593—dc20 94-11102
CIP

This book is printed on acid-free paper.

MARCEL DEKKER, INC.
270 Madison Avenue, New York, New York 10016

Current printing (last digit):
10 9 8 7 6 5 4 3 2 1

PRINTED IN THE UNITED STATES OF AMERICA

Preface

Selective catalytic oxidations mediated by iron porphyrin–containing enzymes form the basis of a great many biosynthetic and biodegradative processes in vivo. The ubiquitous cytochrome P450–dependent monooxygenases, for example, play key roles in the biosynthesis of prostaglandins and steroid hormones and in the detoxification of foreign substances (xenobiotics) in the body, such as hydrocarbons, pesticides, and drugs. The reaction pathways involve a wide variety of synthetically interesting transformations, e.g., alkene epoxidation and the hydroxylation of alkanes and aromatic hydrocarbons, using molecular oxygen (dioxygen) as the oxidant. Many of these transformations are characterized by a high degree of chemo-, regio-, and enantioselectivity.

Catalytic oxidation is also the single most important technology for the conversion of petroleum hydrocarbon feedstocks (alkenes, alkanes, and aromatics) to commodity chemicals. Moreover, as a result of increasing environmental constraints, catalytic oxidation with "clean" oxidants, such as dioxygen, hydrogen peroxide, and alkyl hydroperoxides, is also becoming more important in fine chemicals manufacture.

In the last two decades, therefore, increasing attention has been focused on the use of model metalloporphyrin complexes, e.g., of iron and manganese, as catalysts for selective oxidations. This interest was stimulated partly by a desire for a better understanding of the intricate mechanisms of biochemical pathways and partly by the expectation that relatively simple model systems could emulate the selective oxidations of hydrocarbons observed in vivo. If the latter was feasible it could provide the basis for industrially attractive processes for both bulk and fine chemicals production.

A major aim of this book is to provide an overview of this multifaceted topic. It is intended for both academic and industrial researchers involved with synthetic and/or mechanistic aspects of chemical and biochemical oxidation processes. The book comprises 11 chapters spanning the different aspects of the subject, all written by internationally regarded experts in the field.

The first chapter reviews the long and colorful history of metalloporphyrin chemistry, an area of endeavor that has produced several Nobel Prizes. The second and third chapters deal with mechanistic and theoretical aspects of chemistry involving high-valent oxometalloporphyrin complexes, the putative intermediates in metalloporphyrin-mediated oxidations.

Chapters 4 and 5 review the chemistry of model metalloporphyrin systems related to cytochrome P450 and ligninase, respectively. The latter plays a key role in the biodegradation of lignocellulose. Chapter 6 elaborates the intricate mechanisms of hemoprotein-mediated oxidations involved in the biosynthesis of prostaglandins.

A recent trend in metalloporphyrin research is toward the use of perhalogenated and related "unnatural" porphyrins, which combine high activity with high chemical stability. In order for them to be practically useful, it is essential that efficient methods for their synthesis be available. Hence, synthetic methods are reviewed in Chapter 7.

Chapter 8 deals with asymmetric oxidations catalyzed by chiral metalloporphyrins, an area of increasing interest in connection with the enantioselective synthesis of optically pure drugs and agrochemicals.

Chapter 9 concerns metalloporphyrin-catalyzed autoxidations of hydrocarbons, i.e., using dioxygen as the stoichiometric oxidant. The use of dioxygen brings metalloporphyrin-catalyzed oxidations into the realm of commodity chemicals manufacture. In order to be economically viable, such systems must not only emulate the biocatalysts of Nature but transcend them and become suprabiotic. This interesting concept is further elaborated in Chapter 10, which deals with metalloporphyrin-catalyzed oxidations of light alkanes with dioxygen, via nonclassical pathways and without the requirement for a sacrificial reductant.

Finally, the practical utility of expensive metalloporphyrin catalysts is also dependent on their ease of recovery and recycling. Hence, Chapter 11 deals with the preparation and application of metalloporphyrin catalysts immobilized on solid supports.

The principal literature has been covered up to 1993, and an extensive list of references is included at the end of each chapter. The index has been thoroughly cross-referenced, and the reader is encouraged to seek the various topics under more than one entry.

I express sincere thanks to all my friends and colleagues who have contributed chapters to this book.

Roger A. Sheldon

Contents

Contributors

Pierrette Battioni, Ph.D. Directeur de Recherche C.N.R.S., Laboratoire de Chimie et Biochimie, Université Paris V, Paris, France

Thomas C. Bruice, Ph.D. Professor, Department of Chemistry, University of California, Santa Barbara, Santa Barbara, California

David Dolphin, B.Sc., Ph.D., D.Sc. Professor, Department of Chemistry, University of British Columbia, Vancouver, British Columbia, Canada

Paul E. Ellis, Jr., Ph.D. Senior Staff Scientist, Catalysis Technology Division, Research and Development Department, Sun Company, Inc., Marcus Hook, Pennsylvania

Gong-Xin He, Ph.D. Research Scientist, Department of Organic Chemistry, Gilead Sciences, Inc., Foster City, California

Karl Anker Jørgensen, Ph.D. Professor, Department of Chemistry, Aarhus University, Aarhus, Denmark

John R. Lindsay Smith, B.A.(Oxon), D.Phil.(Oxon) Department of Chemistry, University of York, York, England

James E. Lyons, Ph.D. Senior Principal Scientist, Catalysis Technology Division, Research and Development Department, Sun Company, Inc., Marcus Hook, Pennsylvania

Daniel Mansuy, Ph.D. Directeur de Recherche C.N.R.S., Laboratoire de Chimie et Biochimie, Université Paris V, Paris, France

Bernard Meunier, Ph.D. Laboratoire de Chimie de Coordination, Centre National de la Recherche Scientifique, Toulouse, France

Teresa Mlodnicka, Ph.D., D.Sc. Associate Professor, Institute of Catalysis and Surface Chemistry, Polish Academy of Sciences, Kraków, Poland

Yoshinori Naruta, Ph.D.* Department of Chemistry, Kyoto University, Kyoto, Japan

Dražen Ostović, Ph.D. Research Fellow, Department of Pharmaceutical Research and Development, Merck Research Laboratories, West Point, Pennsylvania

Hans H. Ruf, Ph.D. Biomedical Engineering Department, Fraunhofer Institute, IBMT; and University of Saarland, St. Ingbert, Germany

Roger A. Sheldon, Ph.D. Professor, Laboratory for Organic Chemistry and Catalysis, Delft University of Technology, Delft, The Netherlands

V. Ullrich, Ph.D. Fakultät für Biologie, University of Konstanz, Konstanz, Germany

Tilak P. Wijesekera, Ph.D. Research Scientist, Catalysis Technology Division, Research and Development, Sun Company, Inc., Marcus Hook, Pennsylvania

**Present affiliation*: Associate Professor, Department of Applied Molecular Science, Institute for Molecular Science, Okazaki, Japan

Abbreviations

Br_8TPP	*meso*-tetrakis(2,6-dibromophenyl)porphyrinato
COX	cyclooxygenase
CPCRR	cyclopropyl to homoallyl radical rearrangement
CPO	chloroperoxidase
Cl_8TPP	*meso*-tetrakis(2,6-dichlorophenyl)porphyrinato
$Cl_{20}TPP$	*meso*-tetrakis(pentachlorophenyl)porphyrinato
$Cl_{20}TPP$-β-Cl_8	*meso*-tetrakis(pentachlorophenyl)-β-octachloroporphyrinato
CT complex	charge transfer complex
DDQ	2,3-dichloro-5,6-dicyanobenzoquinone
DDT	1,1-bis(4-chlorophenyl)-2,2,2-trichloroethane
DMF	dimethylformamide
DMSO	dimethylsulfoxide
E,(Z)	Entgegen; designation for *trans*-isomers
EPR = ESR	electron paramagnetic resonance
Et	ethyl
$F_{20}TPP$	*meso*-tetrakis(pentafluorophenyl)porphyrinato
$F_{20}TPP$-β-Br_8	*meso*-tetrakis(pentafluorophenyl)-β-octabromoporphyrinato

$F_{20}TPP$-β-F_8 = $F_{28}TPP$	*meso*-tetrakis(pentafluorophenyl)-β-octafluoro-porphyrinato
FAD	flavine adenine dinucleotide
FADH	flavine adenine dinucleotide reduced
FMN	flavine mononucleotide
Hb	hemoglobin
HHD	12(S)-hydroxy-8(E),10(E)-heptadecadienoic acid
HHT	12(S)-hydroxy-5(Z),8(E),10(E)-heptadecatrienoic acid
HOMO	highest occupied molecular orbital
HRP	horseradish peroxidase
Im	imidazole
kDa	kilodalton
KHT	12-keto-5(Z),8(E),10(E)-heptadecatrienoic acid
LiP	lignin peroxidase
LUMO	lowest unoccupied molecular orbital
MDA	malondialdehyde
Me	methyl
MeNAHN	methyl dihydronicotinamide
$Me_{12}TPP$	*see* TMP
MMP = MMPP	magnesium monoperoxyphthalate
MnP	manganese peroxidase
NAD	nicotinamide adenine dinucleotide
NADH	nicotinamide adenine dinucleotide reduced
NMR	nuclear magnetic resonance
OEP	β-octaethylporphyrinato
$OEP(NO_2)_4$	*meso*-tetranitro-β-octaethylporphyrinato
OMP	β-octamethylporphyrinato
OPP	β-octaphenylporphyrinato
P = por(ph)	general notation for the porphyrinato ligand, i.e., the porphyrin dianion (the free porphyrin is denoted by PH_2)
P450	cytochrome P450
PAA	polyacrylic acid

PGG_2	15(S)-hydroperoxy-9α,11α,epidioxy-5(Z),13(E)-prostadienoic acid
PGH	prostaglandin H synthase
PGH_2	15(S)-hydroxy-9α,11α-epidoioxy-5(Z),13(E)-prostadienoic acid
PGI	prostacyclin synthase
Ph	phenyl
POD	peroxidase
PPIX	protoporphyrin IX
PQQ	2,7,9-tricarboxy-1*H*-pyrrolo[2,3*f*] quinoline-4,5-dione(pyrrolo quinoline quinone)
PS-Im	imidazole-modified polystyrene
PSS	poly(styrene-4-sulfonate)
PVP	poly(4-vinylpyridine)
PVS	poly(vinylsulfonate)
SCE	standard calomel electrode
Si-Im	imidazole-modified silica
TBHP	*tert*-butyl hydroperoxide
TCPP	*meso*-tetrakis(4-carboxyphenyl)porphyrinato
$TCSMCl_8P$	*meso*-tetrakis(2,4,6-trimethyl-3-chloro-5-sulfonatophenyl)porphyrinato
$TDCPBr_8P$	*meso*-tetrakis(2,6-dichlorophenyl)-β-octabromoporphyrinato
$TDCPCl_8P$	*meso*-tetrakis(2,6-dichlorophenyl)-β-octachloroporphyrinato
TDCPP	*see* $Cl_{20}TPP$
TDCSPP = TDCPPS	*meso*-tetrakis(2,6-dichloro-3-sulfonatophenyl)porphyrinato
$TDSMBr_8P$	*meso*-tetrakis(2,4,6-trimethyl-3,5-disulfonatophenyl)-β-octabromoporphyrinato
TDSMPP	*meso*-tetrakis(2,4,6-trimethyl-3,5-disulfonatophenyl)porphyrinato
$TFPBr_8P$	*see* $F_{20}TPP$-β-Br_8
$TFPF_8P$	*see* $F_{20}TPP$-β-F_8
TFPP = TF_5PP	*see* $F_{20}TPP$

$TFPN_4P$	*meso*-tetrakis(pentafluorophenyl)-β-tetranitroporphyrinato
$TFPS_4P$	*meso*-tetrakis(pentafluorophenyl)-β-tetrasulfonatoporphyrinato
THF	tetrahydrofuran
TMP	*meso*-tetrakis(2,4,6-trimethylphenyl)porphyrinato
$TMPBr_8$	*meso*-tetrakis(2,4,6-trimethylphenyl)-β-octabromoporphyrinato
TMPyP = T4MPyP	*meso*-tetrakis(4-N-methylpyridyl)porphyrinato
T2MPyP	*meso*-tetrakis(2-N-methylpyridyl)porphyrinato
TMPS	*meso*-tetrakis(4-sulfonatophenyl)porphyrinato
TPCPP	*see* $Cl_{20}TPP$
TPFPP	*see* TFPP
TPP	*meso*-tetraphenylporphyrinato
$TPPCl_{20}$	*see* $Cl_{20}TPP$
$TPPF_{20}$	*see* $F_{20}TPP$
$TPPF_{28}$	*see* $F_{20}TPP$-β-F_8
$TPPF_{20}$-β-Br_8	*see* $F_{20}TPP$-β-Br_8
TSPP = TPPS	*meso*-tetrakis(4-sulfonatophenyl)porphyrinato
TSMP	*see* TMPS
TTAP	*meso*-tetrakis(4-trimethylammoniumphenyl)porphyrinato
TTP	*meso*-tetrakis(4-methylphenyl)porphyrinato
TXA	thromboxane synthase
Z,(E)	Zusammen; designation for *cis*-isomers

1

Oxidation Catalysis by Metalloporphyrins

A Historical Perspective

Roger A. Sheldon
Delft University of Technology, Delft, The Netherlands

The controlled partial oxidation of hydrocarbons (alkanes, alkenes, and aromatics) by molecular oxygen (dioxygen) is both highly desirable and difficult to achieve [1–3]. The central problem is the same as that encountered in the use of dioxygen in biological systems, namely, how to promote reaction whilst at the same time limiting the damage caused by indiscriminate attack of dioxygen.

I. STRUCTURE AND REACTIVITY OF DIOXYGEN

The complete oxidation of organic materials by dioxygen to carbon dioxide and water is a thermodynamically highly favorable (i.e., exothermic) process. Fortunately for mankind, the kinetic barrier to reaction is large. The ground state of dioxygen is a triplet and consequently its reaction with singlet organic molecules to give singlet products is a spin-forbidden process with a very low rate. One way of circumventing this barrier is via the intermediacy of free radicals. The reaction of a singlet molecule with $^3O_2$ (Reaction 1) to give two doublets (free radicals) is spin-allowed. However, Reaction 1 is highly endothermic for most substrates and occurs readily only with very reactive substrates that form resonance-stabilized radicals, e.g., reduced flavins (Reaction 2). This is a key step in oxidations mediated by flavin-dependent monooxygenases.

$$RH + {}^3O_2 \longrightarrow R\bullet + HO_2\bullet \qquad (1)$$

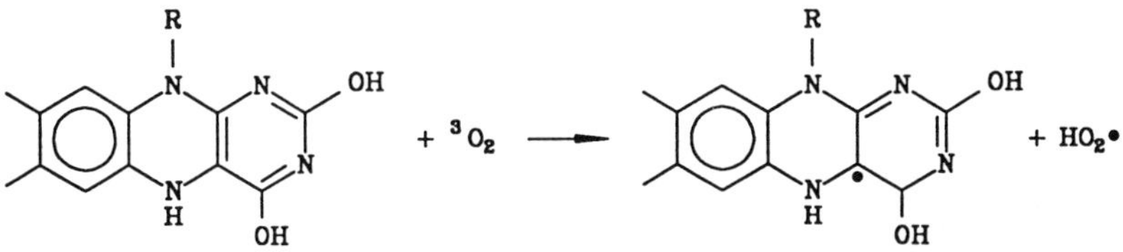 (2)

A second pathway for overcoming the spin conservation barrier is for $^3O_2$ to combine with a paramagnetic transition metal ion. In this way various coordinated oxygen species can be generated (Figure 1) which are capable of selectively oxidizing organic substrates such as hydrocarbons. Nature achieves this by employing metalloenzymes, many of which contain iron porphyrin complexes as the moiety responsible for activating the dioxygen molecule. Interest in oxidations mediated by metalloporphyrins is twofold. On the one hand, it is driven by the quest for a better understanding of vital biological processes and, on the other hand, by the hope that we may succeed in designing model systems capable of emulating these unique oxidation catalysts.

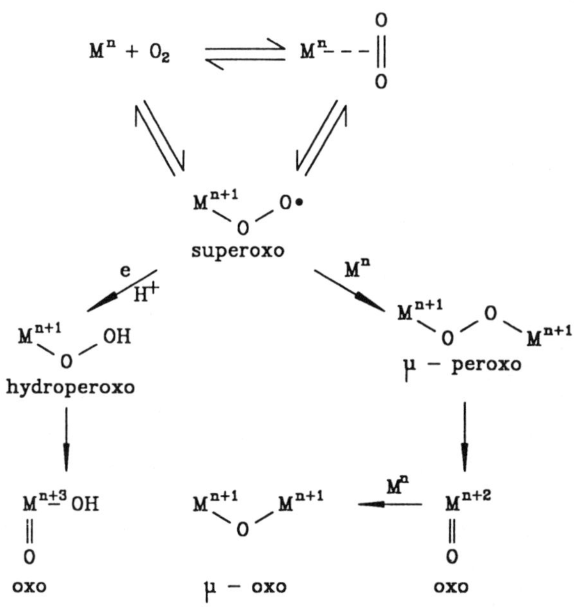

Figure 1 Metal–oxygen species.

II. INTRODUCTION TO HEMOPROTEINS

Macrocyclic tetrapyrrole ligands, exemplified by the porphyrins and the closely related chlorins, play a central role in life processes. Chlorophyll, which mediates the photosynthesis of carbohydrates in green plants, is a magnesium complex of a substituted chlorin. Iron porphyrin complexes,

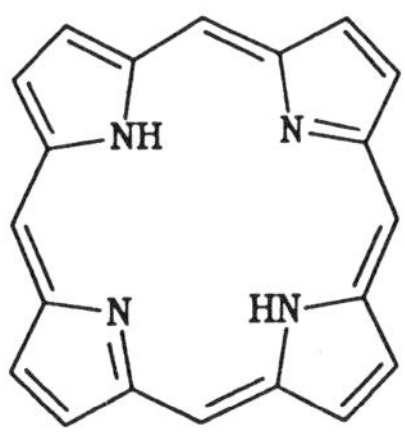

porphyrin

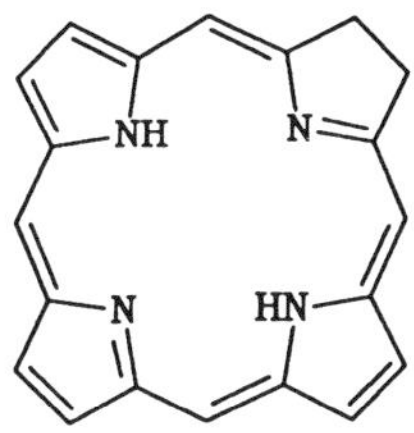

chlorin

on the other hand, constitute the prosthetic group of an important class of proteins and enzymes known collectively as hemoproteins. They control a wide variety of fundamental biological processes involving the complexation of dioxygen. For example, the respiratory proteins hemoglobin and myoglobin are responsible for the transport of oxygen from the lungs to all parts of the organism and the storage of oxygen in muscle tissue, respectively. Structurally related hemoproteins mediate electron transfer (cytochromes b and c), the production of cellular energy by mitochondrial reduction of oxygen to water (cytochrome c oxidase), the biosynthesis of steroids and detoxification of foreign substances (cytochrome P450-dependent monooxygenases), and the destruction or utilization of hydrogen peroxide (catalase and peroxidases, respectively). The prefix heme (or hemo) refers to the iron(II)porphyrin moiety and hemin to the iron(III)porphyrin. Apparently nature chose the iron porphyrin complexes as the most appropriate redox metal–ligand combination for the binding and activation of dioxygen [4].

Hemoproteins differ with regard to the structure of both the heme and the polypeptide component, and the oxidation state of the iron (2+ or 3+). The most widely occurring heme is known as heme b or c or, more commonly, protoheme or protoporphyrin IX iron(II), or simply "heme." Protoporphyrin IX (see Figure 2) is also the progenitor in the biosynthesis of other porphyrin ligands. The iron(III) form is called hemin or protohemin and with chloride as axial ligand chlorohemin or chloroprotoporphyrin(IX)iron(III). Iron(II) has six d electrons and, hence, a four-coordinate iron porphyrin has two vacant coordination sites. When two extra ligands bind to the iron they are constrained by the macrocycle to occupy axial positions, trans with respect

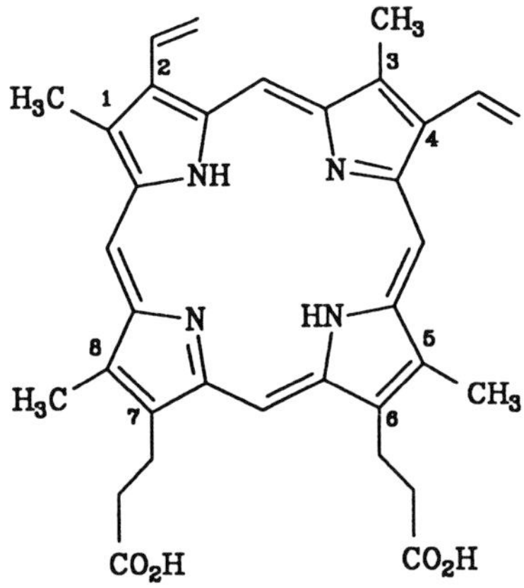

Figure 2 Protoporphyrin IX. Numbers 1–8 are referred to as β positions and α, β, γ, and δ are known as meso positions.

to each other across the porphyrin rings. In a strain-free porphyrin the radius of the porphyrin central cavity is 2.01 Å [4], which is too small to fit a high-spin (S = 2) five-coordinate iron(II). Hence, the high-spin iron(II) lies substantially out of the plane of the ring toward the axial ligand.

In hemoglobin and myoglobin five-coordinate high-spin iron(II) is coordinated to protoporphyrin IX and an imidazole group of a histidine residue in the polypeptide chain occupies an axial position on what is defined as the "proximal" side of the porphyrin ring. The sixth coordination site, defined as the distal site, is free for coordination to dioxygen or other small molecules. Oxygenation leads to the smaller low-spin ($S = 0$) six-coordinate iron complex and the iron moves into the plane of the porphyrin ring. The motion of the iron into the porphyrin ring upon oxygenation is responsible for the changes in the protein conformation that lead to an increased affinity for dioxygen, the so-called cooperative effect [4].

In catalase and horseradish peroxidase a histidine imidazolate forms the axial ligand in a protoporphyrin(IX)iron(III) complex. In chloroperoxidase (EC 1.11.1.10) and cytochrome P450–dependent monooxygenases, in contrast, a cysteine thiolate serves as the proximal ligand in a protoporphyrin(IX)iron(II) complex. By changing the axial ligand from imidazole in hemoglobin and myoglobin to thiolate in cytochrome P450–dependent monooxygenases the function of the hemoprotein changes from reversible binding of dioxygen to dioxygen activation (see below).

III. EARLY HISTORICAL DEVELOPMENT

The early history of metalloporphyrins (see Table 1 for selected landmarks) has been the subject of several reviews [5–8]. The presence of iron in blood was first demonstrated by Menghini in 1747. Lavoisier's explanation of the phenomenon of combustion in 1774 signaled the demise of the phlogiston theory and the advent of the modern era of chemistry. Lavoisier also initiated studies of biological oxidation processes. Based on quantitative studies with humans he concluded that respiration is a slow combustion of various nutrients, such as fats and sugars. Accordingly, he defined oxidation as the addition of oxygen atoms to a substrate and reduction as its removal (Reaction 3).

$$X + {}^{1}/_{2}\, O_2 \underset{\text{reduction}}{\overset{\text{oxidation}}{\rightleftharpoons}} XO \qquad (3)$$

The first allusion to the red pigments we now call porphyrins was by Scherer in 1841. He showed that removal of iron from powdered blood by treatment with sulfuric acid left a substance which afforded a red solution in alcohol. Hoppe-Seyler coined the name hemoglobin in 1864 and in 1871 he isolated porphyrins from blood and showed that they were pyrrole derivatives. He also established (in 1879) the structural similarity between chlorophyll and heme. It took until 1912, however, before Kuster proposed the correct macrocyclic tetrapyrrole structure (which was not accepted) for porphyrins. The first general synthesis of porphyrins was described by Milroy in 1918 [9]. This was followed by the epochal work of Hans Fischer on the synthesis of porphyrins, including octamethylporphyrin (1926) and his crowning achievement, the de novo synthesis of chlorohemin in 1929 [10], thus confirming the tetrapyrrole structure proposed by Kuster 17 years earlier. A year later, in 1930, Fischer was awarded the Nobel Prize for his porphyrin work and the book by Fischer and Orth *Die Chemie des Pyrrols*, published in 1934 [11], became an outstanding classic in the field. The synthesis of *meso*-tetraphenylporphyrin, which was later to form the basis for a plethora of studies with model synthetic porphyrin complexes, was first reported by Rothemund in 1939. In short, by 1940 the structure and biological functions of iron porphyrin complexes were well established. Moreover, it did not take long before the intimate molecular structures of myoglobin [12] and hemoglobin [13] were elucidated by X-ray crystallography, feats which earned Kendrew and Perutz the Nobel Prize in 1962.

As outlined above, the major effort in the first half of this century was focused on the dioxygen-binding proteins, hemoglobin and myoglobin. It is also interesting to follow the development of our understanding of enzymatic

Table 1 Oxidation Catalysis by Metalloporphyrins: Selected Landmarks

1747	Menghini demonstrates the presence of iron in blood
1774	Priestley discovers oxygen
	Lavoisier explains combustion
1777	Lavoisier assigns a role to O_2 in respiration
1841	Scherer isolates the red, iron-free pigment from blood
1864	Hoppe-Seyler names hemoglobin
1871	Hoppe-Seyler isolates porphyrins from blood and demonstrates that they are pyrrole derivatives
1879	Hoppe-Seyler shows structural similarity of chlorophyll and heme
1897	Bertrand coins the term oxidase
1898	Linossier names peroxidase
1901	Loew names catalase
1912	Kuster proposes the correct macrocyclic tetrapyrrole structure for porphyrins (which is not accepted)
1913	Willstatter suggests the wrong structure for porphyrins (which is accepted)
1915	Willstatter awarded Nobel prize for chlorophyll work
1918	Milroy describes the first general synthesis of porphyrins
1925	Keilin names cytochromes and shows hemoprotein structure
1926	Fischer synthesizes octamethylporphyrin
1929	Fischer and Zeile synthesize hemin and confirm the Kuster structure
1930	Fischer awarded Nobel Prize for porphyrin work
1931	Warburg awarded Nobel Prize for respiration work
1934	Publication of *Die Chemie des Pyrrols* by Fischer and Orth
1939	Rothemund synthesizes *meso*-tetraphenylporphyrin (TPP)
1955	Hayaishi and Mason independently discover oxygenases
1960	Kendrew and Perutz report the X-ray structures of myoglobin and hemoglobin, respectively
1962	Kendrew and Perutz awarded Nobel Prize
1963	Estabrook established role of CO-binding pigment in microsomal steroid hydroxylation
1964	Omura and Sato establish that the CO-binding pigment contains protoporphyrin(IX)iron(III) and name it cytochrome P450
1966	Ullrich and Staudinger propose high-valent oxoiron porphyrin intermediate for cytochrome P450
1973	Collman synthesizes "picket fence" iron porphyrins
1975	Baldwin synthesizes "capped" iron porphyrins
1975	Hrycay demonstrates the use of peroxidases and other oxygen donors with cytochrome P450. Ullrich reports the use of iodosylbenzene as oxygen donor
1979	Groves reports first oxygen transfer (peroxide shunt pathway) with model iron porphyrin and iodosylbenzene
	Tabushi describes the use of NaOCl in combination with manganese(III) TPP

Table 1 (Continued)

1980	Meunier and Guilmet report catalytic epoxidation with manganese(III) TPP and NaOCl
1983	Groves reports first in vitro asymmetric epoxidation with a chiral metalloporophyrin
1989	Ellis and Lyons describe low-temperature hydroxylation of light alkanes catalyzed by perhalogenated iron porphyrins

oxidations, particularly those involving oxygen activation by heme-dependent oxidoreductases, during the same period. As early as 1855 Schönbein reported that oxidations of certain organic substrates by dilute hydrogen peroxide are catalyzed by "substances" present in plants and animals. The terms *oxidase, peroxidase,* and *catalase* were introduced around the turn of the century (see Table 1). Subsequently, Bach and Chodat showed that horseradish is a particularly good source of peroxidase and Wilstätter (of chlorophyll fame) succeeded in purifying the enzyme and determining its activity.

Parallel to these developments, observations made in the 19th century linked the deterioration of many organic materials, such as natural oils and fats, to the absorption of dioxygen. Around the turn of the century it was recognized that these processes involved organic peroxide intermediates and subsequent detailed mechanistic studies with simple hydrocarbons led to the free radical chain theory of autoxidation [1].

The intermediacy of iron–dioxygen complexes in enzymatic oxidation processes was first proposed by Warburg in the 1920s. In his classic studies of cellular respiration he demonstrated the heme character of the cellular catalyst which he referred to as *Atmungsferment* or respiratory enzyme. He further reasoned that such enzymatic redox processes involved the interaction of the iron of the heme with dioxygen. Warburg was awarded the Nobel Prize in 1931 for his respiratory studies, a year after Hans Fischer received it for his porphyrin work.

Hence, the scene was already set around 1930 for further elucidation of the role of dioxygen activation by iron porphyrin complexes in many biological oxidations. However, further evolution of these concepts was seriously hampered by the emergence in the late 1920s of the dehydrogenation theory of enzymatic oxidation promulgated by Wieland, a distinguished chemist and Nobel laureate. This theory, which was widely accepted for more than two decades, held that the sole function of dioxygen in enzymatic oxidations was one of electron acceptor, forming water or hydrogen peroxide but never being incorporated in the substrate. This situation changed dramatically in 1955 when Hayaishi [14] and Mason [15] independently demonstrated, by $^{18}O_2$ labeling studies, the direct incorporation of dioxygen into the substrate in Reactions 4 and 5.

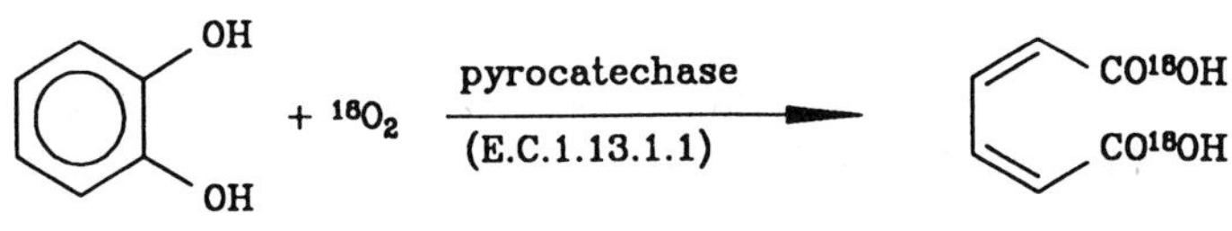

(4)

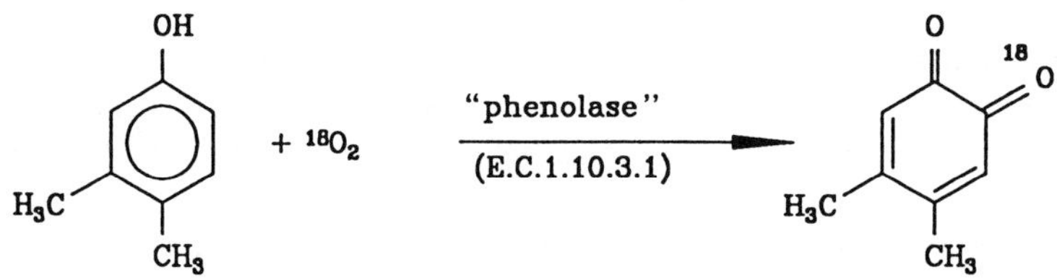

(5)

Hayaishi proposed the generic term oxygenase to describe such enzymes and distinguished two basic types: monooxygenases, which incorporate one oxygen atom of dioxygen into the substrate, and dioxygenases, which incorporate both oxygen atoms (Figure 3). In essence this was a revival of the ideas of Lavoisier, which dated from almost two centuries earlier but had temporarily been buried by Wieland.

IV. CYTOCHROME P450-DEPENDENT MONOOXYGENASES

A. Discovery, Structure, and Function

Although the above-mentioned oxygenases were nonheme enzymes it did not take long before heme-dependent monooxygenases were shown to play

Monooxygenase

external: $X + O_2 + DH_2 \longrightarrow XO + D + H_2O$

internal: $XH_2 + O_2 \longrightarrow XO + H_2O$

Dioxygenase

$X + O_2 \longrightarrow XO_2$

Figure 3 Types of oxygenases.

an important role in many biological oxidation processes. In the late 1950s liver microsomes were shown to contain a previously unnoticed carbon monoxide-binding pigment [16]. It was subsequently shown that this pigment is an essential component of a class of monooxygenases that mediate a variety of important biosynthetic pathways, such as steroid hormone and prostaglandin biosynthesis, and the detoxification of foreign substances in the body including drugs, pesticides, and petroleum hydrocarbons.

Estabrook and coworkers [17] showed that the photochemical action spectrum for reversal of CO inhibition of O_2-dependent microsomal drug and steroid hydroxylation has a maximum at 450 nm. Since the CO difference spectrum of the reduced pigment also has a maximum at 450 nm the metabolic role of this microsomal component was thus established. In 1964 Omura and Sato [18] demonstrated that the CO-binding pigment, which they gave the name P450, was a protoporphyrin IX iron(III)-containing cytochrome which was reduced by both NADH and NADPH, under anaerobic conditions, and rapidly reoxidized by dioxygen. The term cytochrome P450* was subsequently widely adopted to describe this class of monooxygenases. The widespread interest in these enzymes stems from their unique ability to catalyze a variety of regio- and/or stereoselective oxidations, including olefin epoxidation and alkane and arene hydroxylation. One oxygen atom of dioxygen is inserted into the substrate and the other is reduced to water (Reaction 6). The reducing equivalents are provided by the NADH or NADPH cofactor which in vivo is regenerated in a separate step.

$$S + O_2 + NAD(P)H_2 \xrightarrow{\text{cyt-P450}} SO + H_2O + NAD(P) \tag{6}$$

Cytochrome P450-dependent monooxygenases are ubiquitous in nature. They have been isolated from a variety of mammalian tissues, insects, plants, yeasts, and bacteria. The liver contains the highest concentration of mammalian P450 enzymes. The bacterial camphor-hydroxylating enzyme from *Pseudomonas putida* was the first cytochrome P450-dependent monooxygenase to be crystallized and have its structure determined [19]. Its substrate binding site is lined with hydrophobic residues and the camphor is located about 4 Å from the porphyrin plane, directly adjacent to the O_2 binding site. The substrate is protected from the external environment by a reversible hydrophobic lid [19]. The protoporphyrin(IX)iron(III) contains a cysteine thiolate as the proximal ligand. Both the thiolate axial ligand and the hydrophobic binding site appear to be essential for the unique activity and selec-

*A cytochrome is a hemoprotein whose characteristic mode of action is electron and/or proton transport by virtue of a reversible valency change of the iron.

tivity of cytochrome P450. The function of the thiolate may be to facilitate heterolytic cleavage of the O-O bond thereby producing the reactive oxo-iron intermediate (see below).

B. Mechanism of Cytochrome P-450-Mediated Oxidations

In 1954 Udenfriend and coworkers [20] reported that a mixture of Fe(II), EDTA, ascorbic acid, and dioxygen is able to hydroxylate aromatic rings at neutral pH and ambient temperature. The Udenfriend reagent came to be regarded as a model for iron-dependent monooxygenases which it predated by one year. Subsequent studies showed that the ascorbic acid could be replaced by a variety of hydrogen donors [21]. Indeed, one may conclude that in addition to Fe(II) and dioxygen all that is needed is a source of electrons and protons.

In 1964 Hamilton [22] proposed an "oxenoid" mechanism for the hydroxylation of aromatics with Udenfriend's reagent (Figure 4). In hindsight it was only a small step from Hamilton's oxenoid mechanism to the oxometal mechanism that is currently widely accepted for cytochrome P450–dependent monooxygenases. To our knowledge Ullrich and Staudinger [23] were the first (in 1966) to propose such a mechanism, in which a formally oxoiron(V)–protoporphyrin IX complex is the active oxidant. Nowadays it is generally accepted (see Chapter 2) that the porphyrin ligand donates one electron to the iron to give an oxoiron(IV)porphyrin radical cation as the actual intermediate. Moreover, electron density may also be transferred from the thiolate axial ligand to the iron. In the extreme case this would give a complex which

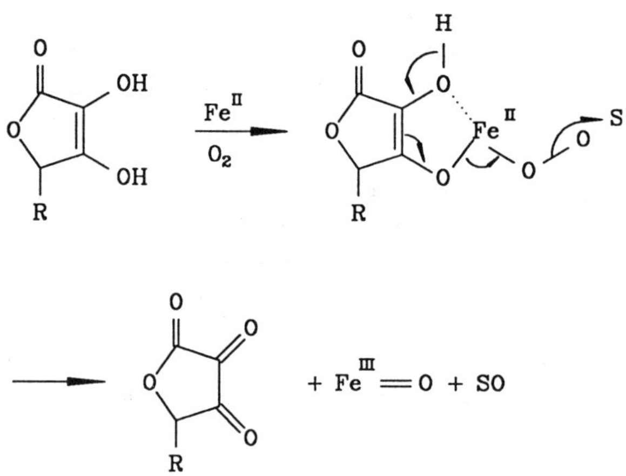

Figure 4 The "oxenoid" mechanism.

may be regarded as [$RS\cdot Fe^{III}$-$O^{-}P^{+}$]. Whatever its precise electronic structure this high-valent oxoiron species is a powerful oxidant which transfers its oxygen to an amazing variety of substrates. At the same time it does not undergo reaction with the protein ligand. This is assumed to result from the fact that the substrate binds to the (hydrophobic) active site prior to activation of the dioxygen. Once it is bound the substrate is ideally placed to react with the oxoiron species as it forms. The catalytic cycle for cytochrome P450–mediated oxidations is illustrated in Figure 5 [16,19]. Initial one-electron reduction of the native iron(III) cytochrome is followed by binding of dioxygen. A second one-electron reduction is postulated to give a (hydro)peroxoiron(III) species which heterolytically cleaves to afford the oxoiron(IV)porphyrin radical cation, which is the active oxidant.

$$S + O_2 + DH_2 \longrightarrow SO + D + H_2O$$

S = substrate DH_2 = NAD(P)H

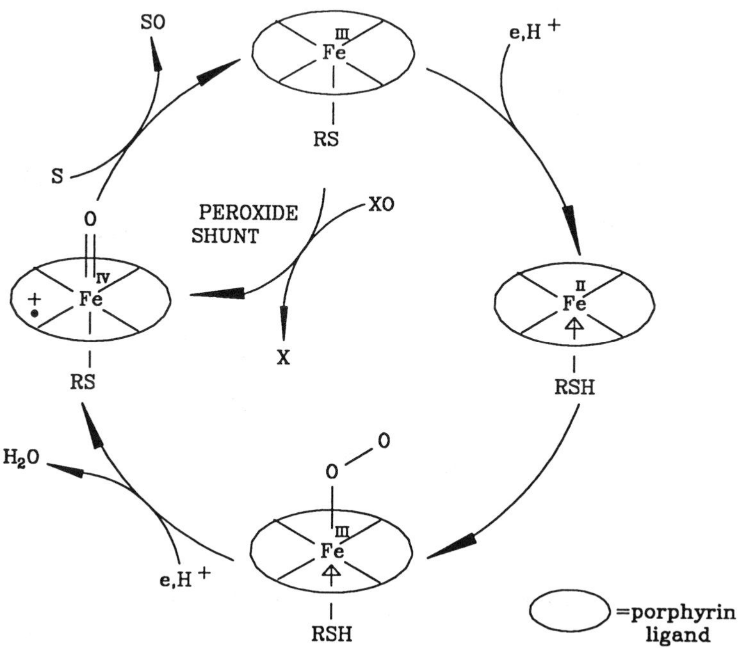

Figure 5 Mechanism of cytochrome P450–catalyzed oxidations.

C. The Peroxide Shunt Pathway

In the mid-1970s various groups demonstrated [24–27] that liver microsomal cytochrome P450 can catalyze the hydroxylation of hydrocarbon substrates by a variety of single-oxygen donors, e.g., H_2O_2, RO_2H, chlorite, periodate, and, particularly, the lipophilic iodosylbenzene (27). This pathway, which later became known as the "peroxide shunt" (see Figure 5), provided a means for circumventing the need for a coreductant (cofactor) in such systems. Moreover, it paved the way for the many model systems which were to follow (see below).

The peroxide shunt pathway is closely related to the mechanism of oxidations mediated by peroxidases [28] such as horseradish peroxidase (HRP). The active oxidant in the latter reactions is also an oxoiron(IV)protoporphyrin IX radical cation, commonly referred to as the HRP-I compound [29]. The essential difference between HRP and cytochrome P450 is that in

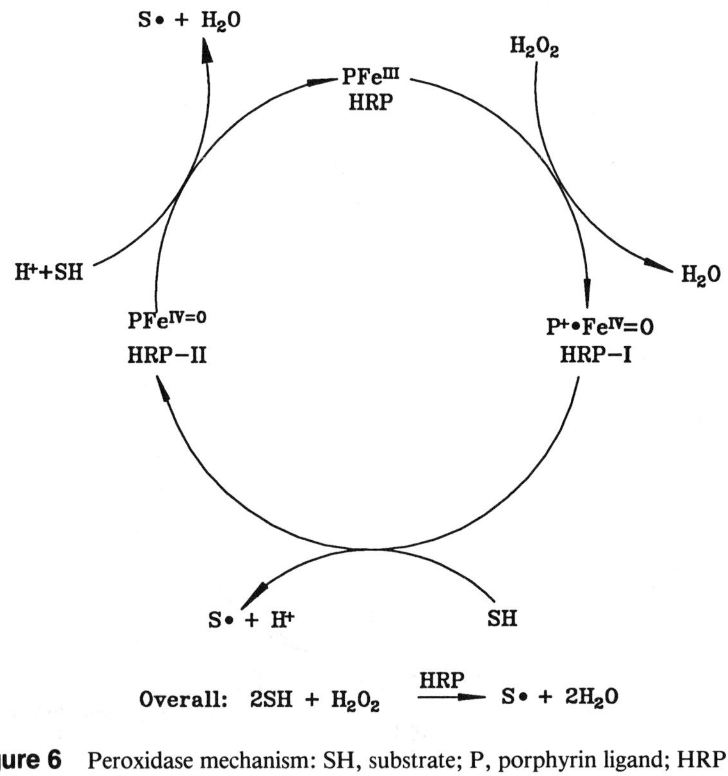

$$\text{Overall: } 2SH + H_2O_2 \xrightarrow{HRP} S\bullet + 2H_2O$$

Figure 6 Peroxidase mechanism: SH, substrate; P, porphyrin ligand; HRP, horseradish peroxidase.

Table 2 Structural Features of Hemoproteins

Enzyme	Axial ligand	Polarity heme pocket	Access to heme: Edge	Access to heme: Iron
Cytochrome P450	Cysteine	−	−	+
Chloroperoxidase	Cysteine	+	+	+
Peroxidase	Histidine	+	+	−
Cytochrome C peroxidase	Histidine	+	+	+
Catalase	Tyrosine	+	−	−
Myoglobin	Histidine	−	+	+

the former there is limited access to the iron center whereas the latter has a substate binding site directly adjacent to the porphyrin nucleus. Consequently, peroxidases generally operate via an outer sphere electron transfer from the substrate to the periphery of the porphyrin radical cation. This is followed by a second electron transfer from the resulting oxoiron(IV)porphyrin which returns the enzyme to its trivalent state (see Figure 6). The differences in axial ligand and degree of access to various hemoprotein-based oxidoreductases are summarized in Table 2. Interestingly, chloroperoxidase (CPO) contains a cysteine thiolate as the proximal ligand and its active site closely resembles that of cytochrome P450. In vivo, CPO catalyzes the oxidation of chloride ion to hypochlorite [30]. However, it has recently been demonstrated that CPO can also catalyze oxygen transfer processes analogously to cytochrome P450. For example, CPO catalyzes the enantioselective oxidation of thioethers to sulfoxides by H_2O_2 [31].

V. MODEL METALLOPORPHYRINS: BIOMIMETIC CHEMISTRY

A. Models for Oxygen Transport Hemoproteins

In the 1970s several groups, e.g., those of Collman, Baldwin, Traylor, and Momenteau [32–35], carried out elegant studies on model metalloporphyrins with the goal of emulating the reversible oxygen binding of hemoglobin and myoglobin. The basic approach of these studies was to prevent further reaction of the iron–dioxygen complex by hindering the irreversible formation of the μ-oxo complex via reaction with a second molecule of the iron(II) porphyrin (Reaction 7). In the hemoprotein the approach of a second molecule of iron(II)porphyrin is prevented by the steric bulk of the protein ligand.

$$\mathrm{PFe^{III}\text{-}O\text{-}O} \xrightarrow{\mathrm{PFe^{II}}} \underset{\mu\text{-peroxo}}{\mathrm{PFe^{III}\text{-}O\text{-}O\text{-}Fe^{III}P}} \longrightarrow$$

$$\mathrm{2PFe^{IV}{=}O} \xrightarrow{\mathrm{PFe^{II}}} \underset{\mu\text{-oxo}}{\mathrm{PFe^{III}\text{-}O\text{-}Fe^{III}P}} \qquad (7)$$

Square-planar iron(II)porphyrins are readily prepared by reduction of the corresponding iron(III)porphyrins in noncoordinating solvents in the absence of dioxygen [4]. In the absence of axial ligands the $d_{z}2$ orbital is of relatively low energy and the four-coordinate complex is in the intermediate spin state, S = 1 (see Figure 7). Five-coordinate high-spin (S = 2) iron(II)porphyrins are more difficult to prepare because the formation of low-spin six-coordinate complexes is favored by crystal field stabilization energy, i.e., $K_2 > K_1$ in Reaction 8.

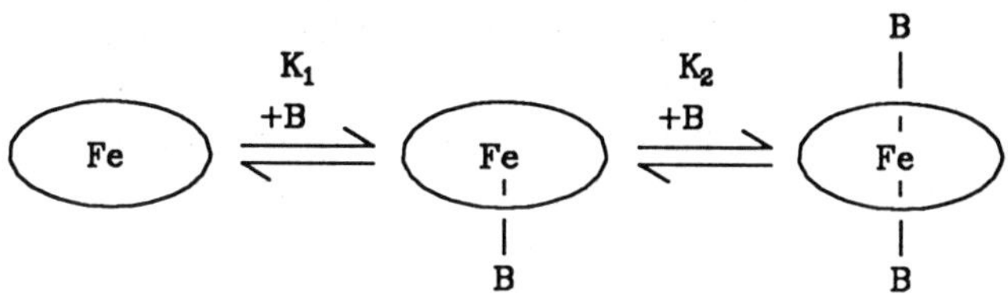

(8)

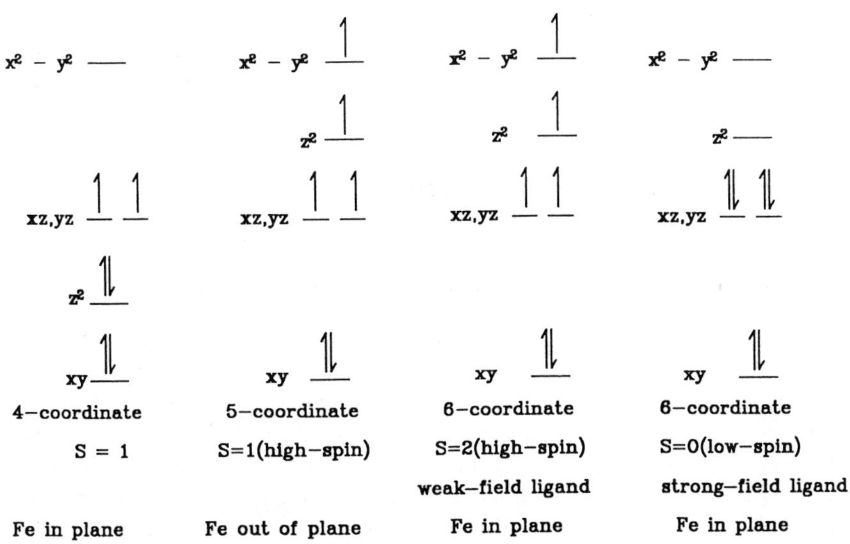

Figure 7 Spin states of iron(II)porphyrins.

Collman and Reed [32] showed that with the sterically hindered 2-methylimidazole as the axial base movement of the iron into the porphyrin ring is sufficiently hampered to allow isolation of a high-spin five-coordinate complex. The sixth coordination position is then free to coordinate to a small molecule such as dioxygen. However, this does not prevent reaction of the six-coordinate iron porphyrin–dioxygen complex with a second molecule of iron(II)porphyrin, resulting in irreversible formation of the iron(III)–μ-peroxo and eventually the iron(III)–μ-oxo complex (Reaction 7). Several approaches have been used to block this irreversible decomposition of the dioxygen complex, all of which involve restricting access to the iron center by constructing a protective pocket around it (Figure 8). For example, Collman and coworkers [33] prepared "picket fence" porphyrins which are able to reversibly bind dioxygen at ambient temperature. The dioxygen complexes exhibited stabilities of the same order of magnitude as those of hemoglobin and myoglobin. Similarly, Baldwin and coworkers [34] and Momenteau and coworkers [35] constructed "capped" and "basket handle" porphyrins, respectively, to achieve the same goal.

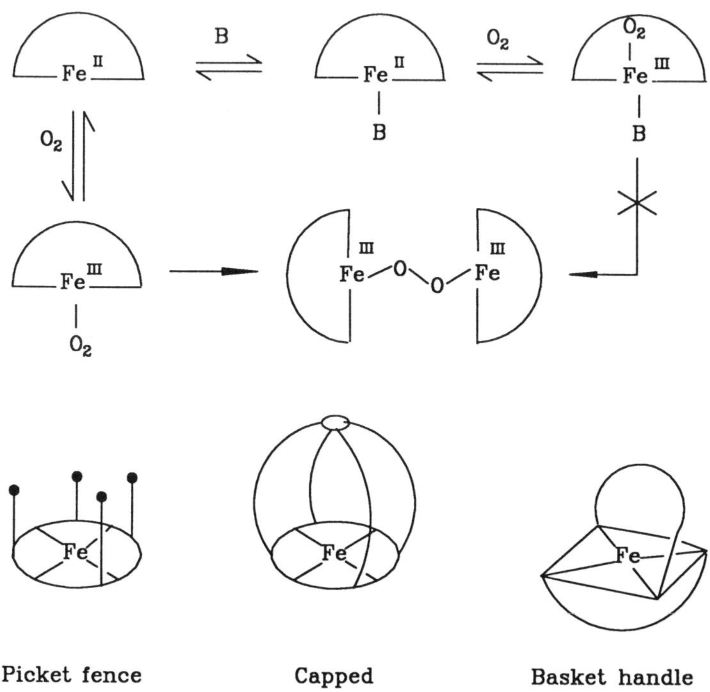

Figure 8 "Protected pocket" porphyrin complexes as O_2 transport models.

Although these elegent studies led to a better understanding of oxygen transport by hemoproteins they do not, as such, constitute models for the activation of dioxygen. As discussed earlier, a source of protons and electrons, i.e., reducing equivalents, is needed to convert the iron–dioxygen complex to the active oxoiron(V)porphyrin oxidant (Figure 5). In vivo these reducing equivalents are provided by NAD(P)H. In order to develop model systems for heme-dependent oxygenases a different approach had to be followed.

B. Models for Cytochrome P450-Dependent Monooxygenases

The discovery of the peroxide shunt (see Section IV.C) provided an important impetus in the development of functional models for cytochrome P450-dependent monooxygenases. In 1979 Groves and coworkers [36] were the first to apply the peroxide shunt pathway to a model system. They described the use of iron(III)*meso*-tetraphenylporphyrin (TPP) chloride in combination with the lipophilic iodosylbenzene, first used in vivo by Ullrich (see above), for the epoxidation of olefins and the hydroxylation of alkanes. Subsequently, chromium [37] and manganese [38,39] complexes of TPP were shown to catalyze oxygen atom transfer reactions from iodosylbenzene to an olefin or an alkane (Figure 9).

Following these seminal studies extensive investigations were carried out in the last decade on metalloporphyrin-catalyzed oxidations of olefins and alkanes with a variety of single-oxygen donors (Figure 9). Manganese porphyrins, which had been shown [38] to be more effective than iron or chromium,

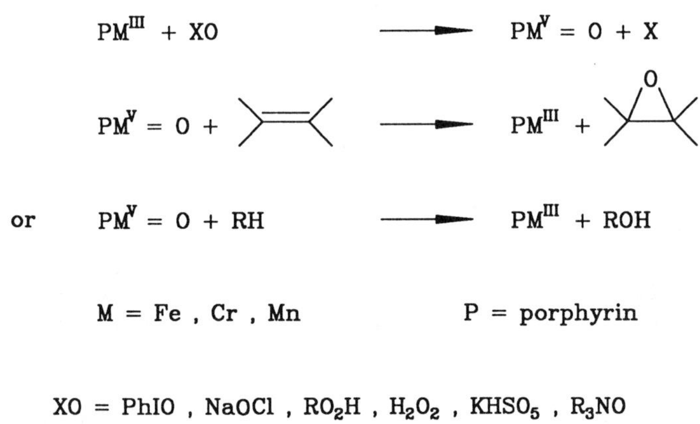

Figure 9 Metalloporphyrin-catalyzed oxygen transfer reactions.

were generally the catalyst of choice. Tabushi and Koga [40] had already described, in 1979, the use of sodium hypochlorite, in combination with TPPMn(III)Cl and a phase transfer catalyst in a two-phase system (CH_2Cl_2-H_2O), for the oxidation of alcohols and cyclohexane. Several groups, e.g., those of Meunier [41,42], Collman [43], Drenth [44], and Montanari [45], subsequently applied this system to the selective epoxidation of olefins [46]. The efficiency of the system is further increased by the addition of a basic axial ligand, such as pyridine [42] or an imidazole group appended to the porphyrin ring [45].

Cumyl hydroperoxide [47] and hydrogen peroxide [48] were introduced by Mansuy in 1984 and 1985, respectively. The presence of imidazole as an axial ligand was essential for good performance, which was attributed to its dual function as a stabilizing ligand and as a base in promoting the heterolysis of the M^{III}-O-OR bond to form the putative $M^{V}=O$ intermediate. *p*-Cyano-*N,N*-dimethylaniline *N*-oxide was introduced by Bruice and Nee in 1982 [49] and $KHSO_5$ by Meunier in 1985 [50]. More recently, Querci and Ricci described [51] the use of the water-soluble magnesium monoperoxyphthalate as an oxygen donor.

Various systems have also been developed for the direct utilization of dioxygen in these systems by employing a coreductant. Tabushi [52], for example, employed H_2 in combination with colloidal platinum. Similarly, the use of zinc and acetic acid [53], ascorbate [54], or sodium borohydride [55] as the reducing agent has been described (see Chapter 9 for more detailed discussion). However, although they constitute interesting biomimetic models, from a practical viewpoint these "reductive oxygen activation" systems do not offer any advantages compared to the oxygen donor systems described above. The lower cost of dioxygen is offset by the need for a "sacrificial" reducing agent.

C. Second- and Third-Generation Metalloporphyrins

A fundamental problem associated with metalloporphyrin-catalyzed oxidations, inherent in all hemoprotein-mediated oxidations, is the vulnerability of the porphyrin ligand to destructive oxidation. Metalloporphyrins have a marked proclivity for self-destruction in oxidizing media. Consequently, parallel to the development of suitable oxygen donors considerable effort has also been focused on improving the stability of the porphyrin ligand.

In 1981 Chang and Ebina [56] described the use of pentafluorophenyl groups instead of unsubstituted phenyl substituents at the oxidation-vulnerable meso positions of the porphyrin ring. Tetrakis(pentafluorophenyl)porphyrinato ($TPPF_{20}$)iron(III)chloride in combination with iodosylbenzene converted cyclohexene to cyclohexene oxide in 95% yield. In 1984, Traylor and Dolphin [57] reported high turnovers and good yields of epoxides using *meso*-tetrakis(2,6-dichlorophenyl)porphyrinato ($TPPCl_8$)iron(III)chloride or *meso*-tetrakis(pentachlorophenyl)porphyrinato ($TPPCl_{20}$)iron(III)chloride in conjunction with pentafluoroiodosylbenzene. The epoxidation of 10,000 mol of norbornene per mol of catalyst in 20 min (eight turnovers per second) was observed.

If tetraphenylporphyrin (TPP) is regarded as the first generation then the more robust $TPPF_{20}$, $TPPCl_8$, and $TPPCl_{20}$ ligands (see Figure 10 for structures) described above may be regarded as the second generation of porphyrin ligands for biomimetic models. However, these ligands still contain oxidation-vulnerable C-H bonds in the pyrrole rings. In third-generation

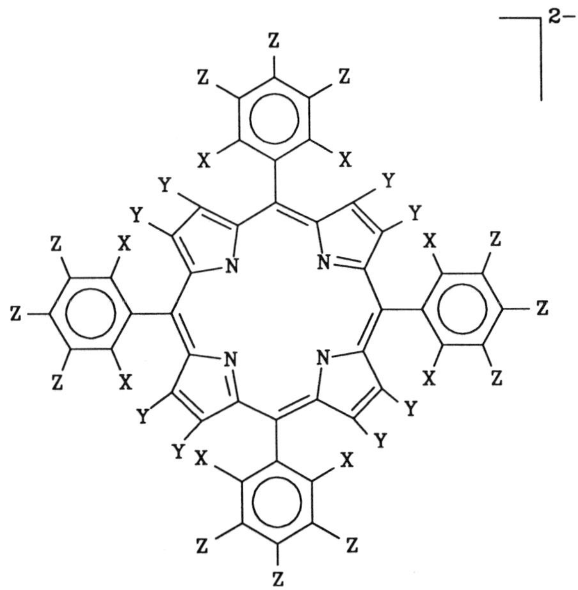

Ligand	X	Y	Z
TPP	8H	8H	12H
$TPPCl_8$	8Cl	8H	12H
$TPPCl_{20}$	8Cl	8H	12Cl
$TPPF_{20}$	8F	8H	12F
$TPPF_{28}$	8F	8F	12F
$TPPF_{20}\beta-Br_8$	8F	8Br	12F

Figure 10 Structures of halogenated porphyrin ligands.

porphyrin ligands the oxidative stability is increased even further by replacing all 28 hydrogens of the TPP macrocycle with halogen. A pertinent example is the "Teflon" porphyrin, *meso*-tetrakis(pentafluorophenyl)-β-octafluoroporphyrin ($TPPF_{28}$) described by Tsuchiya and Seno [58]. Similarly, Lyons and Ellis [59,60] described the synthesis and application of *meso*-tetrakis(pentafluorophenyl)-β-octabromoporphyrinatoiron(III)chloride ($TPPF_{20}\beta$-Br_8). It was found to be the most active catalyst yet studied for the selective oxidation of alkanes with dioxygen (see later and Chapter 10). It is also worth noting that the requirement for modification of the porphyrin ligand in order to enhance oxidative stability has focused attention on the need for efficient methods of synthesis [61] (see Chapter 7).

D. Supported Metalloporphyrin Catalysts

Another problem to be addressed with regard to the development of practically useful oxidations mediated by metalloporphyrins is recovery and reuse of these expensive catalysts. One approach to this problem is to anchor the metalloporphyrin to a solid surface thus combining the advantage of easy recovery of a heterogeneous catalyst with the high activity exhibited by the metalloporphyrin in homogeneous solution. Anchoring the metalloporphyrin to a solid support could provide added benefits arising from steric and electronic effects of the support, analogous to the influence of the polypeptide ligand in hemoproteins. Consequently, considerable attention has recently been focused on the immobilization of metalloporphyrins on both organic and inorganic supports. For example, the tetracationic manganese(III)*meso*-tetra(4-*N*-methylpyridinium)porphyrin chloride adsorbed on silica [62] or in the interlamellar space of montmorillonite clay [63] was shown to be an effective catalyst for alkene epoxidation and alkane hydroxylation with iodosylbenzene.

An added benefit of immobilization in the interlamellar space of clay minerals is the possibility of observing shape selectivity [63]. A variation on this theme comprises the encapsulation of metalloporphyrins within the micropores of zeolites and related molecular sieves: so-called ship-in-the-bottle complexes [64]. For a more detailed discussion of the synthesis and application of supported metalloporphyrin catalysts, see Chapter 11.

E. Asymmetric Oxidations

Another facet of oxidations mediated by cytochrome P450–dependent monooxygenases, which is of more than passing interest to synthetic organic chemists, is the possibility for obtaining high enantioselectivity. In vivo, the asymmetric induction is orchestrated by the protein ligand. In vitro, this can in principle be achieved by building chirality into the porphyrin ligand.

In 1983 Groves and Myers [65] reported the first example of asymmetric induction using a chiral metalloporphyrin catalyst. Epoxidation of styrene with iodosylbenzene in the presence of chiral iron metalloporphyrin catalysts, for example, gave R-styrene oxide in up to 48% ee. In the intervening 10 years effort has continued to intensify on the further development of chiral metalloporphyrin-catalyzed asymmetric epoxidations, alkane hydroxylations, and sulfoxidations. The groups of Groves [66], Collman [67], and Naruta [68] are particularly active in this area. The approaches used bear a strong resemblance to those described earlier for modeling oxygen transport proteins, i.e., via the construction of space-restricted cavities in chiral vaulted [66], capped [67], and picket fence [68] metalloporphyrins. For a more detailed discussion, see Chapter 8.

VI. BEYOND BIOMIMETICS: SUPRABIOTIC CHEMISTRY

The development of efficient methods for the selective partial oxidation of light alkanes is currently a major challenge in the petrochemical industry. The economics of light alkane hydroxylation to alcohols for commodity chemicals or fuel additives restricts the choice of oxidant to dioxygen. Moreover, the stoichiometric addition of electrons and protons, even in its least expensive form, molecular hydrogen, is not economically viable.

In short, a practical oxidation catalyst must have several features that are above and beyond the capabilities of biological systems and their biomimetic models described above. First, it must be capable of generating the required high-valent oxometalporphyrin directly from dioxygen without the prohibitive requirement for a sacrificial reductant. Second, since in vitro no repair mechanisms are possible, it must be resistant to catalytic deactivation both at the metal center and at the ligand.

Ellis and Lyons [59] coined the term "suprabiotic catalyst" to describe this utopian system. An important step forward was their finding that the catalytic activity of metalloporphyrins is greatly enhanced by perhalogenation of the porphyrin macrocycle [59]. In particular, replacement of the β-pyrrole hydrogens by halogens significantly enhances both the stability and the activity of the catalyst. The latter presumably derives from the more pronounced effect of the β-substituents on the redox potential of the metalloporphyrin. As has been pointed out by Dolphin and coworkers [61], the phenyl rings in TPP are sterically constrained to lie perpendicular to the porphyrin ring. Consequently, electronic effects of substituents in the phenyl rings can be only inductively transmitted to the porphyrin ring. When directly substituted in the porphyrin ring, in contrast, electron-withdrawing groups can have a pronounced effect on the porphyrin redox chemistry.

Thus, Ellis and Lyons [59,60] found that $Fe(TPPF_{20}\beta\text{-}Br_8)Cl$ shows unprecedented activity in the oxidation of isobutane with dioxygen to give *tert*-butanol. More than 22% conversion of neat isobutane was observed in 70 hr at room temperature, corresponding to >13,000 turnovers or >190 moles of product per mole of catalyst per hour. A mechanism was proposed [59] in which the active oxidant is an oxoiron(IV)porphyrin, formed via cleavage of the μ-peroxoiron(III)porphyrin (see Figure 11). Steric and electronic effects of the halogen substituents were postulated to stabilize the oxoiron(IV) species with respect to μ-oxodiiron(III) via reaction with the iron(II)porphyrin (see Figure 11). Moreover, the electron-withdrawing halogen substituents stabilize the porphyrin macrocycle and at the same time increase the reactivity of the ferryl intermediate toward the hydrocarbon substrate. Having demonstrated the important rate-enhancing effect of the halogen substituents in the porphyrin ring, one can envisage that the introduction of other strongly electron-withdrawing substituents, e.g., nitro, cyano, sulfonate, into the pyrrole rings may afford even more active catalysts. Hence, the quest for even better suprabiotic metalloporphyrin catalysts continues (see Chapter 10 for a more detailed discussion).

The only other example of a direct oxygenation mediated by a metalloporphyrin, without the requirement for a sacrificial reductant, is the ruthenium

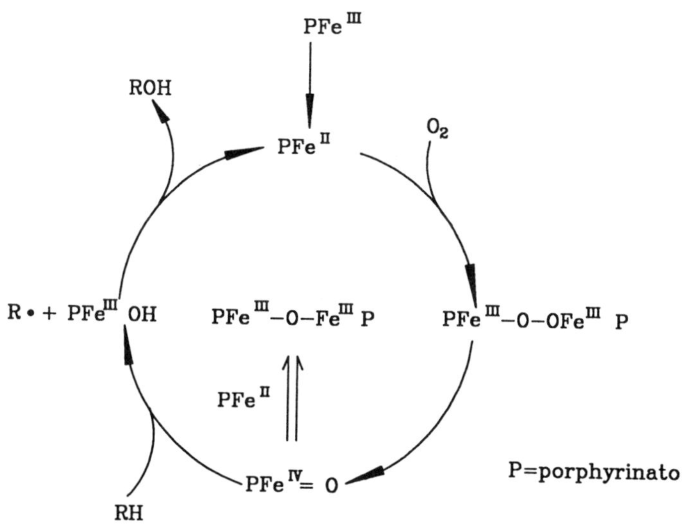

Figure 11 Mechanism of alkane hydroxylation with O_2 and perhalogenated iron porphyrins.

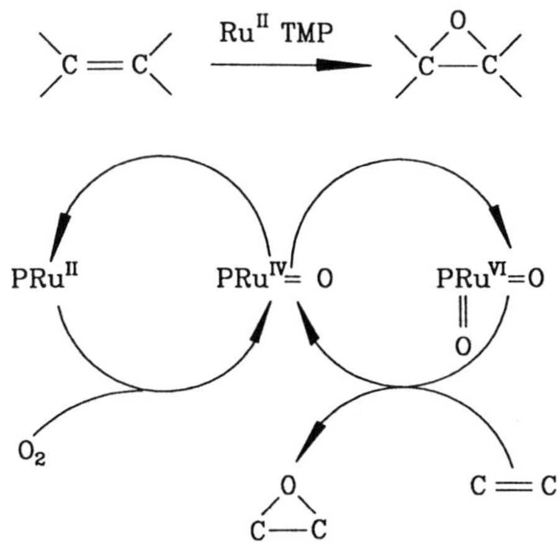

Figure 12 Ru(TMP)-catalyzed epoxidation with O_2.

tetramesitylporphyrin (Ru^{II}TMP)–catalyzed epoxidation of olefins (Figure 12) reported by Groves and Quinn [69]. In order to account for the observed results the authors proposed a mechanism in which a dioxoruthenium(VI)porphyrin is the active oxidant. Reaction of the latter with the olefin affords the epoxide and an oxoruthenium(IV)porphyrin. Disproportionation of the oxoruthenium(IV) species gives dioxoruthenium(VI) and ruthenium(II). The cycle is then completed by reaction of the latter with dioxygen (see Figure 12).

VII. CONCLUDING REMARKS

The development of metalloporphyrin-catalyzed oxidations has come a long way in the last two decades. With regard to practical synthetic applications, two pathways for further development have clearly emerged. First is the design of chiral metalloporphyrins for efficient enantioselective oxidations of prochiral substrates. Second is the design of even more active metalloporphyrin catalysts for suprabiotic chemistry with dioxygen thus bringing the economic viability of metalloporphyrin-catalyzed oxidations into the realm of commodity chemicals. Parallel to these developments much effort continues to be focused on achieving a better understanding of the intricate mechanisms of these reactions, e.g., the mechanism of oxygen transfer from the oxometalloporphyrin to the substrate (see Chapters 2 and 3). In short, oxidation catalysis by metalloporphyrins has a long history and appears to have a bright future.

REFERENCES

1. R.A. Sheldon and J.K. Kochi, *Metal-Catalyzed Oxidations of Organic Compounds*, Academic Press, New York, 1981.
2. L. Simandi, *Catalytic Activation of Dioxygen by Metal Complexes*, Kluwer, Amsterdam, 1992.
3. D.T. Sawyer, *Oxygen Chemistry*, Oxford University Press, Oxford, 1991.
4. J.P. Collman, T.R. Halpert, and K.S. Suslick, O_2 binding to heme proteins and their synthetic analogs, in *Metal Ion Activation of Dioxygen* (T.G. Spiro, ed.), Wiley, New York, 1980, p. 1.
5. D.L. Drabkin, Selected landmarks in the history of porphyrins and their biologically functional derivatives, in *The Porphyrins, Vol. 1, Structure and Synthesis, Part A* (D. Dolphin, ed.), Academic Press, New York, 1978, p. 29.
6. A.D. Adler, Some landmarks in the history of porphyrin related materials to 1950 AD, in *Porphyrin Chemistry Advances* (F.R. Longo, ed.), Ann Arbor Science, Ann Arbor, 1979, p. 1.
7. A.D. Adler and M. Gouterman, An historical survey of porphyrins, their related spectra and their interpretation, in *Optical Properties and Structures of Tetrapyrroles* (A. Blauer and H. Sund, eds.), Walter de Gruyter, Berlin, 1985, p. 3.
8. A. Corwin, The chemistry of the porphyrins, in *Organic Chemistry* (H. Gilman, ed.), Wiley, New York, 1943.
9. J.A. Milroy, Observations on some metallic compounds of hematoporphyrin, *Biochem. J., 12*: 318 (1918).
10. H. Fischer and K. Zeile, Synthese des Hämatoporphyrins, Protoporphyrins und Hämins, *Justus Liebig's Ann. Chem., 486*: 98 (1929).
11. H. Fischer and H. Orth, *Die Chemie des Pyrrols*, Vols. I–III, Akademische Verlagsgesellschaft, Leipzig, 1934.
12. J.C. Kendrew, R.E. Dickerson, B.E. Strandberg, R.G. Hart, D.R. Davies, D.C. Phillips, and V.C. Shore, Structure of myoglobin, *Nature, 185*: 422 (1960).
13. M.F. Perutz, M.G. Rossmann, A.F. Cullis, H. Muirhead, G. Will, and A.C.T. North, Structure of hemoglobin, *Nature, 185*: 416 (1960).
14. O. Hayaishi, M. Katagari, and S. Rothberg, Mechanism of the pyrocatechase reaction, *J. Am. Chem. Soc., 77*: 5450 (1955).
15. H.S. Mason, W.L. Fowlks, and E. Peterson, Oxygen transfer and electron transport by the phenolase complex, *J. Am. Chem. Soc., 77*: 2914 (1955).
16. M.J. Coon and R.E. White, Cytochrome P450, a versatile catalyst in monooxygenation reactions, in *Metal Ion Activation of Dioxygen* (T.G. Spiro, ed.), Wiley, New York, 1980, p. 73.
17. R.W. Estabrook, D.Y. Cooper, and O. Rosenthal, The light-reversible carbon monoxide inhibition of the steroid C-21 hydroxylase system of the adrenal cortex, *Biochem. Z., 338*: 741 (1963).
18. T. Omura and R. Sato, Carbon monoxide-binding pigment of liver microsomes. I. Evidence for its hemoprotein nature, *J. Biol. Chem., 239*: 2370 (1964).
19. M.J. Gunter and P. Turner, Metalloporphyrins as models for the cytochromes P450, *Coord. Chem. Rev., 108*: 115 (1991) and references cited therein.

20. S. Udenfriend, C.T. Clark, J. Axelrod, and B.D. Brodie, Ascorbic acid in aromatic hydroxylation. I. A model system for aromatic hydroxylation, *J. Biol. Chem., 208*: 731 (1954).
21. H. Mimoun, Metal complexes in oxidation, in *Comprehensive Coordination Chemistry* (G. Wilkinson, R.D. Gillard, and J.A. McCleverty, eds.), Pergamon Press, New York, 1987, p. 317.
22. G.A. Hamilton, Oxidation by molecular oxygen. II. The oxygen atom transfer mechanism for mixed-function oxidases and the model for mixed-function oxidases, *J. Am. Chem. Soc., 86*: 3391 (1964).
23. V. Ullrich and H.J. Staudinger, Studies on oxygen activation in model systems for microsomal mixed function oxygenases, in *Biological and Chemical Aspects of Monooxygenases* (K. Bloch and O. Hayaishi, eds.), Maruzen, Tokyo, 1966, p. 235.
24. G.D. Nordblum, R.E. White, and M.J. Coon, Studies on hydroperoxide-dependent substrate hydroxylation by purified liver microsomal cytochrome P450, *Arch. Biochem. Biophys., 175*: 524 (1976).
25. E.G. Hrycay, J.A. Gustafsson, M. Ingelman-Sundberg, and L. Ernster, Sodium periodate, sodium chlorite, organic hydroperoxides, and H_2O_2 as hydroxylating agents in steroid hydroxylation reactions catalyzed by partially purified cytochrome P450, *Biochem. Biophys. Res. Commun., 66*: 209 (1975).
26. E.G. Hrycay and P.J. O'Brien, Cytochrome P450 as a microsomal peroxidase utilizing a lipid peroxide substrate, *Arch. Biochem. Biophys., 147*: 14 (1971).
27. F. Lichtenberger, W. Nastainczyk, and V. Ullrich, Cytochrome P450 as an oxene transferase, *Biochem. Biophys. Res. Commun., 70*: 939 (1975).
28. H.B. Dunford, Horseradish peroxidase: structure and kinetic properties, in *Peroxidases in Chemistry and Biology*, Vol. 2 (J. Everse, K.E. Everse, and M.B. Grisham, eds.), CRC Press, Boca Raton, 1991, p. 1.
29. T.G. Traylor, W.A. Lee, and D.V. Stynes, Model compound studies related to peroxidases. Mechanisms of reactions of hemins with peracids, *J. Am. Chem. Soc., 106*: 755 (1984).
30. B.W. Griffin, Chloroperoxidase: a review, in *Peroxidases in Chemistry and Biology*, Vol. 2 (J. Everse, K.E. Everse, and M.B. Grisham, eds.), CRC Press, Boca Raton, 1991, p. 85.
31. S. Colonna, N. Gaggero, L. Cassella, G. Carrea, and P. Pasta, Chloroperoxidase and hydrogen peroxide: an efficient system for enzymatic enantioselective sulfoxidations, *Tetrahedron Asymm., 3*: 95 (1992).
32. J.P. Collman and C.A. Reed, Syntheses of ferrous-porphyrin complexes. A hypothetical model for deoxymyoglobin, *J. Am. Chem. Soc., 95*: 2048 (1973).
33. J.P. Collman, R.R. Gagne, T.R. Halbert, J.C. Marchon, and C.A. Reed, Reversible oxygen adduct formation in ferrous complexes derived from a "picket fence" porphyrin. A model for oxymyoglobin, *J. Am. Chem. Soc., 95*: 7868 (1973).
34. J. Almog, J.E. Baldwin, and J. Huff, Reversible oxygenation and autoxidation of a "capped" porphyrin iron(II) complex, *J. Am. Chem. Soc., 97*: 227 (1975).
35. M. Momenteau, B. Loock, J. Mispelter, and E. Bisagni, "Basket handle" porphyrins and their ferrous complexes as stable oxygen carriers, *Nouv. J. Chim., 3*: 77 (1979).

36. J.T. Groves, T.E. Nemo, and R.S. Myers, Hydroxylation and epoxidation catalyzed by iron porphyrin complexes. Oxygen transfer from iodosylbenzene, *J. Am. Chem. Soc., 101*: 1032 (1979).
37. J.T. Groves and W. J. Kruper, Preparation and characterization of an oxoporphinatochromium(V) complex, *J. Am. Chem. Soc., 101*: 7613 (1979).
38. C.L. Hill and B.C. Schardt, Alkane activation and functionalization under mild conditions by a homogeneous manganese(III)porphyrin-iodosylbenzene oxidizing system, *J. Am. Chem. Soc., 102*: 6374 (1980).
39. J.T. Groves, W.J. Kruper, and R.C. Haushalter, Hydrocarbon oxidations with oxometalloporphinates: isolation and reactions of a (porphinato)manganese(V) complex, *J. Am. Chem. Soc., 102*: 6375 (1980).
40. I. Tabushi and N. Koga, Synergetic combination of catalysis of the phase transfer—electron transfer type for the oxidation of alcohols of hydrocarbons, *Tetrahedron Lett.,* 3681 (1979).
41. E. Guilmet and B. Meunier, A new catalytic route for the epoxidation of styrene with sodium hypochlorite activated by transition metal complexes, *Tetrahedron Lett., 21*: 4449 (1980).
42. E. Guilmet and B. Meunier, Role of pyridine in the catalytic activation of sodium hypochlorite in the presence of manganese porphyrin, *Tetrahedron Lett., 23*: 2449 (1982).
43. J.P. Collman, T. Kodadek, and J.I. Brauman, Oxygenation of styrene by cytochrome P450 model systems: a mechanistic study, *J. Am. Chem. Soc., 108*: 2588 (1986).
44. J.A.S.J. Razenberg, R.J.M. Nolte, and W. Drenth, Mechanism of olefin epoxidation by a mono-oxygenase model, *Tetrahedron Lett., 25*: 789 (1984).
45. F. Montanari, M. Penso, S. Quici, and P. Vigano, Highly efficient NaOCl olefin epoxidations catalyzed by imidazole or pyridine "tailed" manganese porphyrins under two-phase conditions. Influence of pH and of the anchored axial ligand, *J. Org. Chem., 50*: 4888 (1985).
46. For an excellent review, see B. Meunier, Metalloporphyrins as versatile catalysts for oxidation reactions and oxidative DNA cleavage, *Chem. Rev., 92*: 1411 (1992).
47. D. Mansuy, P. Battioni, and J.P. Renaud, In the presence of imidazole, iron and manganese porphyrins catalyze the epoxidation of alkenes by alkyl hydroperoxides, *J. Chem. Soc., Chem. Commun.,* 1255 (1984).
48. J.P. Renaud, P. Battioni, J.F. Bartoli, and D. Mansuy, A very efficient system for alkene epoxidation by hydrogen peroxide: catalysis by manganese-porphyrins in the presence of imidazole, *J. Chem. Soc., Chem. Commun.,* 888 (1985).
49. M.W. Nee and T.C. Bruice, Use of the N-oxide of p-cyano-N,N-dimethylaniline as an oxygen donor in a cytochrome P450 model system, *J. Am. Chem. Soc., 104*: 6123 (1982).
50. B. de Poorter, M. Ricci, and B. Meunier, Oxone as oxygen donor in the catalytic hydroxylation of saturated hydrocarbons, *Tetrahedron Lett., 26*: 4459 (1985).
51. C. Querci and M. Ricci, Hydroxylation of saturated hydrocarbons by magnesium monoperoxyphthalate catalysed by manganese porphyrins, *Tetrahedron Lett., 31*: 1779 (1990).

52. I. Tabushi and A. Yazaki, P450-type dioxygen activation using H_2/colloidal Pt as an effective electron donor, *J. Am. Chem. Soc., 103*: 7371 (1981).
53. P. Battioni, J.F. Bartoli, P. Leduc, M. Fontecave, and D. Mansuy, A new efficient biomimetic system for hydrocarbon oxidation by dioxygen using manganese porphyrins, imidazole and zinc, *J. Chem. Soc., Chem. Commun.,* 791 (1987).
54. D. Mansuy, M. Fontecave, and J.F. Bartoli, Mono-oxygenase-like dioxygen activation leading to alkane hydroxylation and olefin epoxidation by a Mn(III) porphyrin-ascorbate biphase system, *J. Chem. Soc., Chem. Commun.,* 253 (1983).
55. M.M. Shimizu, H. Orita, T. Hayakawa, and K. Takehira, The oxidation of olefins with O_2 and $NaBH_4$ catalyzed by manganese meso-tetrakis(p-sulfonatophenyl)porphyrin, *J. Mol. Catal., 45*: 85 (1988).
56. C.K. Chang and F. Ebina, NIH shift in haemin-iodosylbenzene-mediated hydroxylations, *J. Chem. Soc., Chem. Commun.,* 778 (1981).
57. P.S. Traylor, D. Dolphin, and T.G. Traylor, Sterically protected hemins with electronegative substituents: efficient catalysts for hydroxylation and epoxidation, *J. Chem. Soc., Chem. Commun.,* 279 (1984).
58. S. Tsuchiya and M. Seno, Novel synthetic method of phenol from benzene catalysed by perfluorinated hemin, *Chem. Lett.,* 263 (1989).
59. P.E. Ellis and J.E. Lyons, Selective air oxidation of light alkanes catalyzed by activated metalloporphyrins—the search for a suprabiotic system, *Coord. Chem. Rev., 105*: 181 (1990).
60. J.E. Lyons and P.E. Ellis, Selective low temperature hydroxylation of isobutane by molecular oxygen catalyzed by an iron perhaloporphyrin complex, *Catal. Lett., 8*: 45 (1991).
61. T. Wijesekera, A. Matsumoto, D. Dolphin, and D. Lexa, Perchlorinated and highly chlorinated meso-tetraphenylporphyrins, *Angew. Chem. Int. Ed. Engl., 29*: 1028 (1990).
62. P. Battioni, J.P. Lallier, L. Barloy, and D. Mansuy, Monooxygenase-like oxidation of hydrocarbons using supported manganese-porphyrin catalysts: beneficial effects of a silica support for alkane hydroxylation, *J. Chem. Soc., Chem. Commun.,* 1149 (1989).
63. L. Barloy, P. Battioni, and D. Mansuy, Manganese porphyrins supported on montmorillonite as hydrocarbon mono-oxygenation catalysts: particular efficiency for linear alkane hydroxylation, *J. Chem. Soc., Chem. Commun.,* 1365 (1990).
64. M. Nakamura, T. Tatsumi, and H. Tominaga, Synthesis and catalytic capability of zeolite-encapsulated iron and manganese tetramethylporphyrin complexes, *Bull. Chem. Soc. Jpn., 63*: 3334 (1990).
65. J.T. Groves and R.S. Myers, Catalytic asymmetric epoxidation with chiral iron porphyrins, *J. Am. Chem. Soc., 105*: 5791 (1983).
66. J.T. Groves and P. Viski, Asymmetric hydroxylation, epoxidation and sulfoxidation catalyzed by vaulted binaphthyl metalloporphyrins, *J. Org. Chem., 55*: 3628 (1990).
67. J.P. Collman, X. Zhang, V.J. Lee, and J.I. Brauman, Asymmetric epoxidation catalysed by an iron "binap capped" porphyrin, *J. Chem. Soc., Chem. Commun.,* 1647 (1992).

68. Y. Naruta and K. Maruyama, Cytochrome P450 modelling oxygenation of olefins within the space-restricted cavity of iron "binap porphyrin": rate enhancement in the presence of imidazole, *Tetrahedron Lett., 28*: 4553 (1987).
69. J.T. Groves and R. Quinn, Aerobic epoxidation of olefins with ruthenium porphyrin catalysts, *J. Am. Chem. Soc., 107*: 5790 (1985).

ADDITIONAL READING

1. P.R. Ortiz de Montellano (ed.), *Cytochrome P450: Structure, Mechanism and Biochemistry*, Plenum Press, New York, 1986.
2. P.W. White, Mechanistic studies and selective catalysis with cytochrome P450 model systems, *Bioorg. Chem., 18*: 440 (1990).
3. M. Momenteau, Synthesis and oxygenation studies of hemoprotein models, *Bull. Soc. Chem. Belg., 100*: 131 (1991).
4. R.E. White and M.J. Coon, Oxygen activation by cytochrome P450, *Ann. Rev. Biochem., 49*: 315 (1980).
5. I. Tabushi, Reductive dioxygen activation by use of artificial P450 systems, *Coord. Chem. Rev., 86*: 1 (1988).
6. G.L. Eichhorn (ed.), *Inorganic Biochemistry*, Vols. 1 and 2, Elsevier, Amsterdam, 1973.
7. J. Reedijk (ed.), *Bioinorganic Catalysis*, Marcel Dekker, New York, 1993.

2

Mechanisms of Reaction of Hypervalent Oxochromium, Iron, and Manganese Tetraphenylporphyrins with Alkenes

Dražen Ostović

Merck Research Laboratories, West Point, Pennsylvania

Gong-Xin He

Gilead Sciences, Inc., Foster City, California

Thomas C. Bruice

University of California, Santa Barbara, Santa Barbara, California

I. INTRODUCTION

It has been widely assumed—but not established—that an enzyme-bound hypervalent iron-oxoprotoporphyrin IX species is the oxidant in cytochrome P450 reactions [1]. From this have ensued investigations directed to the mechanisms of reaction of hypervalent metal-oxoporphyrins with alkenes. We describe here a summary of our studies directed toward the mechanism of epoxidation of alkenes when using iron, manganese, and chromium *meso*-tetraarylporphyrins.

In studies that involve oxidation of metalloporphyrins, chemists stabilize the porphyrin ring by substituting phenyl groups at the 5, 10, 15, and 20 positions: tetraphenylporphyrin (TPP), tetra(2,6-dichlorophenyl)porphyrin (Cl_8TPP), tetra(2,6-dibromophenyl)porphyrin (Br_8TPP), tetra(pentafluorophenyl)porphyrin ($F_{20}TPP$), tetra(2,4,6-trimethylphenyl)porphyrin (TMP). Two $1e^-$ oxidation of the Cr^{III} species with oxygen transfer provides chromium(V)-oxoporphyrin [$(TPP)Cr^V(O)$, $(Cl_8TPP)Cr^V(O)$, $(Br_8TPP)Cr^V(O)$ etc.] while $2e^-$ oxidation above the Fe^{III} state with oxygen transfer generates the iron(IV)-oxoporphyrin π-cation radical [$(+\cdot TPP)Fe^{IV}(O)$, $(+\cdot Cl_8TPP)Fe^{IV}(O)$, $(+\cdot Br_8TPP)Fe^{IV}(O)$, etc.].

II. CHROMIUM PORPHYRINS

A. Kinetics of Alkene Oxidation by Oxochromium(V) Porphyrins

Chromium porphyrins form relatively stable hypervalent oxo complexes, (Porph)Cr^{IV}(O) and (Porph)Cr^{V}(O)(X). Their distinct UV/vis spectral properties allow easy monitoring of alkene oxidation reactions under stoichiometric conditions. The ability to determine the second-order rate constants for the bimolecular oxidation of alkenes provided a unique opportunity to probe the mechanism of the oxidation reaction using conventional kinetic approaches [2,3].

The second-order rate constants for the oxidation of 25 alkenes by oxo-(*meso*-tetrakis(2,6-dibromophenyl)porphinato)chromium(V) (Br_8TPP)Cr^{V}(O)(X) were determined under stoichiometric conditions. The (Br_8TPP)Cr^{V}(O)(X) reagent was prepared by electrochemical oxidation of the stable (Br_8TPP)Cr^{IV}(O). The electrochemical potential for the couple (Br_8TPP)Cr^{IV}(O)/(Br_8TPP)Cr^{V}(O)(X) has been determined by cyclic voltametry to be 0.865 V (vs. SCE). Both literature and experimental values were used for one-electron oxidation potentials of alkenes.

The following kinetic scheme (Scheme 1) was utilized for calculation of the oxidation rate constants.

B. Linear Free-Energy Correlation and the Rate-Limiting Step

A linear correlation between the log of the second-order rate constant (k_1) for the reaction of (Br_8TPP)Cr^{V}(O)(X) with various alkenes and alkene $E_{1/2}$ values is shown in Figure 1 [3]. The log k_1 values for (E)-1,2-bis(*trans*-2-*trans*-3-diphenylcyclopropyl)ethene (8-E), *trans*-stilbene, and *trans-p,p'*-

$$\text{(P)Cr}^{V}\text{(O)(X)} + \text{alkene} \xrightarrow{k_1} \text{(P)Cr}^{III}\text{(O)(X)} + \text{alkene oxidation products}$$

$$\text{(P)Cr}^{V}\text{(O)(X)} + \text{(P)Cr}^{III}\text{(X)} \underset{k_3}{\overset{k_2}{\rightleftharpoons}} \text{(P)Cr}^{IV}\text{(O)} + \text{(P)Cr}^{IV}\text{(X)}_2$$

$$\text{(P)Cr}^{V}\text{(O)(X)} \xrightarrow{k_4} \text{(P)Cr}^{IV}\text{(O)} + X^-$$

$$X^- = Cl^-, ClO_4^-$$

Scheme 1

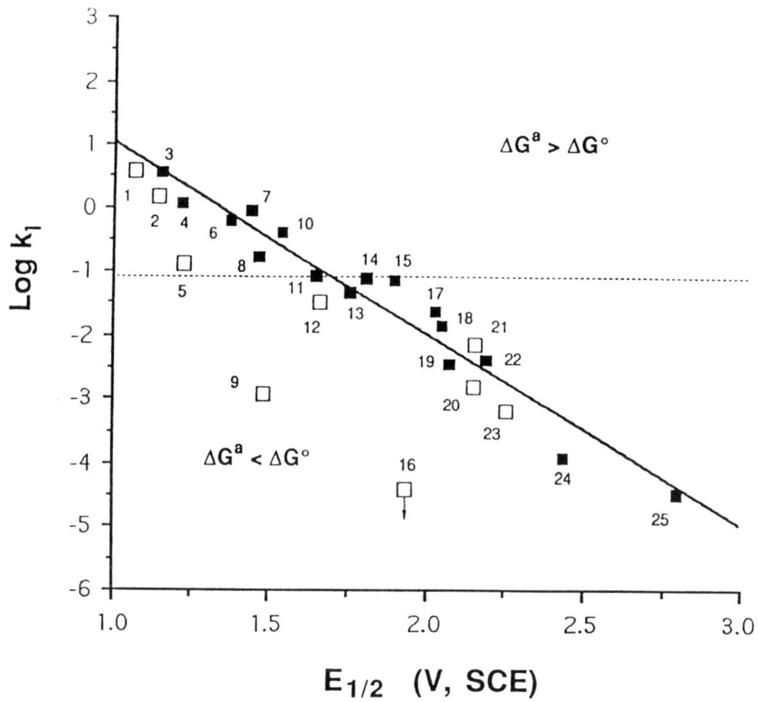

Figure 1 Plot of the log of the second-order rate constant (k_I) for reaction of a series of alkenes with $(Br_8TPP)Cr^V(O)(X)$ vs. the $E_{1/2}$ for $1e^-$ oxidation of the alkenes. The results as shown in open squares are for *trans*-alkenes. Dark squares represent the results for terminal and *cis*-alkenes. They are: (1) *trans-p,p'*-dimethoxystilbene, (2) 1,4-diphenyl-1,3-butadiene, (3) 4-methoxystyrene, (4) 1,1-diphenylethylene, (5) (E)-1,2-bis(*trans*-2, *trans*-3-diphenylcyclopropyl)ethene, (6) 4-methylstyrene, (7) 2,3-dimethyl-2-butene, (8) (Z)-1,2-bis(*trans*-2, *trans*-3-diphenylcyclopropyl)ethene, (9) *trans*-stilbene, (10) *cis*-stilbene, (11) styrene, (12) *trans*-β-methylstyrene, (13) 4-acetoxystyrene, (14) cyclohexene, (15) norbornene, (16) *trans-p,p'*-dicyanostilbene, (17) *cis*-cyclooctene, (18) 4-cyanostyrene, (19) *cis*-2-pentene, (20) *trans*-2-pentene, (21) *trans*-5-decene, (22) cyclopentene, (23) *trans*-2-hexenyl acetate, (24) 1-hexene, and (25) 1-octene. On the intersection of the dashed line and the solid line (linear free energy line) the standard free energy (Δ G°) is equal to the free energy of activation (Δ G^a) for the reaction.

dimethoxystilbene exhibit large negative deviations due to steric effects, which will be discussed later. The values of k_I cover the range of five orders of magnitude, from 3.2×10^{-5} M^{-1} s^{-1} to 4.5 M^{-1} s^{-1}, while the alkene $E_{1/2}$ values range from 1.05 to 2.80 V. The correlation line of Figure 1 is given by Eq. 1.

$$\log k_1 = 2.99E_{1/2} + 4.02 \tag{1}$$

In addition, the second-order rate constants were determined for the oxidation of norbornene by oxo-(*meso*-tetrakis(2,6-dichlorophenyl)porphinato)chromium(V) $(Cl_8TPP)Cr^V(O)(X)$, oxo-(*meso*-tetrakis(2,6-difluorophenyl)porphinato)chromium(V) $(F_8TPP)Cr^V(O)(X)$, oxo-(*meso*-tetrakis(2,4,6-trimethylphenyl)porphinato)chromium(V) $(Me_{12}TPP)Cr^V(O)(X)$, and oxo-(*meso*-tetraphenylporphinato)chromium(V) $(TPP)Cr^V(O)(X)$ [2]. A linear correlation between log of the second-order rate constant (k_1) for the reaction of these chromium oxophorphyrins with norbornene and $E_{1/2}$ values of the chromium oxoporphyrins shown in Figure 2 is given by Eq. 2.

$$\log k_1 = 9.08E_{1/2} - 8.76 \tag{2}$$

Linear correlations of Figures 1 and 2 suggest that an electron transfer process was taking place in the rate-limiting step. But can $1e^-$ transfer be rate limiting?

For a reaction that proceeds via an initial electron transfer three situations are possible: (a) at one extreme the diffusion together of the reactants is rate limiting and log k_1 for $1e^-$ transfer shows no dependence on $\Delta E_{1/2}$ (slope = 0); (b) in the other extreme $1e^-$ transfer is so endothermic that log k_1

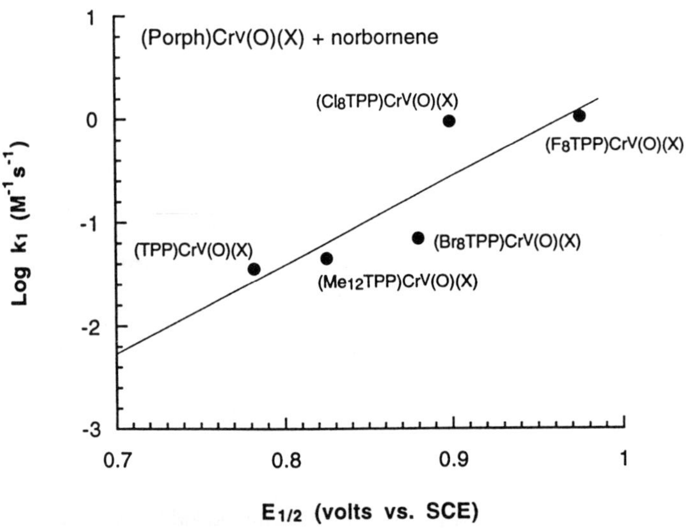

Figure 2 Plot of the log of the second-order rate constants (k_1) for reaction of a series of five $(Porph)Cr^V(O)(X)$ species with norbornene vs. the $E_{1/2}$ for the couple $(Porph)Cr^V(O)(X)/(Porph)Cr^{IV}(O)$.

is directly proportional to the $\Delta E_{1/2}$ for the overall $1e^-$ transfer (slope = $-16.6\ V^{-1}$); and (c) the intermediate region where log k_1 for rate-limiting $1e^-$ transfer is dependent on $\Delta E_{1/2}$ with the slope dependent on the degree of electron transfer in the transition state, changing from zero to $-16.6\ V^{-1}$ [4,5]. When our data were plotted in the form of a modified Brønsted plot log k_1 vs. $E_{1/2}$), a slope of $-2.99\ V^{-1}$ was obtained. In a conventional Brønsted plot (log k_1 vs. log K_{eq}) this corresponds to Brønsted $\alpha = 0.18$. If $1e^-$ transfer was involved in the reaction of (Porph)Cr^V(O)(X) with alkenes, the transition state must be early, such that the electron has undergone only a partial transfer. Such a situation would be equivalent to rate-determining formation of a charge-transfer (CT) complex. Only a late transition state would be in accord with the formation of a solvent-caged /(Br_8TPP)Cr^{IV}(O) + alkene π-cation radical/ pair. Again, a single consistent mechanism involving caged /(Br_8TPP)Cr^{IV}(O) + alkene π-cation radical/ pair with full charge separation may be dismissed. In contrast, the rate-limiting formation of a CT complex would appear feasible.

The observation that the slopes of the free-energy relationships of Eqs. 1 and 2 [k_1 vs. log K_{eq}, when (porph)Cr^V(O)(X) was varied and when alkene was varied] were different indicated the presence of perpendicular (Thornton) contribution to the overall slope. In the first approximation the overall slope (α) is given by Eq. 3, where the

$$\alpha = \chi \pm 0.5(\tau - 1) \tag{3}$$

first term (χ) incorporates the parallel (Leffler–Hammond) contribution, and the second term $0.5(\tau - 1)$ is the perpendicular (Thornton) contribution [6]. The term χ represents the progress of the reaction along the reaction coordinate and τ is the measure of the tightness of the transition state. The limiting values for τ are 0 and 2, representing fully dissociative and fully associative transition states. When τ is 1, the bond order in the transition state is conserved. The term χ has a positive sign when the (electron) acceptor is varied and a negative sign when the (electron) donor is varied in a homologous series of reactions. With two slopes (Eqs. 1 and 2) available, $\alpha = -0.18$ when the alkene (electron donor) is varied, and $\alpha = 0.55$ when porphyrin (electron acceptor) is varied, two equations with two unknowns (χ and τ) were solved. The values obtained were $\chi = 0.37$ and $\tau = 0.63$. Presuming a $1e^-$ transfer mechanism and the validity of Eq. 1, the transition state is located ~37% along the reaction coordinate for the formation of (Br_8TPP)Cr^{IV}(O) + alkene π-cation-radical, and it is moderately dissociative. This approximation of the structure of the transition state is not consistent with a full electron transfer but is reasonable for rate-limiting formation of a CT complex between alkene and (Br_8TPP)Cr^V(O)(X). An initially formed CT complex could collapse to give (Br_8TPP)Cr^{III}(X) + epoxide (or other oxidation products) in one or more rapid sequential or parallel steps (Scheme 2).

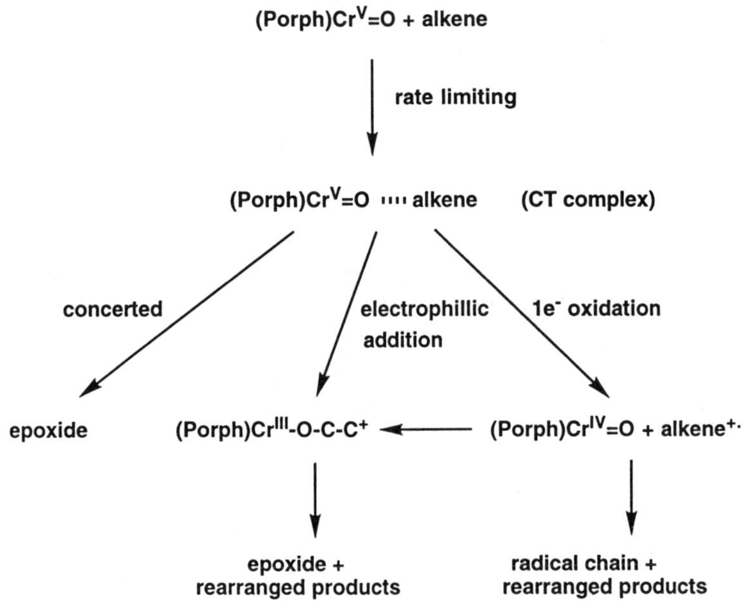

Scheme 2

Initial formation of the alkene derived π-cation radical upon oxidation of alkenes by oxometalloporphyrins has been suggested as a plausible oxidation mechanism [7–11]. This can be dismissed based on the following analysis. The interpretation of the linear free energy of log k_1 to $E_{1/2}$ values as evidence for the intermediacy of $(Porph)Cr^{IV}(O)$ + alkene-derived π-cation radical leads to Eq. 4.

$$(Br_8TPP)Cr^{V}O(X) + e^- \rightleftharpoons (Br_8TPP)Cr^{IV}O + X \qquad P_1 \qquad (4a)$$

$$Alkene^{\cdot +} + e^- \rightleftharpoons alkene \qquad P_2 \qquad (4b)$$

$$(Br_8TPP)Cr^{V}O(X) + alkene \rightleftharpoons (Br_8TPP)Cr^{IV}O + X + alkene^{\cdot +} \qquad (4c)$$

The emf values for $1e^-$ oxidations of various alkenes by $(Br_8TPP)Cr^{V}(O)(X)$ are given as emf $= P_1 - P_2$ and the standard free energies (ΔG°) for the $1e^-$ transfer reaction were calculated from the relationship $\Delta G^\circ = (n = 1) \times F \times$ (emf). The free energies of activation for oxidation of alkenes by $(Br_8TPP)Cr^{V}(O)(X)$ were calculated by use of the Eyring equation [12]. The dashed line in Figure 1 separates those alkenes for which $\Delta G^a < \Delta G^\circ$ and those for which $\Delta G^a > \Delta G^\circ$. The standard free-energy for the formation of the putative $(Br_8TPP)Cr^{IV}(O)$ + alkene π-cation radical intermediates is greater than the observed free energy of activation for the reaction for about half of the alkenes investigated. For example, ΔG° for the formation of $(Br_8TPP)Cr^{IV}(O)$ + 1-octene π-cation radical exceeds the observed ΔG^a by about 20 kcal/mole. An immediate concern is the plausibility of an intermediate solvent-caged/ $(Br_8TPP)Cr^{IV}(O)$ + 1-octene π-cation radical/species, which must be stabilized, relative to the solvent-separated pair by up to 20 + Y kcal/mole (where Y = ΔG^a for reversion of solvent-caged intermediates to solvent-separated alkene + $(Br_8TPP)Cr^V(O)$). A stabilization of this magnitude seems unlikely.

The possibility of a change of a rate-limiting step in mechanism for the oxidation from one involving intermediate formation of solvent-caged/ $(Br_8TPP)Cr^{IV}(O)$ + alkene π-cation ratical/ pair (when $\Delta G^a > \Delta G^\circ$) to a concerted reaction or one involving the intermediacy of a covalent species (when $\Delta G^a < \Delta G^\circ$) does not seem likely since there is no break in the plot of log k_1 vs. $E_{1/2}$ over a wide range of free energies studied. It was therefore concluded that the mechanism involving the formation of an alkene-derived π-cation radical in the rate-limiting step could not be correct.

Electrophilic addition of the metal-bound oxygen to the alkene double bond to provide a carbocation intermediate is a mechanism that is generally consistent with the energetics given by the linear free-energy relationship of Eq. 1. Cationic intermediates have been proposed in order to explain the presence of rearranged products [8–11,13–15]. The proposition has been made that the most likely candidate is an $M\text{-}O\text{-}C\text{-}C^+$ species with the positive charge partially stabilized by interaction with the electron cloud on porphyrin nitrogens. In order to compare this mechanism to the proposed rate-limiting CT mechanism, the logs of rate constants for oxidation of a series of substituted styrenes by $(Br_8TPP)Cr^V(O)(X)$ were plotted vs. substituent σ^+ values. The value of the slope ρ^+ of the resulting Hammet plot is -1.9. The reported ρ^+ values for electrophilic additions to substituted styrenes, which involve the formation of carbocation intermediates, are much greater, i.e., -3.58 (hydration) [16] and -4.8 (bromination) [17]. This indicates that if the electrophilic addition mechanism were operative, an early transition state is necessary to accommodate the observed slope.

Another mechanism that is consistent with the correlation of Eq. 1 is a concerted insertion of oxygen into the alkene double bond. A plot of the log of second-order rate constant for the oxidation of 15 alkenes by *meta*-chloroperbenzoic acid vs. $E_{1/2}$ possesses a slope of $-2.4\ V^{-1}$ [18]. This value can be compared to the slope of $-2.99\ V^{-1}$ of the plot of log k_1 vs. $E_{1/2}$ for the oxidation of alkenes by $(Br_8TPP)Cr^V(O)(X)$. It is obvious that oxidation of alkenes by both *meta*-chloroperbenzoic acid and $(Br_8TPP)Cr^V(O)(X)$ is equally dependent on the electron availability of the alkene. A nonsymmetric transition state for peracid expoxidations and carbene insertions into C-C double bonds has been proposed, wherein one carbon carries a partial positive charge [19,20]. To continue, the following values of ρ have been published for related electrophilic reactions: −0.62 for $:CCl_2$ insertion into the double bond of substituted styrenes using $C_2H_5ZnCH_2I$ [22] and −1.2 for epoxidation of substituted stilbenes with perbenzoic acid [22].

A number of oxidation products other than epoxide are obtained in reactions of oxometalloporphyrins and alkenes (Table 1) [3]. This would tend to argue that if various reaction products are formed in partially rate-limiting steps, the linear free-energy relationships would not hold. However, a simple linear free-energy relationship appears to hold over a wide range of oxidation potentials and alkene structures. It seems reasonable to assume that different oxidation products arise from the common intermediate which is formed in the rate-limiting step. The only mechanism which is fully consistent with these observations is the rate-limiting formation of CT complex between hypervalent oxometalloporphyrins and alkenes. Rate-limiting CT complex formation has been proposed for the reaction of triplet oxygen atoms with alkenes in the gas phase [23]. Bearing in mind the partial triplet "oxenoid" character of the metal-bound oxygen in hypervalent oxometalloporphyrins, a striking similarity becomes evident [24].

C. Carbocation Radical Intermediate of Alkene Oxidation by Oxochromium(V) Porphyrins

The unified mechanism of Scheme 2 has been proposed to account for (a) the various products formed on reaction of alkenes with oxo(*meso*-tetrakis(2,4,6-trimethylphenyl)porphinato)iron(IV) π-cation radical [$(^{+}\cdot TMP)Fe^{IV}(O)$] and oxo(*meso*-tetrakis(2,6-dibromophenyl)porphinato)chromium(V) [$(Br_8TPP)Cr^{V}(O)(X)$]; and (b) the sensitivity of the log of the second-order rate constants for such reactions to the electron density of the alkene ($E_{1/2}$ and σ^+). In the mechanism the rate-limiting step in the oxidation of alkenes by hypervalent metal-oxoporphyrin species is the formation of a CT complex. The CT complex then partitions, dependent upon the alkene and the metal, to

Table 1 Oxidation Products in Reactions of $(Br_8TPP)Cr^{V}(O)(X)$ with Various Alkenes

	Product yield (%)		
Alkene	Predicted from Scheme 1	Observed	
1-Octene	2.8	3.3	(100% 1,2-epoxyoctene)
cis-Stilbene	68	73	(21% *cis*-stilbene oxide, 52% 1,1-diphenyl-acetaldehyde)
Styrene	61	65	(59% styrene oxide, 6% phenylacetaldehyde)
Norbornene	64	63	(100% *exo*-norbornene oxide)

various oxidation products such as epoxide and carbocation radical. From a combination of kinetic and electrochemical studies we have shown that the carbocation radical of **III** cannot be formed *in the rate-limiting step* in the oxidation of alkenes by $(Br_8TPP)Cr^V(O)(X)$. Furthermore, **III** cannot be an intermediate in epoxidations by $(Br_8TPP)Cr^V(O)(X)$ if the $1e^-$ oxidation potential of the alkenes exceeds ~1.7 V (SCE) [2,3]. For electron-rich alkenes, however, $1e^-$ oxidation is possible and carbocation radical (**III**) should be considered as a possible intermediate on the reaction path [3]. (See Chart 1, p. 43, for structures of **I, II, III, IV**, and **V.**)

Oxidation of (*Z*)-1,2-bis(*trans*-2,*trans*-3-diphenylcyclopropyl)ethene (**1**) by $(Br_8TPP)Cr^V(O)(X)$ in the presence of the nonoxidative electrolyte $(n\text{-}Bu)_4NBF_4$ provides as the major product a mixture of the two isomeric (3*E*,5*E*)-1,8-dichloro-1,2,7,8-tetraphenylocta-3,5-diene (referred to as *trans, trans*-**3**) and (3*E*,5*Z*)-1,8-dichloro-1,2,7,8-tetraphenylocta-3,5-diene (referred to as *trans, cis*-**3**) [25]. The same diene mixture is obtained as the major product (and same ratio of the configurational isomers) on $1e^-$ oxidation of **1** by controlled-potential bulk electrolysis (CPBE) in CH_2Cl_2 using the nonoxidative $(n\text{-}Bu)_4NBF_4$ as the supporting electrolyte. This shows that the oxidation of **1** by $(Br_8TPP)Cr^V(O)(X)$ provides both *trans, trans*-**3** and *trans, cis*-**3** by way of an intermediate carbocation radical (Scheme 3). The diene products arise from carbocation radical by opening one cyclopropyl ring via a cyclopropylcarbinyl to homoallyl-carbinyl radical rearrangement (CPCRR), while the other cyclopropyl ring opens in a cyclopropylcarbinyl to homoallylcarbinyl carbocation rearrangement (CPCCR). The intermediate radical and cation species are trapped by the solvent decomposition products chlorine radicals and chloride, respectively. In contrast to the oxidation of **1** by $(Br_8TPP)Cr^V(O)(X)$ in the presence of the nonoxidative electrolyte $(n\text{-}Bu)_4NBF_4$, which provides the mixture of dienes plus $(Br_8TPP)Cr^{IV}(O)$, oxidation of **1** by $(Br_8TPP)Cr^V(O)(X)$ in the presence of the cooxidant $(n\text{-}Bu)_4NClO_4$ provides *trans*-2, *trans*-3-diphenylcyclopropanecarboxyaldehyde as a major product plus the immediate product $(Br_8TPP)Cr^{III}(X)$. A cooxidation of **1** by $(Br_8TPP)Cr^V(O)(X)$ and ClO_4^- is required to explain the high-yield formation of the aldehyde [25,26].

Since the second-order rate constants are much the same for the reaction of **1** with $(Br_8TPP)Cr^V(O)(X)$ in the presence of $(n\text{-}Bu)_4NBF_4$ or $(n\text{-}Bu)_4NClO_4$, it is concluded that the rate-limiting step for reaction of **1** with $(Br_8TPP)Cr^V(O)(X)$ is separate from the product-forming reactions and that the formation of carbocation radical **2** (Scheme 3) is not the rate-limiting step. Also, the point for **1** fits well in the general correlation for the rate-limiting formation of CT complex of Figure 1 despite the fact that the oxidation potential of **1**[1.47 V (SCE)] is such that the full $1e^-$ oxidation by $(Br_8TPP)Cr^V(O)(X)$ is allowed. These results are again consistent with the rate-limiting formation of a CT complex and the general mechanism of Scheme 2.

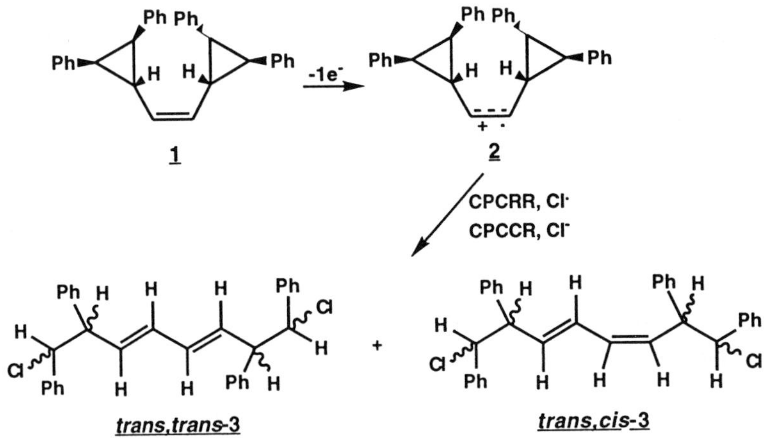

Scheme 3

D. Stereoselectivity of Alkene Oxidation by Oxochromium(V) Porphyrins

Stereoselectivity has been reported for the epoxidation of alkenes catalyzed by metalloporphyrins [14,27–31]. In order to better understand cis vs. trans stereoselectivity in alkene epoxidation we have compared the dynamics, products, and van der Waals closest approach minima in the reaction of oxo(*meso*-tetrakis(2,6-dibromophenyl)porphinato)chromium(V) [$(Br_8TPP)Cr^V(O)(X)$] with a series of terminal, *cis*-, and *trans*-alkenes.

Of considerable assistance to this study is the previous finding that the log of the second-order rate constants (k_1) for *cis*-alkene oxidation by $(Br_8TPP)Cr^V(O)(X)$ is a linear function of the alkene $1e^-$ oxidation potential (Figure 1) [3]. The use of this linear free-energy

$$\log k_1 = -2.99E_{1/2} + 4.02 \qquad (5)$$

relationship allows the quantification of steric factors brought about by a trans disposition of substituent groups through comparison of determined values of k_1 and values (k_{1cal} calculated from Eq. 5 [25]. In Figure 1 the data points for the 16 *cis*- and terminal alkenes are shown as dark squares and for the nine *trans*-alkenes open squares are used [25]. *trans-p,p'*-Dicyanostilbene exhibited no reactivity with $(Br_8TPP)Cr^V(O)(X)$. The data point for this alkene in Figure 1 indicates that the value of k_1 is less than $4 \times 10^{-5}\ M^{-1}\ s^{-1}$.

A finding of considerable interest is that *trans*-alkenes do not generally react with the sterically restricted $(Br_8TPP)Cr^V(O)(X)$ with second-order rate constants (k_1) which are appreciably less than those predicted from electronic effects (k_{1cal}) [25]. Thus (Figure 1), values of k_1 for the oxidation of *trans-p,p'*-dimethoxystilbene, *trans-β*-methylstyrene, *trans*-2-pentene, *trans*-5-decene, and *trans*-2-hexenyl acetate are very close to the values of k_{1cal} anticipated (Eq. 5) from their electron availability. Therefore, there can be little or no steric hindrance for these alkenes in the rate-limiting step. A significant steric effect is found only for *trans*-stilbene, *trans-p,p'*-dicyanostilbene, and (*E*)-1,2-bis(*trans*-2, *trans*-3-diphenylcyclopropyl)ethene. The influence of this steric effect is not only discernable in k_{1cal}/k_1 but also from product analysis. Especially for *trans*-stilbene and *trans-p,p'*-dicyanostilbene no oxidation products could be detected. The fact that *trans*-stilbene cannot be epoxidized was also found for the system using $(Br_8TPP)Fe^{III}(Cl)$ and perfluoroiodosylbenzene [32,33].

The lack of a steric effect (dynamics and products) in the oxidation of *trans-p,p'*-dimethoxystilbene with $(Br_8TPP)Cr^V(O)(X)$ deserves comment because the X-ray structure of *trans-p,p'*-dimethoxystilbene [34] and *trans*-stilbene [35] are comparable. They have a similar planar structure and the same C=C bond length in solid state (1.316 and 1.313 Å for *trans-p,p'*-dimethoxystilbene and *trans*-stilbene, respectively). The potentials P_1 of Eq. 6 are provided in Figure 1 while the potential P_2 (Eq. 7) is equal to 0.88 V(SCE) [3].

$$>C(\cdot)-C(+)< + e^- \rightarrow >C=C< \qquad P_1 \qquad (6)$$

$$(Br_8TPP)Cr^V(O)(X) + e^- \rightarrow (Br_8TPP)Cr^{IV}(O)(X) \qquad P_2 \qquad (7)$$

The value of the standard free energy ($\Delta G°$) for the reaction of Eq. 8 is equal to unity when $P_1 = 0.88$ V. Thus the 1e$^-$ oxidation of the

$$>C=C< + (Br_8TPP)Cr^V(O)(X) \rightarrow >C(\cdot)-C(+)< + (Br_8TPP)Cr^{IV}(O)(X) \qquad (8)$$

alkenes in Figure 1 (Eq. 8) are all endergonic. For *cis*-stilbene and *trans*-stilbene the oxidation by $(Br_8TPP)Cr^V(O)(X)$ is endergonic by 0.76 and 0.63 V, respectively [25]. For *trans-p,p'*-dimethoxystilbene the 1e$^-$ oxidation in Eq. 8 is endergonic by only 0.17 V. We have proposed a unifying mechanism (Scheme 2) to explain the formation of epoxide and other products [3]. In Scheme 2 the formation of a CT complex between alkenes and $(Br_8TPP)Cr^V(O)(X)$ is rate limiting. In Figure 1 for those alkenes whose values of P_1 are to the left of the intersection of the dashed line and linear free energy line $\Delta G° < \Delta G^a$ so that 1e$^-$ transfer (Eq. 8) is allowed along the reaction path to products. For alkenes whose P_1 values are to the right of this intersection $\Delta G° > \Delta G^a$ and 1e$^-$ transfer along the reaction path is not allowed [2,3]. Therefore 1e$^-$ transfer (Eq. 8) along the reaction coordinate is allowed with *cis*- and *trans*-stilbene and particularly *trans-p,p'*-dimethoxystilbene. The distance of approach of reactants is of importance to an electron transfer reaction. The calculated van der Waals energy minimum for the approach of *trans*-stilbene to $(Br_8TPP)Cr^V(O)(X)$ occurs (Figure 3B) at a distance of the center of the C=C bond to the oxygen in Cr^VO moiety of 3.98 Å (vide infra), which is well within the distance for an electron transfer [36,37]. Since the potential P_l for *trans-p,p'*-dimethoxystilbene is favorable for an electron transfer, this may be the mode of its reaction with $(Br_8TPP)Cr^V(O)(X)$. A similar lack of a trans steric effect, which may be attributed to a very favorable electron transfer from alkene to $(Br_8TPP)Cr^V(O)(X)$, is apparent in the comparison of k_l and k_{lcal} for 1,4-diphenyl-1,3-butadiene (Figure 1) [3].

In the unifying mechanism the rate-limiting step has been considered as the formation of a CT complex (Scheme 2) [3,25]. The steric requirement of the formation of the CT complex should be the origin of the stereoselectivity of the epoxidation reaction. For charge transfer to occur the highest occupied molecular orbital (HOMO) of alkenes should be sufficiently close to the lowest unoccupied molecular orbital (LUMO) of the reactive center of the hypervalent oxometalloporphyrin. For alkenes the HOMO is the π orbital of the double bond. For $(Br_8TPP)Cr^V(O)(X)$, as suggested by several theoretical studies [38–40], the LUMO is the two empty Cr-O (d_π-p_π) antibonding orbitals (Scheme 4).

In order to determine the energy minima for approach of alkenes to $(Br_8TPP)Cr^V(O)(X)$ without any restraints, detailed docking studies were carried out in which both the docking distance r and angle α were varied

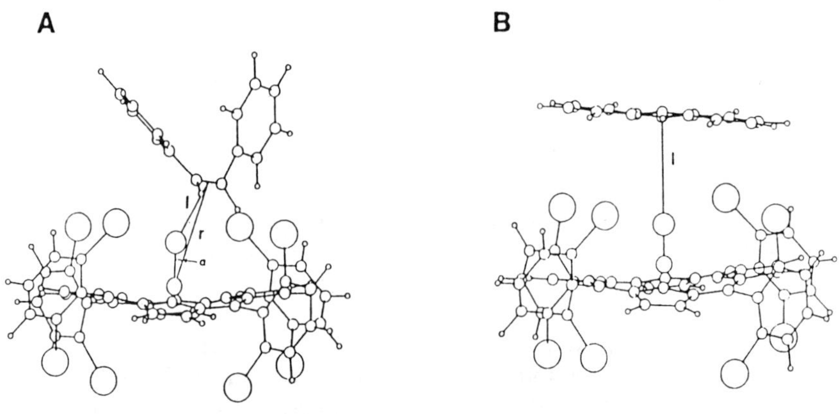

Figure 3 The lowest energy minima positions from the docking experiments: (A) *cis*-stilbene on $(Br_8TPP)Cr^V(O)(X)$ with $\theta = 45°$, $\alpha = 24°$, $r_{min} = 4.10$ Å, and $l_{min} = 2.81$ Å; (B) *trans*-stilbene on $(Br_8TPP)Cr^V(O)(X)$ with $\theta = 0°$, $\alpha = 0°$, $r_{min} = 5.45$ Å, and $l_{min} = 3.88$ Å.

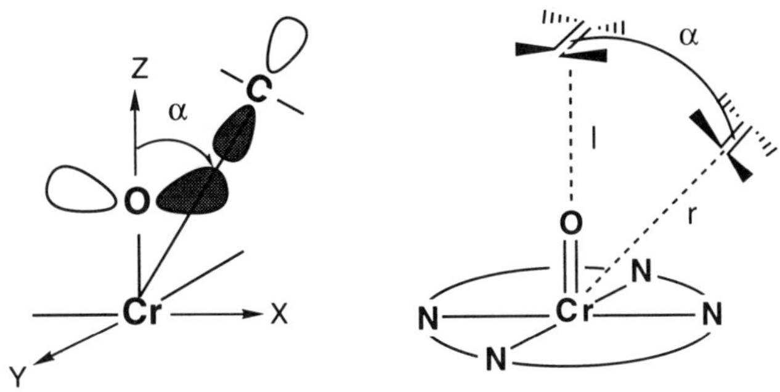

X Y

Scheme 4

(**X** and **Y**). The calculated distance thus obtained between the C=C bond of *cis*- and *trans*-stilbene and Cr in $(Br_8TPP)Cr^V(O)(X)$ at the lowest energy minima positions (l_{min}, r_{min}, α, and θ) are given in Figures 3A and 3B, respectively. In Figure 3A, The C=C bond aligns at a bisect of the 90° angle formed between Cr^V and adjacent pyrrole nitrogens ($\theta = 45°$), and the minimum in potential energy, −7.63 kcal/mole, occurs when *cis*-stilbene lies at 2.81 Å away from the oxygen (l_{min}) with $\alpha = 24°$. For *trans*-stilbene in Figure 3B the C=C bond is parallel with the N-Cr-N fragment ($\theta = 0°$), and the minimum potential energy is equal to −6.34 kcal/mole with $l_{min} = 3.88$ Å and $\alpha = 0°$. Any deviations in distance or angle resulted in higher van der Waals energy. A similar computer docking study was performed for trans-β-methylstyrene. Interestingly, *trans*-β-methylstyrene shows a steric effect similar to that of *cis*-stilbene (l_{min} of 3.13 Å vs. 2.98 Å) rather than *trans*-stilbene (l_{min} = 3.88 Å). This finding is consistent with the results from the kinetic study (Figure 1) and strongly suggests that both configuration and substituent size of the alkenes are important in determining the steric hindrance of the epoxidation reaction.

The closest distance (l_{min}) between the nucleophilic C=C bond and the electrophilic oxygen of $(Br_8TPP)Cr^V(O)(X)$ can also be obtained from the study. For *trans*-stilbene r_{min} is 0.90 Å longer than that for the *cis*-stilbene. This steric difference is even more enhanced when α varies away from 0°. When $\theta = 45°$, *cis*-stilbene can be docked to the porphyrin from a range of angle α with less steric hindrance, while docking of *trans*-stilbene is still restricted at $\alpha = 0°$ with higher van der Waals energy. Therefore, the stereoselectivity of *cis*- over *trans*-stilbene can be considered to be understood. The formation of a CT complex usually requires only a moderate separation of the two interacting moieties. The critical distance for CT complex to occur was reported to be 3.5 Å or less [37], whereas an electron transfer reaction can proceed at a relatively long distance, ~7–9 Å or even more [36]. As shown in Figure 3, the closest distance between *cis*-stilbene and the oxygen in porphyrin (l_{min}) is less than 3 Å, and for *trans*-stilbene it is ~4 Å. From the product analysis, *cis*-stilbene is quantitatively oxidized but *trans*-stilbene is not oxidized. These results are in accord with a mechanism involving rate-limiting formation of a CT complex. The mechanism of Scheme 2 is also applicable to alkene epoxidation catalyzed by iron porphyrins since similar stereoselectivity is found in both iron and chromium systems [14, 27–29].

III. IRON PORPHYRINS

A. Introduction

Higher valent iron-oxoporphyrins are exceedingly reactive species and, unlike higher valent chromium-oxoporphyrins, cannot be prepared for oxidation studies under stoichiometric conditions except at very low temperatures. With iron(III) porphyrins charging of the catalyst is the rate-limiting step, so that the mechanism of epoxidation could not be studied using conventional kinetic approaches at ambient temperatures (Eq. 9).

$$(P)M^{III} + [O] \dashrightarrow (+\cdot P)M^{IV}{=}O \qquad \text{slow} \qquad (9a)$$

$$(+\cdot P)M^{IV}{=}O + S \rightarrow\rightarrow\rightarrow (P)M^{III} + SO \qquad \text{fast} \qquad (9b)$$

A $\rho+$ value reported for the reaction of substituted sytrenes with oxo-(*meso*-tetrakis(2,4,6-trimethylphenyl)porphinato-π-cation radical)iron(IV) is -1.9 [41], indicating that the substitution of chromium for iron does not significantly change the energetics of the rate-limiting step of the reaction with alkenes.

Although our goal of establishing the rate-limiting step in the reaction of alkenes with hypervalent oxometalloporphyrins seemed to be met, the answers pertaining to the mechanism of epoxidation remained unanswered. Epoxidation as well as other oxidation reactions occur after the rate-limiting formation of the CT complex, rendering kinetic analysis useless. However, the intricate steps that follow the formation of CT complex can be elucidated by careful examination of the reaction products. This has been accomplished by studying oxidation of a wide variety of structurally and electronically different alkenes with different hypervalent iron-oxoporphyrins under catalytic conditions.

Deliberations concerning the mechanism of epoxidation have revolved around questions of the intermediacy of **(I)** metallaoxetane [42–44], **(II)** π-radical cation [7–11], **(III)** carbocation [7–11,13–15], and **(IV)** carbon radical [41,45–50] as opposed to **(V)** concerted "oxene" insertion [51–54] (Chart 1).

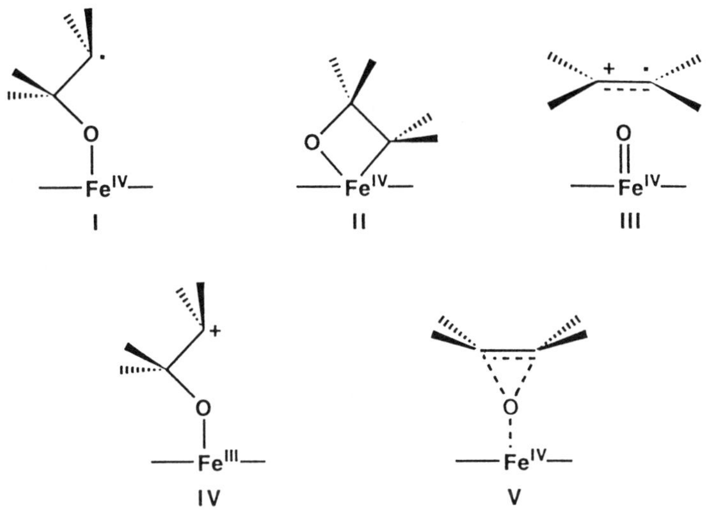

Chart 1

The stereochemistry of epoxide products relates directly to the mechanism of epoxidation. The nature of other products may not be relevant to the mechanism of epoxidation if these side products arise from reactions that are parallel and competing with the epoxidation reaction. However, much of the evidence that has been used in support of the involvement of intermediates **I**, **II**, **III**, and **IV** during epoxidation is based on the structures of products other than epoxide. The numerous products can be accommodated by schemes in which they either are formed from intermediates on the way to epoxide or are the products of reaction paths parallel to and competing with epoxidation.

B. Metallaoxetane Intermediate

Formation of metallaoxetane intermediate (**I**) was proposed to occur by a 2a + 2s cycloaddition reaction and to give way to epoxide plus metal(III) porphyrin via a concerted reductive elimination [42–44]. Though alternate rationales have been advanced for the spectral and kinetic observations offered in support of the accumulation of **I**, the occurrence of **I** as a low-concentration steady-state intermediate remained in contention [55].

Because both formation and breakdown are concerted processes, any loss of stereochemical control has been relegated to a minor parallel reaction involving metallaoxetane ring opening to give an acyclic intermediate

(M-O-C-C). With styrene as substrate, phenylacetaldehyde is obtained as a product. Phenylacetaldehyde has been postulated to result from a metallaoxetane **4b**, which is regioisomeric with the metallaoxetane **4a**, responsible for epoxide formation (Scheme 5). Rearrangement of **4b** to aldehyde is proposed to occur by preferential migration of the *cis*-hydrogen (deuterium) in *cis*- and *trans*-styrene-β-d to give the deuterated phenylacetaldehydes because the conformational preference of the four-membered ring results in the cis C-H bond being better aligned for rearrangement (Scheme 5) [15]. These conclusions were based on the distribution of label in phenylacetaldehyde derived from the oxidation of deuterium-labeled styrenes and the lack of phenyl migration to give acetophenone. The stereoelectronic control required for preferential migration of *cis*-H(D) is not present in a rotationally unrestricted acyclic intermediate. However, a quasiclosed cationic intermediate with a positive charge on the benzylic center stabilized by the electron cloud on porphyrin nitrogens can be used as well as a metallaoxetane to explain the 1-d/2-d phenylacetaldehyde product ratios.

With *cis*-stilbene only one metallaoxetane (**4c**) can be formed. Explicit in the metallaoxetane hypothesis is a much faster collapse of **4a** to epoxide compared to phenyl migration. With **4c** the situation should be different, since the metal–carbon bond in **4c** should have nearly the same polarization as in **4a**. Therefore, *cis*-stilbene should behave in an analogous manner to styrene, and no phenyl migration should occur. Since no hydrogens are in the proper spatial arrangement for their migration, no rearrangement products should be seen. This predicted outcome is not realized, and all catalytic systems we used gave substantial amounts of diphenylacetaldehyde resulting from phenyl migration and deoxybenzoin resulting from hydrogen migration. Significantly, the yield of diphenylacetaldehyde (phenyl migration rearrangement product) is always much greater than the yield of deoxybenzoin (hydrogen migration rearrangement product). This observation is in accord with

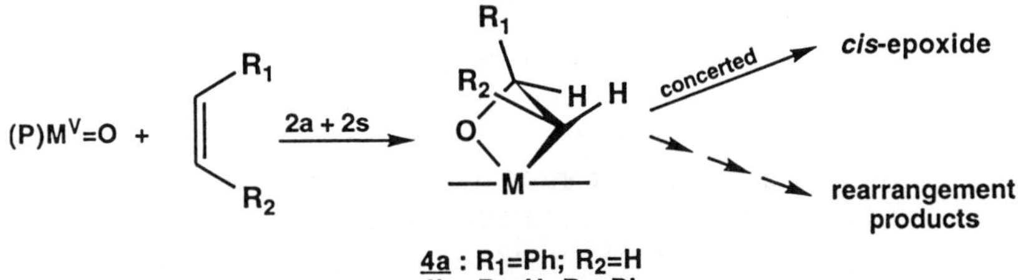

4a : R_1=Ph; R_2=H
4b : R_1=H; R_2=Ph
4c : R_1=R_2=Ph

Scheme 5

the formation of cationic intermediate, since phenyl migration to the carbocation center is favored over hydrogen migration [14].

If acyclic or quasiclosed cationic intermediates (M-O-C-C$^+$) (**IV**) are involved in alkene epoxidation, then the intermediate formed from styrene would be a substituted benzyl carbocation so that phenyl migration would be disfavored. This would explain why phenyl migration to provide acetophenone does not occur during styrene epoxidation. One argument used in favor of a metallaoxetane intermediate and against a carbocation intermediate is based on the lack of substituent effects on the ratio of epoxide to phenylacetaldehyde obtained by hydrogen migration. Although substituent effects would control the rate of formation of the carbocation intermediate, these would have a questionable effect on the partitioning of carbocation to epoxide and phenylacetaldehyde.

The preceding discussion successfully challenged the rationale for obligatory involvement of metallaoxetane intermediate. Most of the results could be rationalized equally well by invoking a carbocation intermediate. Still, it was not clear whether the presence of rearrangement products during epoxidation can be used to dismiss a metallaoxetane intermediate since they may arise from the pathways parallel and competitive to epoxidation.

All *cis*-alkenes, including terminal alkenes, are cleanly epoxidized when using (Br_8TPP)Fe^{III}(Cl) as catalyst with C_6F_5IO as oxidant to generate the (+·Br_8TPP)Fe^{IV}(O) epoxidizing species. Though almost quantitative yields of epoxide were obtained, approach of the alkene to the iron–oxo reaction center is restricted by the bulk of the eight *ortho*-bromo substituents. We have used molecular modeling to determine if formation of **I** is possible for the reaction of alkene with (+·Br_8TPP)Fe^{IV}(O) [32,33].

Two template X-ray structures were used as possible models for (Br_8TPP) Fe^{III}(Cl): (a) *meso*-(tetraphenylporphinato)iron(III) chloride ((TPP)Fe^{III}(Cl)) [56] and *meso*-(tetraphenylporphinato-π-cation radical)iron(III) chloride hexachloroantimonate ((+·TPP)Fe^{III}(Cl)($SbCl_6$)) [57]. Model (a) is justified on the basis of the same spin multiplicity as in the catalase and horseradish peroxidase oxidants and by the fact that metalloporphyrins and porphyrin radical cations are most likely flat in solution. Model (b) was considered because it shows the degree to which the porphyrin ring can deviate from planarity. *o*-Bromo substituents on the phenyl rings were appended with the fixed bond distance of 1.85 Å. Due to the steric interaction of bromine atoms with pyrrole carbons, the phenyl groups can rotate to a minimal dihedral angle of 70° with respect to the plane of the porphyrin ring. Although this relieves some steric interaction with the incoming substrate, considerable steric restrictions are still present. For modeling of the putative metallaoxetane intermediate, two template X-ray structures were considered: (a) iridium azametallacyclobutane (Cp*(PMe_3)Ir($CH_2CMe_2NH_2$+)) [58] (bond lengths in the four-membered ring Ir-C = 2.11 Å, C-C = 1.52 Å, C-N = 1.51 Å, N-Ir = 2.12 Å) and (b) platinum oxametallacyclobutane $[(As(C_6H_5)_3)_2Pt(C_2(CN)_4O)]$ [59] (bond lengths in the four-membered ring Pt-C = 2.10 Å, C-C = 1.58 Å, C-O = 1.36 Å, O-Pt = 2.05 Å). Both structures have similar bond lengths and angles in the relevant four-membered ring, and the slightly more relaxed $Cp^*(PMe_3)Ir(CH_2CMe_2NH_2^+)$ was used as the actual template for the proposed iron metallaoxetane. Structures of metallaoxetane ring systems derived from *cis*-stilbene, *trans*-stilbene, and 2,3-dimethyl-2-butene were constructed by fixing the structure of the four membered ring according to the aforementioned X-ray structure and energy minimizing the orientation of the substituents. Assembly of the putative porphyrin metallaoxetanes was carried out by superimposing the metal of the porphyrin and that of metallaoxetane, followed by readjustment of the substituent positions to provide minimal steric interactions, by rotation to positions of minimal energy [33].

Examination of the constructed structures of putative metallaoxetanes formed from the $(+\cdot Br_8TPP)Fe^{IV}(O)$ moiety and the alkenes *cis*-stilbene and 2,3-dimethyl-2-butene shows (a) for the iron *cis*-diphenylmetallaoxetane porphyrin, one phenyl ring is invaginated into both the porphyrin ring and the *o*-bromo substituents; and (b) for the iron tetramethylmetallaoxetane porphyrin structure, two of the methyl groups are deeply penetrated into the porphyrin plane. In order to avoid these unfavorable steric interactions, the Fe-C and Fe-O bond lengths must be extended to unreasonable values of >3.2 Å. Both *cis*-stilbene and 2,3-dimethyl-2-butene epoxidize in almost quantitative yields. In order to elaborate on this point, an extremely "dished" structure of $(+\cdot Br_8TPP)Fe^{IV}(O)$ was constructed in which the iron was 0.46 Å out of plane, defined by porphyrin pyrrole nitrogens, and 0.99 Å out of plane, defined by pyrrole β carbons. The *cis*-stilbene metallaoxetane moiety was then docked on this porphyrin structure. In this case shorter bond distances were achieved (Fe-C = 2.72 Å and Fe-O = 2.37 Å; this is a minimal Fe-O distance sterically allowed by O-N pyrrole repulsions even when Fe is 0.46 Å out of plane, defined by porphyrin nitrogens; compare to Pt-O = 2.05 Å and Pt-C = 2.10 Å, the distances in the X-ray structure of a known metallaoxetane [59]). Also, with this porphyrin model a *trans*-stilbene porphyrin metallaoxetane with identical geometry of the four-membered ring can be constructed without any additional steric interactions, suggesting that if this model was correct, *trans*-stilbene should also epoxidize via metallaoxetane. This, however, is not observed. The studies described above exclude metallaoxetanes as required intermediates in alkene epoxidation by hypervalent metal-oxoporphyrin species.

C. Radical Intermediate

To explore the intermediacy of the radical species in alkene epoxidation a special substrate was required. The cyclopropyl substituent has been used in many chemical and enzymatic reactions as a radical trap. Rate constants for the cyclopropylcarbinyl to homoallyl radical rearrangement (CPCRR) of a 2° to 1° radical is too small ($2 \times 10^7\ s^{-1}$, 30°C) to trap a radical intermediate that collapses to a product with a rate constant $> 10^{10}\ s^{-1}$ [60]. We established from competitive experiments that the rate constant for the 2° → 2° benzyl CPCRR of the *trans*-2, *trans*-3-diphenylcyclopropylcarbinyl radical exceeds the rate constant for CPCRR of 2° → 1° radical of cyclopropylcarbinyl radical by at least a factor of 10^3 [49].

The alkene (*Z*)-1,2-bis(*trans*-2, *trans*-3-diphenylcyclopropyl)ethene (**1**) (Scheme 6) incorporates the *trans*-2, *trans*-3-diphenylcyclopropyl group as a hypersensitive radical trap into a suitable substrate for epoxidation. To establish that CPCRR will proceed in **1**, the alkene was photocatalytically reduced with *n*-Bu_3SnH. ^1H NMR analysis of the two resulting tin adducts showed that they were diastereomers and were composed of a cyclopropylvinyl and a 1,2-diphenylethyl spin system connected via a SnCH methine

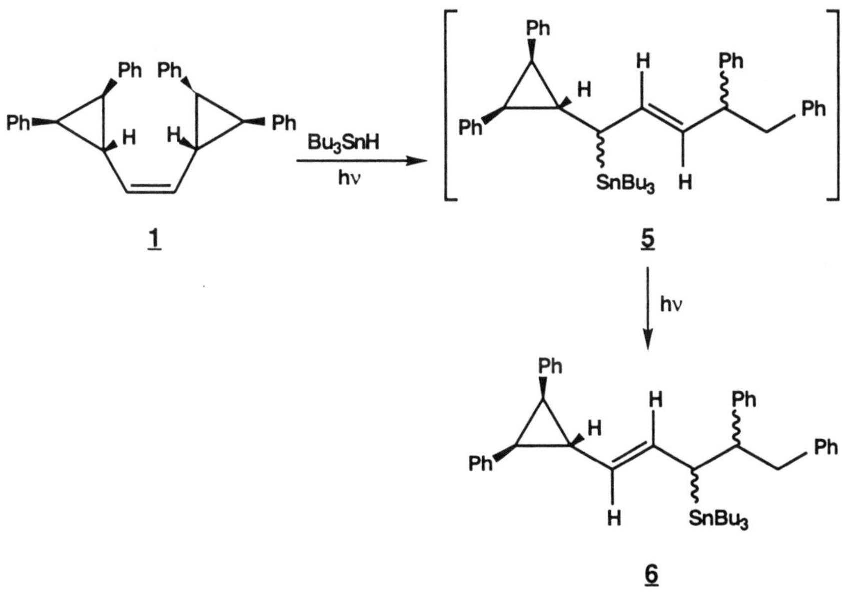

Scheme 6

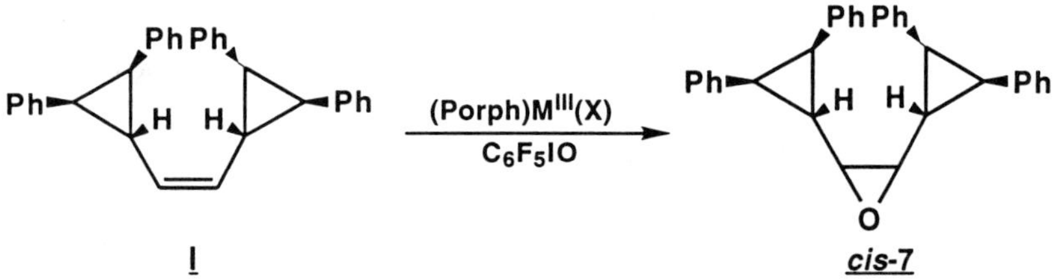

Scheme 7

hydrogen. Thus, one *trans*-2, *trans*-3-diphenylcyclopropyl group remained intact, while the other had undergone a CPCRR to give **5**. Subsequent allyl tin rearrangement of **5** produced **6**.

With C_6F_5IO as oxidant, the catalysts employed in the epoxidation of **1** were $(F_{20}TPP)Fe^{III}(Cl)$, $(Cl_8TPP)Fe^{III}(Cl)$, and $(Cl_8TPP)Mn^{III}(OH)$ (Scheme 7) [50]. Reactions were carried out under anaerobic conditions in CH_2Cl_2. The results are presented in Table 2. They show all systems giving high yields of the *cis*-epoxide (>80% based on reacted **1**). In none of the reactions investigated could either *trans*-alkene or *trans*-epoxide be detected at 0.1% detection limit.

The remaining reacted **1** was found as very nonpolar non–oxygen-containing products at 8–16% yield dependent on catalyst. The structures of two nonpolar products (Scheme 8) were found to be **8** and **9** after the examination of the fragmentation pattern from the mass spectra. The identity of the substituent X could not be established, but it is most likely Cl, since chloride ions are generated from oxidation of the reaction solvent CH_2Cl_2. Pathways that would explain the formation of **8** and **9** from a common cation radical intermediate **2** generated by one-electron oxidation of **1** are given in Scheme 8.

Knowing that the rate constant for CPCRR of the (*trans*-2, *trans*-3-diphenylcyclopropyl)carbinyl radical is $>2 \times 10^{10}\ s^{-1}$ and the yields of non-

Table 2 Epoxide Yields in the Reactions of Alkene **1** with Several Metalloporphyrins as Catalysts

Catalyst	% yield of epoxide **7**[a]
$(F_{20}TPP)Fe^{III}(Cl)$	95
$(Cl_8TPP)Fe^{III}(Cl)$	89
$(Cl_8TPP)Mn^{III}(OH)$	84

[a]Yield is based on reacted **1**.

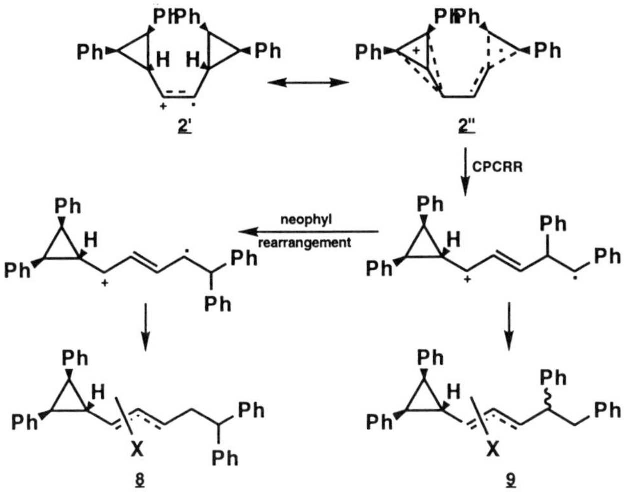

Scheme 8

polar products derived from cation radical **2** are 5–16% (Table 2), an obvious conclusion is that the rate constant for conversion of **2** to epoxide is required to be $>5 \times 10^{10}$ s^{-1} to 4×10^{11} s^{-1} (Scheme 9; Part A). A rate constant of about 10^{11} s^{-1} for the recombination of solvent-caged pair involving **2** to provide epoxide is not unreasonable. However, no products other than the epoxide **8** and **9** were found with (Cl_8TPP)Mn^{III}(OH). Therefore, if products attributable to the neutral radical were formed, their yields must fall below the 0.1% detection limit. Thus, the rate constant for formation of epoxide with (Cl_8TPP)Mn^{III}(OH) catalyst is required to be $>2 \times 10^{12}$ s^{-1}. A rate constant of such magnitude would prevent the detection of the neutral radical as a discrete intermediate. On this basis the neutral radical cannot be a discrete intermediate to epoxide formation with (Cl_8TPP)Mn^{III}(OH) as catalyst. This same analysis applies for (F_{20}TPP)Fe^{III}(Cl) and (Cl_8TPP)Fe^{III}(Cl) as catalysts because neither provided an oxygen-containing product other than **7**. This analysis is appropriate only if **2** is an intermediate in the formation of epoxide. It is possible that the nonpolar products are formed from the cation radical **2** in a reaction that is parallel to epoxidation.

Based on the preceding considerations, the intermediacy of an alkene-derived neutral radical species during the metalloporphyrin-catalyzed epoxidation can be dismissed.

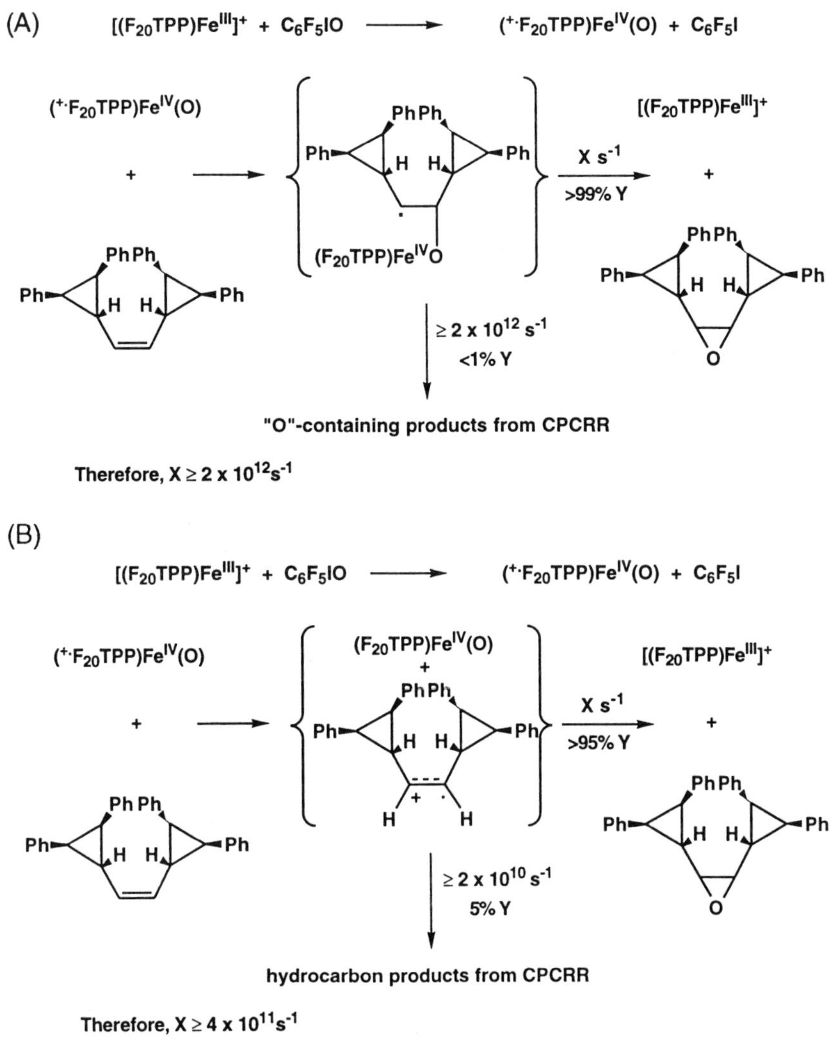

Scheme 9

D. Carbocation Radical and Carbocation Intermediates

The intermediacy of the carbocation radical and the carbocation has been proposed on the basis of the formation of rearranged products during epoxidation [8–11,13–15]. Hypervalent oxometalloporphyrins are strong $1e^-$ oxidants, and isomerization and rearrangement of alkenes through carbocation radicals is reasonable for those alkenes with low oxidation potential that

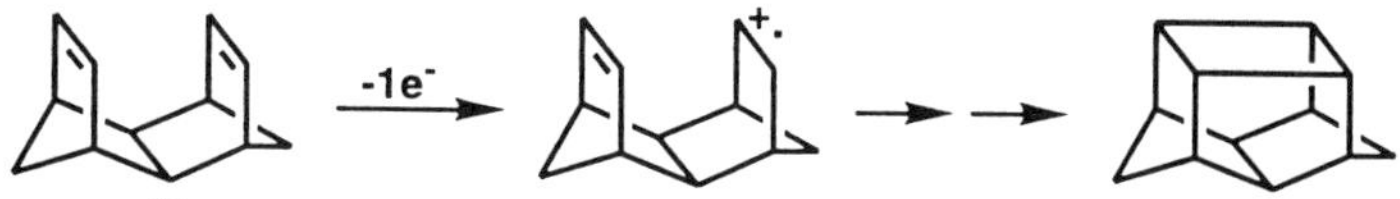

Scheme 10

are susceptible to rearrangement. For example, the rearrangement of hexamethyl Dewar benzene to hexamethylbenzene [9] and the rearrangement of alkene **10** offer good evidence for the intermediacy of a carbocation radical (Scheme 10) [101].

A mechanism that involves both carbocation radical and carbocation was proposed based on the structures of products obtained during epoxidation of norbornene with iodosylarenes using several substituted tetraphenylporphyrin and tetrapyridylporphyrin iron(III) salts as catalysts. *exo*-Norbornene oxide **(13)** and *endo*-norbornene oxide **(14)** were obtained in unusually low exo/endo ratios, along with norcamphor **(15)** and cyclohexene-4-carboxyaldehyde **(16)**. The proposed mechanism involves an initial $1e^-$ transfer from alkene to $(+\cdot\text{Porph})\text{Fe}^{IV}(\text{O})(\text{X})$ to form the intimate ion pair **(11)**, followed by the collapse of the intimate ion pair to exo and endo carbocations **(12)**, which would yield the final products (Scheme 11) [10].

The lack of stereospecificity was used to rule out the direct formation of the acyclic cationic intermediate which would give *exo*-epoxide only. Collapse of the ion pair would, however, be less discriminate and substantial yield of *endo*-epoxide could result. The observed trend was that the lower exo/endo ratios were observed with catalysts of increasing oxidation potential and increasing steric demand of the ortho substituents on the tetraphenylporphyrin moiety. Increased reactivity (oxidation potential) of the hypervalent oxometalloporphyrin in itself can explain lower selectivity. When we employed a $(\text{Br}_8\text{TPP})\text{Fe}^{III}(\text{Cl})/\text{C}_6\text{F}_5\text{IO}$ catalytic system to epoxidize norbornene, norbornene oxides were obtained in 99% combined yield with the exo/endo norbornene oxide ratio of 6.8 [33]. No other products could be detected. Computer graphics docking experiments were used to assess steric factors involving norbornene H-C2 and H-C3 interactions with *ortho*-bromo substituents vs. *syn*-H-C5, *syn*-H-C6 (endo), and *syn*-H-C7 (exo) interactions with iron(IV)-bound oxygen. The interactions of the *syn*-H-C5 and *syn*-H-C6 (endo attack), and *syn*-H-C7 (exo attack) with ortho substituents on the porphyrin phenyl groups are sterically unimportant, and their interactions with oxygen remain invariant, while the steric interactions of ortho substituents on the porphyrin phenyl groups with H-C2 and H-C3 (double-bond hydrogens) become more demanding with increase in size of the substituent

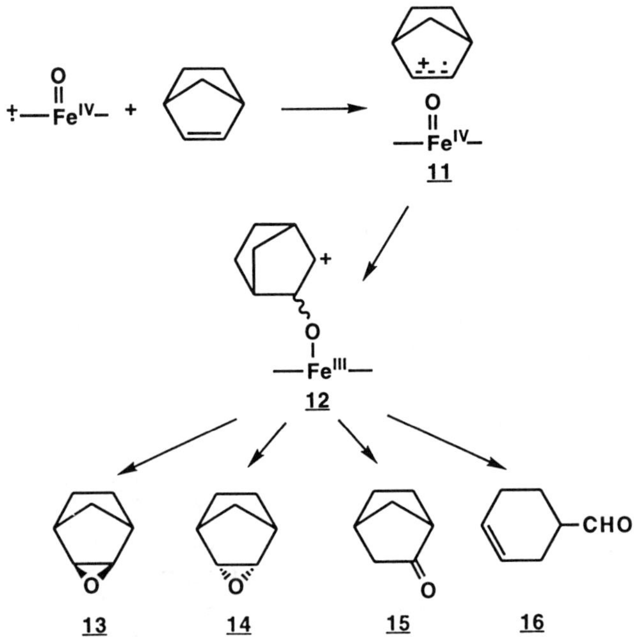

Scheme 11

(Figure 4). Thus, an additional steric factor is present, making the usual exo/endo selectivity less pronounced. The observed change in ratio of *exo*- to *endo*-norbornene oxide can be explained by these electronic and steric effects. Thus, **11** is not required as an intermediate in order to explain experimental results. However, only the intermediacy of $\cdot + Fe^{III}$-O-C-C^+ can explain the formation of the small amounts of norcamphor and cyclohexene-4-carboxyaldehyde.

cis-Stilbene was also utilized as a simple mechanistic probe for a stepwise reaction pathway for epoxidation [14]. Destruction of the π system with formation of an acyclic intermediate can result in loss of stereochemical information and formation of rearrangement products. When using $(F_{20}TPP)Fe^{III}(Cl)$, $(Cl_8TPP)Fe^{III}(Cl)$, and $(Cl_8TPP)Mn^{III}(Cl)$ catalysts and a C_6F_5IO oxygen donor for epoxidation of *cis*-stilbene the following products were obtained: *trans*-stilbene **23**, benzaldehyde **24**, *cis*-stilbene oxide **19**, *trans*-stilbene oxide **20**, diphenylacetaldehyde **21**, and deoxybenzoin **22** (Scheme 12). In addition, *cis*-stilbene destruction was found to exceed the C_6F_5IO oxidant employed. The previously proposed stepwise mechanism for epoxidation which involves carbocation radical and carbocation successfully

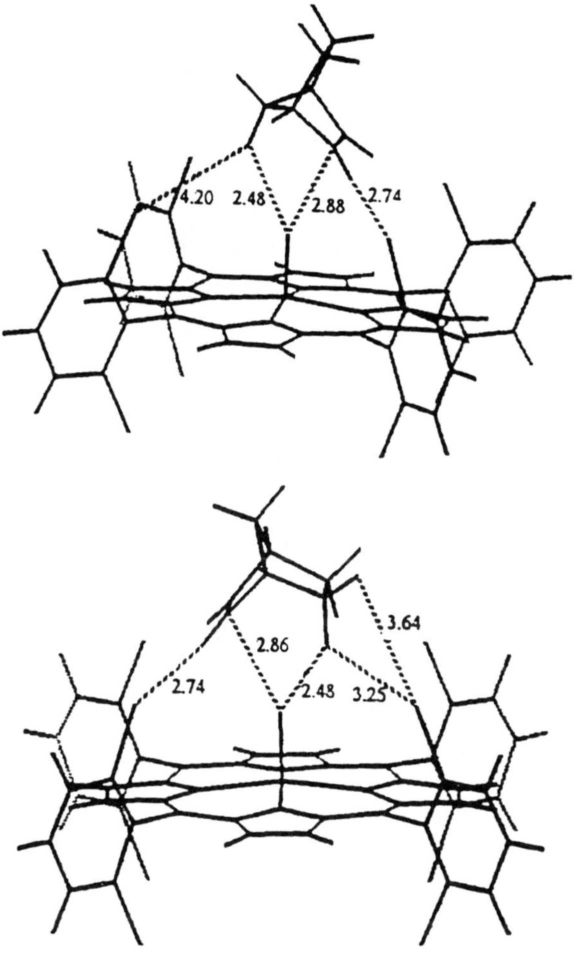

Figure 4 Computer-generated stick models of norbornene docked on $(+\cdot Br_8TPP)Fe^{IV}(O)$: (top) attack of oxygen from the *exo* side (interaction energy = 0.71 kcal/mole); (bottom) attack of oxygen from the *endo* side (interaction energy = 0.69 kcal/mole).

accounted for all of the observed products (Scheme 12) [10]. A caged ion pair consisting of alkene-derived carbocation radical and $(Porph)M^{IV}(O)$ **17** is formed by $1e^-$ transfer from the alkene to the hypervalent oxometalloporphyrin $(Porph+\cdot)M^{IV}(O)$, which then collapses to the acyclic carbocation **18**. Alternatively, a direct electrophilic attack on the alkene can provide an acyclic carbocation that partitions between epoxides **19** and **20** and rearrange-

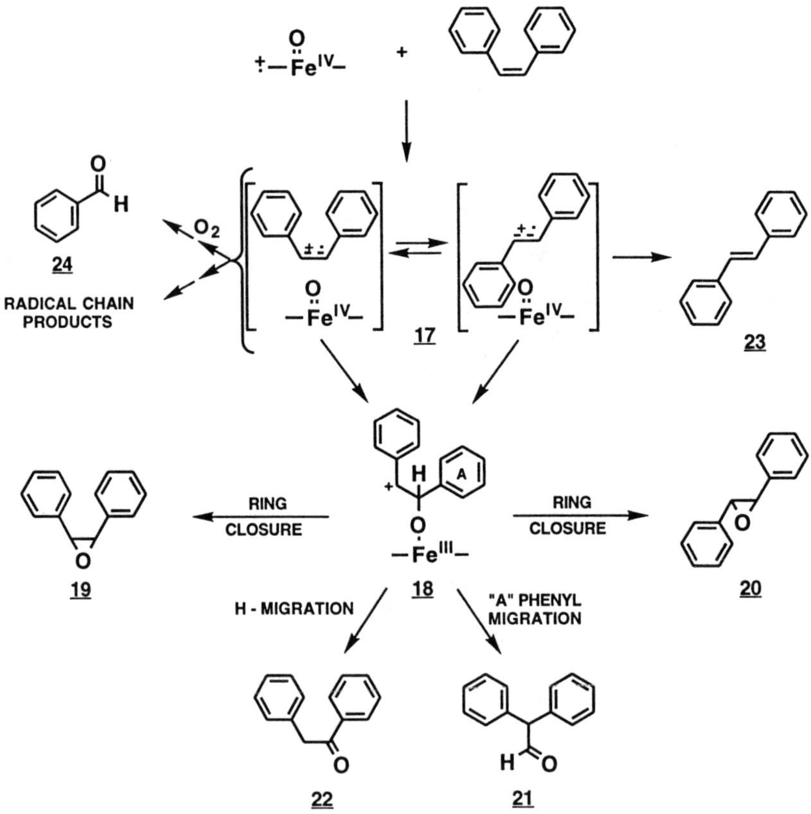

Scheme 12

ment products diphenylacetaldehyde **21** and deoxybenzoin **22**. *trans*-Stilbene **23**, benzaldehyde **24**, and radical chain products are formed from the stilbene carbocation radical that escapes the cage.

However, the finding that a CT complex is formed in the rate-limiting step necessitates that Scheme 12 be rewritten in the more general form of Scheme 2.

The fact that rearranged products resulting from alkene derived carbocation radicals and carbocations are observed during epoxidation of certain alkenes does not necessitate their involvement on the reaction path to epoxide. The evidence for the involvement of a carbocation on the epoxidation reaction path is questionable. It is apparent from the material discussed thus far that the only remaining reasonable alternative to a carbocation intermediate on the epoxidation reaction path is the reaction path that would involve a direct insertion of metalloporphyrin-bound oxygen into the alkene double bond.

E. Concerted Mechanism for Epoxidation

Most of the experimental and theoretical evidence accumulated to date suggests the following. The structure of an oxo-iron porphyrin $2e^-$ oxidized above the iron(III) state is an oxo-iron(IV)(porphyrin π-cation radical) where oxygen px and py orbitals are mixed with d_{xz} and d_{yz} orbitals on the iron [24]. This results in partial unpaired electron density and triplet oxenoid character of the iron-bound oxygen. If this is the case, a direct concerted insertion of iron-bound oxygen into the alkene double bond to form epoxide can be excluded, since it would be required that an intermediate be formed that could collapse to epoxide only after a spin inversion. Thus, for any concerted mechanism it is necessary to assume extremely efficient mixing of triplet and singlet states after the initial CT complex formation. Efficient changes of spin states and the presence of "spin equilibria" are well documented for iron porphyrins [61–63]. This indicates that effective means for interconversion of spin states are available in these systems. It is known that mixing of spin states and spin inversions are accelerated by the presence of other unpaired spins in an unsymmetric environment [64]. The unpaired electron in the porphyrin π-cation radical might just create such an environment at the point where the electron redistribution within the alkene and the oxidant has commenced. This is supported by the fact that $(+\cdot Br_8TPP)Fe^{IV}(O)(X)$, which has the porphyrin π-cation radical structure, epoxidizes alkenes in almost quantitative yields, while $(Br_8TPP)Cr^{V}(O)(X)$ which does not have porphyrin π-cation radical structure provides substantial yields of rearranged products. The most likely mechanism of epoxidation would involve rate-limiting formation of a CT complex between alkene and $(+\cdot Porph)M^{IV}(O)(X)$, followed by a very fast change of spin state, and concerted oxygen insertion into the alkene double bond (Scheme 13).

Asymmetric oxygen insertion, in which the two O-C bonds are formed at different rates but simultaneously, has support in the observation of an inverse secondary isotope effect ($k_H/k_D = 0.93$) for deuterium on α carbon and no isotope effect for deuterium on β carbon in the epoxidation of styrene by cytochrome P450 [54]. After formation of a CT complex and a change of spin state, the ratio of epoxide to carbocation rearrangement products would be determined by the relative rates of (a) concerted oxygen insertion to give epoxide and (b) electrophilic attack of metal-bound oxygen on the alkene to provide a carbocation intermediate (Scheme 13). The latter would

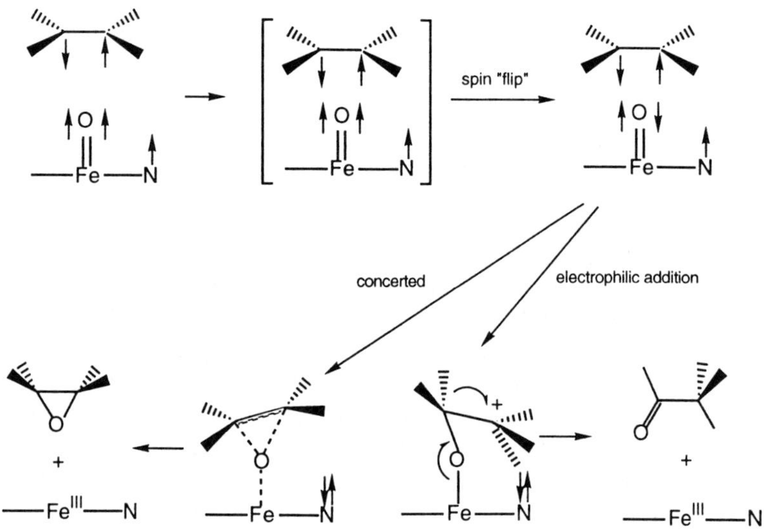

Scheme 13

be greatly favored by the overlap of p orbitals on the alkene and porphyrin nitrogens.

Support for the concerted mechanism may be found in the following findings: Epoxidation of alkenes by higher valent oxometalloporphyrins is stereospecific for the vast majority of alkenes. The ρ^+ value of -0.93, determined in competitive experiments [13], is reported for epoxidation of a series of substituted styrenes by the (TPP)Fe^{III}(Cl)/C_6H_5IO system. This value is similar to the ρ^+ values reported for known concerted processes such as carbene insertions into the double bond ($\rho^+ = -0.62$ to -1.61) [65] or epoxidation with perbenzoic acid ($\rho^+ = -1.2$) [22]. The reported ρ^+ values for known electrophilic additions which involve the formation of carbocation intermediates are generally $\geqslant -3.5$ [16,17]. This would tend to argue against the electrophilic mechanism for epoxidation by ($+\cdot$Porph)M^{IV}(O)(X). Additional support for a concerted mechanism comes from the fact that the most efficient and selective catalyst is sterically hindered (Br_8TPP)Fe^{III}(Cl) [33]. Bulky *o*-bromo substituents can play a dual role: (a) The steric hindrance prevents the interaction of alkenes with the electron density on the porphyrin ring, which can participate in stabilization of a developing positive charge on the alkene and favor electrophilic mechanism and the formation of a carbocation intermediate and the corresponding rearrangement products. (b) Bromines are heavy atoms and as such they can facilitate spin-spin interconversions via spin-orbit coupling, which would create a favorable environment for concerted oxygen insertion into the alkene double bond and the formation of epoxide.

F. Conclusion

During the course of our work a general picture has emerged whereby charge-transfer complex formation is the rate-limiting step during oxidation of alkenes by higher valent oxometalloporphyrins, and whereby what follows (epoxidation vs. various rearrangement pathways) is dependent on the oxidation potentials of the species involved (alkene and the active oxidant) and their steric and electronic structures, as well as the propensity of various substrates to undergo rearrangements. On many occasions the choice of substrates which are susceptible to specific rearrangements was misleading for the understanding of the mechanism of epoxidation. However, the involvement of intermediates such as alkene-derived radical, alkene-derived carbocation-radical, and metallaoxetane on the epoxidation reaction path can be dismissed based on the available data. Of the two remaining proposed mechanisms, (a) electrophilic addition of the metal-bound oxygen to the alkene double bond to form a carbocation intermediate and (b) concerted insertion of the metal-bound oxygen into the alkene double bone to form epoxide directly, the evidence supporting the latter seems more convincing.

IV. MANGANESE PORPHYRINS

A. Epoxidation of Alkenes by Hypervalent Manganese-Oxoporphyrin

Loss of stereospecificity is observed in the epoxidation of *cis*-alkenes when catalyzed by manganese porphyrins [12,15]. For example, the catalytic epoxidation of *cis*-stilbene by $(Cl_8TPP)Mn^{III}(Cl)$ with C_6H_5IO [14], and (TPP) $Mn^{III}(Cl)$ with 2-phenylsulfonyl-3-(*p*-nitrophenyl)oxaziridine [31] provides both *cis*- and *trans*-stilbene oxides. That *trans*-epoxides are obtained from such alkenes as *cis*-stilbene has led to the suggestion of a radical oxidation mechanism. This is not true of epoxidations catalyzed by chromium and iron porphyrins. Another feature is the observation that the stereoselectivity and stereospecificity of alkene epoxidation using manganese porphyrins are remarkably improved by the addition of pyridine or imidazole derivatives [66–70]. This has been shown to be due to the stabilization of the manganese(V)-oxoporphyrin and prevention of formation of the manganese(IV)-oxoporphyrin by the nitrogen base acting as an axial ligand [71–73].

Recently it has been shown that the *trans*-epoxide product resulting from a radical-type oxidation most likely arises from contamination of the manganese(V)-oxoporphyrin by manganese(IV)-oxoporphyrin [26,71]. The appearance of manganese(IV)-oxoporphyrin is from both spontaneous decomposition of manganese(V)-oxoporphyrin and the rapid reaction of manganese(V)-oxoporphyrin with any manganese(III) porphyrin that is present. Both manganese(V)-oxoporphyrin and manganese(IV)-oxoporphyrin are capable of transferring an oxygen atom to alkenes but the alkene epoxidation mechanisms of the two complexes are different.

Manganese(V)-oxoporphyrins are somewhat unstable and have not been isolated at ambient temperatures though solutions or slurries of manganese(V)-oxoporphyrins have been obtained, contaminated with manganese(IV)-oxoporphyrin, at $-80°C$ [71,74,75]. As compared to the manganese(V)-oxoporphyrin, manganese(IV)-oxoporphyrin is relatively stable. Several stable manganese(IV)-oxoporphyrin species generated by oxidation with peroxy acid or electrochemical method have been isolated and fully characterized [71]. Other examples of the stable manganese(IV)-oxoporphyrin include a manganese(IV) bis(iodosylbenzene) adduct [76] and a dimeric (μ-oxo)manganese(IV) bis(iodosylbenzene) species [77]. However, in contrast to the related iron(IV)-oxoporphyrin and chromium(IV)-oxoporphyrins which are inactive in the oxidation of alkenes, these manganese(IV)-oxoporphyrin species are capable of oxidizing organic substrates [71,78,79].

B. Mechanistic Studies of Manganese(V)-Oxo- and Manganese(IV)-Oxoporphyrin Species

Scheme 2 describes a unified mechanism that we have employed in describing reactions of both chromium(V)-oxoporphyrins and iron(IV)-oxo porphyrin π-cation radicals [2,3]. It is most likely that the same conclusions can be reached for alkene oxidations involving manganese(V)-oxoporphyrins. The reactions following the rate-limiting step (epoxidation vs. rearrangement) are dependent on a number of factors that include the oxidation potentials of the alkenes and the active oxidant, the steric and electronic structures of the reactants (steric bulk and geometry of both species and the metal axial ligand of the metalloporphyrins), as well as the propensity of various substrates to undergo rearrangements [2,3,25].

Groves and Stern showed that a manganese(V)-oxoporphyrin could be prepared in CH_2Cl_2 using *m*-CPBA as oxidant in the presence of pyridine or imidazole at $-78°C$. In the absence of pyridine there is produced a substantial amount of the manganese(IV)-oxo species. The manganese(IV)-oxoporphyrin can be isolated by chromatography on alumina at $-78°C$. Manganese(IV)-oxoporphyrins are relatively stable. For example, a CH_2Cl_2 solution of $(Cl_8TPP)Mn^{IV}(O)$ can be kept at $-20°C$ for several days without decomposition. Groves and Stern [80] examined the oxidation of cis-β-methylstyrene using $(TMP)Mn^{V}(O)$ and $(TMP)Mn^{IV}(O)$ and we [26] studied the oxidation of a series of *cis*- and *trans*-alkenes by $(Cl_8TPP)Mn^{IV}(O)$ and *cis*-stilbene by $(Cl_8TPP)Mn^{V}(O)$.

Kinetic studies of the reaction of $(Cl_8TPP)Mn^{IV}(O)$ with *trans*-4-methoxystilbene, *cis*-4-methoxystilbene, 1,4-diphenyl-1,3-butadiene, 4-methoxystyrene, 1,1-diphenylethylene, 4-methylstyrene, 2,3-dimethyl-2-butene, *trans*-stilbene, *cis*-stilbene, styrene, and 4-acetoxystyrene in methylene chloride solution (30°C) in air [26] showed that the *trans*-alkenes are slightly more reactive than their cis isomers. In comparison, the epoxidation of *trans*-alkenes with iron(IV)-oxoporphyrin π-cation radical and chromium(V)-oxoporphyrin species is slower than is the reaction with the cis isomers, especially for the electron-deficient alkenes. The values of log of the second-order rate constants (k_1) correlate well with the $E_{1/2}$ ($1e^-$ oxidation potential of alkene) values. The values of k_1 range from $1.3 \times 10^{-3}\ M^{-1}\ s^{-1}$ to $1.1 \times 10^{-2}\ M^{-1}\ s^{-1}$, while the alkene $E_{1/2}$ values range from 1.05 to 1.76 V. The linear correlation is given by Eq. 10

$$\log k_1 = -0.89E_{1/2} - 1.14 \tag{10}$$

and the slope of $-0.89\ V^{-1}$ corresponds to a Brønsted α of 0.05 in a conventional Brønsted plot of log k_1 vs. log K_a. Thus, in the reaction of $(Cl_8TPP)Mn^{IV}(O)$ with alkenes there is very little electron transfer. Much less so than in the reaction of $(Br_8TPP)Cr^{V}(O)$ [3]. Obviously, it is concluded that a mechanism involving rate-limiting formation of an alkene-derived carbocation radical cannot be correct. Additionally, a linear free-energy correlation of log k_1 and σ^+ for the epoxidation of substituted styrenes by $(Cl_8TPP)Mn^{IV}(O)$ gives the ρ^+ value of -0.99, which is much smaller than the ρ^+ values for known electrophilic additions and for $(Br_8TPP)Cr^{V}(O)(X)$. The low ρ^+ value precludes rate-limiting carbocation formation via electrophilic addition. These observations indicate transition states for rate-limiting steps that involve very little charge separation. If $1e^-$ transfer were involved in the reaction of $(Cl_8TPP)Mn^{IV}(O)$ with alkenes, then the transition state is very early, which is described by the rate-limiting formation of a CT complex. On the other hand, formation of a neutral carbon radical would also show slight sensitivity to polar effects. For example, the radical decomposition of the substituted peresters of $PhCH_2C(O)OOC(CH_3)_3$ yield ρ^+ in the neighborhood of -1 [81,82]. Therefore, formation of the neutral carbon radical species in the rate-limiting step is also possible for the epoxidation of alkenes with manganese(IV)oxoporphyrins.

The aerobic reaction of $(Cl_8TPP)Mn^{IV}(O)$ with *cis*-stilbene produced *cis*-stilbene oxide (7%), *trans*-stilbene (5%), and benzaldehyde (3%), while the reaction with *trans*-stilbene produced *trans*-stilbene oxide (15%) and benzaldehyde (13%) [26]. The production of *trans*-epoxide from *cis*-stilbene

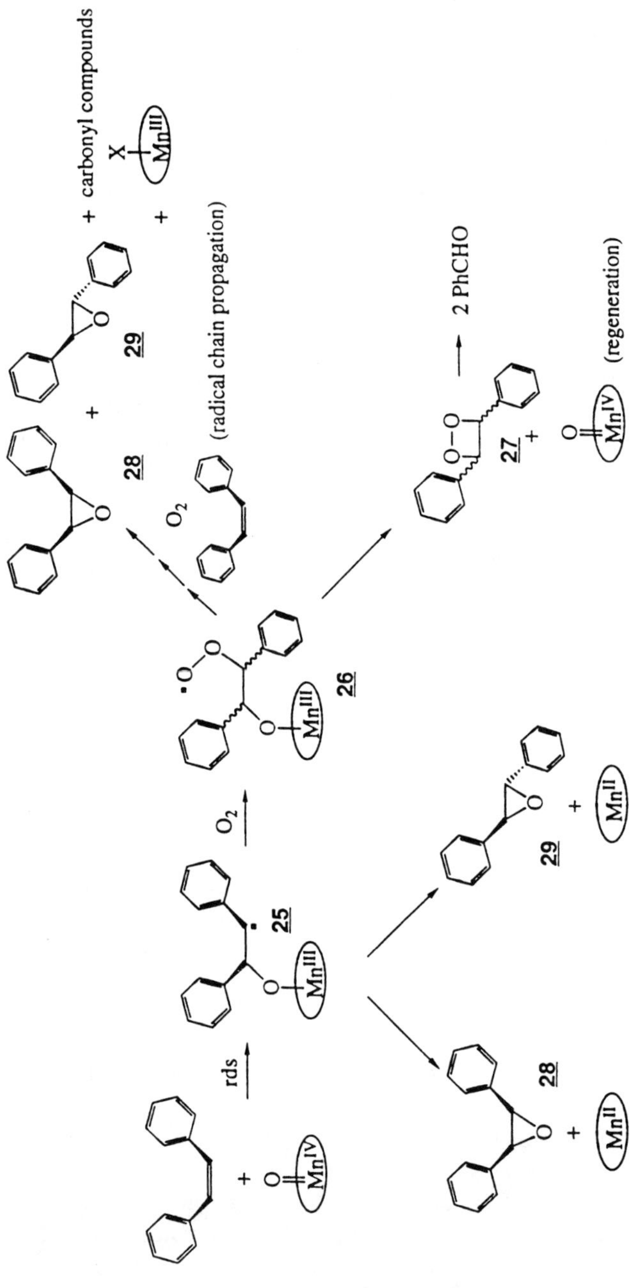
rds
O_2
25
26
27
28
29
+ carbonyl compounds
(radical chain propagation)
2 PhCHO
(regeneration)
Mn^{IV}
Mn^{III}
Mn^{II}
X

Scheme 14

requires that the intermediate leading to its direct formation must undergo a cis/trans inversion. This may be accounted for by a mechanism involving radical intermediates as described by the series of reaction shown in Scheme 14. The *cis*-carbon radical intermediate **25** has a sufficient lifetime to allow partial isomerization by rotation around the alkene C-C bond to provide the more stable *trans*-**25** and *trans*-epoxide. In the presence of molecular dioxygen, however, the carbon radical intermediate can react with O_2 to generate the peroxy radical species **26** which could ring-close to form a dioxetane **27** and regenerate $(Cl_8TPP)Mn^{IV}(O)$. Cleavage of the dioxetane provides PhCHO. Additionally, it has been proposed that peroxy radicals could act as radical chain initiators to generate more alkylperoxyl radicals with O_2, which could oxidize alkenes. When the reactions were carried out in deoxygenated solutions, both *cis*-stilbene oxide **28** (8%) and *trans*-stilbene oxide **29** (3%) were obtained. On the other hand, the aerobic reaction of $(Cl_8TPP)Mn^{IV}(O)$ with *trans*-4-methoxystibene gave 53% of *trans*-epoxide and 290% of carbonyl compounds based on $(Cl_8TPP)Mn^{IV}(O)$. The same aerobic reaction with 2,3-dimethyl-2-butene produced 10% of epoxide and 220% of carbonyl compounds. Such a radical chain reaction mechanism is also supported by the result from a similar epoxidation reaction of $(TMP)Mn^{IV}(O)$ with *cis*-β-methylstyrene in the presence of $^{18}O_2$ [71].

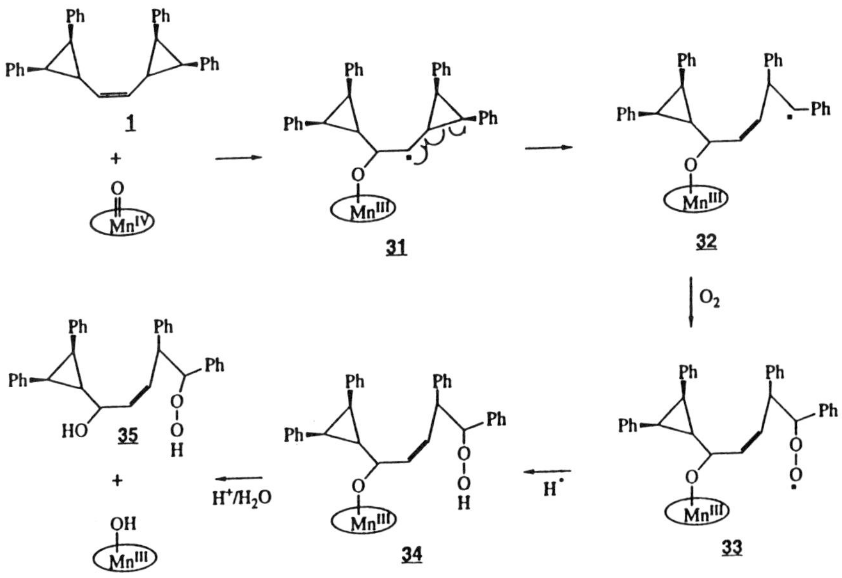

Scheme 15

We have shown (Scheme 15) in studies with $(Cl_8TPP)Mn^{III}(Cl)$ and (*Z*)-1,2-bis(*trans*-2,*trans*-3-diphenylcyclopropyl)ethene **1**, using C_6F_5IO as oxidant, that $(Cl_8TPP)Mn^{V}(O)$ does not provide radical species **II** and **III** (Chart 1) as intermediates in the epoxidation [50]. Formation of radical intermediates on reaction of $(Cl_8TPP)Mn^{IV}(O)$ with **1** was confirmed. The reaction of **1** with $(Cl_8TPP)Mn^{IV}(O)$ provided a polar oxygen-containing product and a very non-polar product. No epoxide product could be detected to 0.1% detection limit [26]. The structure of the polar product was identified as **30** following exam-

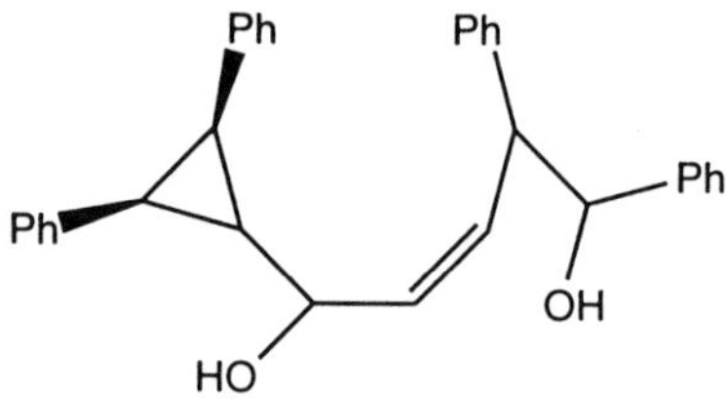

30

ination of the fragmentation patterns from the mass spectra. A pathway that could explain the formation of **30** from **1** is shown in Scheme 15. After generating the radical intermediate **31**, CPCRR of one cyclopropyl ring of **31** gives a benzyl radical **32**. The benzyl radical, which is relatively stable, adds dioxygen and then abstracts a H· to form a hydroperoxide **34**. Hydrolysis of **34** provides **35**, which under MS analysis is detected as **30**.

C. Conclusion

In summary, as with iron(IV)-oxoporphyrin π-cation radical and chromium(V)-oxoporphyrin, epoxidation of alkenes with manganese(V)-oxoporphyrin species gives epoxides with a high yield and stereospecificity through a quasi-concerted oxygen insertion reaction following rate-limiting CT formation (Scheme 2). Manganese(IV)-oxoporphyrin species also epoxidize alkenes, which is in contrast to iron(IV)-oxoporphyrin and chromium(IV)-oxoporphyrin. However, the oxidation of alkenes with manganese(IV)-oxoporphyrin species only provides expoxide in low yields and has no stereospecificity. The mechanism is considered to involve the neutral carbon radical intermediate **IV** of Chart 1 (p. 43).

REFERENCES

1. P. R. Ortiz de Montellano (ed.). *Cytochrome P-450: Structure, Mechanism, and Biochemistry*, Plenum Press, New York, 1986.
2. J.M. Garrison and T.C. Bruice, Intermediates in the epoxidation of alkenes by cytochrome P-450 models. 3. Mechanism of oxygen transfer from substituted

oxochromium(V) porphyrins to olefinic substrates, *J. Am. Chem. Soc., 111*: 191 (1989).
3. J.M. Garrison, D. Ostovic, and T.C. Bruice, Is a linear relationship between the free energies of activation and one-electron oxidation potential evidence for one-electron transfer being rate determining? Intermediates in the epoxidation of alkenes by cytochrome P-450 models. 4. Epoxidation of a series of alkenes by oxo(meso-tetrakis(2,6-dibromophenyl)porphinato)chromium(V), *J. Am. Chem. Soc., 111*: 4960 (1989).
4. D. Rehm and A. Weller, Kinetics of fluorescence quenching by electron and H-atom transfer, *Isr. J. Chem., 8*: 259 (1970).
5. C.P. Andrieux, C. Blocman, J.-M. Dumas-Bouchiat, and J.-M. Saveant, Heterogeneous and homogeneous electron transfers to aromatic halides. Electrochemical redox catalysis study in the halobenzene and halopyridine series, *J. Am. Chem. Soc., 101*: 3431 (1979).
6. M.M. Kreevoy and I.-S. H. Lee, Marcus theory of a perpendicular effect for hydride transfer between NAD analogs, *J. Am. Chem. Soc., 106*: 2550 (1984).
7. G.-X. He, R.D. Arasasingham, G. Zhang, and T.C. Bruice, The rate-limiting step in the one-electron oxidation of an alkene by oxomesotetrakis-(2,6-dibromophenylporphynato)chromium(V) is the formation of a charge-transfer complex, *J. Am. Chem. Soc., 113*: 9828 (1991).
8. T.G. Traylor, T. Nakano, B.E. Dunlap, P.S. Traylor, and D. Dolphin, Mechanisms of hemin-catalyzed alkene epoxidation. The effect of catalyst on the regiochemistry of epoxidation, *J. Am. Chem. Soc., 108*: 2782 (1986).
9. T.G. Traylor and A.R. Mikzstal, Mechanisms of hemin-catalyzed epoxidations. Electron transfer from alkenes, *J. Am. Chem. Soc., 109*: 2770 (1987).
10. T.G. Traylor, T. Nakano, A.R. Mikzstal, and B.E. Dunlap, Transient formation of N-alkylhemins during hemin-catalyzed epoxidation of norbornene. Evidence concerning the mechanism of epoxidation, *J. Am. Chem. Soc., 109*: 3625 (1987).
11. O. Bortolini and B. Meunier, Enhanced selectivity by an open-well effect in a metalloporphyrin-catalyzed oxygenation reaction, *J. Chem. Soc., Perkin. Trans., 2*, 1967.
12. H.J. Eyring, Activated complex in chemical reactions, *J. Chem. Phys., 3*: 107 (1935).
13. J.R. Lindsay-Smith and P.R. Sleath, Model systems for cytochrome P-450 dependent mono-oxygenases. 1. Oxidation of alkenes and aromatic compounds by tetraphenylporphynatoiron(III) chloride and iodosylbenzene, *J. Chem. Soc. Perkin. Trans., 2*: 1009 (1982).
14. A.J. Castellino and T.C. Bruice, Intermediates in the epoxidation of alkenes by cytochrome P-450 models. 1. cis-Stilbene as a mechanistic probe, *J. Am. Chem. Soc., 110*: 158 (1988).
15. J.P. Collman, T. Kodadek, and J.I. Brauman, Oxygenation of styrene by cytochrome P-450 model systems. A mechanistic study, *J. Am. Chem. Soc., 108*: 2588 (1986).
16. W.M. Schubert and J.R. Keefe, The acid-catalyzed hydration of styrenes, *J. Am. Chem. Soc., 94*: 559 (1972).

17. K. Yates, R.S. McDonald, and S.A. Shapiro, Kinetics and mechanisms of electrophilic addition. I. A comparison of second- and third-order brominations, *J. Org. Chem., 38*: 2460 (1973).
18. D.F. Shellhammer, Electrophylic additions to methylene-cyclopropane cyclopropanes, Ph.D. Dissertation, University of California at Santa Barbara, 1974, pp. 168–171.
19. R.P. Hanzlik and G.O. Shearer, Transition-state structure for peracid epoxidation—secondary deuterium-isotope effects, *J. Am. Chem. Soc., 97*: 5231 (1975).
20. D. Seyferth, J.Y.-P. Mui, and R. Damrauer, Halomethyl-metal compounds. XIX. Further studies of the aryl(bromodichloromethyl)mercury-olefin reaction, *J. Am. Chem. Soc., 90*: 6182 (1968).
21. J. Nishimura, J. Furukawa, N. Kawabata, and M. Kitayama, The relative reactivity of olefins in cycloaddition with zinc carbenoid, *Tetrahedron, 27*: 1799 (1971).
22. Y. Ogata and I. Tabushi, Kinetics of the epoxidation of substituted α-methylstilbenes, *J. Am. Chem. Soc., 83*: 3440 (1961).
23. R.J. Cvetanovic, Electrophylic character of oxygen atoms, *Can. J. Chem., 38*: 1678 (1960).
24. G.H. Loew, C.J. Kert, L.M. Hjelmeland, and R.F. Kirchner, Active site models of horseradish peroxidase compound I and a cytochrome P-450. Analog electronic structure and electronic field gradients, *J. Am. Chem. Soc., 99*: 3534 (1977).
25. G.-X. He, H.-Y. Mei, and T.C. Bruice, Evaluation of the selectivity of the epoxidation of cis-alkenes versus trans-alkenes by oxo(meso-tetrakis(2,6-dibromophenyl)porphynato)chromium(V)—dynamics, products, and van der Waals energies, *J. Am. Chem. Soc., 113*: 5644 (1991).
26. R.D. Arasasingham, G.-X. He, and T.C. Bruice, Mechanism of manganese porphyrin catalyzed oxidation of alkenes. Role of manganese(IV)-oxo species, *J. Am. Chem. Soc., 115*: 7985 (1993).
27. J.T. Groves, T.E. Nemo, and R.S. Myers, Hydroxylation and epoxidation catalyzed by iron-porphine complexes. Oxygen transfer from iodosylbenzene, *J. Am. Chem. Soc., 101*: 1032 (1979).
28. T.G. Traylor and F. Xu, Model reactions related to cytochrome P-450. Effects of alkene structure on the rates of epoxide formation, *J. Am. Chem. Soc., 110*: 1953 (1988).
29. J.P. Collman, P.D. Hampton, and J.I. Brauman, Suicide inactivation of cytochrome P-450 model compounds by terminal olefins. 2. Steric and electronic effects in heme N-alkylation and epoxidation, *J. Am. Chem. Soc., 112*: 2986 (1990).
30. J.T. Groves and W.J. Kruper, Jr., Preparation and characterization of an oxoporphyrinatochromium(V) complex, *J. Am. Chem. Soc., 101*: 7613 (1979).
31. L.-C. Yuan and T.C. Bruice, Use of an oxaziridine as an oxene transfer agent to manganese(III) tetraphenylporphyrin chloride, *J. Chem. Soc. Chem. Commun.,* 868 (1985).
32. D. Ostovic and T.C. Bruice, Transition state geometry in epoxidation by iron-oxo porphyrin at the compound I oxidation level. Epoxidation of alkenes cata-

lyzed by a sterically hindered (meso-tetrakis(2,6-dibromophenyl)porphinato)-iron(III) chloride, *J. Am. Chem. Soc., 110*: 6906 (1988).

33. D. Ostovic and T.C. Bruice, Intermediates in the epoxidation of alkenes by cytochrome P-450 models. 5. Epoxidation of alkenes catalyzed by a sterically hindered meso-tetrakis(2,6-dibromophenyl)porphinato)iron(III) chloride, *J. Am. Chem. Soc., 111*: 6511 (1989).
34. C.R. Theocharis, W. Jones, and C.N. Ramachandra Rao, An unusual photo-induced conformational polymorphism. A crystallographic study of bis(para-methoxy)-trans-stilbene, *J. Chem. Soc., Chem. Commun.*, 1291 (1984).
35. J. Bernstein, Refinement of trans-stilbene. Comparison of 2 cyrstallographic studies, *Acta Crystallogr., Sect. B: Struct. Crystallogr. Cryst. Chem., B31*: 1268 (1975).
36. K. Shimada and M. Szwarc, Flexibility of molecular chains studied by electron-spin resonance technique, *J. Am. Chem. Soc., 97*: 3313 (1975).
37. K. Shimada and M. Szwarc, Peculiarities of intramolecular electron-transfer in α-naphthyl-$(CH_2)_3$-α-naphthyl-radical anions and α-naphthyl-$(CH_2)_4$-α-naphthyl-radical anions, *J. Am. Chem. Soc., 97*: 3321 (1975).
38. K. Tatsumi and R. Hoffmann, Metalloporphyrins with unusual geometries. 2. Slipped and skewed bimetallic structures, carbene and oxo complexes, and insertions into metal-porphyrin bonds, *Inorg. Chem., 20*: 3771 (1981).
39. K.A. Jørgensen, On the mechanism of stereoselective epoxidation of alkenes by oxo-iron porphyrins, *J. Am. Chem. Soc., 109*: 698 (1987).
40. K. Yamaguchi, Y. Takahara, and T. Fueno, in *Applied Quantum Chemistry* (V.H. Smith, Jr., et al., eds.), Reidel, Dordrecht, 1986, pp. 155–184.
41. J.R. Groves and Y. Watanabe, On the mechanism of olefin epoxidation by oxo-iron porphyrins. Direct observation of an intermediate, *J. Am. Chem. Soc., 108*: 507 (1986).
42. J.P. Collman, J.I. Brauman, B. Meunier, S.A. Raybuck, and T. Kodadek, Epoxidation of olefins by cytochrome P-450 model compounds—mechanism of oxygen atom transfer, *Proc. Natl. Acad. Sci. USA, 81*: 3245 (1984).
43. J.P. Collman, J.I. Brauman, B. Meunier, T. Hayashi, T. Kodadek, and S.A. Raybuck, Epoxidation of olefins by cytochrome P-450 model compounds. Kinetics and stereochemistry of oxygen atom transfer and origin of shape selectivity, *J. Am. Chem. Soc., 107*: 2000 (1985).
44. J.P. Collman, T. Kodadek, S.A. Raybuck, J.I. Brauman, and L.M. Papzian, Mechanism of oxygen atom transfer from high valent iron porphyrins to olefins. Implications to the biological epoxidation of olefins by cytochrome P-450, *J. Am. Chem. Soc., 107*: 4343 (1985).
45. F.P. Guengerich and T.L. MacDonald, Chemical mechanisms of catalysis by cytochrome P-450. A unified view, *Acc. Chem. Res., 17*: 9 (1984).
46. J.R. Groves, W.J. Kruper, Jr., and R.C. Haushalter, Hydrocarbon oxidations with oxometalloporphyrinates. Isolation and reactions of a (porphynato)manganese(V) complex, *J. Am. Chem. Soc., 102*: 6375 (1980).
47. J.R. Groves and R.S. Myers, Catalytic asymmetric epoxidations with chiral iron porphyrins, *J. Am. Chem. Soc., 105*: 5791 (1983).

48. M. Fontecave and D. Mansuy, Alkene epoxidation by iodosylbenzene catalyzed by porphyrin and non-porphyrin iron complexes. The importance of the porphyrin ligand in cytochrome P-450 and heme model reactions, *J. Chem. Soc., Chem. Commun.*, 879 (1984).
49. A.J. Castellino and T.C. Bruice, Radical intermediates in the epoxidation of alkenes by cytochrome P-450 model systems. The design of a hypersensitive radical probe, *J. Am. Chem. Soc., 110*: 1313 (1988).
50. A.J. Castellino and T.C. Bruice, Intermediates in the epoxidation of alkenes by cytochrome P-450 models. 2. Use of the trans-2,trans-3-diphenylcyclopropyl substituent in a search for radical intermediates, *J. Am. Chem. Soc., 110*: 7512 (1988).
51. T. Watabe and K. Akamatsu, Microsomal epoxidation of cis-stilbene. Decrease in epoxidase activity related to lipid peroxidation, *Biochem. Pharmacol., 23*: 1079 (1974).
52. T. Watabe, Y. Ueno, and J. Imazumi, Conversion of oleic acid into threo-dihydroxystearic acid by rat liver microsomes, *Biochem. Pharmacol., 20*: 912 (1971).
53. P.R. Ortiz de Montellano, B.L.K. Mangold, C. Wheeler, K.L. Kunze, and N.O. Reich, Stereochemistry of cytochrome P-450 catalyzed epoxidation and prosthetic heme alkylation, *J. Biol. Chem., 258*: 4208 (1983).
54. R.P. Hanzlik and G.O. Shearer, Secondary deuterium-isotope effects on olefin epoxidation by cytochrome P-450, *Biochem. Pharmacol., 27*: 1441 (1978).
55. R.W. Lee, P.C. Nakagaki, P.N. Balasubramanian, and T.C. Bruice, Observations and comments on the mechanism of epoxidation of alkenes by manganese(III) porphyrins with hypochlorite, *Proc. Natl. Acad. Sci. USA, 85*: 641 (1988).
56. J.L. Hoard, G.H. Cohen, and M.D. Glick, The stereochemistry of the coordination group in an iron(III) derivative of tetraphenylporphine, *J. Am. Chem. Soc., 89*: 1992 (1967).
57. P. Gans, G. Buisson, E. Duee, J.-C. Marchon, B.S. Erler, W. Scholz, and C.A. Reed, High-valent iron porphyrins. Synthesis, x-ray structures, π-cation radical formation, and notable magnetic properties of chloro(meso-tetraphenylporphynato)iron(III) hexachloroantimonate and bis(perchlorato)(meso-tetraphenylporphynato)iron(III), *J. Am. Chem. Soc., 108*: 1223 (1986).
58. D.P. Klein, J.C. Hayes, and R.D. Bergman, Insertion of (Eta-5-C_5Me_5)(PMe_3)Ir into the C-H bonds of functionalized organic molecules. A C-H activation route to 2-oxametallacyclobutanes and 2-azametallacyclobutanes. Potential models for olefin oxidation intermediates, *J. Am. Chem. Soc., 110*: 3704 (1988).
59. R. Schlodder, J.A. Ibers, M. Lenarda, and M. Graziani, Structure and mechanism of formation of the metallooxacyclobutane complex $Pt[C_2(CN)_4O][As(C_6H_5)_2]_2$, the product of the reaction between tetracyanooxirane and $Pt[As(C_6H_5)_2]_4$, *J. Am. Chem. Soc., 96*: 6893 (1974).
60. A.L.J. Beckwith and G. Moad, Ring-opening of some radicals containing the cyclopropylmethyl system, *J. Chem. Soc. Perkin Trans., 2*: 1473 (1980).
61. H.A. Hill, P.D. Skyte, J.W. Buchler, H. Lueken, M. Tonn, A.K. Gregson, and G. Pellizer, Magnetic properties of hemichrome salts (derivatives of iron(III)

octaethylporphyrin complexes). Presence of a spin-equilibrium, *J. Chem. Soc. Chem. Commun.*, 151 (1979).
62. A.K. Gregson, Quantitative description of the magnetic properties and proton isotropic shifts of hemichrome salts (derivatives of iron(III) octaethylporphyrin complexes). The presence of a spin-equilibrium, *Inorg. Chem., 20*: 81 (1981).
63. R.L. Martin and A.H. White, Nature of the transition between high-spin and low-spin octahedral complexes of the transition metals, *Transition Met. Chem., 4*: 113 (1968).
64. R.A. Harman and H.J. Eyring, The structure of substituted ethylenes and their isomerization, polymerization and "peroxide addition reactions," *J. Chem. Phys., 10*: 557 (1942).
65. R.A. Moss, in *Carbenes* (M. Jones, Jr., and R.A. Moss, eds.), Wiley, New York, 1973, Vol. 1, pp. 268–269.
66. B. Meunier, E. Guilmet, M.E. De Carvalho, and R. Poilblanc, Sodium hypochlorite—a convenient oxygen source for olefin epoxidation catalyzed by (porphynato)manganese complexes, *J. Am. Chem. Soc., 106*: 6668 (1984).
67. E. Guilmet and B. Meunier, Role of pyridine in the catalytic activation of sodium hypochlorite in the presence of manganese porphyrin, *Tetrahedron Lett., 23*: 2449 (1982).
68. E. Guilmet and B. Meunier, Unexpected modification of selectivity with pyridine in the NaOCl Mn(TPP)OAc catalytic epoxidation, *Nouv. J. Chim., 6*: 511 (1982).
69. J.P. Collman, T. Kodadek, S.A. Raybuck, and B. Meunier, Oxygenation of hydrocarbons by cytochrome P-450 model compounds. Modification of reactivity by axial ligands, *Proc. Natl. Acad. Sci. USA, 80*: 7039 (1983).
70. A.W. Van der Made and R.J.M. Nolte, Olefin epoxidation by manganese(III) porphyrin complexes. Effect of pyridine on stereoselectivity, *J. Mol. Catal., 26*: 333 (1984).
71. J.T. Groves and M.K. Stern, Synthesis, characterization, and reactivity of oxomanganese(IV) porphyrin complexes, *J. Am. Chem. Soc., 110*: 8628 (1988).
72. J.T. Groves and Y. Watanabe, Heterolytic and homolytic O-O bond cleavage reactions of (acylperoxo)manganese(III) porphyrins, *Inorg. Chem., 25*: 4808 (1986).
73. M. Schappacher and R. Weiss, Formation of manganese(IV)-oxo-porphyrin derivatives by decomposition of peroxycarbonate complexes, *Inorg. Chem., 26*: 1190 (1987).
74. O. Bortolini and B. Meunier, Isolation of a high-valent oxo-like manganese porphyrin complex obtained from NaOCl oxidation, *J. Chem. Soc. Chem. Commun.*, 1364 (1983).
75. O. Bortolini, M. Ricci, B. Meunier, P. Friant, I. Ascone, and J. Goulon, Isolation, characterization and structural investigation by EXAFS of high valent manganese porphyrin complexes as active species in the NaOCl/Mn(Porphyrin) X-oxygenation system, *Nouv. J. Chim., 10*: 39 (1986).
76. J.A. Smegal and C.L. Hill, Synthesis, characterization, and reaction chemistry of a bis(iodosylbenzene)-metalloporphyrin complex, [PhI(OAc)O_2Mn(IV)TPP—

a complex possessing a 5-electron oxidation capability, *J. Am. Chem. Soc., 105*: 2920 (1983).
77. J.A. Smegal, B.C. Schardt, and C.L. Hill, Isolation, purification, and characterization of intermediate (iodosylbenzene)metalloporphyrin complexes from the (tetraphenylporphynato)manganese(III)—iodosylbenzene catalytic hydrocarbon functionalization system, *J. Am. Chem. Soc., 105*: 3510 (1983).
78. M.J. Camenzind, F.J. Hollander, and C.L. Hill, Synthesis, ground electronic state, and crystal and molecular structure of the monomeric manganese(IV) porphyrin complex dimethoxy(5,10,15,20-tetraphenylporphynato)manganese (IV), *Inorg. Chem., 21*: 4301 (1982).
79. J.T. Groves and T. Takahashi, Activation and transfer of nitrogen from a nitridomanganese(V) porphyrin complex—the aza analog of epoxidation, *J. Am. Chem. Soc., 105*: 2073 (1983).
80. J.T. Groves and M.K. Stern, Synthesis, characterization, and reactivity of oxomanganese(IV) porphyrin complexes, *J. Am. Chem. Soc., 110*: 8628 (1988).
81. P.D. Bartlett and C.J. Ruchardt, Peresters. IV. Substituent effects upon the concerted decomposition of t-butyl phenylacetates, *J. Am. Chem. Soc., 82*: 1756 (1960).
82. J.P. Engstrom and J.C. DuBose, Substituent and solvent effects on the rate of perester decomposition. The case for polar contributions to the transition state, *J. Org. Chem., 38*: 3817 (1973).

3

Theoretical Studies of Oxometalloporphyrins and Their Reactions with Organic Substrates

Karl Anker Jørgensen
Aarhus University, Aarhus, Denmark

I. INTRODUCTION

Oxidation of organic compounds catalyzed by synthetic metalloporphyrin complexes has been one of the "hot areas" in chemistry during the last decade [1]. The metalloporphyrins can catalyze the oxidation of a variety of organic substrates, as e.g., outlined in Scheme 1, with many different terminal oxidants [1,2].

Several different transition metal porphyrin systems have been studied, but the main attention has been focused on manganese and iron systems as oxidative catalysts [1]. The catalytic properties of the transition metal porphyrins are due to the fact that an oxo–transition metal porphyrin intermediate is formed, which can transfer the oxygen atom to a substrate or can accept an electron from the substrate [1]. Many stable oxo–transition metal complexes are known and have been characterized, but several of these complexes are inert and unreactive as oxygen transfer reagents. In the field of oxidation catalyzed by transition metal porphyrins, the oxygen transfer step is the crucial step, and from an experimental point of view a lot of attention has been devoted to the nature of the oxo transition metal bond [3]. The early transition metals form a strong $M=O$ bond at room temperature and no oxygen transfer is achieved, e.g., oxovanadium porphyrins are unreactive [4], and the chromium porphyrins are hardly reactive. If we move to the manganese and iron porphyrins, the $M=O$ bond formed is too unstable to be experimentally fully characterized. It should be mentioned, though, that

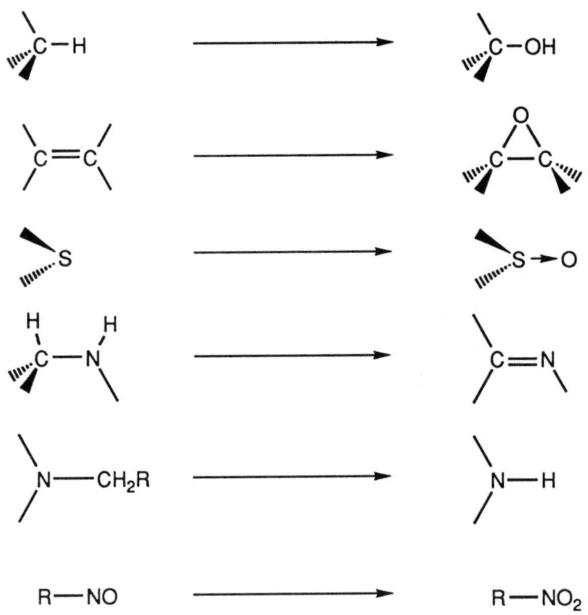

Scheme 1

a stable low-spin d^2 square-pyramidal oxomanganese complex with a macrocyclic tetraamide ligand has been characterized [5]. The Mn$=$O distance was found to be 1.55 Å [5].

The mechanism for the oxygen transfer step from the intermediate oxo-transition metal porphyrin to the substrate or the electron transfer from the substrate to the oxo-transition metal porphyrin has been the subject of numerous studies [6].

This chapter is devoted to a presentation of the theoretical studies of the oxo-transition metal porphyrins and the mechanism for the oxygen transfer step. The intention is to present some of the results obtained by quantum chemical calculations in such a way that the experimentalists (hopefully) will be able to use them in their interpretation of the results. An introduction to the vast number of quantum chemical calculations on transition metal porphyrins was published in a previous review [7].

From a theoretical point of view the electronic structure of the M$=$O bond in the high-valent oxo-transition metal porphyrins and the subsequent oxygen transfer to a substrate is not easy to describe due to the complexity of the system. It might here be appropriate to quote Yamamoto et al. from a study of the electronic structure of the Fe = O bond in highly oxidized iron porphyrins [8]:

> The Fe-O bond is also theoretically interesting because its electronic structure has not been fully understood in detail. This is because its electronic structure is rather complicated and cannot be well described at the Hartree–Fock level. The Fe-O bond is considered to have two triplet-coupled unpaired electrons. The electronic structure is similar to that of the ground state of the O_2 molecule; however, the Fe-O bond is much more complicated than the O-O bond.

Before proceeding to the theoretical work performed on oxo–transition metal porphyrins and their oxygen transfer reactions, a very short and very brief description of the different quantum chemical computation procedures used in the following will be presented. A variety of different programmes are available for calculations of different properties of molecules and reaction paths, and the reliability of these results needs to be discussed together with their interpretation.

II. QUANTUM CHEMICAL METHODS

The validity and accuracy that may be associated with a description of the electronic structure of oxometalloporphyrins and their reactions depend on the level of sophistication of the theoretical method employed. A molecular system is viewed as a collection of electrons and nuclei and its states are determined by quantum mechanics. Our present considerations are limited to the description of electronic states for fixed nuclear positions. The comparison of states for different conformations provides the insight we are aiming at. This is the framework offered by the Born–Oppenheimer picture whereby the nuclear motion in a certain electronic state is determined from an electronic energy surface. Thus we associate an electronic energy value to each nuclear conformation, often through an integral, *the energy expectation value*,

$$E(\mathbf{R}_1,\mathbf{R}_2,\ldots,\mathbf{R}_N) = \int dv\, [W|\Psi|^2 - \sum_{j=1}^{n} \frac{\hbar^2}{2m} |\nabla_j\Psi|^2] \tag{1}$$

for the wave function $\Psi(\mathbf{r}_1,\mathbf{r}_2,\ldots,\mathbf{r}_n;\ \mathbf{R}_1,\mathbf{R}_2,\ldots,\mathbf{R}_N)$ which depends on the electronic coordinates $[\mathbf{r}_j]$ as well as on the nuclear positions $[\mathbf{R}_j]$. The integration in equation (1) is over all electronic coordinates. The wave function is normalized,

$$1 = \int dv |\Psi|^2 \tag{2}$$

and satisfies the conditions imposed by the Pauli principle and electron spin. The last term in the integrand of Eq. 1 represents the electronic kinetic energy, m is the mass of an electron, while the potential W expresses the Coulomb interaction energy between the electrons with charge -e and the nuclei with charges $[eZ_k \mid k = 1,2,\ldots, N]$:

$$W = \frac{e^2}{4\pi\epsilon_0} \sum_{j<k} \frac{Z_j Z_k}{|R_j - R_k|} - \sum_{j,k} \frac{Z_k}{|r_j - R_k|} + \sum_{j<k} \frac{1}{|r_j - r_k|} \tag{3}$$

A variational principle ascertains that the stationary values of the form (1), subject to the auxiliary condition (2), are the electronic energies of the bound states and that the lowest value of the form (1) is an upper bound to the ground state energy. This is the basis for the approximate methods where simplifying limitations on the wave function are imposed.

Many-electron systems, such as are considered here, will invariably be dealt with in a *finite basis set* formulation: the wave function is usually restricted to a finite subspace based on one-electron functions, *the basis orbitals*, and the energy expression (1) becomes a sum over one- and two-electron integrals. These latter ones are numerous and constitute a large dataset; it is common practice in semiempirical methods to replace these integrals by parameters or parameterized forms. The one-electron integrals may similarly be derived by fitting procedures. The many-electron wave function requires further reduction in order to be tractable and the initial form is commonly the *Hartree–Fock* (HF) or *single Slater determinant* form whereby *molecular orbitals* (MOs), which are linear combinations of the basis orbitals, are assigned occupation numbers 0, 1, or 2. This reduces the problem of finding the stationary values of the form (1) to an optimization problem that can be solved with moderate computational effort. An intermediate problem in this procedure is the determination of eigenvalues and eigenvectors of the Fock matrix. Its rows and columns are labeled by the basis orbitals and the eigenvectors give the coefficients in the molecular orbital expansion. The Fock matrix depends on these coefficients through *the density or the charge-and-bond-order matrix.* A *self-consistent "field"* (SCF) is found when the molecular orbitals that come out of the eigenvalue problem equal those that were used to construct the Fock matrix.

General many-electron wave functions can be generated as linear combinations of Slater determinants, each with its own set of occupation numbers for the molecular orbitals. Such forms are called superposition of configurations and the calculation of the energy expression (1) gives a form that accounts for *configuration interaction* (CI). Although efficient algorithms have been developed for the solution of the associated matrix problems, the cost is still prohibitive for all but the smaller systems. One may define an ideal Hartree–Fock energy as the lowest value of the form (1) for a single determinant with a basis that is exhaustive, i.e., no further additions to the basis will improve the result. Similarly there will be a *full* CI limit and the difference between the two is defined as the *correlation energy*. A basis that is not exhaustive in the above sense gives a basis set error in both limits, but the error may well be different and there is no direct way to ascertain that a multideterminant wave function in a small basis gives a better picture than a few-determinant wave function in a more extended basis. The demands on the basis set increase with the quest for detail in the electronic structure. Data management increases with the fourth power of the size of the basis while the configuration interaction space follows the nth power and rapidly exceeds the limits of available computational equipment.

The extended Hückel approach [9a] is a simple quantum chemical calculational procedure which, when applied to organotransition metal complexes such as metalloporphyrins, leads to reasonable results. It has also been used for crystals and surfaces [9b]. Extended Hückel results can be considered as qualitative interpretations with its main strength being in the MO picture that emerges. The energies found are inadequate for the determination of endo- and exothermocities and the paths derived for reactions are only tentative. The results cannot be sufficiently trusted in studies where the formation and cleavage of several bonds are involved simultaneously. No adequate total energy expression is used and thus geometry optimizations are unavailable. The extended Hückel calculations have nevertheless led to many important insights. It is possible to treat large systems and to obtain an MO picture of the bondings of the system studied; furthermore, a picture emerges of the orbitals interacting along a reaction path. The extended Hückel method has the following main characteristics: (a) only the s-, p-, and d-valence orbitals are treated; (b) overlap integrals are calculated from atomic orbital basis functions; (c) geometry optimization is precluded; (d) Fock matrix elements are neglected and for nonbonding atoms integrals are assumed to be zero, except for atoms directly bounded to each other; (e) intra-atomic matrix elements are estimated from the ionization potential of an electron from the same orbital in an isolated atom. The extended Hückel MOs are often used as input for a molecular SCF calculation.

More detailed treatments of electron–electron interactions are offered by semiempirical methods using simplifying assumptions and experimental data to estimate integrals and to avoid explicit evaluations [10]. Two slightly different semiempirical approaches have been applied to study oxometalloporphyrins and their reactions; the MINDO [11] and AM1 [12] and the INDO [13] calculations. The assumptions in these approximate methods are, in principle, that they eliminate all three- and four-center integrals, while the remaining ones are evaluated from empirical data. Optimization of the total energy expression is performed with SCF methods and total energies are obtained. Both closed and open shell systems can be handled in a simplified manner also applicable to radicals. The MINDO and AM1 procedures take only the s- and p-valence orbitals into account and do not apply to calculations where transition metals having d orbitals are involved. The INDO (occasionally also called ZINDO [14]) method can treat first- and second-row transition metals with d orbitals. Both types of calculation are suitable for optimization of the geometric structure of relatively large molecules and can be used to elucidate reaction paths as relatively simple algorithms for transition state calculations are included in the programs. The methods do have deficiencies, as e.g., the INDO procedure is known for its exaggerated stability of ring systems compared with the corresponding open systems [15]. The following points can be made: (a) the MINDO, the AM1, and the INDO calculations cannot handle systems as large as the extended Hückel procedure, and the former methods also require much more computer power than the latter; (b) the MO interpretation is easier with the extended Hückel calculations since the MOs of the other models are more mixed than in the former; (c) radicals having both one and two unpaired electrons can be dealt with by the MINDO, AM1, and INDO procedures; (d) the MINDO, AM1, and INDO calculations can give a more exact picture of a reaction path than the extended Hückel procedure, and are furthermore able to locate transition states.

The ab initio methods evaluate all the terms in Eq. 1 without approximations beyond those originally involved in setting up a framework of an orbital basis [16]. The ab initio methods may be used for the description of the electronic structure for molecules having s, p, and d orbitals, and are all-electron treatments, i.e., both the core and the valence orbitals are included. The ab initio calculations are used to describe the electronic structure as well as reaction paths. For open shell systems, in which electrons are not completely assigned to orbital pairs, the simple orbital theory has been extended in two ways: *the spin-restricted Hartree–Fock* (RHF) *theory* and the *spin-unrestricted Hartree–Fock* (UHF) *theory.* A single set of MOs is used in the former approach, some being doubly occupied and some being singly occupied with electrons of α spin, whereas in the latter approach different spatial orbitals are assigned to α and β orbitals. The optimized UHF energy is below the optimized RHF energy, since a restriction on the total energy is removed. The commonly available ab initio programs are developed to go beyond the "simple Hartree–Fock" approach; they can include a partial treatment of the correlation energy, through wave functions that invoke CI, either in the form of a full CI or a limited CI. Energy optimization is occasionally replaced by the Møller–Plesset perturbation theory. When dealing with molecules having transition metals, calculations using CI should be preferred over calculations using Møller-Plesset perturbation theory, since no convergence can be ascertain for the perturbation series. A variety of more elaborate methods to deal with the correlation problem have also been used as, e.g., *the complete active space* (CAS) *approach* and *the restricted active space* (RAS) *approach*. The ab initio calculations often provide very good algorithms for elucidation of reaction paths and transition states. Ab initio calculations require more computerpower in the form of memory and speed. A supercomputer is desirable for calculations on oxometalloporphyrins because of the complexity in dealing with the theory. In addition, it is often necessary with approximations in the geometric structure in order to reduce the number of geometric variables. The advantages of the ab initio methods are as follows: (a) a variety of different levels of computation are available allowing one to obtain results at different levels of sophistication; (b) geometries are found to be reliable for both small and moderately large molecules; (c) the more refined versions describe the electronic structure of both closed and open shell systems very well; (d) reaction paths and transition states are often well described.

III. GEOMETRIC AND ELECTRONIC STRUCTURES OF OXOMETALLOPORPHYRINS

A. M = O Porphyrin Complexes

The electronic structure of the oxoiron porphyrin has been described using the extended Hückel approach [17,18]. A pseudo-octahedral geometry around the central iron atom was assumed, **1** [17].

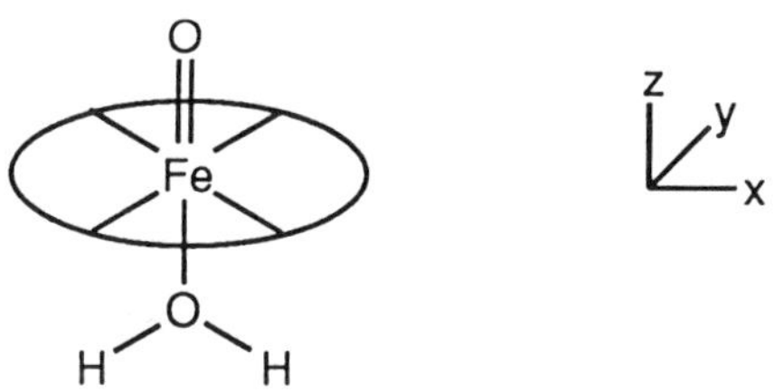

1

The iron atom is located in the porphyrin plane and the Fe = O bond was set to 1.65 Å [17]. The interaction diagram for the Fe(porphyrin)H_2O^{2+} and O^{2-} is outlined in Figure 1. The frontier orbitals of the Fe(porphyrin)H_2O^{2+} are shown to the left in the figure [17]. It appears that iron has two occupied orbitals, i.e., iron is d^4, the highest occupied molecular orbital (HOMO) and the second HOMO are the d_{yz} and d_{xz} orbitals, respectively. Immediately above the HOMO is the lowest unoccupied molecular orbital (LUMO), the iron $d_{x^2-y^2}$ orbital. The remaining two unoccupied iron orbitals are the d_{z^2} orbital, also low in energy but destabilized slightly due to the axial H_2O, and the d_{xy}, high in energy. The interaction with the O^{2-} orbitals, shown to the

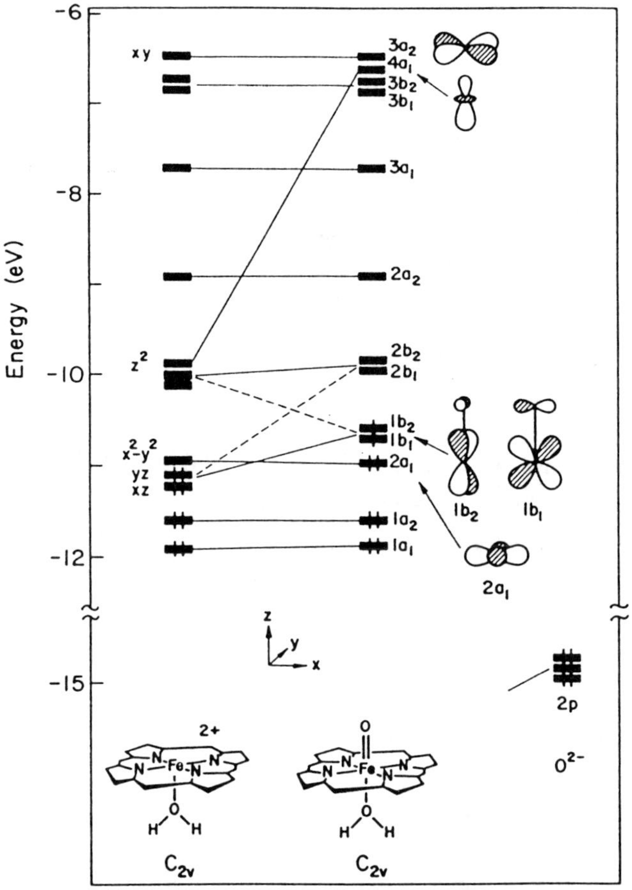

Figure 1 Interaction diagram for Fe(porphyrin)H_2O^{2+} and O^{2-} fragments of the O = Fe(porphyrin)H_2O. (From Ref. 17.)

right in Figure 1, pushes the d_{z^2} orbital upward in energy ($4a_1$), while the two interactions leading to the π orbitals destabilize the d_{yz} ($1b_1$) and d_{xz} ($1b_2$) orbitals slightly. These two π orbitals split due to the axial ligand, but are nearly degenerate (same orbital energy), and are available for the last two electrons of the complex, indicating a triplet ground state [17]. The two orbitals are shown in **2** and **3**:

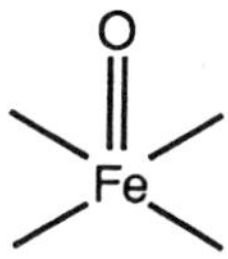

2

3

Based on the electronic structure of the oxoiron bond an interpretation for the unstability of this bond has been presented [17]. The two extra electrons are in the orbitals pictured in **2** and **3**; these orbitals are oxoiron antibonding, causing a substantial weakening of Fe=O bond. The presence of these two extra electrons in the antibonding oxoiron π orbital, leading to the weakening of this bond, is probably the reason for the extraordinary reactivity observed for this intermediate.

The electronic structure of square-pyrimidal oxoiron porphyrin without an axial ligand was also investigated [17]. Removal of H_2O—the axial σ donor—and displacement of the iron atom by 0.5 Å away from the porphyrin plane stabilizes the iron d_{z^2} orbital by nearly 3 eV [17].

The unpaired electron densities for **1**, and the oxoiron porphyrin with no axial ligand, both in an open shell triplet configuration, were also calculated and are presented for the oxoiron part in **4** and **5** [17]:

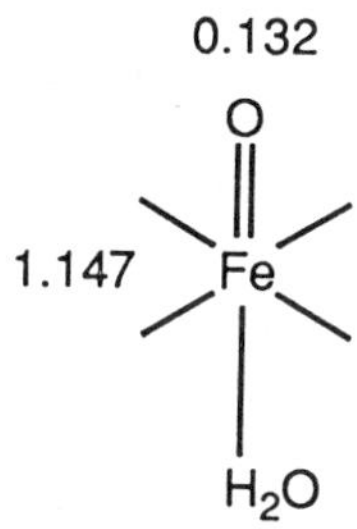

4

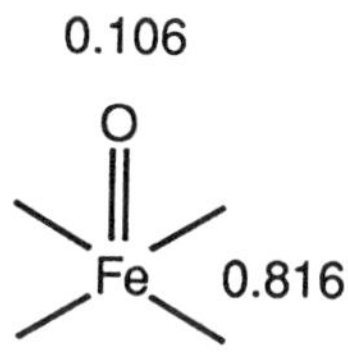

5

These calculations show that most of the unpaired electron density resides on the central iron atom, with only a minor part located on the oxygen and the atoms in the porphyrin ring [17]. It also appears from **4** and **5** that the out-of-plane displacement of the Fe=O group accompanying the removal of the axial ligand only slightly increases the spin density on the porphyrin ring.

Extended Hückel calculations have also been performed for ferrylprotoporphyrin IX with triplet-spin state (S = 1), both with and without an imidazole ligand [19]. The orbital picture which emerges from these calculations is very similar to the one outlined in Figure 1. Special attention was here attributed to the calculation of spin density for different ferryl π cations that were compared with the physical properties reported for compound I of horseradish peroxidase (HRP), chloroperoxidase, and catalase [19]. The ground state of the ferryl π cation was found to be influenced by peripheral substitution and axial ligation [19]. The spin density in HRP I and II has also been calculated using the INDO approach, and the trend observed in both the extended Hückel and the INDO approach is similar [17,20]. It should be mentioned that the INDO calculations show that, in addition to the spin density found at the oxygen and iron ligands, the spin density in the model HRP I a_{2u} π cation resides primarily on the pyrrole nitrogen and on the methine carbons, whereas on the model HRP I a_{1u} π cation the spin is primarily found to be on the α and the methine carbon atoms [20].

Calculation on oxoiron porphyrins using the INDO approach gives a orbital picture nearly similar to the one outlined in Figure 1 obtained by the extended Hückel calculations [21].

Several studies using ab initio calculations on oxometalloporphyrins have been published [8,22–24]. The electronic structure of the oxomanganese porphyrin chloride has been investigated with a minimum basis set [22]. The Mn = O bond was calculated to be 1.436 Å, with the manganese atom located 0.121 Å above the porphyrin plane and the Mn-Cl bond length calculated to be 2.301 Å [22]. The oxygen atom was calculated to be nearly neutral in charge (−0.09) while the manganese atom was found to have a slightly positive charge of 0.40. A relatively high Mn = O bond order was calculated (0.81) indicating a relatively strong Mn = O bond; the high Mn = O bond order is also reflected in the rather short Mn = O bond length. The overlap population in terms of σ and π contributions with respect to the Cl-Mn = O axis revealed that the Cl-Mn bond was predominantly of the σ type, while the oxygen is bound to the manganese atom by a double π bond [22]. The LUMOs of the system was calculated to be degenerate, having an orbital energy of 2.45 eV, and to be of the same types as outlined in **2** and **3**. The HOMO at −4.49 eV is located at the porphyrin ring, while the second HOMO at −4.87 eV is distributed over the porphyrin ring, chloride, and oxygen [22]. The electronic structure of oxoiron porphyrin has also been investigated using relatively simple and more sophisticated ab initio methods [8,23]. For the oxoiron porphyrin (no axial ligand present) with S = 1, the electronic configuration of the Fe(IV) d^4 atom was $(d_{xy})^2(d_{xz})^1(d_{yz})^1$, which has been described as the ground state [23]. With a Fe = O bond length of 1.70 Å it has been found that a stabilization of the system is achieved by an out-of-plane displacement of iron; a displacement of 0.25 Å causes a decrease in total energy for the system by 7.6 kcal/mol relative to the iron atom placed in the porphyrin plane [23]. The electronic state of the oxoiron porphyrin was assigned as $^3A_2$ [23]. The electronic structure has also been calculated for the cation of oxoiron porphyrin assumed with S = 3/2 [23]. This complex can be formed by either an ionization from the a_{1u} or the a_{2u} orbital of the porphyrin ligand. These calculations showed that the state $^4A_1$ with an a_{1u} hole is 9 kcal/mol more stable than the state $^4A_2$ with an a_{2u} hole (iron placed in the porphyrin plane) [23]. The electronic configuration of this for the most stable state was described as $(d_{xy})^2(d_{xz})^1(d_{yz})^1(\text{por } a_{1u})^1$ [23]. The energy difference obtained by ab initio calculations for these two states was found to be very similar to the energy difference obtained by the semiempirical INDO calculations [20]. The Fe = O bond in oxoiron porphyrins with pyridine as the axial ligand has also been studied using RHF and CASSCF ab initio procedures [8]. It was concluded that the RHF approximations were not adequate for the description of the Fe = O bond, while the CASSCF calculations gave satisfactory results [8]. When using the latter type of calculation the minimum for the potential energy curve for the Fe = O bond length was calculated to be 1.66 Å [8], which differs only slightly for the one found by EXAFS (1.64 ± 0.03 Å) [25]. The two unpaired electrons also here occupy the Fe = O π_x* and π_y* antibonding orbitals. The atomic net charge was calculated to be 2.37 for iron and −0.49 for oxygen; besides its nucleophilic nature the oxygen was, due to its biradical nature, also considered to be electrophilic [8]. The spin density for the Fe = O bond was found to be very sensitive to the method for calculation; with the RHF approach iron has a spin density of 1.91 and oxygen 0.06, while the CASSCF approach gives 1.12 and 0.86, respectively. The latter results are said to be consistent with proton nuclear magnetic resonance (NMR) measurements [8]. It was also pointed out that the electronic structure of the Fe = O moiety does not correspond to that of the stable O_2 molecule but to that of the elongated unstable O_2 molecule [8].

B. Oxygen Inserted into the Iron–Nitrogen Bond

Another structural description for the oxoiron porphyrin intermediate has also been presented [17,21,23,26]. The structure of this intermediate where the oxygen is inserted into the iron–nitrogen bond, as illustrated in **6**, is based on the structure of Fe(TPP)[C = C(*p*-ClC$_6$H$_4$)$_2$]Cl where the vinylidene group is inserted into the iron–nitrogen bond, which has been fully characterized [27].

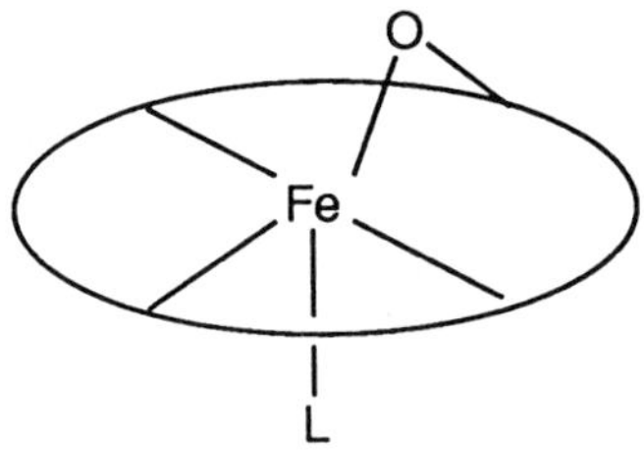

6

If the oxygen atom in **1** is slipped toward one of the pyrrole nitrogens in the porphyrin ring a stabilization of the system is achieved. Figure 2 shows the INDO calculated change in total energy of the oxoiron porphyrin system as a function of the slipping displacement of the oxygen atom toward one of the nitrogens in the porphyrin ligand (x axis) and the distance between the oxygen and the iron porphyrin plane (z axis) [21].

The stability of **6** relative to **1** is calculated within the INDO approach to be about 180 kcal/mol, a high stabilization energy, and probably rather

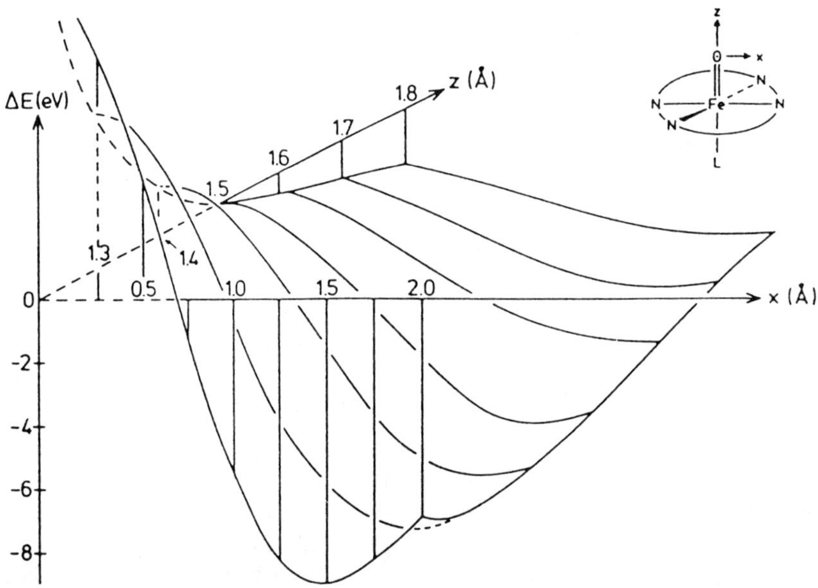

Figure 2 The total energy surface of the oxoiron porphyrin system as a function of the slipping motion of the oxygen as indicated in the upper right corner. The zero energy is taken to be the symmetric form for x = 0 Å and z = 1.5 Å. (From Ref. 21.)

exaggerated due to the deficiency of the method. The Fe-O and N-O bond lengths in **6** are 1.87 and 1.41 Å, respectively, and they correspond well to those found in the nickel analog to **6** [28]. The instability of **1** relative to **6** can be traced to two unpaired electrons in the Fe = O π_x* and π_y* antibonding orbitals (corresponding to the orbitals shown in **2** and **3**). During the slipping displacement of the oxygen toward the nitrogen the antibonding character between iron and oxygen is reduced, and the bonding between oxygen and nitrogen is built up; the overlap population between iron and oxygen diminishes from 0.47 to 0.18 when going from **1** to **6**, whereas the oxygen–nitrogen overlap increases from 0.00 to 0.65 [21].

Both manganese and iron porphyrins with an oxygen inserted into the metal–nitrogen bond have also been the subject of ab initio calculations [23, 26]. For the oxomanganese porphyrin chloride with oxygen inserted into the manganese–nitrogen bond a stabilization of 30 kcal/mol is calculated relative to the high-symmetric complex [26]. The optimized structure of this complex shows that the manganese atom is positioned on the opposite site (0.261 Å) of the oxygen, with Mn-O and N-O bond lengths of 1.840 and 1.464 Å, respectively [26]. The charges on oxygen and manganese have been calculated to be 1.38 and −0.43, respectively, and they are thus different to those in the high-symmetric oxomagnanese porphyrin chloride complex [22]. The Mn-O bond order is also reduced significantly by the insertion of the oxygen into the manganese–nitrogen bond; it is calculated to be 0.48, relative to 0.81 for the Mn = O complex [22,26]. The LUMO of the system at 2.53 eV is located on the porphyrin ligand as a π orbital, at oxygen, manganese, and chloride, while the second LUMO at 3.02 eV is distributed at the metal and the porphyrin ligand [26]; the HOMO is calculated to be −4.87 eV and is located at the porphyrin ligand, while the second HOMO, found at −6.15 eV, is distributed on the porphyrin ligand, and at the oxygen and chloride atoms [26]. Both iron porphyrins (with no axial ligand), with oxygen inserted into the iron–nitrogen bond with S = 1 and S = 3/2, have also been studied [23]; the system with S = 1 is calculated to be about 10 kcal/mol more stable relative to the axial structures, whereas for the cation with S = 3/2 the stabilization is about 28 kcal/mol [23]. The electronic state for the first S = 1 state is $^3A'$ and for the spin state S = 3/2 it is $^4A''$; the electronic configurations for these two spin states are $(d_{xy})^2(d_{yz})^2(d_{x^2-z^2})^1(d_{xz})^1$ and $(d_{yz})^2(d_{xy})^1(d_{x^2-z^2})^1(d_{xz})^1$ [23].

IV. INFLUENCE OF AXIAL LIGANDS (PROXIMAL EFFECTS)

One striking feature of the catalytic properties of the metalloporphyrins is the remarkable improvement of the rate, chemo- and stereoselectivity observed, especially for the manganese porphyrins, by the addition of nitrogen bases, such as pyridine or imidazole, to the reaction mixture [1]. The effect of the nitrogen ligand adding to the axial position (trans to the oxofunctionality) is often called the proximal effect. This effect can be considered as a preliminary approach in the modeling and distal effects involved in the chemical properties of hemoproteins, such as cytochrome P450 and catalase.

The effect of the axial ligand on the properties of the oxometal bond in oxometalloporphyrins has been investigated using methods ranging from extended Hückel to ab initio. Extended Hückel calculations have been mainly performed for the enzyme-like models and discussed mainly in relation to optical and electron spin resonance (ESR) spectra of the compounds [18, 19]. The axial ligand has been found to have an influence on the energy difference between the a_{1u} and a_{2u} orbitals, and the energy increases in the order no ligand < mercaptide << imidazole < mercaptan < imidazolate [19]. Recently, INDO calculations have been performed to investigate the effects of imidazole $H_{\delta 1}$ deprotonation on the electronic structure and reactivity of the oxoiron prophyrin in comparison with the properties of cytochrome P450 [29]. The deprotonation of the imidazole creates a new low-lying imidazole π-cation radical electronic state, in addition to the classical a_{1u} and a_{2u} porphyrin π-cation radical states. In this imidazolate π-radical state, there is one unpaired spin on the imidazole ring, and the negative charge is almost entirely on the porphyrin ring, which has little unpaired spin density [29]. The deprotonation does not affect the spin and charge distribution on iron or on the oxygen atom [29]. The calculations indicated that the deprotonation is a mechanism by which the radical character can be transferred from the porphyrin ring to the imidazole [29].

The effect of the axial ligand on the properties of the oxomanganese and oxoiron bond in oxometalloporphyrins has also been investigated using ab initio calculations [8,30]. The influence of pyridine as axial ligand has been compared with chloride and with having no axial ligand [30]. The system with no axial ligand present is in many respects very similar to the one with chloride as axial ligand; however, replacement of chloride with pyridine causes several changes: (a) the Mn = O bond length is increased from 1.41 Å to 2.08 Å, indicating a weaker bound oxygen; (b) the oxygen becomes more electrophilic as the atomic charge on the oxygen is 0.12 in the complex with pyridine present, while the charge is -0.24 in the complex without axial ligand; (c) the oxygen becomes a better π-electron acceptor [30]. The reason for the increase in Mn = O bond length can be attributed to an interaction of the pyridine π orbital perpendicular to the pyridine plane interacting with the oxomanganese antibonding π ortibal as shown in **7**:

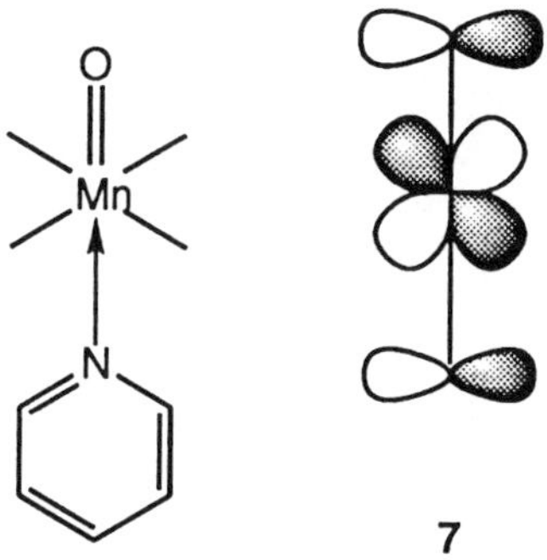

7

By virtue of this interaction pyridine donates electron density (0.27) into the antibonding Mn = O bond thus causing it to weaken [30]. The electronic state of the oxomanganese porphyrin–pyridine complex is $^1A_1$ [30]. The attachment of an axial ligand to the oxoiron porphyrin system has also been suggested to weaken the oxoiron bond [8], as the lone-pair electrons on the imidazole nitrogen hinder the electron donation from oxygen $2p_x$ to iron d_{z^2} [8]. The axial ligand has been suggested to affect the Fe = O bond in the following order [8]:

$$O{=}Fe(TPP) > O{=}Fe(P)(Py) > O{=}Fe(T_{piv}PP)(1\text{-}MeIM)$$

The coordinating pyridine ligand also donates a minor amount of electron density (−0.04) to the remaining complex by the attachment, but due to the difference in sophistication in the calculation procedure a comparison of the axial effect on the Mn = O and Fe = O bonds was not feasible [31].

V. MECHANISM OF OXYGEN TRANSFER TO ALKENES, ALKANES, AND AROMATICS

Compared to the interest devoted to the experimental studies of the mechanism of oxygen transfer from oxometalloporphyrins to substrates, theoretical investigations have not received much attention, and only a limited amount of papers are available dealing with the mechanistic aspects of oxygen transfer reactions.

A. Oxidation of Alkenes

Atomic oxygen has been used as a simple model for oxometalloporphyrins for the study of the epoxidation of alkenes using MINDO/3 calculations [32]. With atomic oxygen in a singlet state (^1S) a concerted symmetric transition state, **8**, is obtained [32]. The carbon–oxygen distance is calculated to be 3.197 Å and the oxygen insertion takes place with retention of the configuration of the alkene [32].

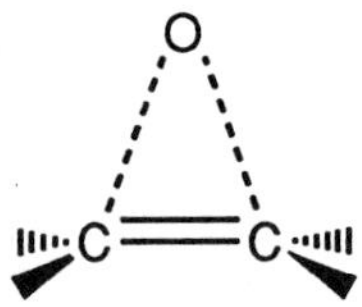

8

This closed shell reaction path has been calculated to have a barrier of 1.4 kcal/mol, while the open shell reaction path with triplet atomic oxygen (^3P) proceeds in two steps with a transition state energy of 16 kcal/mol for the first step leading to an open biradical oxygen adduct, implying a loss of configuration [32]. These calculations thus indicate that two reaction paths are possible for the epoxidation. The most probable one is the concerted [1 + 2] addition because of the lower transition state energy, while the other one, an energetically less favorable process, takes place as a radical mechanism with free rotation around C-C bond, which can lead to a loss of configuration.

The mechanism for the oxygen transfer from oxometalloporphyrins having oxometal bonds to alkenes has been the subject of intense debate and several types of mechanism have been proposed [1,33]; three possible intermediates related to this presentation are shown in **9–11**:

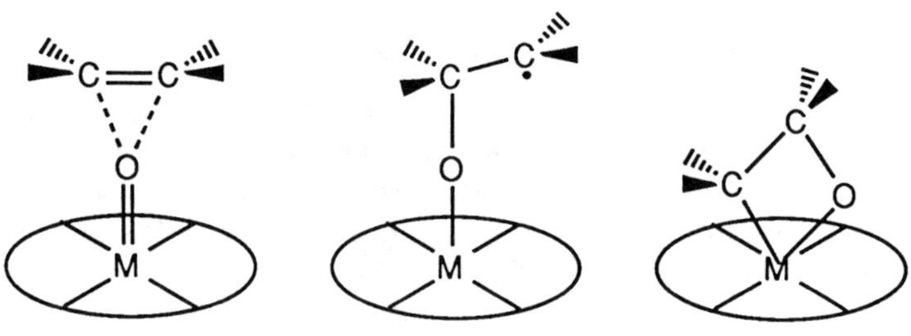

9 **10** **11**

The first approach, **9**, is a symmetric/concerted [1 + 2] interaction of the alkene with the oxygen in the oxometalloporphyrin, maintaining the stereochemistry of the alkene, while the second approach, **10**, is an interaction of only one of the carbons in the alkene with the oxygen of the oxometalloporphyrin, in principle a nonconcerted [1 + 2] addition reaction [1,

33,34]. The oxygen transfer in the third approach, a [2 + 2] addition reaction, **11**, takes place via a metallaoxetane intermediate and the epoxide is formed by a reductive elimination [33–35]. The [2 + 2] addition reaction can take place either in a concerted or nonconcerted manner.

By using extended Hückel calculations augmented by valence bond (VB) analysis, the oxygen transfer from oxoiron porphyrin to an alkene has been investigated [36]. It was shown that the Fe(IV) = O bond bears a substantial radical character and should thus be expected to react like an RO radical species [36]. The VB scheme for the reaction of an alkene with Fe(IV) = O porphyrin shows that the oxygen has an electrophilic character and that the addition is a nonconcerted [1 + 2] addition reaction with an intermediate of the type Fe-O-C-C, **10**, with a covalent radical nature [36]. The reason for this covalent radical-type intermediate is that one electron of a given spin, e.g., β, comes from the π system of the alkene, so that a potentially reactive VB structure of Fe(IV) = O porphyrin must have only one electron of α spin around the oxygen [36]. The spin-freezing of this structure leads to the formation of one covalent bond along the C-O direction and leaves two electrons unpaired, one at the terminal oxygen and one at the iron [36]. Based on this intermediate it was pointed out that the metallaoxetane intermediate was not feasible because of an unfavorable electronic arrangement [36]. These calculations thus suggested that the epoxide is formed as an open intermediate by oxygen transfer from the oxoiron porphyrin to the alkene, but also that along the various reaction paths, one-electron transfer mechanisms are likely to occur [36].

The stereoselective epoxidation of *cis*-alkenes, rather than *trans*-alkenes, by oxoiron porphyrins has also been investigated using INDO calculations, based on the concept that the alkene approaches the oxygen from the side with the oxygen tilted toward the nitrogen in one of the pyrrole rings/inserted into the iron–nitrogen bond, **12** [21]. This reaction can be described as a concerted [1 + 2] addition reaction.

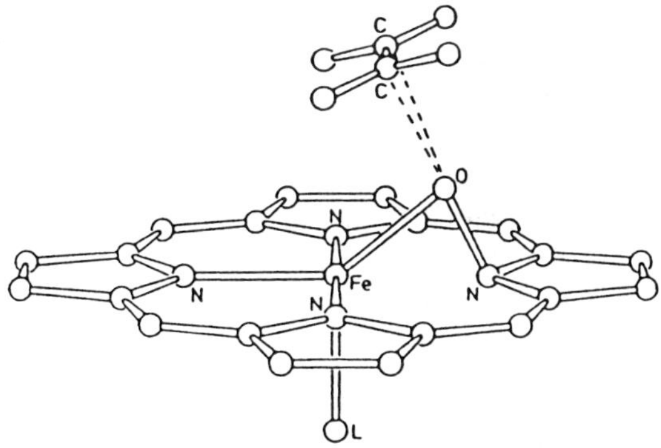

12

The preferred orientation of the alkene in **12** can be attributed to the frontier orbitals of the two interacting systems. The alkene antisymmetric π^* orbital interacts with the oxygen p_y orbital, also antisymmetric with respect to the same plane. This orbital interaction is shown in **13** [21]. Besides this interaction, the alkene π orbital also donates electron density into an empty orbital located at the inserted oxygen [21].

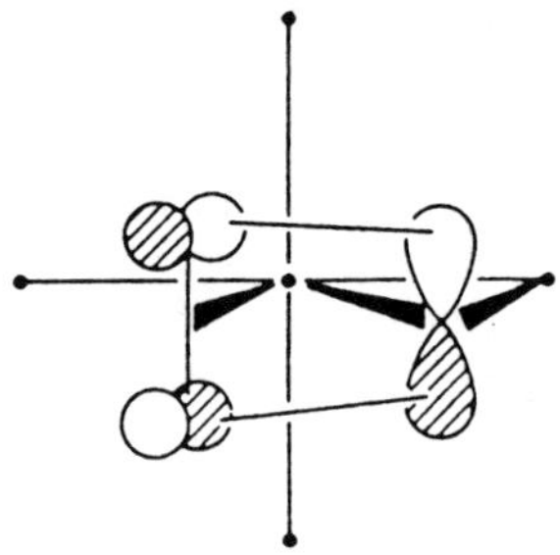

13

The interaction between the alkene and the oxygen in the oxoiron porphyrin shown in **12** might account for the preferred epoxidation of *cis*-alkenes, compared to *trans*-alkenes; one of the substituents in the *trans*-alkene will interact in a sterically repulsive manner with the porphyrin plane, and this is not the case for the *cis*-alkenes [21].

The epoxidation reaction with a six-coordinate oxoruthenium(IV) model complex containing two glyoxal diimines and an ammonia ligand has also been investigated using INDO calculations [37]. The following reaction pathways were considered [37]: a concerted and a nonconcerted [2 + 2] pathway leading to the metallaoxetane (corresponding to **11** as the intermediate), a concerted [1 + 2] oxoruthenium–alkene interaction (corresponding to **9** as the intermediate) and a nonconcerted [1 + 2] oxoruthenium–alkene interaction (corresponding to **10** as the intermediate). Attempts to optimize the geometry for a concerted [2 + 2] cycloaddition where the Ru-O and C-C bonds are paralleled resulted in an intermediate with almost no bonding between the ethylene and the oxoruthenium complex [37]. The nonconcerted [2 + 2] cycloaddition pathway leading to a metallaoxetane by first a [1 + 2] pathway followed by a [2 + 2] pathway was also found to be unfavorable [37]. The reason for the unfavorable [2 + 2] pathway can be traced to a two-orbital, four-electron repulsive interaction between the $\pi_{Ru\text{-}O}$ and $\pi_{C\text{-}C}$ orbitals [37]. The concerted [1 + 2] approach leads to an energy minimization of 9 kcal/mol for $R_{oxo\text{-}C} \approx 2.7$ Å as an ion–molecule complex by which the alkene has donated electron density to the oxoruthenium complex [37]. Proceeding along the [1 + 2] pathway from the shallow ion–molecule intermediate leads to a bound epoxide with $R_{oxo\text{-}C} \approx 2.5$ Å, but the reaction proceeds along a nonconcerted pathway; the reason for the unfavorable concerted pathway arises from a two-orbital, four-electron repulsive interaction between one of the oxygen lone-pair electrons and a $\pi_{C\text{-}C}$ orbital, **14** [37].

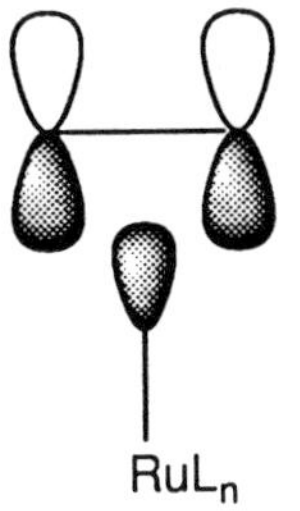

14

15

In the nonconcerted [1 + 2] approach, **10**, the orbital picture changes [37]. Moving the ethylene off the center causes interaction of the filled ethylene π orbital with the empty oxoruthenium π^* orbital, which allows for a second-order mixing of ethylene π and π^* molecular orbitals. This net interaction leads to a bonding and an antibonding interaction, the latter of which is shown in **15**. This interaction is interesting because the species obtained shows increased cation radical character at the terminal carbon [37]. Furthermore, due to the electron donation by the ethylene π molecular orbital there is a reduced electron density of the terminal carbon, which directs the attack of this carbon atom to the electron-rich oxygen atom rather than to the electron-poor, high-valent ruthenium atom [37]. The attack at the oxygen leads to formation of the epoxide, while the attack at the metal leads to the metalla-oxetane. It was concluded that the [1 + 2] nonconcerted addition is the most likely pathway [37].

The ab initio studies of the oxomanganese porphyrins, with both the oxomanganese functionality and with the oxygen inserted into the manganese–nitrogen bond, have also been the subject of a brief discussion on oxygen transfer to an alkene [22,26]; no calculations were performed in an attempt to obtain a quantitative description of the oxygen transfer step. Based on the frontier orbitals of the oxomanganese functionality it was pointed out that the oxo species possesses various low-lying unoccupied antibonding Mn d_π-O p_π combinations of a symmetry proper for interaction with the occupied alkene π orbitals, though it was pointed out that the symmetry of the LUMOs of the oxomanganese porphyrin excludes any symmetric interactions with the alkene π orbital [22]. This proposed intermediate corresponds to the intermediate shown in **10**. For the oxomanganese porphyrin with the oxygen inserted into the manganese–nitrogen bond a carbocation intermediate has been proposed, **16**, based on the unsymmetric orbital interaction outlined in **17**, where the HOMO of the alkene interacts with the LUMO of the active oxomanganese porphyrin complex [26].

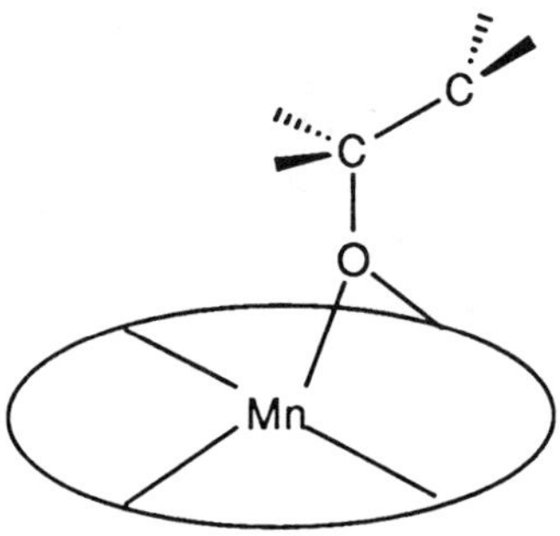

16

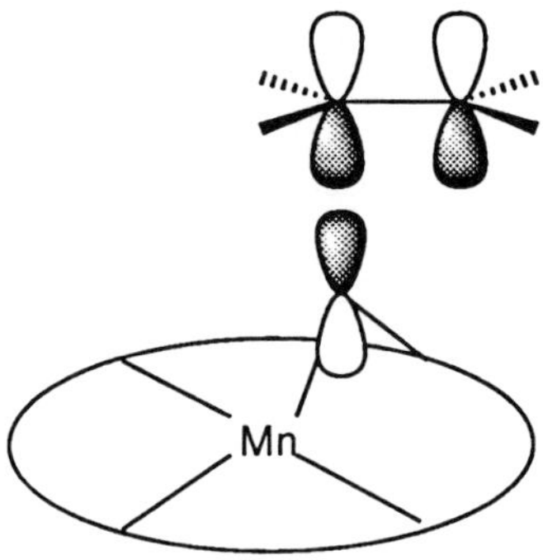

17

The interaction of a terminal oxidant and a metallaporphyrin giving an oxygen atom inserted into the metal–nitrogen bond in relation to oxygen transfer to alkenes has also been discussed from an experimental point of view [38].

It should also be mentioned that the addition of terminal alkenes to oxoiron porphyrin in relation to the putative suicide substrates of cytochrome P450 has been studied using ab initio calculations in combination with molecular mechanics [39].

It is at the present stage very difficult to draw any conclusion about the mechanism for the epoxidation of alkenes by oxometal porphyrins because different calculation procedures are used for the few studies that have been performed. But it seems likely that for several of the systems studied a [1+2] addition reaction is more likely than a [2+2] addition reaction, although a one-electron transfer mechanism might also occur. Furthermore, it also seems difficult to account for the role of the transition metal in these epoxidation reactions as the calculations have been performed with different methods, and these are not directly comparable.

B. Oxidation of C-H Bonds in Alkanes and Arenes

The atomic oxygen model has also been applied for the investigation of oxidation of C-H bonds using MINDO/3 calculation [32]. The reaction of singlet atomic oxygen (^1S) with methane consists, in addition to a reactant-like transition state with O-H and C-O distances of 2.262 and 3.399 Å, respectively, and a transition state energy of 2.5 kcal/mol, also of an "intermediate" saddle point complex with O-H and C-O distances of 0.979 and 2.130 Å, respectively, and with O-H-C found on a straight line with an energy 54.9 kcal/mol lower than the reactants [32]. These two structures are, together with the product for the reaction, methanol, shown in Scheme 2. A hydride abstraction from methane to form OH^- and CH_3^+ does not operate in this reaction [32]. In the transition state structure the partial charges of the atoms are essentially those of the reactants, whereas for the intermediate saddle point a significant negative charge is found at the oxygen (−0.58) and a partial positive charge at the carbon (0.29) [32]. The transition state obtained by the MINDO/3 calculations is claimed to be qualitatively similar to the one obtained by ab initio calculations including CI with an extended basis set including polarization functions [40]; in this study the transition state was found to have O-H and C-H bond lengths of 1.25 and 1.32 Å, respectively.

The reaction path for the triplet atomic oxygen (^3P) with methane indicates a hydrogen abstraction reaction path with no stationary points between the transition state and the product minimum [32]. The activation energy is calculated to be 6.8 kcal/mol [32]. The results for the hydroxylation with atomic oxygen show that the closed shell reaction path can be characterized as a "attachment-rearrangement" that is concerted and should take place with retention of configuration, whereas the open shell reaction path involves the formation of hydroxyl and hydrocarbon radical intermediates, which subsequently can recombine implying a loss of configuration.

The atomic oxygen model has also been used for aromatic oxidation [41]. The reaction of singlet (cationic) and triplet (biradical) atomic oxygen

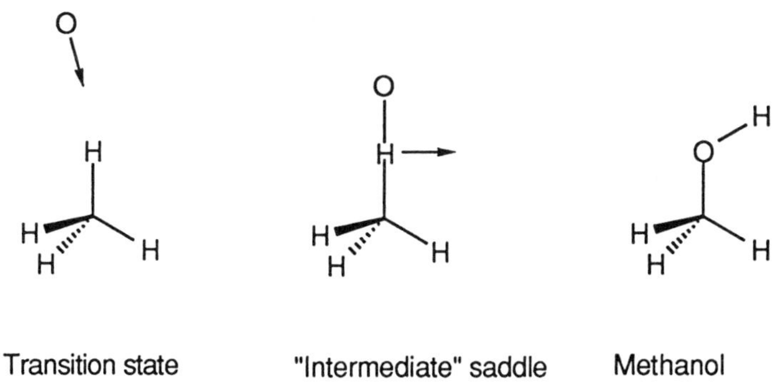

Scheme 2

with benzene and substituted benzenes was studied, and the attention was only focused on the distribution of the products—epoxide, ketone, and phenol—using MNDO calculations [41b]. The starting point for the calculations for the cationic pathway was a protonated species, **18** (Scheme 3), which has a positive charge on the ring. The transition state for the formation of the protonated epoxide, for the protonated ketone, formed by a 1,2 hydrogen shift, and the direct formation of phenol was calculated using **18** as the substrate (Scheme 3) [41b].

Different substituents on the phenyl ring, especially hydroxy and methoxy, were found to stabilize the intermediate and the para-substituted compounds have heats of formation ~ 18 kcal/mol lower in energy than the meta isomers [41b]. The order of the reaction rates for the formation of protonated epoxide, ketone, and direct phenol formation from **18** was phenol > ketone > epoxide [41b].

For the study of the radical pathway the substrate was a system with one unpaired electron on the oxygen, the other one on the ring, and the oxygen bound to the ring in a tetrahedral manner [41b]. The same reaction pathways as for the cationic system were also studied, but the order of the reaction rate was here calculated to be epoxide > ketone > phenol [41b]. The

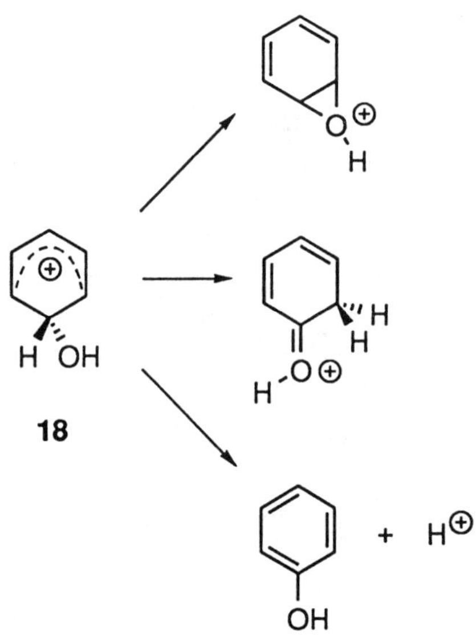

Scheme 3

reaction path for the epoxide formation shows that the triplet state leads to an excited state, while the singlet biradical surface was calculated to be 15 kcal/mol lower in energy [41b].

These semiempirical calculations thus indicated that, depending on the nature of the oxygen, different reaction rates could be expected and it was suggested that both a cationic and a radical pathway might be involved in cytochrome P450–mediated aromatic oxidation [41b]. But no information on the mechanism for the oxidation step from the oxometal porphyrin to the substrate was provided.

The oxidation of the C-H bond in alcohols with the oxoruthenium(IV) model complex with two glyoxal diimines and an ammonia ligand has also been investigated using INDO calculations [42]. Two interactions of the C-H bond in methanol with the oxygen in the oxoruthenium(IV) complex were investigated—a triangular approach, by which the oxygen in the oxoruthenium(IV) complex interacts with the C-H bond as outlined in **19** and a linear one by with the oxygen approaches the C-H bond in a linear fashion as outlined in **20** [42].

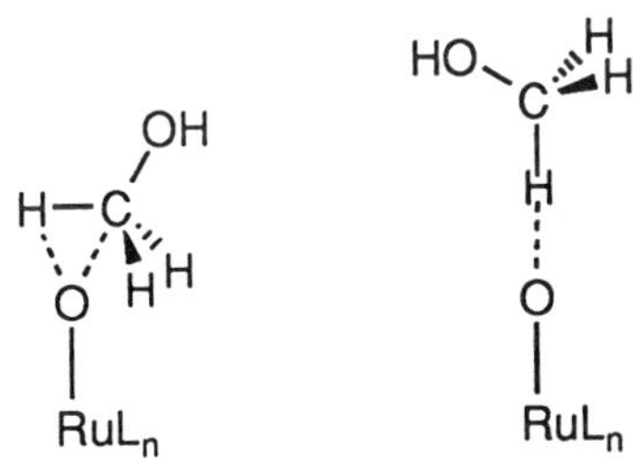

19 **20**

The reaction path, **19**, was found to be unfavorable for all distances with $R_{oxo\text{-}C} < 2$ Å, whereas the linear approach, **20**, was found to be energetically more favorable because of better orbital interaction and less steric hindrance [42]. For $R_{oxo\text{-}C} \approx 1.4$ Å, the latter approach shows a small barrier of 3.1 kcal/mol resulting from repulsive two-orbital, four-electron interaction between the Ru=O σ nonbonding molecular orbital (O $2p_z$) and the methanol C-H σ bond [42]. Decreasing the $R_{oxo\text{-}C}$ to about 1.2 Å results in an energy minimum of 10.6 kcal/mol relative to the separated species, and shortening the $R_{oxo\text{-}C}$ further causes an increase in the total energy of the system [42]. The analysis of different stretching modes for the system gave an estimate of the activation energy of 52 kcal/mol [42]. A third approach of alcohol oxidation by an oxoruthenium(IV) complex, in which the oxygen approaches the metal as outlined in **21**, has also been analyzed [42].

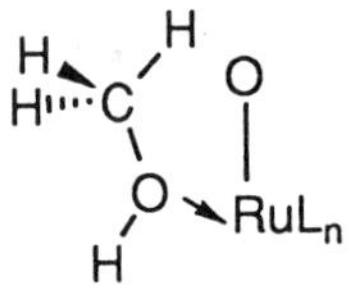

21

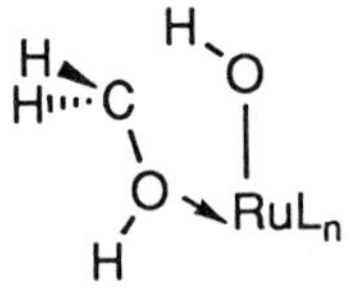

22

The coordination of the substrate to the metal activates it and the calculated barrier for the hydrogen transfer process is 16 kcal/mol; by this hydrogen transfer the system goes from $Ru(N_5)\text{-}O(CH_3OH)^{2+}$ to $Ru(N_5)\text{-}(OH)\text{-}(CH_2OH)^{2+}$, **22** [42]. The hydroxyl group in **22** is then transferred and $CH_2(OH)_2$ is formed. The preferred approach from the site has been suggested to originate from an energetically more favorable interaction of the C-H bond with Ru d_π–O p_π located higher in energy and thus more reactive than the $\sigma_{Ru\text{-}O}$ bond, which is involved in the linear approach [42]. The latter approach is thus the most likely from an energetic point of view.

C. Miscellaneous Reactions

Two types of reactions will be briefly discussed here, the oxidation of CCl_4 and $CHCl_3$ and some amines by atomic oxygen; this work is also related to the oxidation of these substrates catalyzed by cytochrome P450 [43,44]. In both studies MNDO calculations were applied [43,44]. The reaction of triplet oxygen (^3P) with CCl_4 was calculated to take place first by a chlorine abstraction followed by a recombination [43]. The chlorine abstraction has been calculated to be exothermic, in contrast to the experimental results, with an activation energy of 29.5 kcal/mol [43]. The chlorosyl radical (OCl) has the unpaired electron exclusively on the oxygen and is thus similar to the hydroxyl radical. The recombination step therefore takes place between the oxygen atom in OCl and the carbon atom in CCl_3. The reaction path for the radical atomic oxygen reaction with $CHCl_3$ has two possibilities: a hydrogen and a chlorine abstraction. The transition mode for the hydrogen abstraction is calculated to be a "linear" transition state, whereas for the chlorine abstraction a bent OClC, is found; the latter resembles that for the chlorine abstraction from CCl_4 [43]. The transition state energy for the hydrogen abstraction has been calculated to be 37.4 kcal/mol, while the one for chlorine abstraction is calculated to be 33.4 kcal/mol [43], indicating that the latter might be slightly more feasible than the former. However, one should note that reaction of R-O· with $CHCl_3$ results in hydrogen abstraction.

The oxidation of amines (ammonia and mono-, di-, and trialkylamines) by the singlet and triplet atomic oxygen model for cytochrome P450 has been studied [44]. The reaction of singlet oxygen with amines takes place as a two-step addition-rearrangement mechanism leading to both *N*-hydroxyl-amine and *N*-methoxyamine products via an *N*-oxide intermediate [44]. The interaction of the aliphatic amines with singlet oxygen leading to the *N*-oxides is calculated to be exothermic (57–77 kcal/mol) and have relatively small energy barriers (10–15 kcal/mol, highest for trimethylamine) [44]. The rearrangement to the hydroxylamines requires a barrier of up to 49 kcal/mol, with an unsymmetric transition state where the migrating part is closer to nitrogen than oxygen [44]. For the reaction of triplet oxygen the reaction can take place as an N-addition, an N-H abstraction, or an α-C oxidation by a hydrogen abstraction [44]. The calculations showed that triplet oxygen does not add to the nitrogen forming stable *N*-oxides, but undergoes a simultaneous loss of either hydrogen or a methyl radical [44]. The activation parameters for these processes have been found to be 33–46 kcal/mol, which is comparable to the energy required for α-C oxidation by a hydrogen abstraction [44]. It was pointed out that the obtained results were consistent with a radical mechanism for the oxidation of alkylamines by cytochrome P450 [44]. But it should also be mentioned that electron transfer reactions, which have been observed experimentally [2,45], were not considered.

VI. COMMENTS

The intention with the present chapter was to try to present some of the theoretical work performed on oxometalloporphyrins and the subsequent oxygen transfer reactions. The calculations carried out on the different problems related to the chemistry of oxometalloporphyrins do not all give answers that will allow one to understand the experimental results. There are several reasons for this lack. Dealing with oxometalloporphyrins in a satisfactory manner, i.e., doing reliable calculations, is a tough task and the coming generation of computer programs will hopefully include software that will make it possible both to treat the oxometal bonds in the oxometalloporphyrins in a satisfactory manner and locate transition states and intermediates along a reaction path.

If one looks at what has been done to date, one begins to understand the oxometal bond in, say, oxoiron porphyrin, but our understanding of the oxygen transfer from these types of species is in its infancy, and much more work in the field of theoretical calculations is necessary to explain all the interesting results that experimentalists are producing.

ACKNOWLEDGMENTS

Thanks are expressed to Professors Jan Linderberg and Peter Swanstrøm for helpful comments.

NOTES AND REFERENCES

1. B. Meunier, Metalloporphyrins as versatile catalysts for oxidation reactions and oxidative DNA cleavage, *Chem. Rev., 92*: 1411 (1992).
2. J. Larsen and K.A. Jørgensen, A facile oxidation of secondary amines to imines by iodosobenzene or by a terminal oxidant and manganese or iron porphyrins and manganese salen as the catalysts, *J. Chem. Soc. Perkin Trans., 2*: 1213 (1992).
3. See, for example, Ref. 1 for a review about the experimental investigations of the characterization of the high-valent oxometal porphyrin complexes.
4. It should be noted that oxovanadium porphyrin in the presence of iodosylbenzene has been shown to catalyze oxygen transfer to an alkene at about 80°C, but that the oxygen transfer ability is significantly lower than the one found for the later transition metal complexes. C. Sams and K.A. Jørgensen, unpublished results.
5. T.J. Collins, R.D. Powell, C. Slebonick, and E.S. Uffelman, A water-stable manganese(V)-oxo complex; definitive assignments of a $\nu_{Mn(V)\equiv O}$ infrared vibration, *J. Am. Chem. Soc., 112*: 899 (1990).
6. See, for example, Ref. 1 and references therein.
7. A. Dedieu, M.M. Rohmer, and A. Veillard, Ab initio calculations of metalloporphyrins, *Adv. Quant. Chem., 16*: 43 (1982).
8. S. Yamanoto, J. Teraoka, and H. Kashiwagi, Ab initio and CASSCF studies of Fe-O bond in high-valent iron-oxo-porphyrins, *J. Chem. Phys., 88*: 303 (1988).
9. (a) R. Hoffmann, An extended Hückel theory. I. Hydrocarbons, *J. Chem. Phys., 39*: 1397 (1963). (b) R. Hoffmann, *Solids and Surfaces: A Chemist's View on Bonding in Extended Structures*; VCH, New York (1988).
10. G.A. Segal (ed.), *Semiempirical Methods of Electronic Structure Calculations*, Parts A and B, Plenum Press, New York, 1977.
11. R.D. Bingham, M.J.S. Dewar, and D.H. Lo, Ground states of molecules. XXV. MINDO/3. An improved version of the MINDI semiempirical SCF-MO method, *J. Am. Chem. Soc., 97*: 1285 (1975) and the following papers.
12. (a) M.J.S. Dewar, E.G. Zoebisch, E.F. Haley, and J.J.P. Steward, AM1.: A new general purpose quantum chemical molecular model, *J. Am. Chem. Soc., 107*: 3902 (1985).
    (b) J.J.P. Steward, Optimization of parameters for semiempirical methods I. Method, *J. Comp. Chem., 10*: 209 (1989).
13. (a) J.A. Pople, D.L. Beveridge, and P.A. Debosh, Approximate self-consistent molecular-orbital theory. V. Intermediate neglect of differential overlap, *J. Chem. Phys., 47*: 2026 (1967).
    (b) J.A. Ridley and M.C. Zerner, An intermediate neglect of different overlap technique for spectroscopy: pyrrole and azines, *Theoret. Chim. Acta (Berl.), 32*: 111 (1973).
    (c) M.C. Zerner, G.H. Loew, R.F. Kirscher, and U.T. Mueller-Westerhoff, An intermediate neglect of differential overlap technique for spectroscopy of transition-metal complexes. Ferrocene, *J. Am. Chem. Soc., 102*: 589 (1980).

(d) W.P. Anderson, T.R. Cundanari, R.S. Drago, and M.C. Zerner, Utility of the semiempirical INDO/1 method for the calculations of the geometry of second-row transition-metal species, *Inorg. Chem., 29*: 1 (1990).

14. ZINDO—an abbreviation for *Z*erners *I*ntermediate *N*eglect of *D*ifferential *O*verlap.
15. M. Zerner, personal communication.
16. See W.J. Hehre, L. Radon, P.v.R. Schleyer, and J.A. Pople, *Ab Initio Molecular Orbital Theory*, Wiley, New York, 1986 for an excellent book about ab initio theory.
17. K. Tatsumi and R. Hoffmann, Metalloporphyrins with unusual geometries. 2. Slipped and skewed bimetallic structures, carbenes and oxo complexes, and insertions into metal-porphyrin bonds, *Inorg. Chem., 20*: 3771 (1981).
18. G.H. Loew, C.J. Kert, L.M. Hjelmeland, and R.F. Kirshner, Active site models of horseradish peroxidase compound I and a cytochrome P450 analogue: electronic structure and electronic field gradients, *J. Am. Chem. Soc., 99*: 3534 (1977).
19. (a) L.K. Hansson, C.K. Chang, M.S. Davis, and J. Fajer, Electronic pathways in catalase and peroxidase enzymic catalysis. Metal and macrocycle oxidations of iron porphyrins and chlorins, *J. Am. Chem. Soc., 103*: 663 (1981).
    (b) I. Fujita, L.K. Hanson, F.A. Walker, and J. Fajer, Models for compounds I of peroxidases: axial ligand effects, *J. Am. Chem. Soc., 105*: 3296 (1983).
20. G.H. Loew and Z.S. Herman, Calculated spin densities and quatrupole splittings for model horseradish peroxidase compound I: evidence for iron(IV) porphyrin ($S = 1$) π cation radical electronic structure, *J. Am. Chem. Soc., 102*: 6173 (1980).
21. (a) K.A. Jørgensen, Does an unusual structure of oxo-iron porphyrins cause its stereoselective properties?, *Acta Chem. Scand., 40*: 512 (1986).
    (b) K.A. Jørgensen, On the mechanism of stereoselective epoxidation of alkenes by oxo-iron porphyrins, *J. Am. Chem. Soc., 109*: 698 (1987).
22. D.H.W. den Boer, J.H. van Lenthe, and A.W. van der Made, Ab initio calculations on oxomanganese porphyrin chloride; relevance to its oxidizing properties, *Recl. Trav. Chim. Pays-Bas, 107*: 256 (1988).
23. A. Strich and A. Veillard, Ab initio calculations on oxo-iron porphyrins relevance to the structure of the active compounds of horseradish peroxidase and catalase, *New J. Chem., 7*: 347 (1983).
24. K. Yamaguchi, Y. Takahara, and T. Fueno, Ab initio molecular orbital studies of structure and reactivity of transition-metal compounds, in *Applied Quantum Chemistry* (V.H. Smith, H.F. Schaefer, and K. Morokuma, eds.), D. Reidel, Dordrecht, 1986, p. 155.
25. J.E. Penner-Hahn, K.S. Elbe, T.J. McMurry, M. Renner, A.L. Balch, J.T. Groves, J.H. Dawson, and K.O. Hodgson, Structural characterization of horseradish peroxidase using EXAFS spectroscopy. Evidence for Fe=O ligation in compound I and II, *J. Am. Chem. Soc., 108*: 7819 (1986).
26. D.H.W. den Boer, A.W. van der Made, R. Zwaans, and J.H. van Lenthe, Ab initio calculations on oxomanganese porphyrin chloride. II. An explanation of its reactions with alkenes, *Recl. Trav. Chim. Pays-Bas, 109*: 123 (1990).

27. B. Chevrier, R. Weiss, M. Lange, J.C. Chottard, and D. Mansuy, An iron (III)–porphyrin complex with a vinylidene group inserted into an iron–nitrogen bond: relevance to the structure of the active oxygen complex of catalase, *J. Am. Chem. Soc., 103*: 2899 (1981).
28. A.L. Bach, Y.-W. Chan, and M.O. Olmstead, A complex containing a Ni-O unit in the centre of a porphyrin. The X-ray structure of the nickel(II) complex of octaethylporphyrin N-oxide dianion, *J. Am. Chem. Soc., 107*: 6510 (1985).
29. G.H. Loew, F.U. Axe, J.R. Collins, and P. Du, Quantum-chemical studies of model peroxidase compounds I complexes, *Inorg. Chem., 30*: 2291 (1991).
30. K.A. Jørgensen and P. Swanstrøm, On the proximal effect of the nitrogen ligands on the oxomanganese porphyrin system, *Acta Chem. Scand., 43*: 822 (1988).
31. See also M.J. Gunter and R. Turner, The role of the axial ligand in meso-tetra-arylmetalloporphyrin models of the P-450 cytochromes, *J. Mol. Catal., 66*: 121 (1991) for a discussion of the frontier orbitals of oxometalloporphyrin–axial ligand systems.
32. A.T. Pudzianowski and G.H. Loew, Quantum-chemical studies of model cytochrome P450 hydrocarbon oxidation mechanisms. 1. A MNDO/3 study of hydroxylation and epoxidation pathways for methane and ethylene, *J. Am. Chem. Soc., 102*: 5443 (1980).
33. D. Ostovic and T.C. Bruice, Mechanism of alkene epoxidation by iron, chromium, and manganese higher valent oxometalloporphyrins, *Acc. Chem. Res., 25*: 314 (1992).
34. See, for example, K.A. Jørgensen, Transition-metal catalyzed epoxidations, *Chem. Rev., 89*: 431 (1989).
35. For a discussion of metallaoxetanes as intermediate in oxygen transfer reactions, see, for example, K.A. Jørgensen and B. Schiøtt, Metallaoxetanes an intermediate in oxygen-transfer reactions—reality or fiction?, *Chem. Rev., 9*: 1483 (1990).
36. A. Sevin and M. Fontecave, Oxygen transfer from iron oxo porphyrin to ethylene. A semiempirical MO/VB approach, *J. Am. Chem. Soc., 108*: 3266 (1986).
37. T.R. Cundari and R.S. Drago, Molecular orbital investigation of ruthenium-oxo-catalyzed epoxidations, *Inorg. Chem., 29*: 487 (1990).
38. See, for example, J.T. Groves and Y. Watanabe, Reactive iron porphyrin derivatives related to the catalytic cycles of cytochrome P-450 and peroxidase. Studies of the mechanism of oxygen activation, *J. Am. Chem. Soc., 110*: 8443 (1988).
39. B.T. Luke, J.R. Collins, G.H. Loew, and A.D. McLean, Theoretical investigations of terminal alkenes as putative suicide substrates of cytochrome P-450, *J. Am. Chem. Soc., 112*: 8686 (1990).
40. S.P. Walch and T.H. Dunning, Calculated barrier to hydrogen abstraction from CH_4 by $O(^3P)$, *J. Chem. Phys., 72*: 3221 (1980).
41. (a) A.T. Pudzuanowski and G.H. Loew, Mechanistic studies of oxene reactions with organic substrates: Reaction paths on MNDO enthalpy surfaces—models for cytochrome P450 oxidations, *Int. J. Quantum Chem., 23*: 1257 (1983).

(b) K. Korzekwa, W. Trager, M. Gouterman, D. Spangler, and G.H. Loew, Cytochrome P450 mediated aromatic oxidation. A theoretical study, *J. Am. Chem. Soc., 107*: 4273 (1985).

42. (a) T.R. Cundari and R.S. Drago, A molecular orbital investigation of Ru-oxo-catalyzed alcohol oxidations, *Int. J. Quantum Chem., Symp., 23*: 489 (1989). (b) T.R. Cundari and R.S. Drago, Oxidation of alcohols by a six-coordinate Ru(IV)-O complex, *Inorg. Chem., 29*: 3904 (1990).

43. A.Z. Pudzianowski, G.H. Loew, B.A. Mico, R.V. Branchflower, and L.R. Pohl, A molecular orbital study of model cytochrome P450 oxidation of CCl_4 and $CHCl_3$, *J. Am. Chem. Soc., 105*: 3434 (1983).

44. A. Goldblum and G.H. Loew, Quantum chemical studies of model cytochrome P450 oxidations of amines. 1. MNDO pathways for alkylamine reactions with singlet and triplet oxygen, *J. Am. Chem. Soc., 107*: 4265 (1985).

45. (a) J.R. Lindsay Smith and D.N. Mortimer, Model systems for cytochrome P450-dependent monooxygenases. 5. Amine oxidation. 17. Oxidative N-dealkylation of tertiary amines by metalloporphyrin-catalyzed model systems for cytochrome P450 monooxygenases, *J. Chem. Soc. Perkin Trans., 2*: 1734 (1986). (b) K. Fujimoto, T. Takata, S. Fujiwara, O. Kikiuchi, and S. Oae, Intervention of N,N-dimethylaniline cation radical in the Polononovski type reaction of N,N-dimethylaniline N-oxide catalyzed by meso-tetraphenylporphyrin/imidazole, *Tetrahedron Lett., 27*: 1617 (1988).

4
Cytochrome P450 Model Systems

Daniel Mansuy and Pierrette Battioni
Université Paris V, Paris, France

I. INTRODUCTION

A. The Multigene Family of Cytochrome P450

Cytochromes P450 are widely distributed in living organisms and have been found, for instance, in mammals, insects, fish, yeasts, and plants [1]. In mammals they exist in several tissues and organs with a maximum level in the liver. About 200 cytochromes P450 from various species have already been cloned and sequenced [2]. Most of them are part of multienzymatic systems called monooxygenases which catalyze the activation of dioxygen and the transfer of one of its oxygen atom into substrates. However, in recent years it has been found that some cytochromes P450 catalyze quite different reactions such as the isomerization of the prostaglandin endoperoxide PGH_2 into thromboxane and prostacycline (see Ref. 3 and following chapter in this book), the reduction of NO_2^- into N_2O [4] or the dehydration of a fatty acid hydroperoxide into an allene oxide [5]. The corresponding cytochromes P450 do not use dioxygen and are not linked (at least for some of them) to electron transfer proteins. Therefore the only common property of all cytochromes P450 described so far is the distinctive position around 450 nm of the UV-visible Soret peak of their P450–Fe(II)–CO complex. This red-shifted peak, unusual for heme protein–Fe(II)–CO complexes, most of which absorb around 420 nm, has only been found for cytochromes P450 and chloroperoxidases. It is due to the presence of an axial ligand of the iron which is a cysteinate coming from the P450 polypeptide chain and which leads to a marked charge transfer to the heme [6]. The X-ray structure of a cytochrome P450, P450cam, which is present in the microorganism *Pseudomonas putida* and which catalyzes the regioselective hydroxylation of camphor at position 5-exo, has been reported [7]. It shows that the P450–camphor complex involves a pentacoordinate state of the heme Fe(III) with cysteinate 357 as the only axial ligand. The camphor molecule is held in a position very close to the heme mainly by a hydrogen bond between the camphor keto group and a tyrosine residue from the protein located in the distal part of the heme. In the absence of its substrate, P450cam involves a hexacoordinate Fe(III) state for the iron with two axial ligands, the proximal cysteinate 357 and a water molecule. Although no other X-ray data are published presently for other cytochromes P450, many spectroscopic data show that all cytochromes P450 described so far contain iron(III)protoporphyrin IX (Figure 1) at their active site and an iron–cysteinate axial bond. Most cytochromes P450 described so far derive from the binding of iron(III)protoporphyrin IX to a single polypeptide chain with a molecular mass around 50 kDa. However, recently, more complex cytochromes P450 have been reported. These enzymes are constituted by a much longer polypeptide chain bound not only to the heme but also to other flavin cofactors. This is the case of the enzyme present in *Bacillus megaterium*, which contains flavin and heme binding sites on the same protein and is able to catalyze both the electron transfer from NADPH to the heme and the oxygenation of substrates [8]. Even more recently, nitric oxide synthases, the enzymes responsible for the biosynthesis of nitric oxide from oxidation of arginine in living organisms, have been reported to be cytochrome P450–like enzymes [9–12]. They are large proteins (molecular mass around 130 kDa) which bind not only iron(III)protopor-

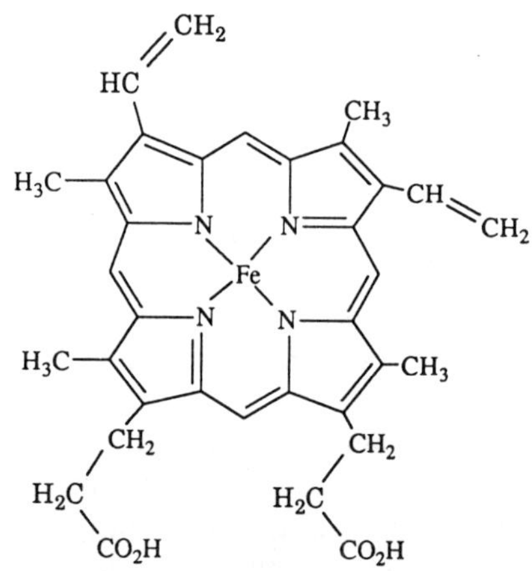

Figure 1 Structure of iron(III)protoporphyrin IX, the cofactor present in all cytochromes P450, as well as in many other heme proteins.

phyrin IX but also several other cofactors necessary for their catalytic activity, two flavins (FAD and FMN) and tetrahydrobiopterin.

B. Different Roles of Cytochrome P450–Dependent Monooxygenases

Most cytochromes P450 found so far in living organisms are part of multienzymatic systems called monoxygenases which catalyze the transfer of an oxygen atom from O_2 into substrates according to Eq. 1.

$$RH + O_2 + 2e^- + 2H^+ \longrightarrow ROH + H_2O \quad (1)$$

Cytochrome P450–dependent monooxygenases catalyze oxidation steps involved in the biosynthesis and biodegradation of endogenous compounds such as steroids, fatty acids, or prostaglandins. For instance, many steps involved in the biosynthesis of steroid hormones from cholesterol are catalyzed by cytochromes P450. In fact, it seems that living organisms that have been faced with a difficult oxidation of an inert part of a molecule (the hydroxylation of an unactivated C-H bond, for instance) have solved this problem very often by using cytochromes P450. Monooxygenases involved in the biosynthesis of important endogenous compounds are generally highly specific for their substrate and catalyze highly regioselective and stereoselective oxidations. A second class of cytochrome P450–dependent monooxygenases play another important role in living organisms, the oxidative biotransformation of exogenous compounds from the environment such as drugs, pesticides, or solvents. The oxidative biotransformation of these compounds generally renders them more hydrophilic and easier to eliminate by renal excretion. The cytochromes P450 that seem in charge of the degradation and elimination of foreign compounds, called xenobiotics, are generally less selective as they accept a broad variety of substrates. As a consequence, these cytochromes P450 are generally less regioselective in their oxidation of exogenous compounds. Because of their key role in the biodegradation of an extraordinarily wide range of environmental compounds, these cytochromes P450 are very important for the adaptation of living organisms in their ever-changing environment and their exposure to new substances.

C. Why Build Up Chemical or Biotechnological Models for Mammalian Cytochromes P450?

Cytochrome P450–dependent monoxygenases should be very important tools for two main present challenges: (a) the finding of new potent catalysts for important reactions in industrial and fine chemistry, like the selective hydroxylation of alkanes, the enantioselective epoxidation of alkenes, and the regioselective hydroxylation of aromatic compounds, and (b) the study and prediction of the oxidative metabolism of new biologically active molecules such as drugs. In fact, as far as the second challenge is concerned, it becomes more and more important for a company that has found a new active compound to answer as soon as possible the following questions:

1. What are the main oxidized metabolites (generally formed by cytochromes P450) of this new compound in man or in laboratory animals?
2. How to prepare these metabolites rapidly?
3. During the oxidative metabolism of this compound, is there formation of reactive intermediates that could be at the origin of toxic effects?
4. Which cytochromes P450 are mainly responsible for the metabolism of this compound in man?

In order to answer these questions, animal or human tissues or organs containing various cytochromes P450 are used. Of course, this is only possible for reactions performed on small amounts of substrates and leads to difficult problems of technical and ethical origin. In fact, cytochromes P450 involved in drug metabolism are membrane-bound enzymes whose purification is difficult. After purification, their catalytic activity can be obtained only after reassociation with a cytochrome P450 reductase in a lipid environment. Moreover, at least under in vitro conditions, they generally undergo an oxidative denaturation after 100–1000 turnovers.

For all these reasons, model systems for cytochromes P450 of chemical or biotechnological origin should be oxidation catalysts easier to handle for preparative purposes. They would also lead to alternative methods not using human (or laboratory animal) tissues or organs to answer questions related to drug (or xenobiotic) metabolism.

During this last decade, three kinds of catalysts able to mimic cytochromes P450 have been obtained. The first ones are homogeneous, purely chemical, catalysts based on Fe(III) [or Mn(III)] porphyrins associated with oxygen atom donors. The second ones are supported catalysts consisting of a metalloporphyrin bound to organic or inorganic polymers. Because of the fixed environment of the metalloporphyrin provided by the polymer, these supported catalysts could lead to a specific recognition of the substrate and to its regioselective oxidation. The third class of catalysts have been recently obtained by using the tools and concepts of molecular biology. They are microorganisms in which mammalian cytochromes P450 have been expressed under a catalytically active form. By insertion of the gene coding for a given cytochrome P450, microorganisms have been made able to catalyze all the reactions of this cytochrome. Such genetically modified microorganisms may be used as specific catalysts and are especially interesting for reproducing the activity of human cytochromes P450.

This chapter will describe the catalytic cycle of cytochrome P450 by comparison with that generally admitted for another class of hemeproteins, the peroxidases. The different kinds of oxidations catalyzed by cytochromes P450 will be presented with a discussion of their mechanisms. Then, a brief paragraph will be devoted to the new biotechnological catalysts obtained by expression of human P450s in microorganisms in order to illustrate the potential interests of these catalysts. A brief survey of the history of the development of homogeneous oxidation catalysts based on Fe(III) and Mn(III) porphyrins will then be presented and followed by a more detailed description of the homogeneous and supported model catalysts reported during the last 5 years.

II. OXIDATIONS CATALYZED BY CYTOCHROME P450-DEPENDENT MONOOXYGENASES

A. Catalytic Cycle of Substrate Oxidation by Cytochrome P450: Comparison with That of Peroxidases

Cytochromes P450 may catalyze the monooxygenation of substrates either by O_2 with consumption of NADPH (or NADH) (Eq. 1) or by artificial single-oxygen-atom donors like H_2O_2, alkylhydroperoxides, PhIO, or peracids (Eq. 2) [1]. Cytochrome P450-dependent oxidations of substrates by these single-oxygen-atom donors are observed in vitro with cytochrome P450 alone without any need of cytochrome P450 reductase or of any other electron transfer protein and without consumption of NADPH.

$$\mathrm{RH + AO \xrightarrow{P450} ROH + A} \tag{2}$$

On the contrary, cytochrome P450-catalyzed oxidations of substrates by O_2, which occur under physiological conditions, require a stoechiometric consumption of NADPH (NADH) and the presence of all the components of the monooxygenase, i.e., proteins that catalyze the transfer of electrons from NADPH (or NADH) to cytochrome P450. All P450-dependent monooxygenases involve a flavoprotein called cytochrome P450 reductase and some of them involve an additional electron transfer protein (generally an iron-sulfur protein).

All cytochromes P450 described so far have an identical catalytic site consisting of Fe(III)-protoporphyrin IX bound to the protein by a cysteinate ligand. They differ mainly in the nature of the protein environment, which is close to the heme and which is able to more or less selectively bind substrates and maintain them in close proximity to the heme. Despite the diversity of substrates that are oxidized and of the reactions performed, all cytochromes P450 appear to involve a common catalytic cycle of dioxygen activation (Figure 2).

In their resting state, cytochromes P450 exist under the form of two ferric complexes in equilibrium: a hexacoordinate low-spin iron complex with two axial ligands which are a cysteinate from the protein and an OH-containing residue (H_2O as in P450cam [7] or an alcohol residue from an amino acid) and a pentacoordinate high-spin complex with the cysteinate as only axial ligand. Binding of the substrate occurs on a protein site close to the heme and generally leads to a shift of this equilibrium toward the pentacoordinate complex. This high-spin P450–substrate complex very often exhibits a higher redox potential and is more easily reduced by an electron coming from NADPH or NADH via the electron transfer chain of the monooxygenase. The iron(II) complex derived from this reduction is a high-spin pentacoordinate complex with a readily accessible position for ligands in trans position to the cysteinate. Not only CO but many other ligands such as imidazoles, phosphines, or isocyanides may bind to P450 Fe(II) leading to complexes characterized by Soret peaks around 450 nm. Binding of dioxygen leads to a low-spin hexacoordinate Fe(II) complex that differs from oxyhemoglobin, HbFe(II)-O_2, by the nature of its axial ligand trans to O_2 (a cysteinate instead of an imidazole in the case of Hb). It is noteworthy that

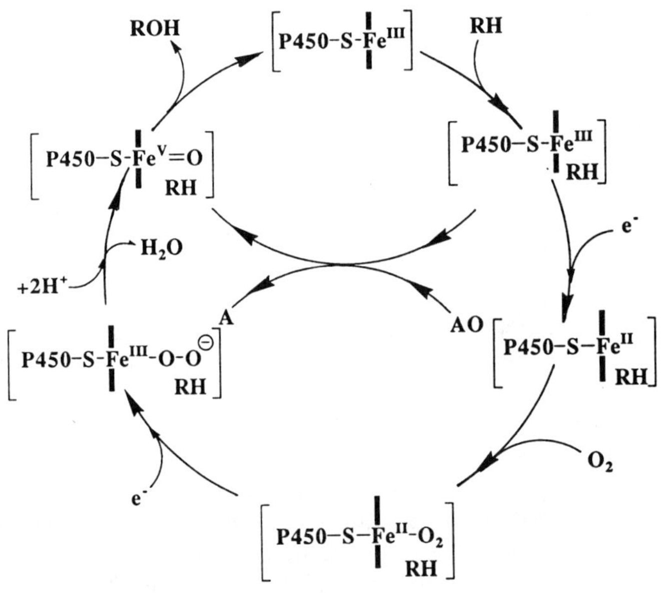

Figure 2 Catalytic cycle of monooxygenation of substrates by cytochrome P450.

the detailed structure and coordination and spin state of these first four intermediates of the catalytic cycle have been determined not only because of the X-ray structures obtained for some of them in the case of P450cam [7] and many spectroscopic studies on the other P450s [6], but also because of a comparison of the spectral properties of these complexes with those of model iron porphyrin complexes that have been completely characterized by X-ray analysis (for reviews, see Refs. 13–15).

The intermediate of the catalytic cycle that is able to transfer an oxygen atom from O_2 (the "active oxygen complex") is not the P450–Fe(II)–O_2 complex but a compound derived from its one-electron reduction. This active oxygen complex is so short-lived that it has never been detected by any spectroscopic technique. Our knowledge of its structure and reactivity is thus mainly based on (a) data coming from mechanistic studies of its reactions with various substrates, (b) a comparison with analogous complexes obtained with iron porphyrin model systems, and (c) a comparison with analogous complexes formed between other heme proteins and single-oxygen-atom donors [16]. In that respect, the ability of cytochrome P450 to catalyze monooxygenation of substrates either by NADPH and O_2 or by single-oxygen-atom donors was also a crucial argument in favor of an "Fe = O" structure for the active oxygen complex. This high-valent iron–oxo complex would be formed either by a heterolytic cleavage of the O-O bond of a possible Fe(III)-OOH intermediate derived from a one-electron reduction of P450 Fe(II)-O_2 (long catalytic cycle of Figure 2), or by direct transfer of an oxygen atom from single-oxygen-atom donors AO to P450 Fe(III) (short catalytic cycle of Figure 2). The detailed electronic structure of this P450 active oxygen complex is not known presently. However, formally at least, it is an equivalent of the active species formed by reaction of heme-containing peroxidases (such as horseradish peroxidase, HRP) in their Fe(III) state with their natural cofactor H_2O_2.

Peroxidases like HRP catalyze the oxidation of electron-rich aromatic compounds by H_2O_2 [17]. Their active site contains Fe(III)-protoporphyrin IX bound to the protein by an imidazole axial ligand. The catalytic cycle of HRP [18] is depicted in Figure 3. Its first step is the formation of an intermediate called compound I by transfer of an oxygen atom of H_2O_2 to HRP Fe(III). This "dehydration" of H_2O_2 is possible because of the presence of amino acid residues of the distal part of the heme pocket able to catalyze the removal of a proton of H_2O_2 and the protonation of an oxygen atom of Fe(III)-OOH to facilitate the removal of H_2O. HRP compound I is more stable than its P450 analog and has been studied by various spectroscopic techniques. It is a (porphyrin radical cation) Fe(IV) = O complex [18,19]. The use of iron porphyrin models has been of great help in understanding the nature of the high-valent iron–oxo intermediates formed in the catalytic

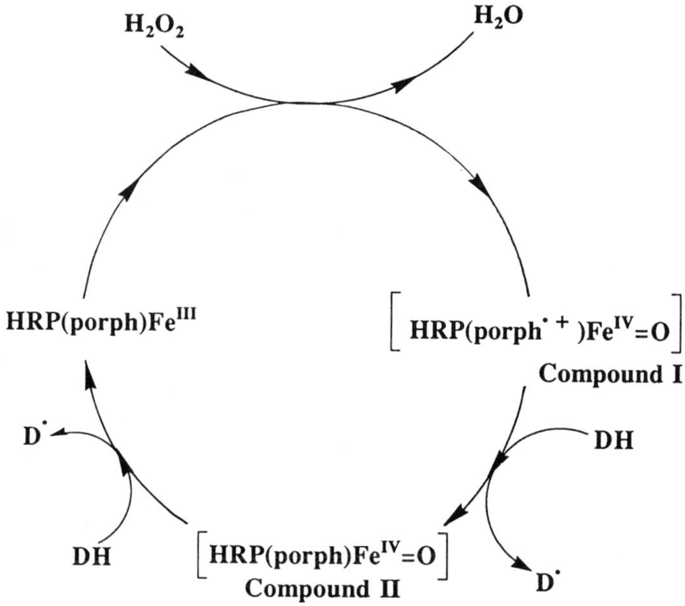

Figure 3 Catalytic cycle of horseradish peroxidase.

cycles of peroxidases, catalases, and cytochromes P450. Such complexes have been prepared by reaction of Fe(III) porphyrins with oxygen atom donors like peracids at low temperature and have been studied by ^{1}H nuclear magnetic resonance, Raman, electron proton resonance, Mössbauer, and EXAFS spectroscopy [20–33]. Their characteristics are generally compatible with a (porphyrin radical cation) Fe(IV) = O structure similar to that of HRP compound I. In fact, in a more general manner, the structure of the complex formed between a heme protein–Fe(III) and H_2O_2 or other oxygen atom donors should depend on the environment of the heme provided by the protein. It could be an Fe(V) = O species, although such a structure has never been proved so far for heme proteins, a (porphyrin radical cation) Fe(IV) = O species as in HRP compound I, or a (porphyrin) Fe(IV) = O complex with a free radical derived from the one-electron oxidation of a protein amino acid residue close to the heme, as in the case of cytochrome c peroxidase [34,35] or of prostaglandin synthase [36–38]. The second step of the catalytic cycle of HRP is the one-electron oxidation of HRP substrates by compound I, which leads to HRP Fe(IV) = O (so called compound II) and a free radical derived from the substrate. Finally, a one-electron reduction of HRP compound II by the electron-rich substrates of HRP regenerates HRP in its resting iron (III) state (Figure 3).

Coming back to the catalytic cycle of cytochrome P450, and whatever the detailed structure of its active oxygen complex (equivalent to HRP compound I) may be, the last step of this cycle is the transfer of the oxygen atom of the Fe = O species to the substrate. In that respect, a major difference between cytochromes P450 and HRP is their very different access of substrates to the Fe = O species. In cytochromes P450, the substrates are held in close proximity of the heme and very easily receive an oxygen atom from the Fe = O species (monooxygenation reaction) whereas the Fe = O active species of HRP is not accessible to substrates. Therefore, HRP Fe = O cannot transfer its oxygen to most of its substrates and can only transfer electrons via the periphery of the heme (peroxidase reaction).

B. Different Oxidations Catalyzed by Cytochrome P450-Dependent Monooxygenases

It is now generally admitted that most of the oxidations of substrates by cytochromes P450 are performed by the high-valent iron–oxo intermediate of its catalytic cycle. This electrophilic and highly reactive species is able to oxidize almost any organic compound. Generally, this results in the transfer of its oxygen atom to the substrate. The different oxidations performed by the P450 active oxygen complex [1,39] are depicted in Figure 4. From a chemical point of view, one of the most interesting properties of cytochromes P450 is their ability to hydroxylate unactivated C-H bonds of alkanes, fatty acids, or xenobiotics. Evidence accumulated from many experiments during these last 15 years favors a mechanism in two steps for this reaction (for reviews on this point, see, for instance, Refs. 1 and 40): (a) a hydrogen abstraction by the P450 active oxygen complex that exhibits a free radical like reactivity, and (b) an oxidation of the free radical derived from the substrate by the P450 Fe(IV)-OH species that results in the transfer of its OH ligand to the substrate free radical (Eq. 3). Main arguments in favor of this mechanism are (a) a partial loss of stereochemistry observed during the hydroxylation of C-H bonds of some substrates, which is due to an epimerization of the intermediate carbon radical before its oxidation by P450 Fe(IV)-OH, (b) some rearrangements that were observed during the hydroxylation of allylic C-H bonds or of C-H bonds in α position to a cyclopropane ring, and (c) large internal isotope effects (k_H/k_D) observed during the hydroxylation of C-H bonds of various substrates. A possible mechanism for the formation of the carbon–oxygen bond during the hydroxylation of substrates could be an oxidative OH ligand transfer from P450Fe(IV) to the carbon radical with regeneration of P450 iron in its ferric state. Another mechanism involves a

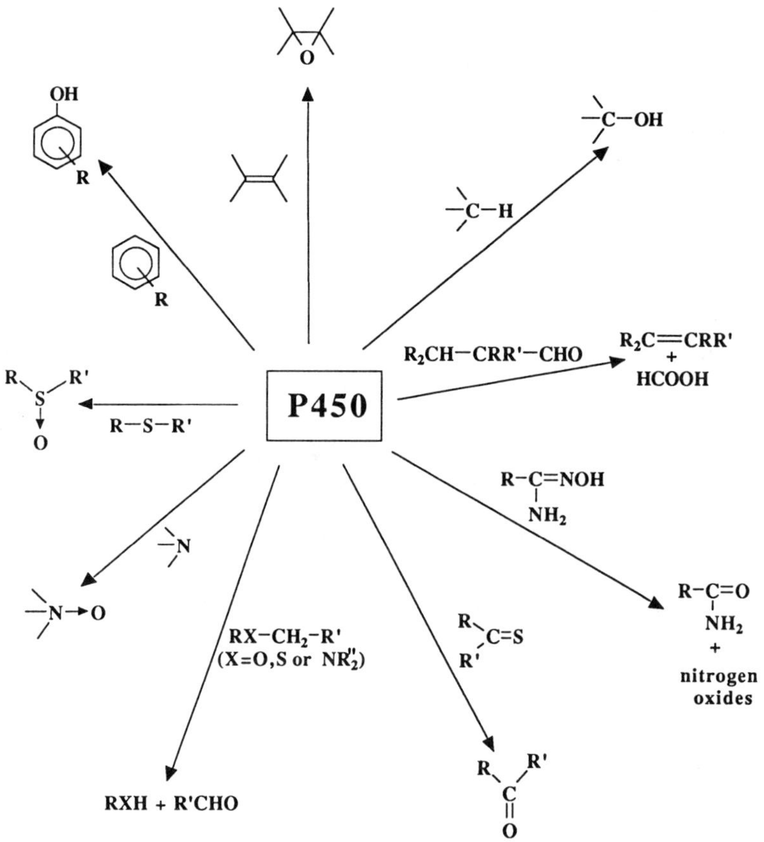

Figure 4 Main classes of oxidations performed by cytochrome P450-dependent monooxygenases.

combination of the carbon radical with P450-iron leading to a P450 σ-alkyl complex followed by a reductive elimination of the σ-alkyl and OH ligands in cis position on the iron (Eq. 4) [41,42].

$$\text{P450 Fe(V)=O} + \text{RH} \longrightarrow \left[\text{P450 Fe(IV)-OH} + \text{R}^{\bullet}\right] \longrightarrow \text{P450 Fe(III)} + \text{ROH} \qquad (3)$$

$$\left[\text{P450 Fe(IV)-OH} + \text{R}^{\bullet}\right] \longrightarrow \left[\text{P450 Fe}\begin{smallmatrix}\diagup \text{R} \\ \diagdown \text{OH}\end{smallmatrix}\right] \longrightarrow \text{P450 Fe(III)} + \text{ROH} \qquad (4)$$

Such a formation of intermediate P450 iron–carbon bonds would explain the formation of inhibitory P450 iron–carbene complexes during the oxidative metabolism of benzodioxole derivatives [42–44].

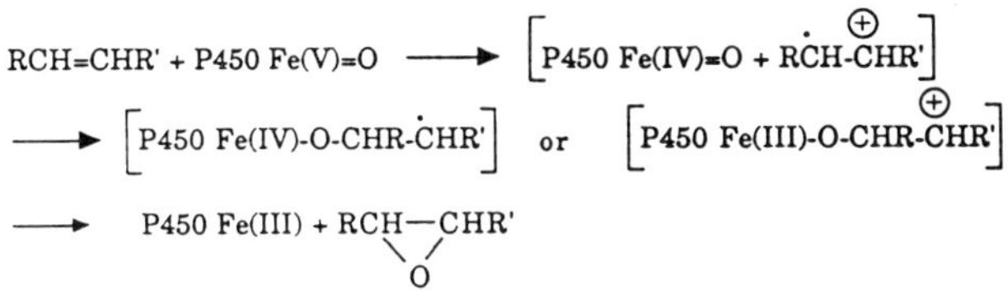

(5)

The second main reaction catalyzed by cytochromes P450 is the epoxidation of compounds containing a double bond like alkenes or arenes. Many mechanisms have been proposed for this reaction. Its first step could be an electron transfer between the P450 oxygen active species and the double bond leading to a radical cation derived from the substrate and P450 Fe(IV) = O, followed by a recombination of the two intermediates that would lead to a substrate-derived free radical or cation linked to the iron by an Fe-O-C-C moiety. Identical intermediates could be formed by direct addition of the P450 active oxygen species to the double bond (Eq. 5). The last step of the reaction would be an intramolecular transfer of the oxygen atom to the intermediate free radical or cation with the eventual formation of an epoxide. Cytochromes P450 not only catalyze the epoxidation of alkenes but also the epoxidation of aromatic rings, which is very often the first step in the hydroxylation of arenes [45]. Phenols are then formed by isomerization of these very reactive arene oxides (Eq. 6).

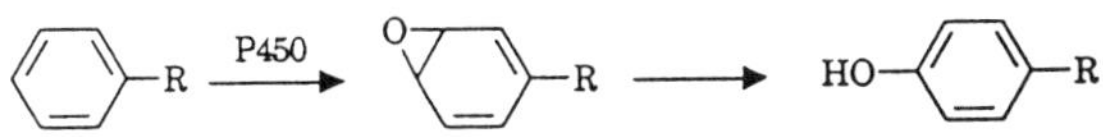

(6)

Because of the highly electrophilic nature of their active oxygen species, cytochromes P450 are very prone to react with electron-rich parts of their substrates, like double bonds or heteroatom-containing functions. Tertiary amines and nitrogen-containing aromatic rings are oxidized to the corresponding *N*-oxides, and thioethers to the corresponding sulfoxides and sulfones [39]. In that respect, it was recently reported that cytochromes P450 are able to oxidize the poorly reactive sulfur atom of thiophene and thiophenes substituted by an ArCO group with formation of the corresponding thiophene-*S*-oxides, which are electrophilic reactive metabolites [46,47]. This result underlines the high reactivity of the P450 oxygen active species. Amines, thioethers and ethers containing C-H bonds in an α position to the heteroatom may undergo P450-dependent oxidative N-dealkylations [39] (Figure 4). These N-dealkylations are due either to a hydroxylation of C-H bonds in an α position to the heteroatom followed by the hydrolysis of intermediate compounds that are of the acetal type (Eq. 7), or to the fate of cation radicals derived from a one-electron transfer from amines or thioethers to the P450 Fe(V) = O species (Eq. 8).

$$\text{R-X-CH}_2\text{R}' \xrightarrow{\text{P450}} \text{RX-CHOHR}' \xrightarrow{\text{H}_2\text{O}} \text{RXH+R}'\text{CHO} \qquad (7)$$

(X = O, S or NR'')

$$\text{R}_2\text{N-CH}_2\text{R}' + [\text{P450 Fe(V)=O}] \longrightarrow \left[\text{R}_2\overset{+\cdot}{\text{N}}\text{-CH}_2\text{R}'\text{+P450 Fe(IV)=O}\right]$$

$$\xrightarrow{-\text{H}^+} [\text{R}_2\overset{+}{\text{N}}\text{=CHR}'\text{+P450 Fe(III)}] \xrightarrow{\text{H}_2\text{O}} \text{R}_2\text{NH+R}'\text{CHO} \qquad (8)$$

Cytochromes P450 also catalyze the oxidative cleavage of C = S, C = N, and C-C bonds. Their ability to oxidize the C = S bond of thioamides [39] with formation of the corresponding amides was reported long ago although the detailed mechanism of this reaction is not yet completely understood. The oxidative cleavage of N^{ω}-hydroxy-L-arginine with formation of citrulline and nitrogen oxides including NO was reported very recently [48] (Eq. 9). This reaction should be related to the formation of NO from L-arginine which is catalyzed by NO synthases in mammals, as N^{ω}-hydroxyarginine is an intermediate in this reaction and as NO synthases are cytochrome P450-like heme proteins [9–12]. In a more general manner, it was recently reported that arylamidoximes are oxidized by cytochromes P450 with the formation of the corresponding arylamides and nitrogen oxides including NO [49] (Figure 4).

$$\text{RNH-}\underset{\text{N-OH}}{\underset{\|}{\text{C}}}\text{-NH}_2 \xrightarrow[\text{NADPH,O}_2]{\text{P450}} \text{RNH-}\underset{\text{O}}{\underset{\|}{\text{C}}}\text{-NH}_2 + \text{NO} \qquad (9)$$

R = $HOOCCH(NH_2)CH_2(CH_2)_2$

A particular class of oxidations catalyzed by cytochromes P450 has been reported to involve a P450 active oxygen species different from P450 Fe(V) = O. This concerns the oxidative cleavage of C-C bonds in aldehydes like cyclohexane carboxyaldehyde [50,51]. The P450-dependent oxidation of this aldehyde leads to cyclohexene and formic acid. It could be performed by NADPH and O_2 (long P450 catalytic cycle) but not by oxygen atom donors like PhIO or alkylhydroperoxides [50]. However, P450 and H_2O_2 perform the reaction indicating that the active oxygen complex involved in that case could be the Fe(III)-O-O$^-$ intermediate of the catalytic cycle (Figure 2). Reaction between this intermediate and the aldehyde should lead to a peroxyhemiacetal-like adduct (Eq. 10). A heterolytic or homolytic cleavage of the O-O bond of this adduct associated to a C-C bond cleavage would lead to the formation of cyclohexene and formic acid. Similar mechanisms would be involved in the P450-catalyzed demethylations of steroids [52], like the

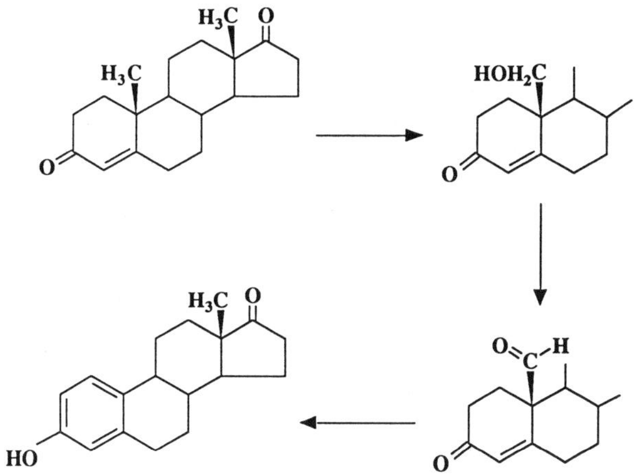

Figure 5 Formation of estrone catalyzed by aromatase.

demethylation of androstenedione which leads to estrone in three steps catalyzed by the same P450 called aromatase (Figure 5): (a) the C19 hydroxylation of androstenedione, (b) the oxidation of the alcohol formed in this first step to the aldehyde, and (c) the oxidation of this aldehyde with formation of estrone and HCOOH.

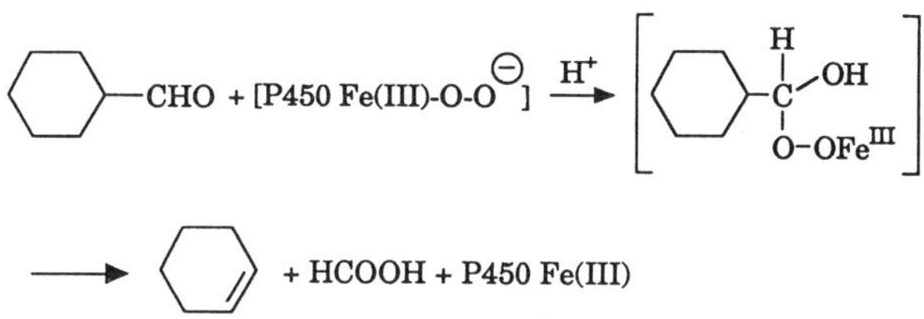

(10)

III. BIOTECHNOLOGICAL CATALYSTS BASED ON YEASTS EXPRESSING HUMAN LIVER CYTOCHROMES P450

Many systems have been used during these last 10 years to express different cytochromes P450 in various kinds of cells or microorganisms [53]. This section describes some recent results obtained by the use of yeasts expressing some P450 isozymes of human liver. *Saccharomyces cerevisiae* is a particularly interesting host microorganism for several reasons. First, it is a biotechnological tool that is very easy to handle and that can be grown on large scales. Second, yeast contains all the components that are necessary for the expressed P450 apoprotein to become catalytically active, i.e., the ability to incorporate the heme into the expressed apoprotein, the existence of the electron transfer proteins, like cytochrome P450 reductase, necessary for the transfer of electrons to cytochrome P450, and the presence of membranes analogous to those of human liver endoplasmic reticulum which are required for the organization of the proteins of the monooxygenase cluster. Several human liver P450s belonging to the 1A [54–56], 2C [57,58], 2D [59], and 3A [60–62] subfamilies have been successfully expressed in yeast to give

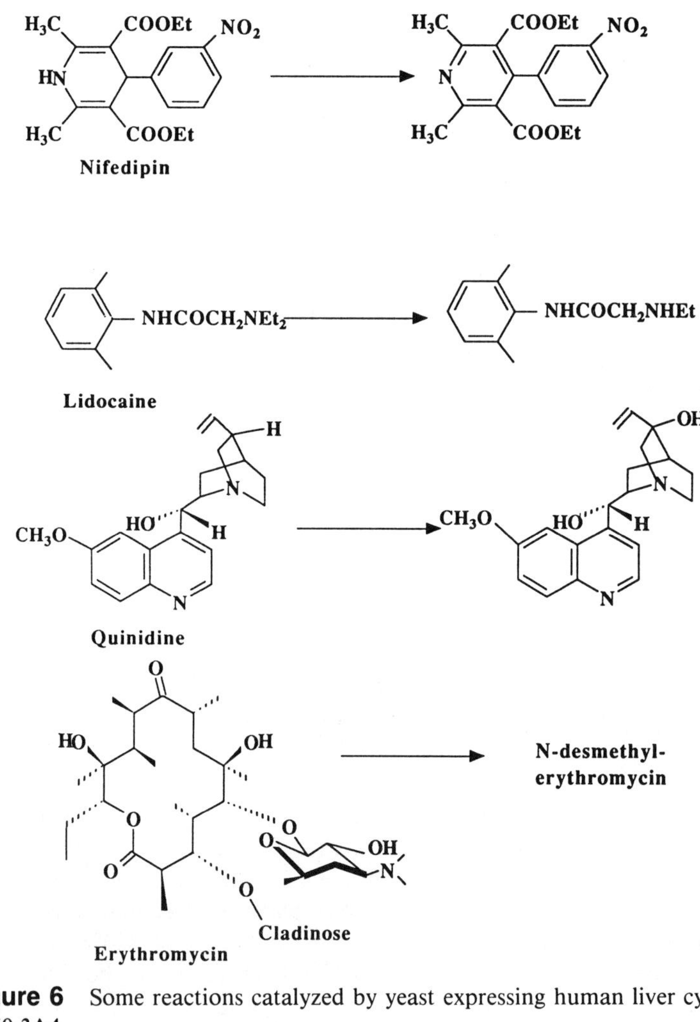

Figure 6 Some reactions catalyzed by yeast expressing human liver cytochrome P450 3A4.

a catalytically active form. The corresponding yeasts are very interesting tools for regioselective oxidations of complex molecules and for studies of drug metabolism. For instance, the gene coding for cytochrome P450 3A4, which is the major P450 of human liver, was expressed in yeast at a level of ~0.1 nmol P450 per mg microsomal yeast protein [61]. This level is high enough to be easily measured in yeast microsomes by visible spectroscopy of the P450–Fe(II)–CO complex, which exhibits a characteristic Soret peak at 450 nm. In a more general manner, this permits a direct study, by visible spectroscopy, of the interactions of this well-defined P450 with its possible substrates like dihydroergotamine and inhibitors like miconazole. The binding constants of these substrates or inhibitors for P450 3A4 may be drawn easily from such spectral studies [61]. Yeasts expressing P450 3A4 or its microsomes are able to catalyze the classical reactions of this cytochrome in human liver. Some oxidations performed by yeast expressing P450 3A4 are shown in Figure 6. Lidocaine, an anesthetic drug, undergoes an oxidative N-deethylation of its tertiary amine function and nifedipin, a cardiovascular drug, is oxidized to the corresponding pyridine derivative with respective rates of 2 and 0.5 turnovers per min, which are comparable to those previously reported for human liver microsomes. Large polyfunctional compounds like the alkaloid quinidine or the antibiotic erythromycin are oxidized in a highly regioselective manner [60–62]. The regioselectivities observed with yeast microsomes, corresponding to an allylic hydroxylation in the case of quinidine and an oxidative N-dealkylation in the case of erythromycin, are identical to those previously found in human liver microsomes.

More recently, much higher monooxygenase activities were obtained with yeast expressing P450 3A4 and coexpressing some other components of the P450-dependent monooxygenase like cytochrome P450 reductase and(or) cytochrome b_5 [62]. Thus yeasts expressing well-defined P450 isozymes from human liver are good functional models for studying the binding capacities and catalytic activities of each human liver P450. Some examples of their application for the study of the oxidative metabolism of drugs will be described in the following sections.

IV. HOMOGENEOUS METALLOPORPHYRIN SYSTEMS MIMICKING CYTOCHROME P450

A. The Three Generations of Metalloporphyrins

Several recent review articles [40,41,63–74] have been devoted to the huge effort of research made by many groups during these last 13 years to develop efficient model systems based on metalloporphyrin catalysts in order to mimic either the short catalytic cycle or the long catalytic cycle of cytochrome P450 (Figure 2). Therefore, the main objective of this section is not to provide another review of all the model systems described so far but to illustrate the most recent results in this field.

The first system, described by J. T. Groves et al. [75], mimicked the short catalytic cycle of cytochrome P450 and used iodosylbenzene (PhIO) as oxygen atom donor in the presence of the simple catalyst Fe(TPP = tetraphenylporphyrin)Cl (Figure 7) which is easily prepared by reaction between pyrrole and benzaldehyde. This system was able to reproduce most reactions performed by cytochromes P450 at least from a qualitative point of view. The main problem encountered with similar systems using PhIO or other single-oxygen-atom donors, like ClO^-, and such simple catalysts, like Fe(TPP)Cl or Mn(TPP)Cl, that will be called first-generation catalysts in the following, is a fast oxidative degradation of their porphyrin ring under the strong oxidizing conditions used for the reactions. Much better results were obtained with a second generation of Fe(III) and Mn(III) *meso*-tetraarylporphyrins whereby electron-withdrawing substituents have been introduced at each *meso*-aryl position [76–78]. Some representatives of this second generation of metalloporphyrins are shown in Figure 7. Both porphyrins, $TFPPH_2$ and $TDCPPH_2$, contain a great number of halogen substituents on their *meso*-aryl rings, 20 F for $TFPPH_2$ and 8 Cl for $TDCPPH_2$. The presence of these electron-withdrawing substituents in the Fe or Mn complexes of those porphyrins should render their corresponding metal-oxo active species much more electrophilic and reactive. Moreover, in the case of Fe (or Mn) (TDCPP)Cl,

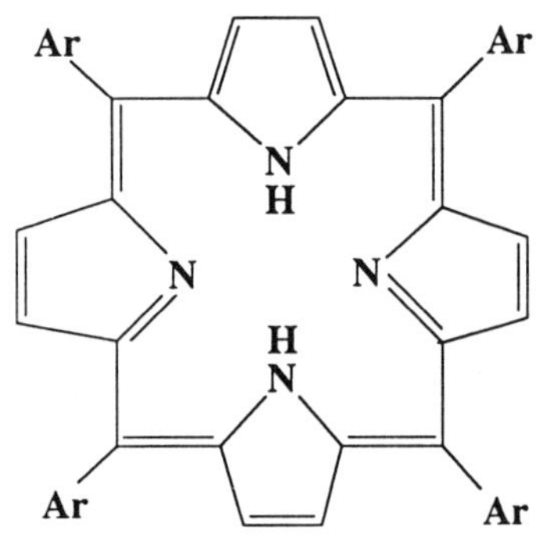

Ar = phenyl	**$TPPH_2$**
Ar = pentafluorophenyl	**$TFPPH_2$**
Ar = 2,6-dichlorophenyl	**$TDCPPH_2$**

Figure 7 Main porphyrins used in first- and second-generation metalloporphyrin catalysts mimicking cytochrome P450.

the *ortho*-Cl substituents of TDCPP provide a steric protection of the *meso*-carbons of the porphyrin ring and prevent rapid destruction of the porphyrin during oxidation catalysis [77]. The two following examples illustrate the interest of the second generation of Fe or Mn porphyrins as oxidation catalysts. Fe(TDCPP)Cl was a very efficient catalyst for epoxidation of alkenes by C_6F_5IO with initial rates as high as 300 turnovers per second [79]. Mn(TDCPP)Cl was a much better catalyst than Mn(TFPP)Cl or Mn(TPP)Cl for the oxidation of hydrocarbons by H_2O_2 [80]. In fact, in the presence of Mn(TDCPP)Cl and imidazole, which was shown to act both as an Mn ligand and as a base catalyst, alkenes are epoxidized by H_2O_2 in high yields and in a stereospecific manner [80,81]. The Mn(TDCPP)Cl–imidazole–H_2O_2 system also performs the hydroxylation of alkanes [81,82] and of aromatic compounds like anisole or naphthalene with satisfactory yields [83].

In a more general manner, the use of metalloporphyrins of the first and second generation in the presence of various oxygen atom donors has led during this last decade to many systems reproducing most oxidations catalyzed by cytochromes P450. Although chromium, cobalt, and osmium porphyrins have been reported to catalyze alkene epoxidation, the metalloporphyrins that gave by far the best results for P450-type oxidations were iron and manganese porphyrins [74]. Various oxygen atom donors like iodosylarenes, NaOCl, alkylhydroperoxides, H_2O_2, amine oxides, peracids, $NaClO_2$, O_3, and oxaziridines have been used successfully for the epoxidation of alkenes and/or hydroxylation of alkanes, in the presence of Fe(III) or Mn(III) porphyrin catalysts. Several systems using these catalysts—O_2 as an oxygen atom source and a reducing agent like a borohydride, H_2 in the presence of colloidal platinum, NADPH, ascorbic acid, Zn powder or amalgam, or electrons coming from an electrode—have also been described [67,74,84]. In that regard one should mention the remarkable result first described in 1985 by Groves et al. [85] which showed that $Ru(TMP)(O)_2$ could catalyze the epoxidation of alkenes at room temperature by O_2 alone without any consumption of a reducing agent.

Olefin epoxidation and alkane hydroxylation were the most often studied reactions with these systems and the use of chiral metalloporphyrins led to asymmetric epoxidations with an enantiomeric excess up to 89% and asymmetric hydroxylations with an enantiomeric excess up to 72%. The biomimetic systems were also used for oxidative N-dealkylation of tertiary amines, oxidation of thioethers to sulfoxides and sulfones, as well as for the oxidation of more complex molecules such as drugs, lignin models, and halogenated aromatic pollutants [74].

With poorly reactive substrates such as linear alkanes, the yields of hydroxylation observed with most catalysts, even those of the second generation, remained below 40% and there was a considerable degradation of the catalyst during the oxidation of those substrates. More recently, even more robust polyhalogenated metalloporphyrins bearing electron-withdrawing substituents (Br, Cl, F, or SO_3H) on their β-pyrrole positions have been prepared and found as very efficient oxidation catalysts. The following section describes the preparation of these third-generation metalloporphyrins and illustrates their particular catalytic properties.

B. Preparation and Catalytic Properties of Third-Generation Metalloporphyrins

The preparation of the first β-octahalogenoporphyrin was reported by Traylor and Tsuchiya in 1987 [86]. *meso*-Tetrakis(2,6-dichlorophenyl)-β-octabromoporphyrin ($TDCPBr_8PH_2$) (Figure 8) was obtained after bromination of Zn (TDCPP) by *N*-bromosuccinimide in refluxing CCl_4 following a method described previously for partial bromination of $TPPH_2$ [87]. The corresponding β-octachloroporphyrin, $TDCPCl_8PH_2$ (Fig. 8), was prepared by chlorination of Fe(TDCPP)Cl either by Cl_2 [88,89] or by reaction of Zn(TDCPP) with *N*-chlorosuccinimide [90]. β-Perhalogenated porphyrins of the $TFPPH_2$ series were similarly obtained by reaction of Zn(TFPP) with *N*-bromosuccinimide in the presence of CF_3COOH [91] or with Br_2 in CCl_4 [92]. In the case of $TMPH_2$, bromination of Zn(TMP) by *N*-bromosuccinimide in refluxing methanol provided Zn($TMBr_8P$) in good yield [93,94]. However, an analogous treatment of Zn(TMP) with *N*-chlorosuccinimide led not only to the perchlorination of the β-pyrrole positions but also to chlorination of the meta positions of the *meso*-2,4,6-trimethylphenyl groups [94]. Finally, $TPBr_8PH_2$ was also obtained by bromination of $TPPH_2$ by a similar procedure [95].

The synthesis of β-octafluoroporphyrins appears more difficult and only one preliminary paper reporting the prepar;ation of a perfluorinated porphyrin containing 28 F substituents, *meso*-tetrakis(pentafluorophenyl)-β-octafluoroporphyrin, $TFPF_8PH_2$, by reaction of $TFPPH_2$ with cobalt or silver fluoride has appeared [96].

Another approach to *meso*-tetraaryl-β-octahalogenoporphyrins bearing various functions, which involves the selective substitution of the *para*-fluorines of $TFPBr_8PH_2$ by some oxygen-, nitrogen-, or sulfur-containing nucleophiles, was described recently [97].

Three other kinds of electron-withdrawing substituents were introduced on the β-pyrrole positions of porphyrins. The water-soluble *meso*-tetra(pentafluorophenyl)-β-tetrasulfonatoporphyrin ($TFPS_4PH_2$), was obtained by sulfonation of $TFPPH_2$ by oleum [98–100]. In fact, this porphyrin exists as a mixture of the regioisomers derived from the introduction of one SO_3H group per pyrrole ring. In a similar manner, nitration of $TFPPH_2$ by concentrated HNO_3 very recently led to a mixture of the regioisomers of *meso*-

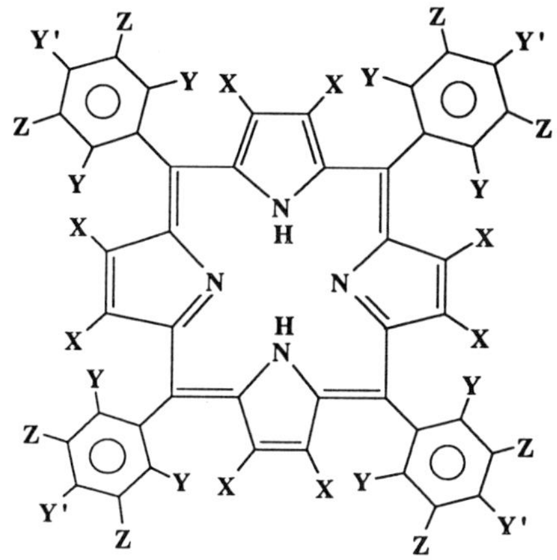

Y = Cl	Z = Y'= H	X = Br	$TDCPBr_8PH_2$
Y = Cl	Z = Y'= H	X = Cl	$TDCPCl_8PH_2$
Y = Y'= CH_3	Z = H	X = H	$TMPH_2$
Y = Y'= CH_3	Z = H	X = Br	$TMBr_8PH_2$
Y = Y'= Z = F		X = Br	$TFPBr_8PH_2$
Y = Y'= Z = F		X = Cl	$TFPCl_8PH_2$
Y = Y'= Z = F		X = F	$TFPF_8PH_2$

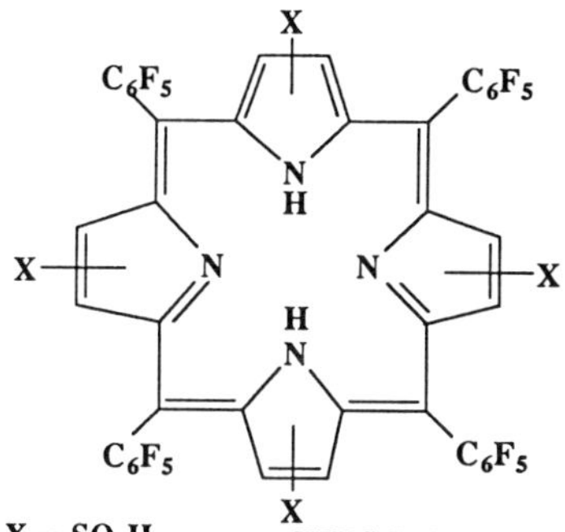

X = SO_3H $TFPS_4PH_2$

X = NO_2 $TFPN_4PH_2$

Figure 8 Formula of third-generation porphyrins used in model systems for cytochrome P450.

tetra(pentafluorophenyl)-β-tetranitroporphyrin ($TFPN_4PH_2$), bearing one nitro group per pyrrole ring (J. F. Bartoli, R. De Foor, P. Battioni, and D. Mansuy, in preparation). Very recently, the possible preparation of β-polynitro- and β-polycyanoporphyrins was indicated in two patents [101,102].

As far as β-octasubstituted porphyrins are concerned, it is noteworthy that β-octaphenylporphyrins [103] and several β-octaalkylporphyrins [104] have been prepared and completely characterized.

It is noteworthy that X-ray structure analyses made on several β-octahalogenoporphyrins, including $TMBr_8PH_2$, Ni($TMBr_8P$), and Ni($TFBr_8P$), showed that these porphyrins are not planar but exist in a saddle-shaped structure for their tetrapyrrole ring, each pyrrole being tilted by about 30° [91]. Recent studies using molecular mechanics calculations and semiempirical quantum calculations [105] showed that although *meso*-tetraarylporphyrins exhibit a planar porphyrin ring and four *meso*-aryl groups perpendicular to this ring, the most stable structure of $TDCPBr_8PH_2$ is of the saddle-shaped type with each *meso*-aryl group very much tilted in order to minimize the strong steric interactions between the *meso*-aryl and β-Br groups.

Fe(III)($TDCPBr_8P$)Cl and Fe(III)($TDCPCl_8P$)Cl were much better catalysts than the second-generation catalyst Fe(TDCPP)Cl for the hydroxylation of linear alkanes [106] or norbornane [86,107] by PhIO. For instance, the yield obtained for hydroxylation of heptane was much higher with Fe($TDCPCl_8P$)Cl than with Fe(TDCPP)Cl (80% instead of 36%) [106].

$$\text{heptane+PhIO} \xrightarrow{\text{Fe(TDCPCl}_8\text{P)Cl}} \text{heptanols+heptanones (80\%)} \tag{11}$$

Interestingly, the regioselectivity of hydroxylation of heptane by Fe ($TDCPCl_8P$)Cl was very different from that obtained with Fe(TDCPP)Cl. The ratio of heptanol-2, 3, and 4 observed with Fe($TDCPCl_8P$)Cl was close to that expected from a statistical reactivity. This could be explained by a much higher reactivity of the Fe = O active species of these third-generation porphyrins and/or to a different accessibility of this species because of the saddle-shaped porphyrin structure [106]. Mn($TDCPCl_8P$)Cl was also much better than Mn(TDCPP)Cl for heptane hydroxylation by PhIO [106]. However, this was not true for alkane hydroxylation by H_2O_2 as Mn($TDCPCl_8P$)Cl was more rapidly destroyed and less efficient than Mn(TDCPP)Cl [90]. In a similar manner, Mn($TMBr_8P$)Cl was found to be a bad catalyst for oxidations of hydrocarbons by H_2O_2 although it gave very good results for the epoxidation of alkenes and the hydroxylation of adamantane and cyclohexane by $KHSO_5$ or magnesium perphthalate (with turnover rates up to 6 cycles/sec for adamantane hydroxylation) [93]. On the other hand, Mn($TDCPBr_8P$)Cl gave good results for the H_2O_2-dependent hydroxylation of anisole [83] and Fe($TFPF_8P$)Cl was used successfully for the hydroxylation of benzene to phenol [96] (Eq. 12).

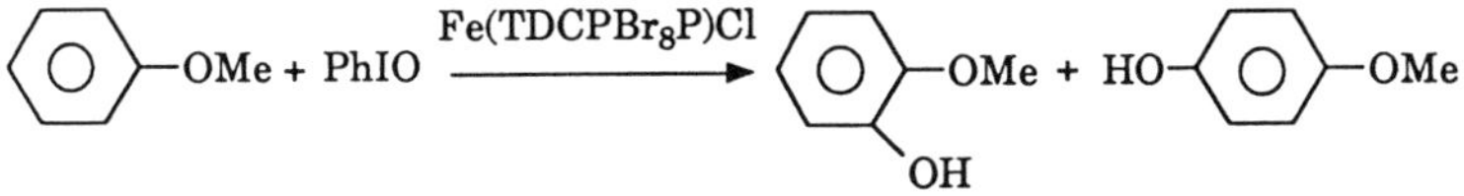

(12)

Some iron porphyrins of the third generation are able to catalyze the oxidation by O_2 at 80°C of alkanes like propane and isobutane [92] or like cyclohexane under photoirradiation at 20°C [108], without any consumption of a reducing agent. The use of Fe(III)(TFPBr$_8$P) allowed the oxidation of isobutane to tertiary butanol at room temperature [92], and the use of Fe(TDCPCl$_8$P) allowed the oxidation of cyclooctane by O_2 at 80°C with the major formation of cyclooctanone [109].

Several sulfonated iron and manganese porphyrins have been used as ligninase models [74]. The water-soluble third-generation catalyst Fe(III) (TFPS$_4$P) also gave good results in the oxidation of lignin dimer models [100]. This catalyst was found particularly good for the oxidation of methoxyarenes to *para*- or *ortho*-quinones [98]. Thus, several dimethoxyarenes were oxidized to the corresponding quinones by a water-soluble oxygen atom donor Mg-monoperoxyphthalate (MMP), in the presence of this iron porphyrin in a CH_3CN-tartrate buffer (pH 3), with yields between 50% and 95% [98] (Eq. 13). This reaction was applied successfully to the preparation of the natural product methoxatin (or PQQ) by oxidation of a methoxyarene precursor [98] (Eq. 14). Formation of quinones by the Fe(TFPS$_4$P)–MMP system only occurred on dimethoxyarenes bearing an electron-donating substituent.

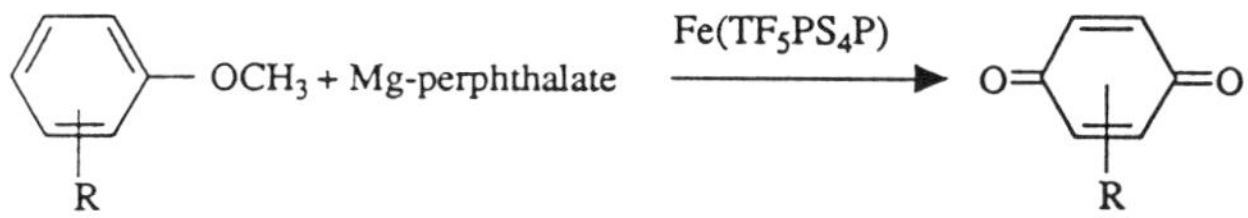

(13)

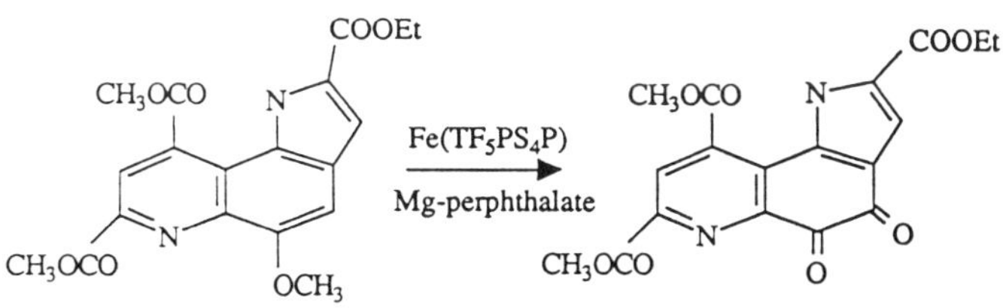

(14)

In contrast, oxidation of dimethoxyarenes bearing an electron-withdrawing substituent by this system led to the selective formation of muconic dimethyl esters derived from the cleavage of the ring C-C bond bearing the two methoxy substituents [99] (Eq. 15). This reaction provides a new access to functional muconic diesters with yields around 40%. Interestingly, sulfonated iron porphyrins of the first and second generation failed to give any muconic diester under identical conditions. The formation of quinones or muconic diesters from dimethoxyarenes involve several steps, i.e., the insertion of two oxygen atoms coming from H_2O or the oxidant, the cleavage of a C-O or C-C bond, and the consumption of 2 moles of oxidant per mole of transformed substrate. The most likely first step of the reaction is a one-electron transfer from the dimethoxyarene substrate to the iron-oxo active species [99,100]. The high selectivity observed for the formation of quinones or muconic diesters in the presence of $Fe(TFPS_4P)$ should be due to a very good control of the substrate-derived free radical intermediates by the iron during the whole catalytic cycle [99,100].

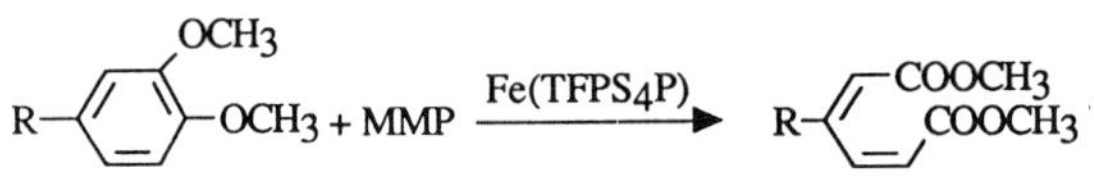

(15)

V. BIOMIMETIC OXYGENATIONS CATALYZED BY SUPPORTED METALLOPORPHYRINS

Various supported metalloporphyrin oxidation catalysts have recently been developed. Metalloporphyrins have been simply adsorbed on silica or alumina or bound to silica bearing imidazole residues linked to its surface. They have been encapsulated into zeolites, immobilized on clays, strongly adsorbed by ionic interactions on ion exchange resins, or covalently linked to polypeptides. The use of such supported metalloporphyrins in biomimetic oxygenations was described in a recent review [74]. Moreover, the preparation and properties of supported metal porphyrin catalysts is the subject of a following chapter of this book (see chapter by J. R. Lindsay-Smith). Therefore, our objective in the following section is not to give a detailed review on these supported catalysts but only to illustrate possible strategies that have been recently used in that field and possible interests of supported catalysts for the selection between substrates of different shapes and for regioselective hydroxylation.

A relatively general method for the binding of polyfluorinated metalloporphyrins of the second or third generation to various supports was recently found [110]. It is based on the property of $TFPPH_2$ to react with several nucleophiles with selective substitution of the four F substituents in the para position of the *meso*-aryl groups [97] (Figure 9). This reaction was used to prepare polymeric polyhalogenated metalloporphyrins [111]. For instance, treatment of Fe(III)(TFPP) or Mn(III)(TFPP) with sodium sulfide leads to the corresponding metalloporphyrin polymers, which are insoluble in most organic solvents. A suspension of the Mn porphyrin polymer in CH_2Cl_2 catalyzes the hydroxylation of cyclohexane by C_6F_5IO to cyclohexanol with a

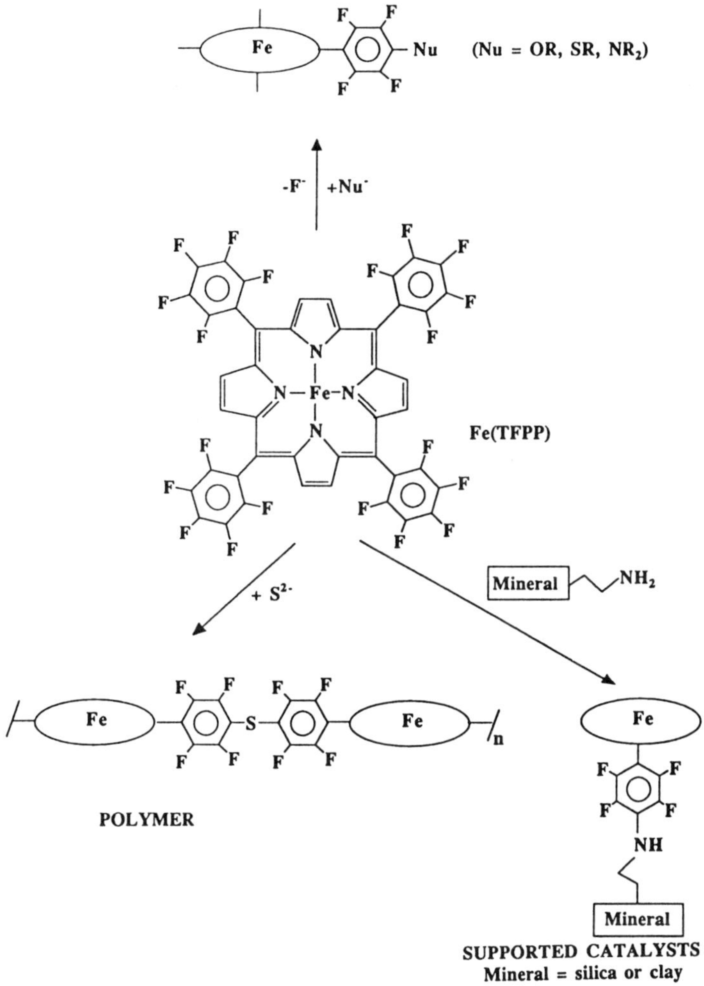

Figure 9 Preparation of soluble and immobilized metalloporphyrins by selective substitution of the para-F substituents of Fe(TFPP) by nucleophiles.

70% yield and with a very minor formation of cyclohexanone [111]. The selective substitution of the *meso*-aryl *para*-F substituents of $TFPPH_2$ mentioned above [97] was also used for the covalent binding of $TFPPH_2$ and $TFPBr_8PH_2$ and their metal complexes to functionalized silica or montmorillonite [110]. For instance, direct reaction of Fe(TFPP) or Mn(TFPP) with silica- or montmorillonite-bearing $(CH_2)_3NH_2$ groups led to a very easy access to supported polyfluorinated metalloporphyrin catalysts in one step. A suspension of the corresponding Mn(III) catalysts in a CH_2Cl_2/CH_3CN mixture catalyzed the epoxidation of cyclooctene by PhIO with nearly quantitative yields. Interestingly enough, Mn(III)(TFPP) bound to montmorillonite was a much better catalyst than Mn(III)(TFPP) bound to silica for the hydroxylation of heptane by PhIO at room temperature (70% yield based on PhIO instead of 30%) [110].

Another approach to supported metalloporphyrins exhibiting a particular fixed environment of the porphyrin ring is to intercalate tetracationic or tetra-anionic metalloporphyrins in the interlayer space of layered minerals [112–115]. For instance, the tetracationic Mn(III)(TMPyP = tetra-4-*N*-methylpyridiniumylporphyrin) was intercalated into montmorillonite by simple ion exchange with cations already present in the interlayer space [113]. An X-ray study of the obtained material indicated that most of the Mn porphyrin has been intercalated, the interlayer distance being 4.2 Å. An intriguing property of this supported Mn porphyrin is its particular ability to efficiently catalyze the hydroxylation of heptane or pentane. In fact, it gave a better yield for heptane hydroxylation (50%) than the same Mn porphyrin simply adsorbed on silica (40%) and the Mn porphyrin alone used as a homogeneous catalyst (3%) [113]. The origin of this particular efficacy of Mn(III)porphyrins supported on montmorillonite [either by intercalation as Mn(TMPyP) or by covalent binding as Mn(TF_5PP)] as catalysts for the hydroxylation of linear alkanes remains to be determined. Another interesting property of Mn(TMPyP) intercalated in montmorillonite is its ability to select substrates of different shapes. This was shown by experiments on the oxidation of two competitive substrates—adamantane, which is very reactive because of its tertiary CH bonds but relatively bulky, and pentane, which is one of the least reactive substrates for the electrophilic Fe(V)=O or Mn(V)=O species but is much less bulky than adamantane [113,114]. Oxidation of a pentane/adamantane (2:1) mixture by PhIO in the presence of Mn(TMPyP) itself or of Mn(TMPyP) adsorbed on silica led to the almost exclusive oxidation of adamantane (adamantanols + adamantanone/pentanols + pentanones = 95:5). On the contrary, oxidation of this mixture by PhIO in the presence of Mn(TMPyP) intercalated into montmorillonite led to a clearly different adamantanols/pentanols ratio of 70:30, which is more in favor of pentane. This is presumably due to the easier access of pentane, which is much less bulky than adamantane, to the Mn(V)=O active species buried in the interlayer space [113, 114]. This is a first encouraging result toward the use of such supported oxidation catalysts in order to choose between substrates of different shapes.

VI. CONCLUSION AND PERSPECTIVES

Cytochrome P450–dependent enzymes are widely distributed in living organisms. They are playing various functions in the biosynthesis or biodegradation of important endogenous compounds like steroid hormones or nitric oxide, and in the oxidation and elimination of xenobiotics from living organisms. They catalyze an extraordinarily wide variety of reactions including the hydroxylation of inert C-H bonds and of aromatic rings, the epoxidation of unsaturated compounds, the N-oxidation of amines and S-oxidation of thioethers, the oxidative cleavage of the C_α-C_β bond of aldehydes and of C=S and C=NOH bonds of thioamides and amidoximes. The use of iron porphyrins was of great help in order to understand the nature and reactivity of the active intermediates of the P450 catalytic cycle and the detailed mechanisms of their reactions with the P450 substrates. Major advances have been made toward efficient catalytically active model systems during this last decade. Chemical systems based on Fe(III) and Mn(III) porphyrins and various oxygen atom donors have been found to reproduce the cytochrome P450 chemistry. Thanks to the use of metalloporphyrins that are very robust toward oxidative degradation, these chemical systems have become highly efficient for obtaining oxidized metabolites at the preparative level. More practically useful supported catalysts have been prepared by incorporation of metalloporphyrins into mineral matrices. At least so far, most chemical model catalysts do not involve an equivalent of the cytochrome P450 protein active site, which can specifically recognize the substrate and allows its regioselective hydroxylation. Therefore, these chemical catalysts should be useful a priori to obtain all the possible oxidized metabolites of a given substrate, which could be an interesting goal for drug metabolism prediction. In that regard, the recently reported biotechnological systems based on yeasts expressing each cytochrome P450 of human liver provide a complementary approach, as they give access to individual metabolites with a very high regio- and stereoselectivity. The complementarity of these two kinds of systems, i.e., the purely chemical models and the yeasts expressing each human liver cytochrome P450, for drug (or other xenobiotic) metabolism study in humans is illustrated in the following example [116]. Lidocaine is a local anesthetic agent that exhibits three main primary oxidized metabolites in humans: a compound derived from the N-deethylation of lidocaine, a phenol derived from the hydroxylation of its aromatic ring, and a metabolite formed by hydroxylation of one of its methyl groups (Figure 10). These three metabolites were believed to derive from oxidations catalyzed by liver cytochromes P450 of the 3A, 1A, and 2B subfamilies, respectively. In fact, oxidation of lidocaine by microsomes from yeast expressing human liver cytochrome P450 3A4 selectively led to its N-deethylated metabolite and CH_3CHO [61,62]. Oxidation of lidocaine by the classical Mn(TDCPP)Cl-H_2O_2-imidazole system [81] also led to products derived from oxidation of its tertiary amine function, the N-deethylated metabolite and a product coming from the combination of this metabolite and CH_3CHO [116] (Figure 10). Interestingly, a different model system operating in a buffer pH 3 and using Mn(III) [TDCSPP

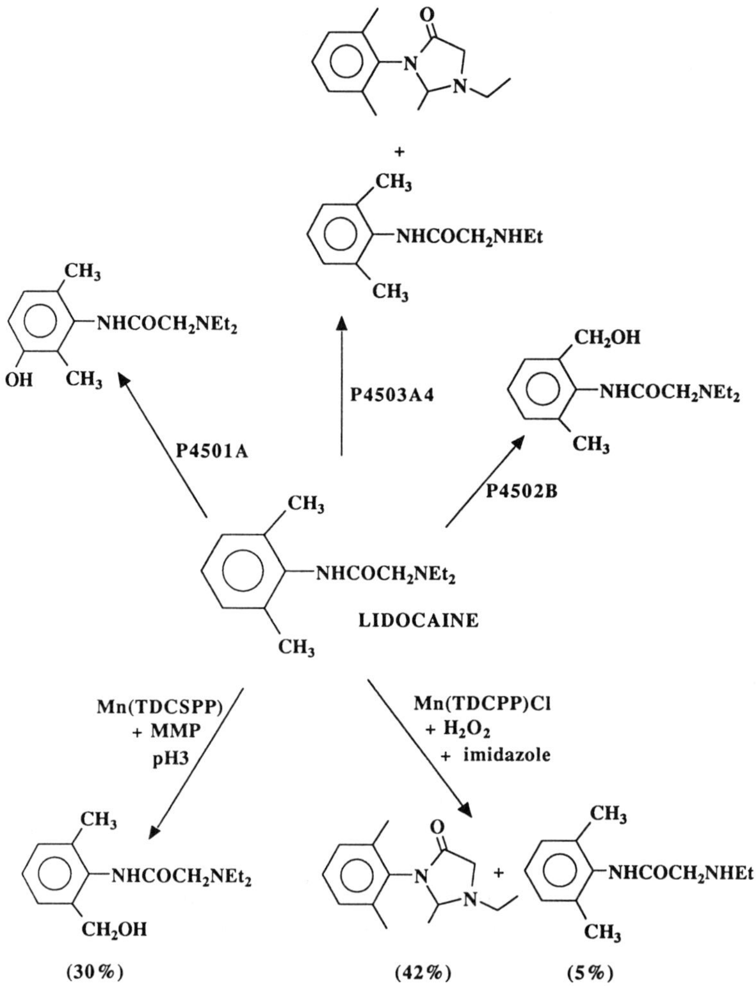

Figure 10 Oxidation of lidocaine by liver cytochromes P450 and model systems. TDCSPP is for *meso*-tetra(2,6-dichloro-3-sulfonatophenyl)porphyrin and MMP for magnesium monoperoxyphthalate.

= *meso*-tetra(2,6-dichloro-3-sulfonatophenyl)porphyrin] as catalyst and magnesium monoperoxyphthalate as oxidant selectively led to the alcohol derived from the hydroxylation of a benzylic methyl group of lidocaine [116]. It is thus possible to prepare most of the oxidized metabolites of lidocaine by a proper choice of the chemical or biotechnological model systems of cytochromes P450 [116].

Possible future improvements of these chemical and biotechnological model systems could be to render the chemical systems more selective by modifying the metalloporphyrin environment and to improve the practical use of the biotechnological systems by increasing the cytochrome P450 expression level and catalytic activity. Besides the use of these systems for drug metabolism studies, they should be very useful for the oxidative degradation of pollutants, the degradation of lignins (see following chapter by B. Meunier), and the selective cleavage of nucleic acid polymers [74].

REFERENCES

1. P.R. Ortiz de Montellano (ed.), *Cytochrome P450, Structure, Mechanism and Biochemistry*, Plenum Press, New York, 1986.
2. D.R. Nelson, T. Kamataki, D.J. Waxman, F.P. Guengerich, R.W. Estabrook, R. Feyereisen, F.J. Gonzalez, M.J. Coon, I.C. Gunsalus, O. Gotoh, K. Okuda, and D.W. Nebert, The P450 superfamily: update on new sequences, gene mapping, accession numbers, early trivial names of enzymes and nomenclature, *DNA Cell Biol., 12*: 1 (1993).
3. M. Haurand and V. Ullrich, Isolation and characterization of thromboxane synthase from human platelets as a cytochrome P450 enzyme, *J. Biol. Chem., 260*: 15059 (1985).
4. H. Shoun and T. Tanimoto, Denitrification by the fungus *Fusarium oxysporum* and involvement of cytochrome P450 in the respiratory nitrite reduction, *J. Biol. Chem., 266*: 11078 (1991).
5. W.C. Song and A.R. Brash, Purification of an allene oxide synthase and identification of the enzyme as a cytochrome P450, *Science, 253*: 781 (1991).
6. J.H. Dawson and M. Sono, Cytochrome P450 and chloroperoxidase: thiolate-ligated heme enzymes. Spectroscopic determination of their active site structures and mechanistic implications of thiolate ligation, *Chem. Rev., 87*: 1255 (1987).
7. T.L. Poulos, B.C. Finzel, and A.J. Howard, High resolution crystal structure of cytochrome P450 cam, *J. Mol. Biol., 195*: 687 (1987).
8. R.T. Ruettinger, L.-P. Wen, and A.J. Fulco, Coding nucleotide, 5′ regulatory and deduced aminoacid sequence of $P450_{BM3}$, a single peptide cytochrome P450: NADPH P450 reductase from *Bacillus megaterium, J. Biol. Chem., 264*: 10987 (1987).
9. K.A. White and M.A. Marletta, Nitric oxide synthase is a cytochrome P450 type hemoprotein, *Biochemistry, 31*: 6627 (1992).
10. D.J. Stuehr and M.I. Saito, Spectral characterization of brain and macrophage nitric oxide synthases: cytochrome P450–like hemeproteins that contain a flavin semiquinone radical, *J. Biol. Chem., 267*: 20547 (1992).
11. P. Klatt, K. Schmidt, and B. Mayer, Brain nitric oxide synthase is a haemoprotein, *Biochem. J., 288*: 15 (1992).
12. K. McMillan, D.S. Bredt, D.J. Hirsh, S.H. Snyder, J.E. Clark, and B.S.S. Masters, Cloned, expressed rat cerebellar nitric oxide synthase contains stoichio-

metric amounts of heme which binds carbon monoxide, *Proc. Natl. Acad. Sci. USA, 89*: 11141 (1992).

13. D. Mansuy, Use of model systems in biochemical toxicology: heme models, in *Reviews in Biochemical Toxicology* (E. Hodgson, J.R. Bend, and R.M. Philpot, eds.), Elsevier, New York, 1981, p. 283.
14. D. Mansuy, Coordination chemistry of cytochrome P450 and heme models, in *The Coordination Chemistry of Metalloenzymes in Hydrolytic and Oxidative Process* (I. Bertini, R.S. Drago, and C. Luchinat, eds.), Reidel, Dordrecht, 1982, p. 343.
15. M. Schappacher, R. Weiss, R. Montiel-Montoya, A. Trautwein, and A. Tabard, Formation of an iron(IV)-oxo "picket-fence" porphyrin derivative via reduction of the ferrous dioxygen adduct and reaction with carbon dioxide, *J. Am. Chem. Soc., 107*: 3736 (1985).
16. P.R. Ortiz de Montellano, Oxygen activation and transfer, in *Cytochrome P450: Structure, Mechanism and Biochemistry* (P.R. Ortiz de Montellano, ed.), Plenum Press, New York, 1986, p. 217.
17. J. Everse, K.E. Everse, and M.B. Grisham (eds.), *Peroxidases in Chemistry and Biology*, CRC Press, Boca Raton, 1991.
18. H.B. Dunford and J.S. Stillman, On the function and mechanism of action of peroxidases, *Coord. Chem. Rev., 19*: 187 (1976).
19. J.E. Roberts, B.M. Hoffman, R. Rutter, and L.P. Hager, Electron nuclear double resonance of horse radish peroxidase compound I, *J. Biol. Chem., 256*: 2118 (1981).
20. J.T. Groves, R.C. Haushalter, M. Nakamura, T.E. Nemo, and B.J. Evans, High-valent iron-porphyrin complexes related to peroxidase and cytochrome P450, *J. Am. Chem. Soc., 103*: 2884 (1981).
21. B. Boso, G. Lang, T.J. Mc Murry, and J.T. Groves, Mössbauer effect study of tight spin coupling in oxidized chloro-5,10,15,20-tetra (mesityl) porphyrinato-iron (III), *J. Chem. Phys., 79*: 1122 (1983).
22. J.E. Penner-Hahn, T.J. McMurry, M. Renner, L. Latos-Grazinski, K. Smith-Eble, I.M. Davis, A.L. Balch, J.T. Groves, J.H. Dawson, and K.O. Hodgson, X-ray absorption spectroscopic studies of high valent iron porphyrins, *J. Biol. Chem., 258*: 12671 (1983).
23. A.L. Balch, The reactivity of spectroscopically detected peroxy complexes of iron prophyrins, *Inorg. Chim. Acta, 297*: 198 (1992).
24. J.T. Groves and Y. Watanabe, On the mechanism of olefin epoxidation by oxo-iron porphyrins. Direct observation of an intermediate, *J. Am. Chem. Soc., 108*: 507 (1986).
25. J.E. Penner-Hahn, K.S. Eble, T.J. McMurry, M. Renner, A.L. Balch, J.T. Groves, J.H. Dawson, and K.O. Hodgson, Structural characterization of horse-radish peroxidase using EXAFS spectroscopy. Evidence for Fe = O ligation in compounds I and II, *J. Am. Chem. Soc., 108*: 7819 (1986).
26. S. Hashimoto, Y. Tatsuno, and T. Kitagawa, Observation of the $Fe^{IV} = O$ stretching raman band for a ferryl porphyrin π cation radical, *J. Am. Chem. Soc., 109*: 8096 (1987).

27. J.T. Groves and Y. Watanabe, Reactive iron porphyrin derivatives related to the catalytic cycles of cytochrome P450 and peroxidase. Studies of the mechanism of oxygen activation, *J. Am. Chem. Soc., 110*: 8443 (1988).
28. J.R. Kincaid, A.J. Schneider, and K.-J. Paeng, The resonance raman spectrum of a ferrylporphyrin cation radical and its photodegradation in the presence of methanol, *J. Am. Chem. Soc., 111*: 735 (1989).
29. A. Gold, K. Jayaraj, P. Doppelt, R. Weiss, E. Bill, X.-Q. Ding, E.L. Bominaar, A.X. Trautwein, and H. Winkler, Oxoferryl (2,6-dichlorophenyl) porphyrin cation radical: evidence for weak metal-radical spin coupling, *New J. Chem., 13*: 169 (1989).
30. E. Bill, X.-Q. Ding, E.L. Bominaar, A.X. Trautwein, H. Winkler, D. Mandon, R. Weiss, A. Gold, K. Jayaraj, W.E. Hatfield, and M.L. Kirk, Evidence for variable metal-radical spin coupling in oxoferrylporphyrin cation radical complexes, *Eur. J. Biochem., 188*: 665 (1990).
31. L.M. Proniewicz, I.R. Paeng, and K. Nakamoto, Resonance raman spectra of two isomeric dioxygen adducts of iron II porphyrins and π-cation radical and nonradical oxoferryl porphyrins produced in dioxygen matrices: simultaneous observation of more than seven oxygen isotope sensitive bands, *J. Am. Chem. Soc., 113*: 3294 (1991).
32. S. Hashimoto, Y. Mizutani, Y. Tatsuno, and T. Kitagawa, Resonance raman characterization of ferric and ferryl porphyrin π cation radicals and the FeIV = O stretching frequency, *J. Am. Chem. Soc., 113*: 6542 (1991).
33. D. Mandon, R. Weiss, K. Jayaraj, A. Gold, J. Terner, E. Bill, and A.X. Trautwein, Models for peroxidase compound I: generation and spectroscopic characterization of new oxoferryl porphyrin π cation radical species, *Inorg. Chem., 31*: 4404 (1992).
34. B.M. Hoffman, J.E. Roberts, C.H. Kang, and E. Margoliash, Electron paramagnetic and electron double resonance of the hydrogen peroxide compound of cytochrome c peroxidase, *J. Biol. Chem., 256*: 6556 (1981).
35. L.A. Fishel, J.E. Villafranca, J.M. Mauro, and J. Kraut, Yeast cytochrome c peroxidase: mutagenesis and expression in *Escherichia coli* show tryptophan-51 is not the radical site in compound I, *Biochemistry, 26*: 351 (1987).
36. R. Karthein, R. Dietz, W. Nastainczyk, and H.H. Ruf, Higher oxidation states of prostaglandin H synthase. EPR study of a transient tyrosyl radical in the enzyme during the peroxidase reaction, *Eur. J. Biochem., 71*: 313 (1988).
37. R. Dietz, W. Nastainczyk, and H.H. Ruf, Higher oxidation states of prostaglandin H synthase. Rapid electronic spectroscopy detected two spectral intermediates during the peroxidase reaction with prostaglandin G_2, *Eur. J. Biochem., 71*: 321 (1988).
38. H. Suanyu and H.B. Dunford, Prostaglandin H. Synthase kinetics, *J. Biol. Chem., 267*: 17649 (1992).
39. F.P. Guengerich and T.L. MacDonald, Mechanisms of catalysis by cytochromes P450: a unified view, *Acc. Chem. Res., 17*: 9 (1984).
40. D. Mansuy and P. Battioni, Alkane functionalisation by cytochromes P450 and by model systems using O_2 or H_2O_2, in *Activation and Functionalization of Alkanes* (Craig L. Hill, ed.), Wiley, New York, 1989, p. 195.

41. D. Mansuy, P. Battioni, and J.P. Battioni, Chemical model systems for drug-metabolizing cytochrome P450 dependent monooxygenases, *Eur. J. Biochem., 184*: 267 (1989).
42. D. Mansuy, J.C. Chottard, M. Lange, and J.P. Battioni, On the nature and reactivity towards substrates of the active oxygen hemoprotein complexes involved in the reaction of peroxidases and cytochrome P450 with two electron oxidants, *J. Mol. Catal., 7*: 215 (1980).
43. V. Ullrich, The mechanism of cytochrome P450 action, in *Biological Reactive Intermediates* (D.J. Jollow, J.J. Kocsis, R. Snyder, and H. Vaino, eds.), Plenum Press, New York, 1977, p. 192.
44. D. Mansuy, J.P. Battioni, J.C. Chottard, and V. Ullrich, Preparation of a porphyrin-iron-carbene model for the cytochrome P450 complexes obtained upon metabolic oxidation of the insecticide synergists of the 1.3-benzodioxole series, *J. Am. Chem. Soc., 101*: 3971 (1979).
45. D.M. Jerina and J.W. Daly, Arene oxides: a new aspect of drug metabolism, *Science, 185*: 573 (1974).
46. D. Mansuy, P. Valadon, I. Erdelmeier, P. Lopez-Garcia, C. Amar, J.P. Girault, and P. Dansette, Thiophene S-oxides as new reactive metabolites: formation by cytochrome P450 dependent oxidation and reaction with nucleophiles, *J. Am. Chem. Soc., 113*: 7825 (1991).
47. P.M. Dansette, D.C. Thang, H. El Amri, and D. Mansuy, Evidence for thiophene-S-oxide as a primary reactive metabolite of thiophene in vivo: formation of a dihydrothiophene sulfoxide mercapturic acid, *Biochem. Biophys. Res. Commun., 186*: 1624 (1992).
48. J.L. Boucher, A. Genet, S. Vadon, M. Delaforge, Y. Henry, and D. Mansuy, Cytochrome P450 catalyzes the oxidation of N^{ω}-hydroxy-L-arginine by NADPH and O_2 to nitric oxide and citrulline, *Biochem. Biophys. Res. Commun., 187*: 880 (1992).
49. V. Andronik-Lion, J.L. Boucher, M. Delaforge, Y. Henry, and D. Mansuy, Formation of nitric oxide by cytochrome P450–catalyzed oxidation of aromatic amidoximes, *Biochem. Biophys. Res. Commun., 185*: 452 (1992).
50. A.D.N. Vaz, E.S. Roberts, and M.J. Coon, Olefin formation in the oxidative deformylation of aldehydes by cytochrome P450. Mechanistic implications for catalysis by oxygen-derived peroxides, *J. Am. Chem. Soc., 113*: 5886 (1991).
51. E.S. Roberts, A.D.N. Vaz, and M.J. Coon, Catalysis by cytochrome P450 of an oxidative reaction in xenobiotic aldehyde metabolism: deformylation with olefin formation, *Proc. Natl. Acad. Sci. USA, 88*: 8963 (1991).
52. D.E. Stevenson, J.N. Wright, and M. Akhtar, Mechanistic consideration of P450 dependent enzymic reactions: studies on oestriol biosynthesis, *J. Chem. Soc. Perkins Trans I, 1*: 2043 (1988).
53. F.J. Gonzalez, The molecular biology of cytochrome P450s, *Pharmacol. Rev., 40*: 243 (1989).
54. H.P. Eugster, C. Sengstag, U.A. Meyer, A. Hinnen, and F.E. Würgler, Constitutive and inducible expression of human cytochrome P450IA1 in yeast *Saccharomyces cerevisiae*: an alternative enzyme source for in vitro studies, *Biochem. Biophys. Res. Commun., 172*: 737 (1990).

55. M.S. Ching, M.S. Lennard, G.T. Tucker, H.F. Woods, D.E. Kelly, and S.L. Kelly, The expression of human cytochrome P450IA1 in the yeast *Saccharomyces cerevisiae, Biochem. Pharmacol., 42*: 753 (1991).
56. C. Gautier, P. Urban, P. Beaune, and D. Pompon, Engineered yeast cell as model to study coupling between human xenobiotic metabolizing enzymes, *Eur. J. Biochem., 211*: 63 (1993).
57. T. Yasumori, N. Murayama, Y. Yamazoe, A. Abe, Y. Nogi, T. Fukasawa, and R. Kato, Expression of a human P450IIC gene in yeast cells using galactose inducible expression system, *Mol. Pharmacol., 35*: 443 (1989).
58. W.R. Brian, P.K. Srivastava, D.R. Umbenhauer, R.S. Lloyd, and F.P. Guengerich, Expression of a human liver cytochrome P450 protein with tolbutamide hydroxylase activity in *Saccharomyces cerevisiae, Biochemistry, 28*: 4993 (1989).
59. S.W. Ellis, M.S. Ching, P.F. Watson, C.J. Henderson, A.P. Simula, M.S. Lennard, G.T. Tucker, and H.F. Woods, Catalytic activities of human liver cytochrome P450 (CYP2D6) expressed in yeast, *Biochem. Pharmacol., 44*: 617 (1992).
60. W.R. Brian, M.A. Sari, M. Iwasaki, T. Shimada, L.S. Kaminsky, and F.P. Guengerich, Catalytic activities of human liver cytochrome P450IIIA4 expressed in *Saccharomyces cerevisiae, Biochemistry, 29*: 11280 (1990).
61. J.P. Renaud, C. Cullin, D. Pompon, P. Beaune, and D. Mansuy, Expression of human liver cytochrome P450IIIA4 in yeast: a functional model for the hepatic enzyme, *Eur. J. Biochem., 194*: 889 (1990).
62. M.A. Peyronneau, J.P. Renaud, G. Truan, P. Urban, D. Pompon, and D. Mansuy, Optimization of yeast-expressed human liver P4503A4 catalytic activities by coexpressing NADPH-P450 reductase and cytochrome b_5, *Eur. J. Biochem., 207*: 109 (1992).
63. T.J. McMurry and J.T. Groves, Metalloporphyrin models of cytochrome P450, in *Cytochrome P450: Structure Mechanism and Biochemistry* (P.R. Ortiz de Montellano, ed.), Plenum Press, New York, 1986, p. 1.
64. B. Meunier, Metalloporphyrin catalysed oxygenation of hydrocarbons, *Bull. Soc. Chim. Fr., 4*: 578 (1986).
65. T.C. Bruice, Chemical studies pertaining to the chemistry of cytochrome P450 and the peroxidases, *Ann. N.Y. Acad. Sci., 471*: 83 (1986).
66. D. Mansuy, Cytochrome P450 and synthetic models, *Pure Appl. Chem., 59*: 759 (1987).
67. I. Tabushi, Reproductive dioxygen activation by use of artificial P450 systems, *Coord. Chem. Rev., 86*: 1 (1988).
68. F. Montanari, S. Banfi, and S. Quici, Oxidations catalyzed by stable metallotetraaryl porphyrins under aqueous-organic two-phase conditions, *Pure Appl. Chem., 61*: 1631 (1989).
69. T. Okamoto, K. Sasaki, and M. Tachibana, Recent progress of metalloporphyrin-catalyzed oxidation and oxygenation, *Bull. Instit. Chem. Res. Kyoto University, 97*: 169 (1989).
70. D. Mansuy, Biomimetic catalysts for selective oxidation in organic chemistry, *Pure Appl. Chem., 62*: 741 (1990).
71. T.C. Bruice, Reactions of hydroperoxides with metallotetraphenylporphyrins in aqueous solutions, *Acc. Chem. Res., 24*: 243 (1991).

72. T.G. Traylor, Kinetics and mechanism studies in biomimetic chemistry: metalloenzyme model systems, *Pure Appl. Chem., 63*: 265 (1991).
73. M.J. Gunter and P. Turner, Metalloporphyrins as models for the cytochrome P450, *Coord. Chem. Rev., 108*: 115 (1991).
74. B. Meunier, Metalloporphyrins as versatile catalysts for oxidation reactions and oxidative DNA cleavage, *Chem. Rev., 92*: 1411 (1992).
75. J.T. Groves, T.E. Nemo, and R.S. Myers, Hydroxylation and epoxidation catalyzed by iron-porphyrin complexes. Oxygen transfer from iodosylbenzene, *J. Am. Chem. Soc., 101*: 1032 (1979).
76. C.K. Chang and F. Ebina, NIH shift in haemin-iodosylbenzene-mediated hydroxylations, *J. Chem. Soc. Chem. Commun.,* 778 (1981).
77. P.S. Traylor, D. Dolphin, and T.G. Traylor, Sterically protected hemins with electronegative substituents: efficient catalysts for hydroxylation and epoxidation, *J. Chem. Soc. Chem. Commun.,* 279 (1984).
78. B. De Poorter and B. Meunier, Catalytic epoxidation of aliphatic terminal olefins with sodium hypochlorite, *Tetrahedron Lett., 25*: 1895 (1984).
79. T.G. Traylor, J.C. Marsters, T. Nakano, and B.E. Dunlap, Kinetics of iron (III)porphyrin catalyzed epoxidations, *J. Am. Chem. Soc., 107*: 5537 (1985).
80. J.P. Renaud, P. Battioni, J.F. Bartoli, and D. Mansuy, A very efficient system for alkene epoxidation by hydrogen peroxide: catalysis by manganese porphyrins in the presence of imidazole, *J. Chem. Soc., Chem. Commun.,* 888 (1985).
81. P. Battioni, J.P. Renaud, J.F. Baroli, M. Reina-Artiles, M. Fort, and D. Mansuy, Monooxygenase-like oxidation of hydrocarbons by H_2O_2 catalyzed by Mn-porphyrins and imidazole: selection of the best catalytic system and nature of the active oxygen species, *J. Am. Chem. Soc., 110*: 8462 (1988).
82. P. Battioni, J.P. Renaud, J.F. Bartoli, and D. Mansuy, Hydroxylation of alkanes by hydrogen peroxide: an efficient system using manganese porphyrins and imidazole as catalysts, *J. Chem. Soc., Chem. Commun.,* 341 (1986).
83. M.N. Carrier, C. Cheer, P. Gouvine, J.F. Bartoli, P. Battioni, and D. Mansuy, Biomimetic hydroxylation of aromatic compounds: hydrogen peroxide and manganese-polyhalogenated porphyrins as a particular good system, *Tetrahedron Lett., 31*: 6645 (1990).
84. D. Mansuy and P. Battioni, Dioxygen activation at heme centers in enzymes and synthetic analogs, in *Bioinorganic Catalysis* (J. Reedjik, ed.), Marcel Dekker, New York, 1993, p. 422.
85. J.T. Groves and R. Quinn, Aerobic epoxidation of olefin with ruthenium porphyrin catalysts, *J. Am. Chem. Soc., 107*: 5790 (1985).
86. T.G. Traylor and S. Tsuchiya, Perhalogenated tetraphenylhemins stable catalysts of high turnover catalytic hydroxylations, *Inorg. Chem., 26*: 1338 (1987).
87. H.J. Callot, Bromuration de la meso-tetra-phenylporphyrine. Structure et réactivité des produits, *Bull. Soc. Chim. Fr.,* 1492 (1974).
88. D.H. Dolphin, T. Nakano, T.K. Kirk, T.E. Maroni, R.L. Farrell, and T.P. Wijesekera, Porphyrins, their syntheses and uses thereof, Patent PCT Int. Appl. WO 88/07988 (1988).
89. T. Wijesekera, A. Matsumoto, D. Dolphin, and D. Lexa, Perchlorinated and highly chlorinated meso-tetraphenylporphyrins, *Angew. Chem. Int. Ed. Engl., 29*: 1028 (1990).

90. A.M. d'A. Rocha Gonsalves, R.A.W. Johnstone, M.N. Pereira, J. Shaw, and A.J.F. do N. Sobral, Metal-assisted reactions. 22. Synthesis of perhalogenated porphyrins and their use as oxidation catalysts, *Tetrahedron Lett., 32*: 1355 (1991).
91. D. Mandon, P. Ochsenbein, J. Fischer, R. Weiss, K. Jayaraj, R.N. Austin, A. Gold, P.S. White, O. Brigaud, P. Battioni, and D. Mansuy, β-halogenated-pyrrole porphyrins. Molecular structures of 2,3,7,8,12,13,17,18-octabromo-5,10,15,20-tetramesitylporphyrin and nickel (II) 2,3,7,8,12,13,17,18-octabromo-5,10,15,20-tetramesitylporphyrin, and nickel (II) 2,3,7,8,12,13,17,18-octabromo-5,10,15,20-tetrakis (pentafluorophenyl)porphyrin, *Inorg. Chem., 31*: 2044 (1992).
92. P.E. Ellis and J.E. Lyons, Selective air oxidation of light alkanes catalyzed by activated metalloporphyrins. The search of a suprabiotic system, *Coord. Chem. Rev., 105*: 181 (1990).
93. P. Hoffman, G. Labat, A. Robert, and B. Meunier, Highly selective bromination of tetramesitylporphyrins M-Br_8TMP and M-Br_8TMPS, *Tetrahedron Lett., 31*: 1991 (1990).
94. P. Hoffman, A. Robert, B. Meunier, Preparation and catalytic activities of the manganese and iron derivatives of Br_8 TMP and Cl_{12} TMP, two robust porphyrin ligands obtained by halogenation of tetramesitylporphyrins, *Bull. Soc. Chim. Fr., 129*: 85 (1992).
95. P. Bhyrappa and U. Krishnan, Octabromotetraphenyl porphyrin and its metal derivatives: electronic structure and electrochemical properties, *Inorg. Chem., 30*: 241 (1991).
96. S. Tsuchiya and M. Seno, Novel synthetic method of phenol from benzene catalysed by perfluorinated hemin, *Chem Lett.*, 263 (1989).
97. P. Battioni, O. Brigaud, H. Desvaux, D. Mansuy, and T.G. Traylor, Preparation of functionalized polyhalogenated tetraarylporphyrins by selective nucleophilic substitution of parafluorine groups of meso-tetra-(pentafluorophenyl) porphyrins, *Tetrahedron Lett., 32*: 2893 (1991).
98. I. Artaud, K. Ben Aziza, C. Chopard, and D. Mansuy, A new easy access to quinones from iron porphyrin-catalysed oxidations of methoxyarenes by magnesium mono peroxyphthalate, *J. Chem. Soc., Chem. Commun.*, 31 (1991).
99. I. Artaud, H. Grennberg, and D. Mansuy, Selective oxidative cleavage of 1,2-dimethoxyarenes to muconic diesters catalyzed by an iron β-sulfonated-tetra-(pentafluorophenyl)porphyrin, *J. Chem. Soc., Chem. Commun.*, 1036 (1992).
100. I. Artaud, K. Ben Aziza, and D. Mansuy, Iron porphyrin-catalyzed oxidation of 1,2-dimethoxyarenes: discussion of the different reactions involved and the competition between the formation of methoxy-quinones or muconic dimethylesters, *J. Org. Chem., 58*: 3373 (1993).
101. P.E. Ellis and J.E. Lyons, Nitrated metalloporphyrins as catalysts for alkane oxidation, Patent, UST U.S. Appl. 5120882 (1992).
102. P.E. Ellis and J.E. Lyons, Cyano- and polycyanometalloporphyrins as catalysts for alkane oxidation, Patent, UST U.S. Appl. 5118886 (1992).
103. S. Tsuchiya, Specific emission spectra observed in dodecaphenylporphyrin, *Chem. Phys. Lett., 169*: 608 (1990).

104. L.D. Sparks, C.J. Medforth, M.S. Park, J.R. Chamberlain, M.R. Ondrias, M.O. Senge, K.L. Smith, and J.A. Shelnutt, *J. Am. Chem. Soc., 115*: 581 (1993).
105. O. Brigaud, P. Battioni, D. Mansuy, and C. Giessner-Prettre, Structure of meso-tetraaryl-β-octahalogeno-porphyrins: a semi-empirical quantum mechanical investigation, *New. J. Chem., 16*: 1031 (1992).
106. J.F. Bartoli, O. Brigaud, P. Battioni, and D. Mansuy, Hydroxylation of linear alkanes catalysed by iron porphyrins: particular efficacy and regioselectivity of perhalogenated porphyrins, *J. Chem. Soc., Chem. Commun.*, 440 (1991).
107. T.G. Traylor, K.W. Hill, W.-P. Fann, S. Tsuchiya, and B.E. Dunlap, Aliphatic hydroxylation catalyzed by iron III porphyrins, *J. Am. Chem. Soc., 114*: 1308 (1992).
108. A. Maldotti, C. Bartocci, R. Amadelli, E. Polo, P. Battioni, and D. Mansuy, Oxidation of alkanes by dioxygen catalysed by photoactivated iron porphyrins, *J. Chem. Soc., Chem. Commun., 1487* (1991).
109. D. Mansuy, Activation of alkanes: the biomimetic approach, *Coord. Chem. Rev., 125*: 129 (1993).
110. P. Battioni, J.F. Bartoli, D. Mansuy, Y.S. Byun, and T.G. Traylor, An easy access to polyhalogenated metalloporphyrins covalently bound to polymeric supports as efficient catalysts for hydrocarbon oxidation, *J. Chem. Soc., Chem. Commun.,* 1051 (1992).
111. T.G. Traylor, Y.S. Byun, P.S. Traylor, P. Battioni, and D. Mansuy, Polymeric polyhalogenated metalloporphyrin catalysts for hydroxylation of alkanes and epoxidation of alkenes, *J. Am. Chem. Soc., 113*: 7821 (1991).
112. H. Kameyama, H. Suzuki, and A. Amano, Intercalation of CoII meso-tetrakis(1-methyl-4-pyridyl) porphyrin into montmorillonite, *Chem. Lett.,* 1117 (1988).
113. L. Barloy, P. Battioni, and D. Mansuy, Manganese porphyrins supported on montmorillonite as hydrocarbon monooxygenation catalysts: particular efficacy for linear alkane hydroxylation, *J. Chem. Soc., Chem. Commun.,* 1365 (1990).
114. L. Barloy, J.P. Lallier, P. Battioni, D. Mansuy, Y. Piffard, M. Tournoux, J.P. Valim, and W. Jones, Manganese porphyrins adsorbed or intercalated in different mineral matrices: preparation and compared properties as catalysts for alkene and alkane oxidation, *New J. Chem., 16*: 71 (1992).
115. K.A. Carrado, P. Thiyagarajan, R.E. Winaus, and R.E. Botto, Hydrothermal crystallization of porphyrin-containing layer silicate, *Inorg. Chem., 30*: 794 (1991).
116. M.N. Carrier, P. Battioni, and D. Mansuy, Studying drug metabolic oxidation with biomimetic metalloporphyrin systems: problems and solutions in the case of lidocaine, *Bull. Soc. Chim. Fr., 130*: 405 (1993).

5
Ligninase and Related Model Systems

Bernard Meunier
Laboratoire de Chimie de Coordination, Centre National de la Recherche Scientifique, Toulouse, France

I. INTRODUCTION

Lignin, cellulose, and hemicellulose are the main polymers of plants. The growing interest in the use of renewable materials derived from wood has stimulated a large number of studies on the enzymatic and/or the chemical treatment of wood or straw [1–3]. A major problem in the cellulose industry is the expensive and polluting pretreatment of wood pulp to produce cellulose materials free of lignin. The currently used chlorine oxidation of lignin in white paper manufacturing produces 4–5 kg of chlorinated and poorly biodegradable phenol residues per metric ton of treated wood pulp [4,5]. In the middle of the 1970s, several groups decided to investigate the possible industrial development of enzymatic degradation of lignin in order to avoid the large chlorinated organic discharges in the environment and also to consider lignin as a renewable source of aromatics which can be considered as raw materials in the future postoil period of chemical industries. Until now, no efficient enzymatic treatment was possible but the recent purification and characterization of two-heme containing peroxidases produced by a ligninolytic filamentous fungus, *Phanerochaete chrysosporium*, have stimulated modeling studies in the field. One of these peroxidases is ligninase, or LiP [6], and the second one is manganese peroxidase, or MnP [7]. The present review on the biomimetic aspects of lignin degradation is organized as follows: after a short section on the biosynthesis of lignin, the properties of the two peroxidases of *Phanerochaete chrysosporium*, LiP and MnP, are described. This is followed by a discussion of recent advances on the modeling of these two heme enzymes by synthetic metalloporphyrins.

II. LIGNIN: BIOSYNTHESIS AND STRUCTURE

Lignin is an essential component of the woody stems of arborescent plants where it is present in amounts ranging from 15% to 36%. This biopolymer is also found as an integral cell wall constituent in all vascular plants including the herbaceous varieties. Because of the large variety of lignin structures, "lignins" is more appropriate than lignin. However, like many authors, we will use the term lignin to cover all different types of this polyphenolic polymer. Besides its structural role in plants (resistance to impact, compression, and bending), lignin has other functions essential for the life of plants: (a) to decrease of the water permeation across the cell walls, (b) to distribute water, nutrients, and metabolites, and (c) to help in the plant's defense against insects and microorganisms [1,2].

A. Biosynthesis of Lignin

This tridimensional polymer results from the free radical polymerization of *p*-hydroxycinnamic alcohols of three different types: *p*-coumarylic (**a**), coniferylic (**b**), and sinapylic (**c**). All three alcohols are biosynthetic products of phenylalanine, produced according to the general scheme indicated in Figure 1. The precise structure of the lignin depends on the ratio of these three cinnamic alcohols present during the polymerization step. Lignin of conifers (gymnosperms) consists of monomers (**G**) based only on coniferyl alcohol polymerization, whereas lignin of trees having leaves (angiosperms) is made

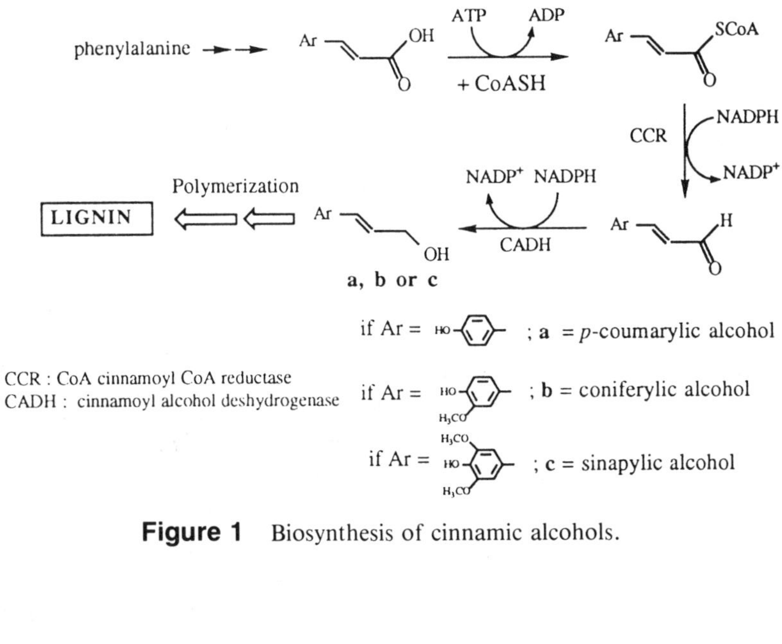

Figure 1 Biosynthesis of cinnamic alcohols.

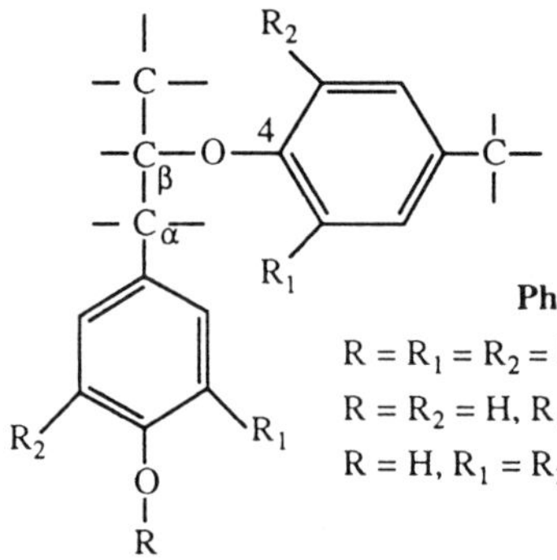

Phenylpropanoid units
R = R_1 = R_2 = H *p*-hydroxyphenyl type (**H**)
R = R_2 = H, R_1 = OCH_3 guaiacyl type (**G**)
R = H, R_1 = R_2 = OCH_3 syringyl type (**S**)

Figure 2 Structure of phenylpropanoids units with arylglycerol-β-arylether (β-O-4) linkages.

of the two monomers (**G** and **S**) (see Figure 2). Herbaceous lignin comprises subunits **G**, **S**, and **H**. These different monomers are involved in the reticulation of lignin via the formation of C-C bonds and arylglycerol-β-arylether bonds, the so-called β-O-4 linkages (Figure 2) [8]. Phenoxy radicals of coumarylic, coniferylic, and sinapylic alcohols produced by the action of peroxidases or laccases polymerize in a tridimensional network producing lignin with a general structure reported in Figure 3.

Several physico-chemical methods have been used to establish lignin structures as precisely as possible (for nuclear magnetic resonance studies on lignin, see Ref. 9). Three main types of linkages have been identified and quantified for beech tree lignin [8]: arylglycerol-β-arylether bonds (the most prevalent linkage in lignin: 48–55% of all different links), biphenyl bonds (9–11%), and phenylcoumaran links (9–12%).

B. Lignin Biodegradation

Lignin is not easily hydrolyzed by natural processes or by chemicals. It can be isolated from wood by treatment with sulfuric acid: polysaccharides are hydrolyzed and lignin is recovered as an insoluble material. Various microorganisms are able to degrade lignin by oxidation, the ultimate products being CO_2 and H_2O. Bacteria [10] play a minor role in delignification compared to that played by fungi [11]. Three different categories of fungi have a ligninolytic activity:

White rot fungi (basidiomycetes) oxidize lignin to CO_2 and H_2O.
Brown rot fungi (basidiomycetes) are able to degrade polysaccharides and, to a lesser extent, lignin, leaving the aromatic rings unchanged.
Soft rot fungi (ascomycetes) are less efficient in lignin oxidation.

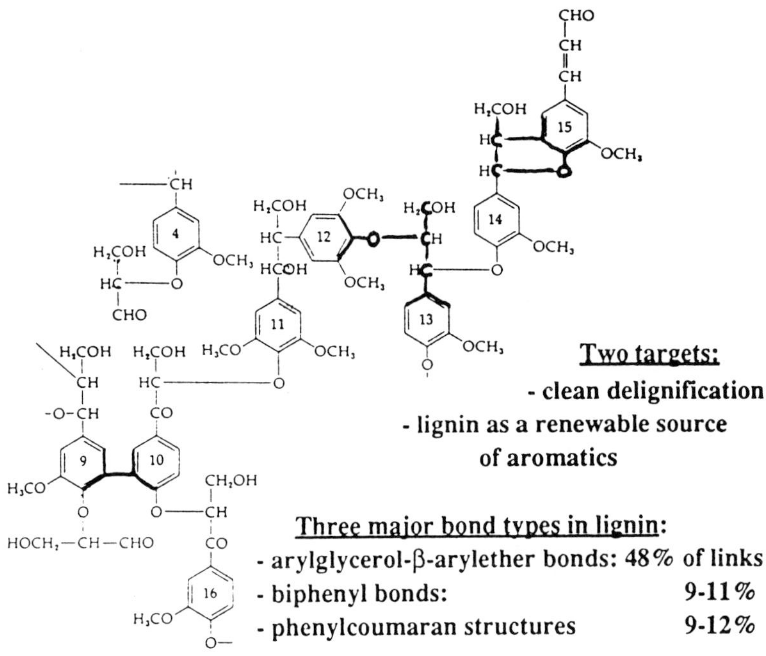

Figure 3 Schematic representation of lignin with the three main types of linkages between aromatic subunits: arylglycerol-β-arylether, biphenyl, and phenylcoumaran links.

Among white rot fungi *Phanerochaete chrysosporium* [12–14] has the highest ligninolytic activity compared to *Pycnoporus cinnabarinus* [15] or *Phlebia radiata* [16]. Several key factors control the ligninolytic activity of *Phanerochaete chrysosporium*. For example, the hydrogen peroxide oxidant necessary for the reaction is produced by microbial oxidation of glucose and this process is inhibited by agitation. This was attributed to low oxygen levels within the pellets containing the fungus. Lignin is not degraded during the primary growth stages of the fungus, metabolism being observed only when growth takes place under conditions of carbon or nitrogen limitation [17]. Three main reactions mediated by the fungus are observed: (a) the cleavage of C_α-C_β bonds of phenylpropanoid units, (b) the cleavage of C-O bonds in the arylglycerol-β-arylether moieties, and (c) the demethoxylation of aromatic rings. This latter reaction was not well documented for peroxidases prior to a study on the metabolism of 9-methoxyellipticine by horseradish peroxidase (HRP) [18]. Two extracellular enzymes of *Phanerochaete chrysosporium* have been purified and identified as H_2O_2-dependent peroxidases: ligninase (LiP) and manganese peroxidase (MnP) [6 and 7, respectively].

III. ENZYMATIC SYSTEMS

A. Ligninase (or LiP)

The purification of ligninase by the groups of Kirk [6a] and Gold [6b] largely contributed to the understanding of the relationships between lignin degradation and reactions catalyzed by this enzyme. The enzyme activity of the concentrated extracellular medium can be monitored by using ^{14}C-labeled dimers **1** and **2** as lignin models (Figure 4) [6a]. Ligninase contained in the extracellular fluid of ligninolytic cultures of *Phanerochaete chrysosporium* has been purified by gel permeation chromatography and six isozymes have been characterized. The main one (H8) has been used for physicochemical analyses. This peroxidase, which has a molecular weight of 42,000 Da, is composed of 15% by weight of carbohydrates and contains one iron-protoporphyrin IX entity as the prosthetic group [6]. Hydrogen peroxide is the cofactor. Early studies on LiP based on ^{18}O incorporation into oxidation products when performed under ^{18}O-labeled molecular oxygen led to the classification of ligninase as an oxygenase [20]. Additional studies with $H_2{}^{18}O$, in contrast, indicated that labeled products were formed confirming the peroxidase na-

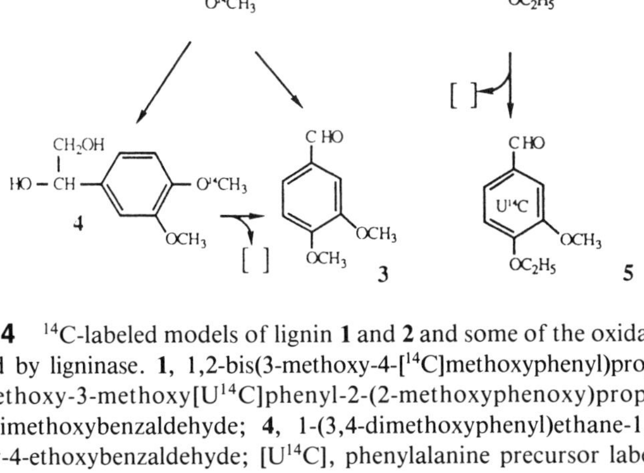

Figure 4 ^{14}C-labeled models of lignin **1** and **2** and some of the oxidation products catalyzed by ligninase. **1**, 1,2-bis(3-methoxy-4-[^{14}C]methoxyphenyl)propane-1,3-diol; **2**, 1-(4-ethoxy-3-methoxy[$U^{14}C$]phenyl-2-(2-methoxyphenoxy)propane-1,3-diol; **3**, 3,4-dimethoxybenzaldehyde; **4**, 1-(3,4-dimethoxyphenyl)ethane-1,2-diol; **5**, 3-methoxy-4-ethoxybenzaldehyde; [$U^{14}C$], phenylalanine precursor labeled with ^{14}C; [ ], other oxidation products.

ture of LiP, the crucial step catalyzed by ligninase being the formation of an initial radical cation on methoxylated aromatic rings [21,22].

The resting state of the enzyme is a high-spin iron(III)porphyrin with a histidine as proximal residue [23,24]. Low-temperature (2°C) electron paramagnetic resonance (EPR) data suggested that a water molecule is present as the sixth ligand [23] but absent at room temperature as confirmed by Raman resonance spectroscopy [24]. In fact there is an equilibrium between the hexa- and pentacoordinated forms between 2 and 25°C. Single crystals of the enzyme have been obtained by T. L. Poulos [25] and coworkers and preliminary data confirmed the presence of a proximal histidine and a large pocket on the heme edge as suggested by Ortiz de Montellano [26]. cDNA of ligninase is now available [27].

As for other peroxidases, a high-valent iron–oxo entity, compound I, is generated by the addition of hydrogen peroxide (see Figure 5 for the catalytic cycle of LiP). This green compound (Soret band at 408 nm) has the same UV-visible characteristics as HRP–compound I that contains an iron

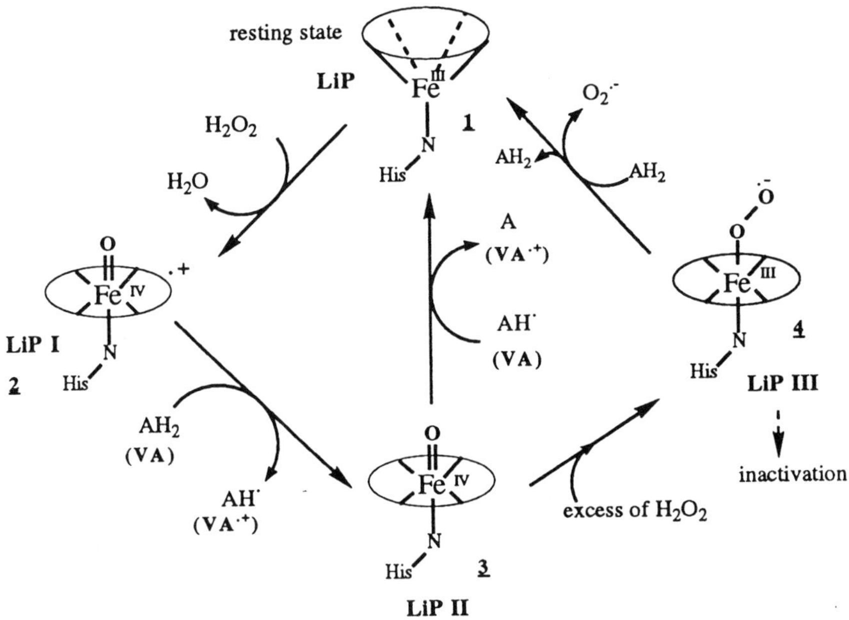

Figure 5 Catalytic cycle of ligninase. **1**, resting state of LiP, Fe^{III} high spin; **2**, $[P^{\cdot+}]Fe^{IV}=O$, LiP–compound I; **3**, $[P]Fe^{IV}=O$, LiP–compound II; **4**, $[P]Fe^{III}O_2^{\cdot -} \rightleftarrows [P]Fe^{II}\ O_2$ LiP–compound III. [P], porphyrin cycle; (VA), veratryl alcohol; $(VA^{\cdot+})$, radical cation of VA.

(IV)oxo species associated with a porphyrin π-radical cation [6c,28]. The addition of 0.5 equivalent of a two-electron reducing agent like veratryl alcohol generates LiP–compound II (Soret band at 420 nm), which corresponds to the iron(IV)oxo without an oxidized porphyrin ligand [6c,28]. Unexpectedly, LiP is a fragile enzyme when exposed to more than 20 equivalents of hydrogen peroxide [19]: besides the two catalytically active forms (compounds I and II), two inactive forms LiP III* and LiP III have been obtained in the presence of an excess of hydrogen peroxide or by addition of molecular oxygen to the reduced form of LiP (ferrous state) [29]. The difference between these two inactive forms is due to the presence of an extra molecule of H_2O_2, probably in the vicinity of the prosthetic group in the case of LiP III*. Any adventitious electron transfer from the ferrous state of the enzyme to this extra hydrogen peroxide molecule would generate a hydroxyl radical, which creates an irreversible inactivation of LiP [30]. The fragility of LiP in the presence of an excess of hydrogen peroxide is the limiting factor for the industrial development of ligninase use in delignification or pollutant oxidation.

Veratryl alcohol is a secondary metabolite in delignification mediated by *Phanerochaete chrysosporium* and is also a good substrate for ligninase and enhances the rate of oxidations catalyzed by LiP [31,32]. It has been proposed that veratryl alcohol, via the formation of its radical cation, is a chemical mediator in the oxidative degradation of lignin. The veratryl alcohol radical cation mediates the transfer of electrons between the two large macromolecules, the enzyme and the lignin. Many oxidation products generated from veratryl alcohol by LiP/H_2O_2 have been identified and quantified by CG-MS methods. Veratraldehyde, two quinones (2-hydromethyl-5-methoxy-1,4-benzoquinone and 2-methoxy-1,4-benzoquinone), and three lactones have been isolated [6d]. Complete identification of the oxidation products of lignin model dimers with an arylglycerol-3-arylether moiety has also been performed [33,35]. The C_α-C_β bond cleavage is mediated by the initial formation of a radical cation on one of the two aromatic rings. In the case of 1-(3,4-dimethoxyphenyl)-2-(2-methoxyphenoxy)propane-1,3-diol, the main oxidation products generated by LiP/H_2O_2 are veratraldehyde, guaiacol, and glycolaldehyde [36,37]. Experiments performed under $^{18}O_2$ and/or with $H_2{}^{18}O$ provided details on the mechanisms of formation of these different oxidation products [38].

B. Manganese Peroxidase (MnP)

The culture medium of *Phanerochaete chrysosporium* contains another extracellular peroxidase. This enzyme is manganese-dependent and has been purified by Gold et al. in 1984 [39]. Four isozymes have been identified, but none of them are able to catalyze the oxidation of veratryl alcohol in the presence of H_2O_2.

Manganese peroxidase is a heme glycoprotein of 46,000 Da and requires manganese(II) for its activity [7]. The primary amino acid sequence of MnP has been deduced from its cDNA [40,41]. This peroxidase exhibits a proximal histidine as well as distal histidine and arginine that catalyze the heterolytic cleavage of the peroxide O-O bond. The presence of the proximal histidine was confirmed by Raman resonance and EPR studies [7a].

The native manganese peroxidase is oxidized by hydrogen peroxide to a compound I species similar to HRP–compound I. Mn^{II} ion, which binds to the enzyme and reduces MnP–compound I to MnP–compound II. This latter oxidized form of MnP is only reduced to native MnP by manganese(II) but not by phenols. The oxidation of Mn^{II} to Mn^{III} by MnP is dependent on the presence of chelating agents in the reaction mixture. These chelating agents, which are α-hydroxy acids (lactate, succinate, malonate), have probably two different roles: they stimulate the catalytic activity by facilitating the dissociation of the enzyme–manganese complex by chelating Mn^{III} and they stabilize manganese at the oxidation state III. Mn^{III} salts have a tendency to quickly dismute to Mn^{II} and Mn^{IV} [42]. However, pyrophosphate, a strong chelating agent of manganese salts, inhibits the MnP activity, suggesting that the enzyme-mediated oxidation of manganese is operated at the level of nonchelated manganese salts. The binding site of manganese is probably on the heme edge close to the δ-*meso* carbon of the prosthetic group [7c].

Manganese ions cannot be replaced by iron(III), copper(II), or cobalt(II) salts. Finally, the addition of a large excess of H_2O_2 to the ferric state of the enzyme results in the formation of a compound III species as in the case of HRP (see Figure 6 for the catalytic cycle of MnP).

Since Mn(III) chelates are only able to oxidize phenol derivatives, the main targets of MnP on lignin are the phenolic residues. Fifteen percent of phenylpropane units in lignin have a hydroxyl group at the *para* position [43].

A detailed study on a lignin model dimer with a phenol function, such as 1-(3,5-dimethoxy-4-hydroxyphenol)-2-(4-methoxy)propane-1,3-diol, indicated that such a dimer was cleaved by $MnP/H_2O_2/Mn^{II}$ to the following monomers: 2,6-dimethoxybenzoquinone, 2-hydroxy-4′-methoxyacetophenone and 1-(4-methoxyphenyl)ethane-1,2-diol. Labeling studies with $^{18}O_2$ and $H_2{}^{18}O$ confirmed that the key reaction was the initial formation of a phenoxy radical [44]. Partial depolymerization of lignin by MnP has been observed in vitro [45].

The ligninolytic activity of *Phanerochaete chrysosporium* is dependent on both enzymes, LiP and MnP, and the presence of molecular oxygen in the culture medium to facilitate the propagation of free radical chain reac-

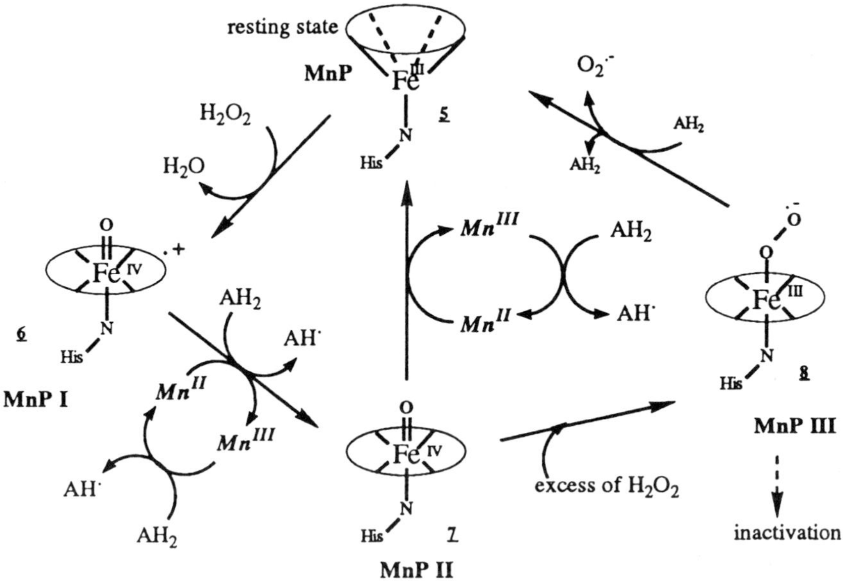

Figure 6 Catalytic cycle of manganese peroxidase MnP. **5**, resting state of MnP, Fe^{III} high spin; **6**, $[P^{\cdot+}]$ $Fe^{IV}=O$, MnP-compound I; **7**, [P] $Fe^{IV}=O$, MnP-compound II; **8**, [P] Fe^{III} $O_2^{\cdot-}$ $\rightleftarrows$ [P] Fe^{II} O_2, MnP-compound III. [P], porphyrin cycle; AH_2, phenolic substrate.

tions. Otherwise, anaerobic conditions favor the recombination of lignin-centered radicals to generate new C-C or C-O bonds within lignin. Such a process would lead to further reticulation of lignin, not to its degradation.

C. Oxidation of Pollutants Mediated by Ligninase

One of the most promising properties of ligninase, which is not substrate-specific like manganese peroxidase, is its capacity to oxidize various recalcitrant organic molecules. Microorganisms in soils and water convert many man-made chemicals into inorganic products [46]. However, some molecules are recalcitrant to biotransformation and are subject to biomagnification and accumulation in plants or animal tissues.

Among the compounds resisting microbial degradation, DDT and lindane are two classical examples [47]. Both are slowly transformed by microbial reduction [48] and few extracellular enzymes are able to catalyze their oxidation. In order to be effective, these enzymes must have higher redox capabilities than horseradish peroxidase (HRP). *Phanerochaete chrysosporium* is able to oxidize organohalide pollutants [49,50]. This oxidation is catalyzed by ligninase and can be extended to other polycyclic aromatics like benzo[a]pyrene and 9,10-dimethylanthracene [49,51].

Ligninase also catalyzes the oxidation of chlorinated phenols. For example, 2,4,6-trichlorophenol, a major pollutant in paper mill effluents [50], is converted to 2,6-dichlorobenzoquinone [49b,50]. 2,3,7,8-Tetrachlorodibenzo-*p*-dioxine is also degraded by LiP [52].

IV. MODEL SYSTEMS OF LIGNINASE

This survey of model systems is limited to those based on metalloporphyrins. Early models of ligninase have been developed with iron protoporphyrin IX in combination with *tert*-butylhydroperoxide [53,54] or Fe(TPP)Cl in combination with O_2 and a dihydropyridine as electron source [55]. However, these metalloporphyrins are not water-soluble and are easily destroyed under oxidizing conditions.

A. Water-Soluble Metalloporphyrin Complexes

Water-soluble sulfonated metalloporphyrins were found to be more suitable catalysts in ligninase model studies [56–59]. In one case, sodium hypochlorite was used as oxygen donor [56]. However, this oxidant cannot be developed in aqueous reaction mixtures, since hypochlorite only exists at elevated pH values and is an efficient oxygen atom donor in hydrophobic medium [60]. Aromatic chlorinations are the main reactions with hypochlorite solution below pH 8–9. In this respect, potassium monopersulfate seems to be a more suitable oxidant in ligninase modeling [57–59]. The best oxidant, i.e., the "clean oxidant," is of course hydrogen peroxide. However, the activation of H_2O_2 by synthetic metalloporphyrins in the absence of cofactors (imidazole derivatives) usually leads to the homolytic cleavage of the peroxide O-O bond [60a]. The produced hydroxy radicals are strong inhibitors and/or bleaching agents of metalloporphyrin catalysts. The activity of the "$KHSO_5$/sulfonated metalloporphyrin" ligninase model has been checked with the usual lignin model molecules: veratryl alcohol and 1-(3,4-dimethoxyphenyl)-2-(2-methoxyphenoxy)propane-1,3-diol. We found that iron(III)tetrakis(4-sulfonatophenyl)porphyrin (FeTPPS) in combination with potassium monopersulfate in a mixture acetonitrile/pH 3 buffer is an efficient catalyst for veratryl alcohol oxidation to veratraldehyde and 2-hydroxymethyl-5-methoxy-1,4-benzoquinone, the same two oxidation products generated by ligninase [59] (see Figure 7).

The dimer with one arylglycerol-β-arylether linkage was also cleaved at the C_α-C_β bond to generate veratraldehyde and 3-methoxy-1,4-benzoquinone. Catalytic activities ranged from 1 to 5 cycles/sec [58,59]. The cor-

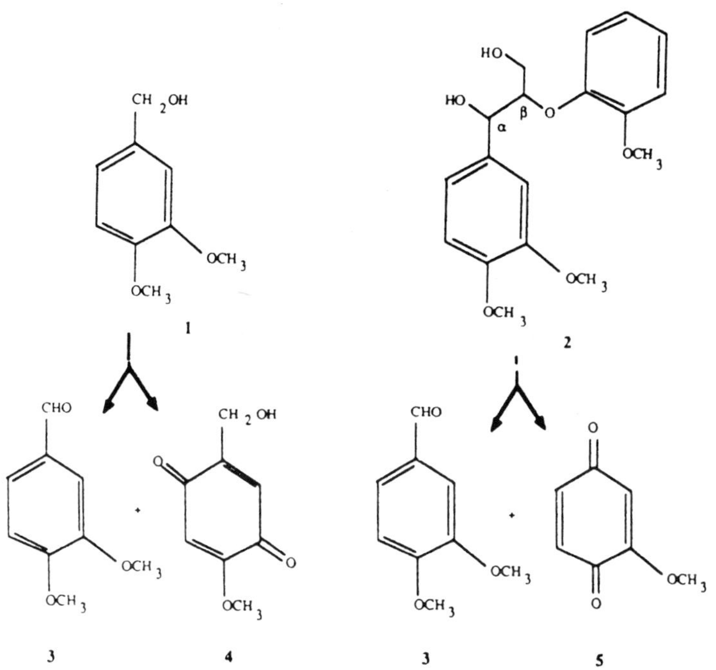

Figure 7 Oxidative degradation of lignin model compounds by the "$KHSO_5$/sulfonated metalloporphyrin" system. (Reproduced with permission from Ref. 59. Copyright 1989, American Chemical Society.)

responding manganese complex, MnTPPS, is less active but the optimum pH of the catalytic activity is around 6 [59].

Robust sulfonated iron and manganese porphyrins have also been used in ligninase modeling, e.g., M-TMPS [61],* M-TDCPPS [62], and M-Br_8TMPS [61] (see Figure 8 for catalyst structures of the first two catalysts). These sterically hindered water-soluble complexes are the most promising catalysts. The degradation products of lignin models are similar to those obtained in ligninase-mediated oxidations. These iron catalysts are able to decolorize Poly B-411, a blue dye used to monitor the activity of ligninase [62,63].

*TMPS, Br_8TMPS, and TDCPPS stand for the dianions of *meso*-tetrakis(3,5-disulphonatomesityl)porphyrin, *meso*-tetrakis(3,5-disulfonatomesityl)-β-octabromoporphyrin, and *meso*-tetrakis(3-sulfonato-2,6-dichlorophenyl)porphyrin, respectively.

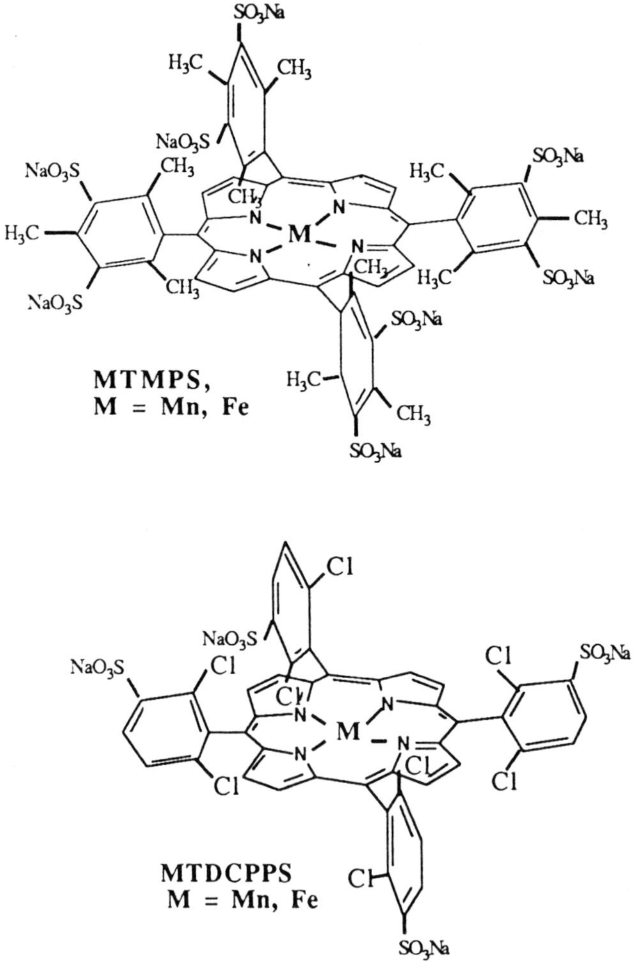

MTMPS,
M = Mn, Fe

MTDCPPS
M = Mn, Fe

Figure 8 Structures of robust sulfonated metalloporphyrins MTMPS and MTDCPPS.

B. Supported Metalloporphyrins

These sulfonated metalloporphyrins are also efficient catalysts in ligninase modeling when supported on ion exchange resins [58,59]. FeTPPS or MnTPPS strongly interacts with the nitrate form of the cationic Amberlite IRA 900, an ion exchange resin derived from polyvinylbenzene and bearing ammonium residues, to give solid catalysts that are more stable than the corresponding soluble catalysts [58,59].

The catalytic activity of MnTPPS-Amberlite used in a second run was still 95% of that measured in the first run [58]. The main problem with supported sulfonated manganese porphyrins is the necessary presence of pyridine in the reaction mixture in order to increase the catalyst activity (for the role of pyridine and imidazole derivatives to boost the catalytic activity of metalloporphyrins by proximal effect, see the review article indicated in Ref. 60a). With a polyvinylpyridine (PVP) support, in contrast, it is possible to take advantage of the proximal effect of a pyridine ligand while avoiding the presence of free pyridine in the reaction mixture. Such supported manganese catalysts are easily prepared from the commercially available poly(4-vinylpyridinium) crosslinked with divinylbenzene [64] (see Figure 9 for a schematic representation of supported manganese porphyrin on PVP). Reticulated poly(4-vinylpyridinium) is first washed with NaOH to yield free pyridine units. Then sulfonated manganese porphyrin is adsorbed on PVP via the coordination of pyridine as axial ligand. This coordination of pyridine

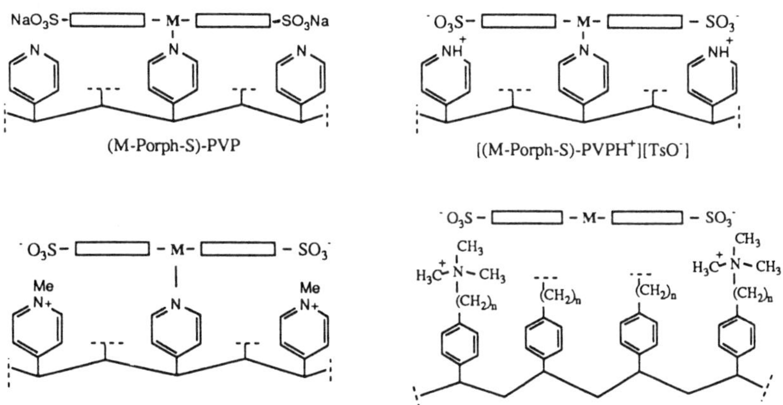

Figure 9 Schematic representation of immobilized sulfonated manganese porphyrins on polyvinylpyridinium polymers with a pyridine of the support acting as proximal ligand.

arising from PVP to the metalloporphyrin can be monitored by UV-visible spectroscopy [65]. The only axial interaction is not sufficient to avoid leaching of the MnTMPS from the PVp during catalytic oxidations. But an additional treatment of MnTMPS-PVP by an acid or a methylating agent regenerates the pyridinium units and reinforces the catalyst–polymer interactions by additional strong electrostatic interactions between peripheral sulfonato groups of the porphyrin ligand and pyridinium units of the cationized PVP. Ninety percent of the initial run activity was found when MnTMPS-protonated PVP was recycled for the third time in the $KHSO_5$ oxidation of 1-(3,4-dimethoxyphenyl)-2-(2-methoxyphenoxyl)propane-1,3-diol [64]. Such supported metalloporphyrins on poly(vinylpyridinium) polymers can also be used as catalysts in olefin epoxidation or alkane hydroxylation [65]. In addition, MnTMPS-protonated PVP was found to be a good biomimetic model of chloroperoxidase, e.g., it is able to catalyze the oxidative chlorination of dimedone [66].

C. Oxidation of Pollutants by Ligninase Models

Sulfonated iron or manganese porphyrin catalysts in combination with $KHSO_5$ or H_2O_2 are highly efficient reagents for the oxidative dechlorination of 2,4,6-trichlorophenol to 2,6-dichlorobenzoquinone (Table 1). Catalytic activities up to 20 cycles/sec have been observed [67]. In addition, it should be noted

Table 1 Catalytic Oxidation of 2,4,6-Trichlorophenol to the Corresponding *p*-Benzoquinone

Run	Catalyst	% Catalyst per substrate	pH	Oxidant	Yield of 2 [%] within: 1 min	15 min
1	FeTPPS	0.3	3.0	$KHSO_5$	95[a]	—
2	FeTPPS	0.3	3.0	H_2O_2	11	68
3	FeTPPS	0.1	3.0	$KHSO_5$	90	92
4	FeTMPS	0.3	3.0	$KHSO_5$	95	—
5	FeTMPS	0.3	6.0	$KHSO_5$	25[b]	10[b]
6	FeTDCPPS	0.3	3.0	$KHSO_5$	92	—
7	FeTPPS-Ad	10	3.0	$KHSO_5$	11	30
8	FeTMPS-Ad	10	3.0	$KHSO_5$	15	36
9	MnTPPS	10	6.0	$KHSO_5$	25[b]	—

[a]89% of 2 in 15 s: catalytic activity = 20 cycles per second.
[b]2 polymerizes at pH 6.0.
Source: Ref. 67. Copyright 1990 Angewandte Chemie.

that exchanging potassium monopersulfate for hydrogen peroxide leads to a small decrease in the catalytic activity. These biomimetic catalysts are also able to oxidize DDT and lindane indicating that high-valent oxometal species generated during the catalytic cycle exhibit a very high oxidation capacity. A classical peroxidase such as HRP is unable to catalyze the oxidation of these pollutants. By using various methoxylated aromatics it has been possible to establish a comparative scale for catalytic oxidations with HRP, LiP, and FeTPPS (Table 2) [67,68]. While HRP is unable to catalyze the oxidation of organic substrates having a $E_{1/2}$ potential value above 1.2 V (Ag/AgCl), LiP is active up to 1.4 V and "FeTPPS/$KHSO_5$" up to 1.7 V. These peroxidase models based on sulfonated metalloporphyrins are also efficient catalysts in pollutant oxidations when immobilized onto ion exchange resins. They may have some future in the removal of micropollutants in water treatment facilities.

D. Manganese Peroxidase Models

The modeling of a nonspecific peroxidase like ligninase is simple compared to the design of metalloporphyrin catalysts able to selectively oxidize manganese(II) salts, in the presence of chelating agents and of easily oxidized phenolic substrates. MnP has a manganese binding site, on the edge of the prosthetic group, that controls the selective oxidation of manganese(II) salts in the presence of phenols. How can we reproduce the same selective oxidation with a synthetic metalloporphyrin without building around the catalytic site an appropriate cavity able to selectively accommodate Mn(II)? Modeling of MnP is obviously more challenging than that of LiP.

Cui and Dolphin reported in 1990 the role of manganese sulfate in the decolorization of Poly B-411 dye catalyzed by a sterically hindered sulfonated

Table 2 The FeTPPS/$KHSO_5$ System Has a Higher Redox Activity Compared to Peroxidases LiP or HRP in the Catalytic Oxidation of Methoxybenzene Derivatives

			Oxidation by		
Run	Substrate	$E_{1/2}$[a]	FeTPPS/$KHSO_5$[b]	LiP	HRP
1	Anisole	1.69	+	–	–
2	1,3,5-Trimethoxybenzene	1.45	+	–	–
3	1,2-Dimethoxybenzene	1.36	+	+	–
4	1,4-Dimethoxybenzene	1.25	+	+	–
5	Guaiacol	1.23	+	+	+

[a]$E_{1/2}$ in volts (Ag/AgCl).
[b]0.3% catalyst/substrate citrate-phosphate buffer at pH = 3.0
(+) Oxidation; (–) no activity.

iron porphyrin, FeTDCPPS, in the presence of *m*-chloroperbenzoic acid [69]. In the absence of $MnSO_4$, dye decolorization stopped after 1 min because of the bleaching of the iron catalyst. Addition of $MnSO_4$ improved the catalytic reaction over a period of 7 min. The catalytic oxidation of Mn(II) chelates, i.e., the key step performed by MnP, has been recently mimicked using synthetic metalloporphyrins [70]. Oxidation of Mn(II)pyrophosphate to the corresponding Mn(III) chelate was performed with catalytic amounts (1–2% based on Mn^{II}) of iron- or manganese-sulfonated porphyrins. Sterically hindered iron complexes were the most active catalysts. We observed catalytic activities of 8 and 3 cycles/min for FeTDCPPS and FeTMPS, respectively.

The catalytic activities of these biomimetic catalysts are well below the activity of MnP itself (8200 cycles/min according to Ref. 71), but this difference of activity between the enzyme and the models is probably related to the nature of the interactions between the substrate and the catalytic system. In the case of MnP, Mn(II) has an affinity for the edge of the prosthetic site of the enzyme. This is not the case for sulfonated metalloporphyrins. With these catalysts having several negatively charged sulfonato groups at the periphery of the ligand, the substrate $[Mn(II)(H_2P_2O_7)_2]^{2-}$, which is also negatively charged, has, in contrast to the enzyme situation, a tendency to be repelled from the catalytic site. A better situation will be obtained with sterically hindered cationic metalloporphyrins; however, such complexes are not yet available. A striking feature of these metalloporphyrin-catalyzed oxidations of manganese pyrophosphate is the improvement of the catalytic activity by the presence of limited amounts of methoxylated benzene derivatives acting as cosubstrates: 1,2-dimethoxybenzene, 3,4-dimethoxybenzyl alcohol (veratryl alcohol), or 1,2,4-trimethoxybenzene (see above for data on oxidations of methoxylated benzenes catalyzed by ligninase or by a FeTPPS; see also Ref. 68). The catalytic activity of FeTMPS is substantially increased by 2–5% of one of these methoxylated benzenes (but not by anisole). This enhancement is dependent on the amount of cosubstrate. The influence of 2%, 3%, and 13% of 1,2,4-trimethoxybenzene with respect to Mn pyrophosphate on the oxidation of manganese pyrophosphate is shown in Figure 10.

In the presence of both FeTMPS and a di- or trimethoxylated benzene compound, oxidation yields of Mn pyrophosphate are above 50–80%. Two reasons for these improved oxidations of Mn pyrophosphate can be envisaged: (a) an inhibition of the bleaching of the metalloporphyrin catalyst by the presence of an easily oxidized substrate in the reaction mixture (80–90% of the catalyst is still present after 10 min with 3% of one of these di- or trimethoxylated benzene derivatives instead of 20% of remaining catalyst in their absence) and, more important, (b) a role of redox mediator for these methoxylated benzene derivatives, as reported in the oxidation of Mn salts

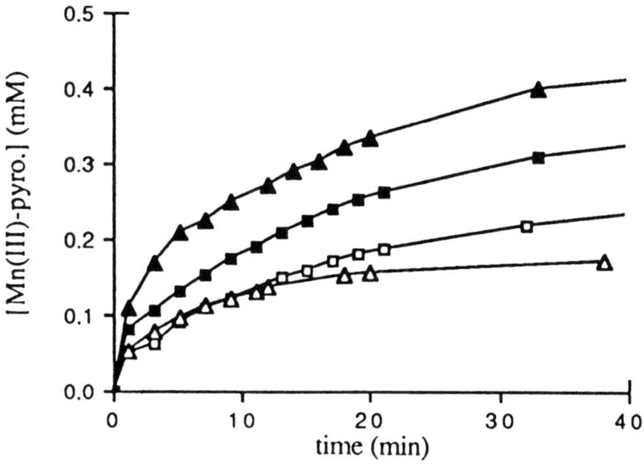

Figure 10 Enhancement of the FeTMPS-catalyzed Mn^{II} pyrophosphate oxidation by limited amounts of 1,2,4-trimethoxybenzene. Open triangles: without; open squares: 2%; black squares: 3%; black triangles: 13%. Molar ratios of trimethoxybenzene are based on manganese salt.

by lignin peroxidase. (Fifty percent of the starting organic mediator is still present in the reaction medium as evidenced by high-performance liquid chromatography. In the absence of manganese pyrophosphate, the organic mediator is completely oxidized within a few minutes.) The generation of the radical cation form of veratrole, veratryl alcohol, or 1,2,4-trimethoxybenzene by the high-valent oxometalloporphyrin complex might behave as a strong oxidant with respect to Mn(II) pyrophosphate and enhance its oxidation. These results with a biomimetic model system further illustrate the key role of low molecular weight organic mediators in the oxidative degradation of lignin by fungus peroxidases.

V. CONCLUDING REMARKS

Iron and manganese complexes having a water-soluble sterically hindered porphyrin ligand in combination with potassium monopersulfate, an efficient oxygen atom donor to metalloporphyrins that is soluble in water at pH 7, are efficient peroxidase models. Low molecular weight molecules used as lignin models are oxidized by these biomimetic catalysts that behave as efficient ligninase models. These metalloporphyrin catalysts can be supported on regular ion exchange resins and even better on polyvinylpyridinium polymers. In the latter case enhancement of the catalytic activity by the proximal effect results from the coordination of pyridine units of the polymer itself to the supported metalloporphyrins. They are also efficient catalysts for the oxidation of pollutants and might have a future in the oxidation of micropollutants in water treatment. These porphyrin catalysts are able to oxidize manganese(II) chelates, the first step in the oxidation of phenolic residues of lignin by manganese peroxidase. However, such an oxidation route is disfavored when performed in the presence of phenolic substrates. The modeling of the complete catalytic cycle of manganese peroxidase still remains an important challenge. Finally, we note that in order to design model porphyrin systems capable of substrate selection the question of emulating the effect of the protein ligand will also have to be addressed.

ACKNOWLEDGMENTS

The author is particularly indebted to the work of two coworkers: Gilles Labat and Sabine Defrance. Among the colleagues, Jean-Louis Séris (Centre de Recherche de Lacq, Elf-Aquitaine) is gratefully acknowledged for all the fruitful discussions on ligninase and manganese peroxidase over a period of 6 years. Research work of my group in this field was supported by Elf-Aquitaine and CNRS.

REFERENCES

1. T.K. Kirk, T. Higuchi, and H.M. Cheng (eds.), *Lignin Biodegradation: Microbiology, Chemistry, and Potential Applications*, Vols. 1–2, CRC Press, Boca Raton, 1980.
2. W.A. Wood and S.T. Kellog (eds.), *Biomass, Part B, Lignin, Pectin, and Chitin* (Methods in Enzymology, Vol. 161), Academic Press, New York, 1988.
3. (a) G. Labat and B. Meunier, Etudes de la dégradation enzymatique de la lignine: évolutions récentes, *Bull. Soc. Chim. Fr., 127*: 553 (1990). (b) H.E. Schoemaker, On the chemistry of lignin biodegradation, *Rec. Trav. Chim. Pays-Bas, 109*: 255 (1990).
4. V.B. Huynh, H.M. Chang, T.W. Joyce, and T.K. Kirk, Dechlorination of chloro-organics by a white-rot fungus, *Tappi J., 68*: 98 (1985).
5. H.M. Chang, T.W. Joyce, A.G. Campbell, E.D. Gerrard, V-B. Huynh, and T.K. Kirk, Fungal decolorization of bleach plant effluents, in *Recent Advances in Lignin Biodegradation Research* (T. Higuchi, H.M. Chang, and T.K. Kirk, eds.), UNI, Tokyo, 1983, p. 257.
6. (a) M. Tien and T.K. Kirk, Lignin-degrading enzyme from the hymenomycete *Phanerochaete chrysosporium* Burds, *Science, 221*: 661 (1983). (b) J.K. Glenn, M.A. Morgan, M.B. Mayfield, M. Kuwahara, and M.H. Gold, An extracellular H_2O_2-requiring enzyme preparation involved in lignin biodegradation by the white rot basidiomycete *Phanerochaete chrysosporium*, *Biochem. Biophys. Res. Commun., 114*: 1077 (1983). (c) V. Renganathan and M.H. Gold, Spectral character-

ization of the oxidized states of lignin peroxidase, an extracellular heme enzyme from the white rot basidiomycete *Phanerochaete chrysosporium, Biochemistry, 25*: 1626 (1986). (d) H.W.H. Schmidt, S.D. Haemmerli, H.E. Schoemaker, and M.S.A. Leisola, Oxidative degradation of 3,4-dimethoxybenzyl alcohol and its methyl ether by the lignin peroxidase of *Phanerochaete chrysosporium, Biochemistry, 28*: 1776 (1989).

7. (a) Y. Mino, H. Wariishi, N.J. Blackburn, T.M. Loehr, and M.H. Gold, Spectral characterization of manganese peroxidase, an extracellular heme enzyme from the lignin-degrading basidiomycete, *Phanerochaete chrysosporium, J. Biol. Chem., 263*: 7029 (1988). (b) M.D. Aitken and R.L. Irvine, Characterization of reactions catalyzed by manganese peroxidase from *Phanerochaete chrysosporium, Arch. Biochem. Biophys., 276*: 405 (1990). (c) R.Z. Harris, H. Wariishi, M.H. Gold, and P.R. Ortiz de Montellano, The catalytic site of manganese peroxidase, *J. Biol. Chem., 266*: 8751 (1991). (d) L. Banci, I. Bertini, A.E. Pease, M. Tien, and P. Turano, ^{1}H NMR investigation of manganese peroxidase from *Phanerochaete chrysosporium*. A comparison with other peroxidases, *Biochemistry, 31*: 10009 (1992).
8. H. Nimz, Beech lignin-proposal of a constitutional scheme, *Angew. Chem. Int. Ed. Engl., 13*: 313 (1974).
9. C.L. Chen and D. Robert, Characterization of lignin by ^{1}H and ^{13}C NMR spectroscopy, in Ref. 2, p. 137.
10. T.K. Kirk and R.L. Farrell, Enzymatic "combustion": the microbial degradation of lignin, *Annu. Rev. Microbiol., 41*: 465 (1987).
11. H. Janshekar and A. Fiechter, Lignin: biosynthesis, application, and biodegradation, *Adv. Biochem. Eng. Biotechnol., 27*: 119 (1983).
12. K. Lundquist, T.K. Kirk, and W.J. Connors, Fungal degradation of kraft lignin and lignin sulfonates prepared from synthetic ^{14}C-lignin, *Arch. Microbiol., 112*: 291 (1977).
13. T.K. Kirk, E. Schultz, W.J. Connors, L.F. Lorentz, and J.G. Zeikus, Influence of the culture parameters on lignin metabolism by *Phanerochaete chrysosporium, Arch. Microbiol., 117*: 277 (1978).
14. M. Tien and T.K. Kirk, Lignin peroxidase of *Phanerochaete chrysosporium*, in Ref. 2, p. 238.
15. A.I. Hattaka and A.K. Uusi-Rauva, Degradation of ^{14}C-labelled poplar wood lignin by selected white rot fungi, *Eur. J. Appl. Microbiol. Biotechnol., 18*: 350 (1983).
16. J.G. Zeikus, Lignin metabolism and the carbon cycle. Polymer biosynthesis, biodegradation and environmental recalcitrance, in *Advances in Microbial Ecology*, Vol. 6 (K.C. Marshall, ed.), 1981, p. 277.
17. M.S.A. Leisola and A. Fiechter, New trends in lignin biodegradation, *Adv. Biotechnol. Proc., 5*: 59 (1985).
18. (a) G. Meunier and B. Meunier, Evidences: for an efficient demethylation of methoxyellipticine derivatives catalyzed by a peroxidase, *J. Am. Chem. Soc., 107*: 2558 (1985). (b) G. Meunier and B. Meunier, Peroxidase-catalyzed *O*-demethylation reactions. Quinone-imine formation from 9-methoxyellipticine derivatives, *J. Biol. Chem., 260*: 10576 (1985).

19. (a) H. Wariishi and M.H. Gold, Lignin peroxidase compound III. Formation, inactivation, and conversion to the native enzyme, *FEBS Lett., 243*: 165 (1989). (b) H. Wariishi and M.H. Gold, Lignin peroxidase Compound III, *J. Biol. Chem., 265*: 2070 (1990).
20. M.H. Gold, M. Kuwahara, A.A. Chui, and J.K. Glenn, Purification and characterization of an extracellular H_2O_2-requiring diarylpropane oxygenase from the white rot basidiomycete, *Phanerochaete chrysosporium, Arch. Biochem. Biophys., 234*: 353 (1984).
21. P.J. Kersten, M. Tien, B. Kalyanaraman, and T.K. Kirk, The ligninase of *Phanerochaete chrysosporium* generates cation radicals from methoxybenzenes, *J. Biol. Chem., 260*: 2609 (1985).
22. H.E. Schoemaker, P.J. Harvey, R.M. Bowen, and J.M. Palmer, On the mechanism of enzymatic lignin breakdown, *FEBS Lett., 183*: 7 (1985).
23. A.L. Andersson, V. Renganathan, A.A. Chui, T.M. Loehr, and M.H. Gold, Spectral characterization of diarylpropane oxygenase, a novel peroxide-dependent, lignin-degrading heme enzyme, *J. Biol. Chem., 260*: 6080 (1985).
24. D. Kuila, M. Tien, J.A. Fee, and M.R. Ondrias, Resonance raman spectra of extracellular ligninase: evidence for a heme active site similar to those of peroxidases, *Biochemistry, 24*: 3394 (1985).
25. S.L. Edwards, R. Raag, H. Wariishi, M.H. Gold, and T.L. Poulos, Crystal structure of lignin peroxidase, *Proc. Natl. Acad. Sci. USA, 90*: 750 (1993).
26. G.D. De Pillis and P.R. Ortiz de Montellano, Substrate oxidation by the heme edge of fungal peroxidases. Reaction of coprinus macrorhizus peroxidase with hydrazines and sodium azide, *Biochemistry, 28*: 7947 (1989).
27. (a) M. Tien and C.P.D. Tu, Cloning and sequencing of a cDNA for a ligninase from *Phanerochaete chrysosporium, Nature, 326*: 520 (1987). (b) T.L. Smith, H. Schalch, J. Gaskell, S. Covert, and D. Cullen, Nucleotide sequence of a ligninase gene from *Phanerochaete chrysosporium, Nucleic Acids Res., 16*: 1219 (1988).
28. L.A. Andersson, V. Renganathan, T.M. Loehr, and M.H. Gold, Lignin peroxidase: resonance raman spectral evidence for compound II and for a temperature-dependent coordination-state equilibrium in the ferric enzyme, *Biochemistry, 26*: 2258 (1987).
29. M. Mylrajan, K. Valli, H. Wariishi, M.H. Gold, and T.M. Loehr, Resonance raman spectroscopic characterization of compound III of lignin peroxidase, *Biochemistry, 29*: 9617 (1990).
30. B. Meunier, Characterization of compound III of lignin peroxidase by resonance raman spectroscopy, *Chemtracts Inorg. Chem., 3*: 199 (1991).
31. P.J. Harvey, H.E. Schoemaker, and J.M. Palmer, Veratryl alcohol as a mediator and the role of radical cations in lignin biodegradation by *Phanerochaete chrysosporium, FEBS Lett., 195*: 242 (1986).
32. S.D. Haemmerli, M.S. A. Leisola, D. Sanglard, and A. Fiechter, Oxidation of benzo(a)pyrene by extracellular ligninases of *Phanerochaete chrysosporium, J. Biol. Chem., 261*: 6900 (1986).
33. K. Miki, V. Renganathan, and M.H. Gold, Mechanism of β-aryl ether dimeric lignin model compound oxidation by lignin peroxidase of *Phanerochaete chrysosporium, Biochemistry, 25*: 4790 (1986).

34. K.E. Hammel, B. Kalyanaranan, and T.K. Kirk, Substrate free radicals are intermediates in ligninase catalysis, *Proc. Natl. Acad. Sci. USA, 83*: 3708 (1986).
35. K.E. Hammel, M. Tien, B. Kalyanaranan, and T.K. Kirk, Mechanism of oxidative C^{α}-C^{β} cleavage of a lignin model dimer by *Phanerochaete chrysosporium* ligninase, *J. Biol. Chem., 260*: 8348 (1985).
36. T.K. Kirk, M. Tien, J.A. Fee, and M.R. Ondias, Ligninase of *Phanerochaete chrysosporium*: mechanism of the degradation of the non-phenolic arylglycerol β-aryl ether substructure of lignin, *Biochem. J., 236*: 279 (1986).
37. T. Higuchi, Biodegradation of lignin, *Wood Res., 73*: 58 (1983).
38. K. Miki, R. Kondo, V. Renganathan, M.B. Mayfield, and M.H. Gold, Mechanism of aromatic ring cleavage of a β-biphenylyl ether dimer catalyzed by lignin peroxidase of *Phanerochaete chrysosporium, Biochemistry, 27*: 4787 (1988).
39. M. Kuwahara, J.K. Glenn, M.A. Morgan, and M.H. Gold, Separation and characterization of two extracellular H_2O_2-dependent oxidases from ligninolytic cultures of *Phanerochaete chrysosporium, FEBS Lett., 169*: 247 (1984).
40. D. Pribnow, M.B. Mayfield, V.J. Nipper, J.A. Brown, and M.H. Gold, Characterization of a cDNA encoding a manganese peroxidase, from the lignin-degrading basidiomycete *Phanerochaete chrysosporium, J. Biol. Chem., 264*: 5036 (1989).
41. E.A. Pease, A. Andrawis, and M. Tien, Manganese-dependent peroxidase from *Phanerochaete chrysosporium, J. Biol. Chem., 264*: 13531 (1989).
42. H. Wariishi, H.B. Dunford, I.D. MacDonald, and M.H. Gold, Manganese peroxidase from the lignin-degrading basidiomycete *Phanerochaete chrysosporium, J. Biol. Chem., 264*: 3335 (1989).
43. K.V. Sarkanen and C.H. Ludwig (eds.), *Lignins*: *Occurrence, Formation, Structure and Reactions*, Wiley, New York, 1971.
44. (a) H. Wariishi, K. Valli, and M.H. Gold, Oxidative cleavage of a phenolic diarylpropane lignin model dimer by manganese peroxidase from *Phanerochaete chrysosporium, Biochemistry, 28*: 6107 (1989). (b) U. Tuor, H. Wariishi, H.E. Schoemaker, and M.H. Gold, Oxidation of phenolic arylglycerol β-aryl ether lignin model compounds by manganese peroxidase from *Phanerochaete chrysosporium*: oxidative cleavage of an α-carbonyl model compound, *Biochemistry, 31*: 4986 (1992).
45. H. Wariishi, K. Valli, and M.H. Gold, In vitro depolymerization of lignin by manganese peroxidase of *Phanerochaete chrysosporium, Biochem. Biophys. Res. Commun., 176*: 269 (1991).
46. (a) M. Alexander, Biodegradation of chemicals of environmental concern, *Science, 211*: 132 (1981). (b) A.H. Neilson, The biodegradation of halogenated organic compounds, *J. Appl. Bacteriol., 69*: 445 (1990). (c) R.A. Hites, Environmental behavior of chlorinated dioxins and furans, *Acc. Chem. Res., 23*: 194 (1990).
47. (a) DDT and lindane are 1,1,1-trichloro-2,2-bis(*p*-chlorophenyl)ethane and $1\alpha,2\alpha,3\beta,4\alpha,5\alpha,6\beta$-hexachlorocyclohexane, respectively.
48. (a) R. Kallenborn, H. Hühnerfuss, and W.A. König, Enantioselective metabolism of ($\pm$)-α-1,2,3,4,5,6-hexachlorocyclohexane in organs of the eider duck, *Angew. Chem. Int. Ed. Engl., 30*: 320 (1991). (b) C.J. Gantzer and L.P. Wack-

ett, Reductive dechlorination catalyzed by bacterial transition-metal coenzymes, *Environ. Sci. Technol., 25*: 715 (1991).

49. J.A. Bumpus, M. Tien, D. Wright, and S.D. Aust, Oxidation of persistent environmental pollutants by a white rot fungus, *Science, 228*: 1434 (1985).
50. K.E. Hammel and P.J. Tardone, The oxidative 4-dechlorination of polychlorinated phenols is catalyzed by extracellular fungal lignin peroxidases, *Biochemistry, 27*: 6563 (1988).
51. S.D. Haemmerli, M.S.A. Leisola, D. Sanglard, and A. Fiechter, Oxidation of benzo(a)pyrene by extracellular ligninases of *Phanerochaete chrysosporium*, *J. Biol. Chem., 261*: 6900 (1986).
52. J.A. Bumpus and S.D. Aust, Biodegradation of environmental pollutants by the white rot fungus *Phanerochaete chrysosporium*: involvement of the lignin degrading system, *BioEssays, 6*: 166 (1987).
53. T. Habe, M. Shimida, T. Okamoto, B. Panijpan, and T. Higuchi, Incorporation of dioxygen into the hydroxylated product during the C-C single bond cleavage of 1,2-bis(*p*-methoxyphenyl)propane-1,3-diol catalysed by hemin. A novel model system for the hemoprotein ligninase, *J. Chem. Soc., Chem. Commun.*, 1323 (1985).
54. A. Paszczynski, R.L. Crawford, and R.A. Blanchette, Delignification of wood chips and pulps by using natural and synthetic porphyrins: models of fungal decay, *Appl. Environ. Microbiol., 54*: 62 (1988).
55. T. Okamoto, K. Sasaki, and S. Oka, Biomimetic oxidation with molecular oxygen. Selective carbon-carbon bond cleavage of 1,2-diols by molecular oxygen and dihydropyridine in the presence of iron-porphyrin catalysts, *J. Am. Chem. Soc., 110*: 1187 (1988).
56. D. Dolphin, T. Nakano, T.E. Maione, T.K. Kirk, and R. Farrell, in *Lignin Enzymic and Microbiological Degradation* (INRA, Ed.), Paris, 40: 157 (1987). *Chem. Abstr., 109*: 69402u (1988).
57. B. Meunier, G. Labat, and J.L. Seris, French Patent 88 09 169 (1988). Oxidation of benzyl alcohols and analogs using biomimetic metalloporphyrin catalysts, *Chem. Abst., 113*: 152001r (1990).
58. G. Labat and B. Meunier, Efficient ligninase models based on water-soluble manganese- and iron-porphyrin complexes associated to potassium monopersulfate, *New J. Chem., 13*: 801 (1989).
59. Same authors, Factors controlling the reactivity of a ligninase model based on the association of potassium monopersulfate to manganese and iron porphyrin complexes, *J. Org. Chem., 54*: 5008 (1989).
60. (a) B. Meunier, Metalloporphyrins as versatile catalysts for oxidation reactions and oxidative DNA cleavage, *Chem. Rev., 92*: 1411 (1992). (b) B. Meunier, E. Guilmet, M.E. De Carvalho, and R. Poilblanc, Sodium hypochlorite: a convenient oxygen atom source for olefin epoxidation catalyzed by porphyrinatomanganese complexes, *J. Am. Chem. Soc., 106*: 6668 (1984).
61. P. Hoffmann, G. Labat, A. Robert, and B. Meunier, Highly selective bromination of tetramesitylporphyrin: an easy access to robust metalloporphyrins, M-TBr_8TMP and M-TBr_8TMPS, *Tetrahedron Lett., 31*: 1991 (1990).

62. F. Cui and D. Dolphin, Biomimetic studies in lignin degradation, *ACS Symp. Ser., 399*: 519 (1989).
63. J.K. Glenn and M.H. Gold, Decolorization of several dyes by the lignin degradating basidiomycete *Phanerochaete chrysosporium, Appl. Environ. Microbiol., 45*: 1741 (1983).
64. G. Labat and B. Meunier, Facile preparation of supported metalloporphyrin catalysts with a proximal ligand arising from the support istself, *C. R. Acad. Sci. Paris, Part. II, 3*(11): 625 (1990).
65. S. Campestrini and B. Meunier, Olefin epoxidation and alkane hydroxylation catalyzed by robust sulfonated manganese and iron porphyrin supported on cationic ion-exchange resins, *Inorg. Chem., 31*: 1999 (1992).
66. G. Labat and B. Meunier, First example of a chloroperoxidase-type chlorination of dimedone using a supported manganese porphyrin catalyst, *J. Chem. Soc., Chem. Commun.,* 1414 (1990).
67. G. Labat, J.L. Seris, and B. Meunier, Comparison of lignin peroxidase, horseradish peroxidase and laccase in the oxidation of methoxybenzenes, *Angew. Chem. Int. Ed. Engl., 29*: 1471 (1990).
68. P.J. Kersten, B. Kalyanaraman, K.E. Hammel, B. Reinhammar, and T.K. Kirk, Comparison of lignin peroxidase, horseradish peroxidase and laccase in the oxidation of methoxybenzenes, *Biochem. J., 268*: 475 (1990).
69. F. Cui and D. Dolphin, Veratryl alcohol as a mediator in lignin model compound biodegradation, *Holzforschung, 45*: 31 (1991).
70. S. Defrance, J.L. Séris, and B. Meunier, Metalloporphyrin-catalyzed oxidation of manganese pyrophosphate. Methoxylated benzene derivatives as mediators in a chemical model of manganese peroxidase, *New J. Chem., 16*: 1015 (1992).
71. J.K. Glenn, L. Akileswaran, and M.H. Gold, Mn(II) oxidation is the principal function of the extracellular Mn-peroxidase from *Phanerochaete chrysosporium, Arch. Biochem. Biophys., 251*: 688 (1986).

6
Heme Proteins in Prostaglandin Biosynthesis

V. Ullrich
University of Konstanz, Konstanz, Germany

Hans H. Ruf
Fraunhofer Institute, IBMT; and University of Saarland, St. Ingbert, Germany

I. INTRODUCTION

During evolution the photosynthetic activity in plants generated molecular oxygen with its unique potential for aerobic metabolism and energy conservation but also with its high toxic potential through the pronounced tendency for free radical chain reactions. It was already outlined in the previous chapters as to how nature succeeded in handling dioxygen by the use of transition metal complexes from which the iron porphyrins play a pivotal role. The tetradentate porphyrin ligand separates two coordination sites of the hexacoordinated iron leaving no possibility for ligand exchange reactions and thus providing a unique means for steroeselective catalysis at the sixth free ligand position of the iron–porphyrin complex. The remaining fifth ligand position offers a handle to bind to a polypeptide chain through appropriate amino acid side chains and together with additional electrostatic forces the heme prosthetic group can be tightly fixed to the active site.

In the context of the present chapter it is essential to point out that the resulting hemoproteins when functioning only as an electron shuttle between the ferrous and ferric state are called *cytochromes*.

When used as oxidation catalysts higher oxidation states of the porphyrin iron must be generated since two negatively charged ligands from the porphyrin lower the redox potential of the Fe^{II}/Fe^{III} couple and lend very limited oxidizing capacities to the ferric forms. Without exceptions the required

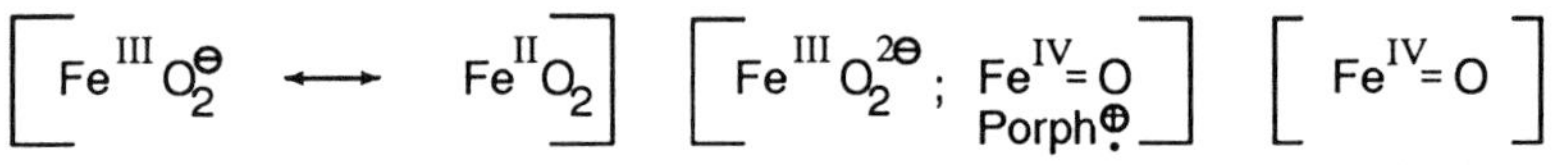

COMPOUND III COMPOUND I COMPOUND II

Scheme 1

oxidizing equivalents are provided by dioxygen through its three one-electron reduced states; the superoxide anion, hydrogen peroxide and the hydroxyl radical forming the perferryl species (compound III), the peroxo or oxo species (equivalent to compound I), and the ferryl species (compound II), respectively (Scheme 1).

All species have oxidizing properties with different redox potentials but generally no oxygen transfer reactions are observed. Such reactions are limited to the above-described group of cytochrome P450 enzymes that catalyze the transfer of one oxygen atom from O_2 to their dusbstrate with concomitant reduction of the second oxygen atom to water (monooxygenases). As outlined above such reactions are not catalyzed by cytochromes and hence all monooxygenases containing heme have been renamed "heme thiolate" proteins since a thiolate ligand from cysteine is a common structural feature of this group of monooxygenases [1,2].

In the previous chapter D. Mansuy pointed out the biochemistry and chemistry of the heme thiolate monooozygenases and has presented the available evidence for the "oxenoid" structure of the active oxygen intermediate. Although the precise role of the thiolate ligand is not known, it seems to be crucial for the "oxene" transfer properties of these hemoproteins. According to our hypothesis thiolate as an axial ligand interferes with the back bonding of the oxygen atom and thus will favor a resonance structure with a radical-type oxygen atom [3] (Scheme 2). It is irrelevant for the purpose of this chapter whether this hypothesis is indeed correct, since we shall be dealing with another group of heme thiolate enzymes in which a second function of the thiolate ligand will be discussed. In the following we shall describe biosynthetic pathways for arachidonic acid in which this molecule is dioxygenated by prostaglandin H (PGH) synthase (EC 1.14.99.1; for review, see [4]). This

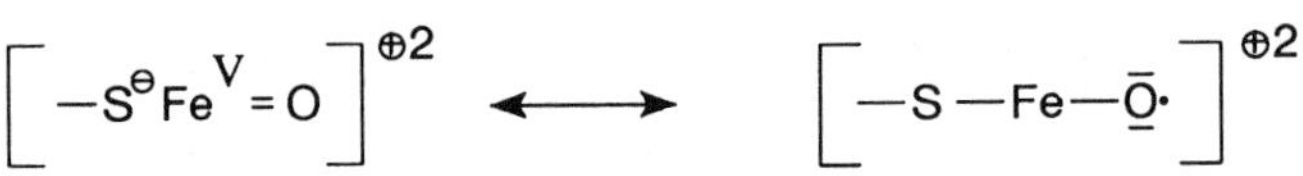

Scheme 2

enzyme belongs to the group of lipoxygenases which abstract hydrogen atoms from the labile allylic positions of unsaturated fatty acids. The dioxygen molecule then adds to the carbon-centered radical and in several steps forms an endoperoxide and a hydroperoxide via substrate activation (arachidonate) and not by dioxygen activation. However, as we shall point out, the H-abstracting species in PGH synthase is not a ferric ion as in the typical lipoxygenases (for review, see [5]) but a higher valence state of the hemoprotein.

Arachidonic acid [5(*Z*),8(*Z*),11(*Z*),14(*Z*)-eicosatetraenoic acid] occurs esterified in the phospholipids of biological membranes and is released by phospholipases as part of signal transduction pathways in the cellular communication network [6]. Having four cis (or Z) double bonds in the 20C skeleton it possesses several allylic positions which can undergo hydrogen atom abstraction. This leads to the formation of the various hydroperoxy derivatives via isomerization and shift of a double bond. The variety of 5-, 8-, 12-, or 15-lipoxygenases form the so-called lipoxygenase pathway.

A second pathway involves the prostaglandin PGH synthase which forms 15(*S*)-hydroxy-9α,11α-epidioxy-5(*Z*),13(*E*)-prostadienoic acid (PGH_2) via reduction of the initially formed 15-hydroperoxy derivative (PGG_2) [4,7]. PGH_2 is the precursor of the four main prostaglandins and of thromboxane A_2, and hence is the key intermediate of the "cyclooxygenase pathway" (Scheme 3). All of the prostaglandins are potent intercellular mediators usually operating within a given organ ("autacoids"). Their participation in inflammation, autoimmune diseases, and shock syndromes, but also in sleep and growth regulation, has made them preferred pharmacological targets. Hence their biosynthetic pathways are of interest as primary points of attack for pharmaceutical drugs. It should be mentioned that a third pathway has been postulated which consists of monooxygenase reactions at the arachidonate molecule leading to epoxides and hydroxy derivatives [8].

In this chapter we concentrate on the mechanism of action of PGH synthase as well as on prostacyclin (PGI) synthase (EC 5.3.99.4) and thromboxane (TxA) synthase (EC 5.3.99.5). All three enzymes are hemoproteins and the latter two enzymes belong to the isomerases but differ from PGE or PGD isomerase by catalyzing an interesting rearrangement of the endoperoxide under formation of oxygen-containing heterocycles.

Undoubtedly, such reactions must involve activated oxygen intermediates since oxygen–carbon bonds are newly formed. This suggested the participation of a transition metal and since hydroperoxides can be cleaved by heme thiolate-dependent monooxygenases we assumed a heme thiolate catalysis for these hitherto unknown synthases [9]. This working hypothesis could be verified in recent years so that today the chemical role of the heme thiolate linkage in the catalytic mechanism has become a most relevant question.

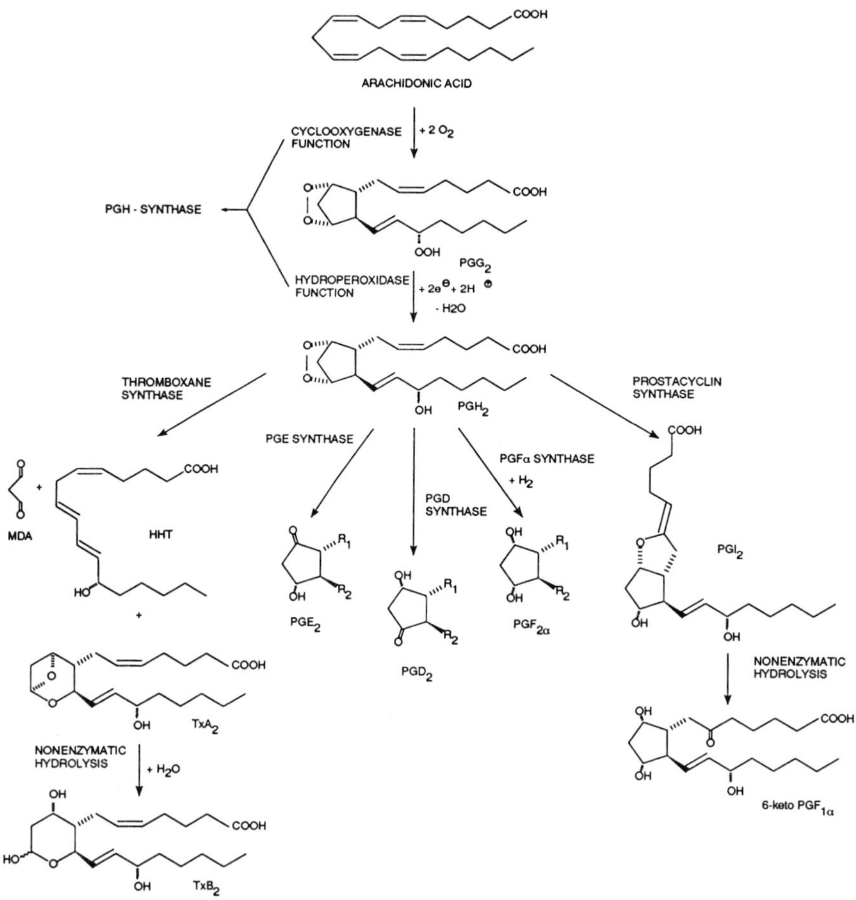

Scheme 3

II. PROSTAGLANDIN ENDOPEROXIDE SYNTHASE

As already mentioned, prostaglandin endoperoxide synthase (prostaglandin H synthase, PGH synthase, cyclooxygenase; EC 1.14.99.1) catalyzes a key reaction in the arachidonic acid cascade, i.e., the oxidation of arachidonic acid to prostaglandin H_2, which is the last common intermediate in the biosynthesis of prostaglandins and thromboxanes. This reaction is considered to be the crucial rate-limiting step in the biosynthesis of prostaglandins and thromboxanes [5].

The formation of PGH_2 is a two-step reaction. In the first step, the cyclooxygenase (COX) reaction, two molecules of O_2 are introduced into the substrate in a bisdioxygenase reaction under formation of the substituted cyclopentane ring which is the typical structural element of the prostanoids (see Scheme 3). The incorporation of molecular oxygen was demonstrated by $^{18}O_2$ experiments [10]. The first O_2 leads to the 9,11-endoperoxide. This introduction of oxygen at C-9 and C-11 provides the important substituents of the cyclopentane ring, which are further modified by the subsequent synthases that specifically produce the individual prostaglandins D, E, F, I, and thromboxanes. The second O_2 incorporated forms the 15-hydroperoxy group of PG_2, which is the first intermediate isolated during the PGH synthase reaction. The second step of the PGH synthase reaction is a hydroperoxidase (POD) reaction which reduces the 15-hydroperoxy group of PG_2 to the 15-hydroxy group of PGH_2. Both steps, cyclooxygenase and peroxidase, are catalyzed by the same enzyme PGH synthase, a heme protein of 72 kDa. In this chapter we focus on the mechanism of the specific oxidation of arachidonic acid by PGH synthase with special emphasis on the role of the heme group. Recent reviews cover various aspects of PGH synthase [4,11].

A. Protein Structure

Prostaglandin biosynthesis and hence PGH synthase are ubiquitous in mammals. Ovine PGH synthase was predominantly isolated from ram seminal vesicles which contain an extraordinarily high activity of PGH synthase [12–18]. The isolated enzyme has an apparent molecular mass of 72 kDa in sodium dodecyl sulfate-polyacrylamide gel electrophoresis (SDS-PAGE) and a specific cyclooxygenase activity on the order of 50–100 μmol O_2 min^{-1} mg^{-1} protein at 30°C. Specific peroxidase activity is higher by about an order of magnitude. Therefore cyclooxygenase is the rate-limiting step under steady-state conditions.

During the isolation, heme can dissociate from the enzyme, and PGH synthase is predominantly isolated as the apoprotein, the activity of which is spontaneously reconstituted upon the addition of ferric heme. There were controversial reports on the stoichiometry between heme and apoprotein with values between 0.5 and 2 [16–20]. A binding stoichiometry of 1 heme/polypeptide of 72 kDa could be clearly shown by electron paramagnetic resonance (EPR) since the EPR spectra of specifically bound (= catalytic) heme and nonspecifically bound (= inactive) heme are resolved [20]. Like other solubilized membrane proteins, isolated PGH synthase tends to form oligomers, predominantly homodimers in detergent solutions, and full cyclooxygenase activity was reported for a dimer with only one heme [19], i.e., with a stoichiometry of 0.5 heme/polypeptide. However, a radiation inactivation study showed that PGH synthase is inactivated with a target size which corresponds to the monomer of 72 kDa both in microsomes and in the purified enzyme [21]. This is strong evidence that the monomer of 72 kDa with one heme is the active enzyme which catalyzes both cyclooxygenase and peroxidase activity.

The primary structure of PGH synthase was revealed by cDNA sequencing for the ovine [22–24], murine [25], and human [26] enzymes which show about 90% sequence identity. After posttranslational removal of a 24-amino-acid signal peptide, ovine PGH synthase contains 576 amino acids with a calculated molecular mass of 65.5 kDa. The difference to the molecular mass of 72 kDa as determined by SDS-PAGE is attributed to glycosylation [4].

Recently, a second form of PGH synthase was discovered and termed COX-2 (the form described above is now termed COX-1) [27–29]. Both forms are encoded by different genes and underly different regulation of gene expression. COX-1 is a constitutive form whereas COX-2 is inducible by mitogens which stimulate certain cells. There is a high sequence identity of 62% between COX-1 and COX-2 [27]. An identical catalytic mechanism is expected for both forms. The small differences in the protein structures can be exploited for the development of inhibitors specific for COX-2. Since COX-2 is responsible for an enhanced prostanoid synthesis during pathological processes such as inflammation, whereas COX-1 has been related to the physiological prostanoid formation, specific inhibition of COX-2 will have great pharmacological and therapeutic relevance.

The three-dimensional structure of PGH synthase has been resolved recently (see Note Added in Proof, p. 185). The growth of suitable crystals of the membrane protein for X-ray diffraction is difficult [30–32], but the recent progress in the resolution of the three-dimensional structure of PGH synthase was achieved by R. M. Garavito and coworkers. The lack of the resolved three-dimensional structure of PGH synthase supplements the assignment of residues to the active site(s) using site-directed mutagenesis, as is described on p. 185.

B. Catalytic Mechanism

The heme in PGH synthase is directly involved in the peroxidase reaction, following a mechanism analogous to other heme-containing peroxidases such as horseradish peroxidase or cytochrome c peroxidase. The elucidation of the mechanism takes advantage of the wealth of spectral information which is available for the heme center. The peroxidase mechanism will be described first. We know much less about the cyclooxygenase mechanism. One of its most challenging features is the activation of the cyclooxygenase reaction by the peroxidase. This will be described as a second point, and finally the question of how the heme generates a strongly oxidizing species and how this species is transferred to and utilized by the cyclooxygenase catalytic center will be discussed.

1. Peroxidase Mechanism

The classical peroxidase mechanism consists of the reaction of the resting ferric enzyme with hydroperoxide under formation of higher oxidation states and the reduction of these states by one-electron transfers from a reducing cosubstrate (see Chapter 2). There is a close analogy between PGH synthase and other heme peroxidases. We will briefly describe the states and transitions of the heme center of PGH synthase.

Resting Enzyme. The heme in samples of purified resting native state of PGH synthase is in its ferric, predominantly high-spin state as can be seen from the electronic [12,20] EPR [20,33,34] and Mössbauer spectra (R. Karthein, E. Bill, A. X. Trautwein, and H.H. Ruf, unpublished results). The electronic absorption spectrum shows the Soret band at 410 nm and bands in the visible at 505 and 633 nm [20]. The Soret band's extinction coefficient of 230,000 M^{-1} cm^{-1} as given by Kulmacz [12] appears to be very high when compared to the values around 100,000 M^{-1} cm^{-1} of other ferric high-spin heme proteins. According to EPR and Mössbauer spectroscopy the enzyme contains ferric heme in its high-spin state with a rhombic distortion of its axial symmetry [20,34]. Deviating results [33] may be due to partially inactivated enzyme or nonspecifically bound heme as also known for cytochrome c peroxidase [35,36].

The coordination sphere of the ferric heme in resting PGH synthase is now definitely known. The high-spin states implies a five coordination or a six coordination with a very weak sixth ligand. This is reasonable for a peroxidase reaction where oxygen from the hydroperoxide enters the inner coordination sphere of the heme iron. Two candidates were suggested for the fifth ligand, tyrosine or histidine. Tyrosine was suggested from the similarity of the rhombicity in EPR with that of catalase [20], the lack of a superhyperfine interaction with an imidazole nitrogen in the EPR spectrum of the ferrous NO adduct of PGH synthase [20] and the chemical modification of PGH synthase by *N*-acetylimidazole [15]. Histidine was inferred from the bisimidazole coordination of the low-spin derivative of PGH synthase [33] and the homology of a decapeptide in PGH synthase with a sequence in thyroid peroxidase and myeloperoxidase [25] which contains the coordinated histidine residue in those enzymes. Site-directed mutagenesis of PGH synthase revealed three histidines as essential for cyclooxygenase and peroxidase activities and the binding of heme [37]. His-309 was suggested as the fifth ligand [37] (the numbering of amino acids corresponds to the ovine enzyme including the signal peptide [22]). Since myeloperoxidase is an example of a heme protein with imidazole coordination [38] exhibiting an EPR signal of unusually high rhombicity ($g = 6.9–6.7, 5.4–5.3$, and 2.0) [39,40], the high rhombicity of the EPR signal of PGH synthase does not rule out histidine as the fifth ligand in PGH synthase. Now X-ray crystallographic structure identifies His-388 as the proximal axial iron ligand. Two distal residues,

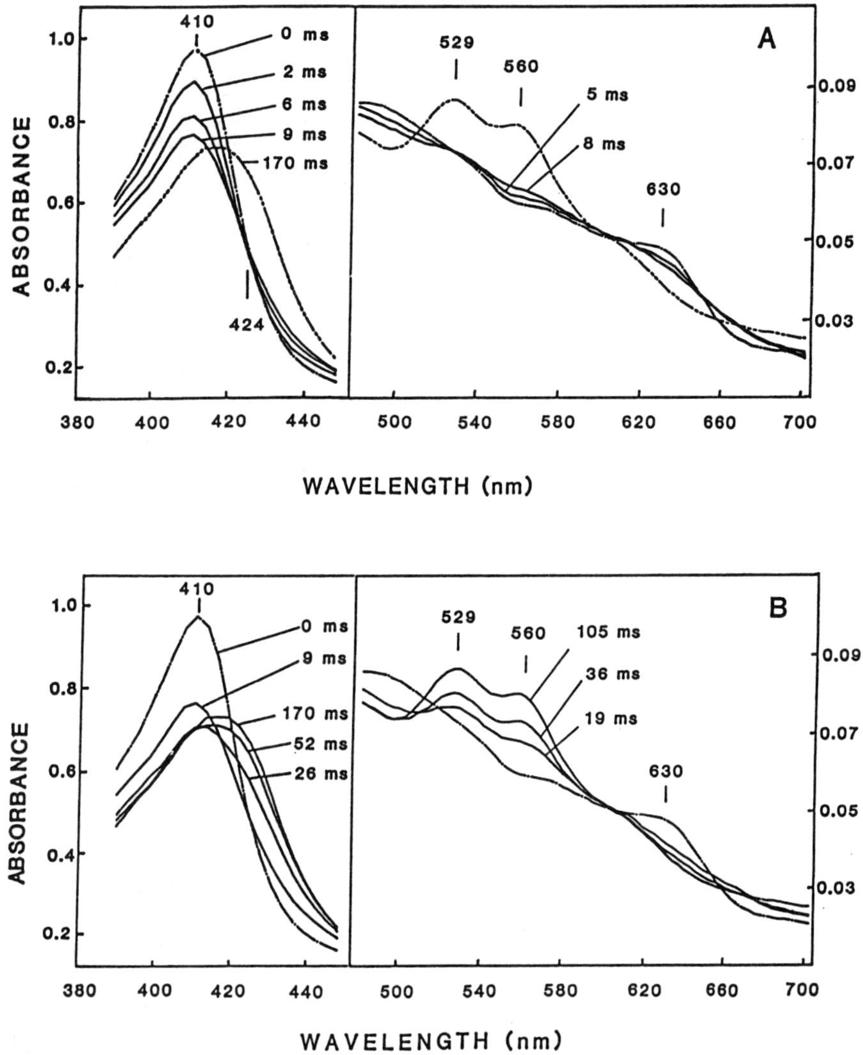

Figure 1 Electronic absorption spectra of higher oxidation states of PGH synthase during the reaction with PGG_2. Panel A shows the formation of intermediate 1 [$(PPIX)^{\cdot +}Fe^{IV}O$] within 10 ms reaction time and panel B the subsequent formation of intermediate II [$(PPIX)Fe^{IV}O$]. The reaction mixture contained 5 μM PGH synthase and 10 μM PGG_2. The reaction was carried out at 1°C in a stopped-flow apparatus with a diode array photometer. (From Ref. 41.)

Gln-203 and His-207, approach the heme iron to approximately 5 Å but do not coordinate it (see Note Added in Proof, p. 185).

Reaction with Hydroperoxide. Hydroperoxides react with resting PGH synthase to give higher oxidation states of the enzyme which can be detected spectroscopically in the absence of exogenous electron donors (Figure 1). The highest rate constants, e.g., $1.4 \times 10^7\ M^{-1}\ s^{-1}$ for PGG_2 at 1°C [41], are achieved with PGG_2 or long-chain fatty acid hydroperoxides whereas hydrogen peroxide reacts slower by three orders of magnitude [41–43]. This indicates that the distal side of the heme in PGH synthase is quite specific for the natural substrate PGG_2.

Spectral Intermediates and Tyrosyl Radical. The first detectable spectral transient (Figure 1A) [41,42] was termed intermediate I, and its electronic spectrum resembles that of compound I of horseradish peroxidase. Hence the structure $[(PPXI)^{\cdot+}\ Fe^{IV}O]$ was suggested for intermediate I of PGH synthase [41,42]. Intermediate I is not stable. Even in the absence of exogenous electron donors, an intermediate II is formed with a rate constant of 65 s^{-1} at 1°C (Figure 1B) [41]. The spectrum of intermediate II resembles that of compound II of horseradish peroxidase or complex ES of cytochrome c peroxidase indicating a structure $[(PPIX)\ Fe^{IV}O]$. A doublet EPR signal was detected concomitantly with intermediate II (Figure 2) and assigned to a tyrosyl radical [44]. The remarkable feature of PGH synthase is that both intermediates represent states with two oxidation equivalents, one at iron Fe^{IV} and the other moving from the porphyrin π cation radical to a tyrosyl residue of the protein. This intramolecular electron transfer produces the tyrosyl radical with the doublet EPR signal [44]. At longer reaction times the doublet signal changes into a "narrow" singlet signal which is also assigned to a tyrosyl radical, but with an altered conformation [45]. A "broad" singlet signal is observed at intermediate reaction times. This signal does not represent a third radical species as suggested [46] but rather a superposition of the doublet and narrow singlet signals [47]. Despite this confusing controversy, the tyrosyl radical with the doublet EPR signal is the only radical species observed which is formed at a rate high enough to have catalytic competence in the cyclooxygenase reaction (see below).

Reactivity of Higher Oxidation States of PGH Synthase. Higher oxidation states of heme can be reduced by two general mechanisms: electron transfer or oxygen transfer. Electron transfer occurs in the classical peroxidase reaction whereas cytochrome P450 is prone to oxygen transfer reactions. In PGH synthase electron transfer in two sequential one-electron steps is most commonly observed. A wide variety especially of phenolic compounds can serve as exogenous electron donors. As physiological donors, uric acid and glutathione were suggested but not identified definitely [4]. Dunford

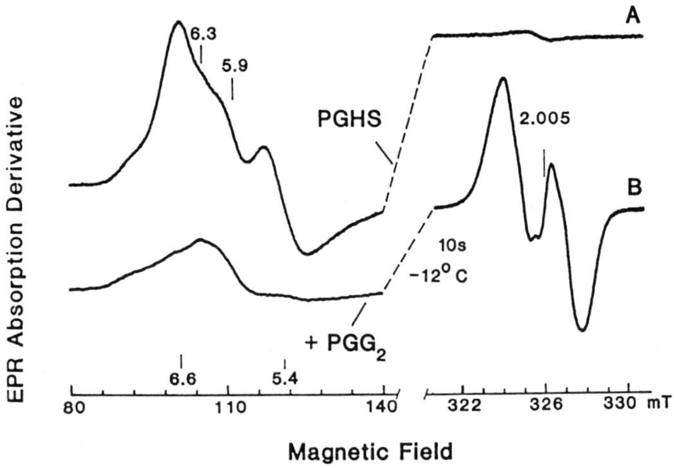

Figure 2 EPR spectra of native PGH synthase and its derivative after the reaction with PGG_2. The reaction with PGG_2 decreased the high-spin ferric heme signal of native PGH synthase at g = 6.6 and 5.4 (curve A) and produced a doublet signal at g = 2.005 (curve B) which was assigned to a transient tyrosyl radical. PGH synthase (98 μM, curve A) was incubated with PGG_2 (300 μM) for about 10 s at −12°C and frozen in liquid nitrogen (curve B). EPR was measured at 15 K. (Adapted from Ref. 44.)

and coworkers determined the reaction rates of electron donors with the higher oxidized intermediates of PGH synthase [48–52].

The one-electron oxidation of electron donors produces free radicals which can initiate subsequent oxidation reactions of additional compounds. By this cooxidation PGH synthase can convert a variety of endogenous or exogenous chemicals to toxic/carcinogenic derivatives [53]. This activation by PGH synthase may be important in tissues with low levels of cytochromes P450.

Oxygen transfer from higher oxidation states of PGH synthase was observed only with alkylaryl sulfides where ^{18}O-labeling studies indicate that the sulfoxide oxygen is quantitatively derived from the hydroperoxide [54]. In the oxygenation of alkylaryl sulfides electronic and steric effects are observed [55]. The chemical reactivity against para-substituted alkylaryl sulfides is comparable to other peroxidases suggesting a similar electronic structure although a spectroscopic characterization of the ferryl–oxo complex of PGH synthase is still lacking. From steric effects the size of the substrate binding pocket at the heme in PGH synthase is estimated to be intermediate between those of horseradish peroxidase (small) and cytochrome P450 (large) [55].

In summary, the peroxidase reaction of PGH synthase shows two special features in comparison with other heme peroxidases: (a) the occurrence of two states with two oxidizing equivalents and (b) the involvement of the peroxidase reaction in the activation of the cyclooxygenase reaction (see below).

2. Cyclooxygenase Mechanism

Hamberg and Samuelsson [56] proposed a reaction mechanism in 1967 which was subsequently substantiated [57–61]. Arachidonic acid binds to the cyclooxygenase center in a bent conformation: A rotation with the C-9/C-10 single bond produces the conformation of arachidonic acid which leads to the formation of trans aliphatic side chains on the cyclopentane ring to be formed. The first step of the reaction is the stereospecific abstraction of the pro-S hydrogen from the allylic C-13. The resulting carbon radical undergoes a series of radical reaction steps: rearrangement to C-11, attack by O_2, formation of the 9,11-endoperoxide and the C-8/C-12 bond, radical rearrangement to C-15, and attack of the second O_2. The reduction of the 15-peroxyl radical to the hydroperoxide of PGG_2 completes the cyclooxygenase reaction cycle. In this model, the enzyme has two main functions: (a) the binding of the correct conformation of the substrate and (b) the stereospecific hydrogen abstraction. In addition, the enzyme might favor the correct radical rearrangements and shield the intermediate radicals to prevent side reactions [59].

With respect to the oxygenation mechanism, PGH synthase does not activate O_2 but rather activates the substrate by the abstraction of hydrogen and formation of a carbon-centered substrate radical. The ferric heme cannot directly abstract this hydrogen since the addition of arachidonic acid under anaerobic conditions does not alter the EPR signal of the resting ferric heme [44]. In contrast, the ferric nonheme iron of lipoxygenases directly abstracts hydrogen from allylic carbon atoms of polyunsaturated fatty acids as a first step of the lipoxygenase reaction. A model for the substrate activation in PGH synthase has to consider the formation of a stronger oxidizing species than ferric heme. A species derived from a higher oxidation state of the peroxidase could provide this activation of arachidonic acid.

Relation Between Cyclooxygenase and Peroxidase. There is longstanding evidence that the cyclooxygenase activity requires both the presence of hydroperoxide and an active peroxidase reaction center [4,62]. The cyclooxygenase reaction shows a characteristic activation phase, typically observed as a lag phase in the O_2 uptake recordings after the start of the assay. If either the hydroperoxide concentration in the assay is diminished or the peroxidase activity is inhibited, a prolonged activation phase or even absence of the cyclooxygenase reaction is observed. There is a mutual interdependence—under decreased peroxidase activity, a higher concentration of hydroperoxide is required for cyclooxygenase activation. For example, MnPPIX-substituted PGH synthase shows only 1% of the peroxidase activity of Fe-PGH synthase but full cyclooxygenase activity although with a longer activation phase and higher requirement for hydroperoxides [63,64]. The Mn-PGH synthase provides a good opportunity to study the activation mechanism of the cyclooxygenase reaction. The results with the manganese enzyme support the evidence accumulated with the iron enzyme [4,65] that the reaction of a hydroperoxide with the metal is the basic step in the activation of the cyclooxygenase reaction [64].

It is important to note that aspirin or indomethacin (and other related compounds) selectively inhibits the cyclooxygenase without affecting the peroxidase activity [4,64]. Partial reduction of the peroxidase activity leaves the maximal rate of the cyclooxygenase activity essentially unchanged although a longer activation phase is observed (see above). The substrates of the cyclooxygenase and peroxidase reaction—arachidonic acid and PGG_2, respectively—do not exhibit a mutual inhibition of the peroxidase and cyclooxygenase reactions [66]. These results suggest distinct separate binding sites

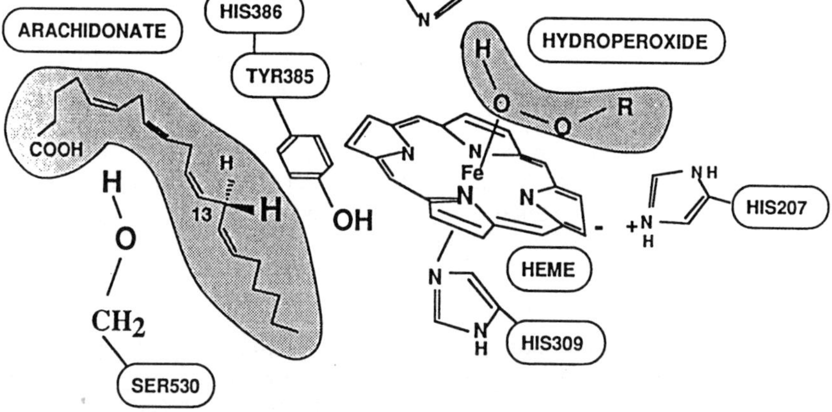

Figure 3 Schematic model for the cyclooxygenase and peroxidase active sites of ovine PGH synthase. This model was suggested by Smith and coworkers on the basis of site-directed mutagenesis studies. Note the differences between this and the X-ray structure described in the Note Added in Proof (p. 185). (From Ref. 65.)

for arachidonic acid and PGG_2 for which several models have been designed. In combination with site-directed mutagenesis studies, Smith et al. [37,65] proposed a model for the active site of PGH synthase (Figure 3) which incorporates the binding site for hydroperoxide at the distal side of the heme, a proximal histidine, and the arachidonic acid binding site separate from the heme. The arachidonic acid binding site contains the Ser-530 the acetylation of which by aspirin inhibits the cyclooxygenase due to steric hindrance of arachidonic acid binding [25].

Reaction Model for the Hydroperoxide-Dependent Activation of Cyclooxygenase. Ruf and coworkers proposed a model ("Ruf model") for the activation of the cyclooxygenase reaction which requires as a first step of the peroxidase reaction the reaction of heme with hydroperoxide to produce higher oxidation states [41,44]. The detection of the tyrosyl radical with the doublet signal [44] prompted the idea that a tyrosyl radical could be the group which abstracts the hydrogen from C-13 of arachidonic acid. The activation phase of the cyclooxygenase reaction is explained by the generation of the tyrosyl radical as a consequence of the reaction of hydroperoxides with the heme. During the steady state of the cyclooxygenase reaction, the tyrosyl radical will be regenerated by the last step of the cyclooxygenase reaction cycle, the reduction of the 15-peroxyl radical to PGH_2 (see Figure 4).

In a first step, from the resting enzyme [Fe(III)] a two-electron oxidized species is formed, formally Fe(V) with the tentative structure $[(PPIX^{\cdot+})FeO]$, which generates the tyrosyl radical by a rapid intramolecular electron transfer. The rate of the reaction of the metal with hydroperoxide is inversely related to the duration of the activation phase of the cyclooxygenase reaction. When Fe-PPIX is substituted with Mn-PPIX in PGH synthase, the reaction rate with hydroperoxides is decreased by three orders of magnitude and the activation is slowed down [63,64]. Substitution with Zn-PPIX or Ga-PPIX (S. Strieder, H.-J. Scherer, U. Raab, and H.H. Ruf, unpublished data), both non–redox-active metals which do not react with hydroperoxides, abolished both peroxidase and cyclooxygenase activities completely. Thus the redox activity of the metal with hydroperoxide seems to be an essential condition also for the cyclooxygenase reaction in contrast to other hypotheses [65,68].

The Ruf model, as described above, suggests that a tyrosyl radical participates in the transfer of the oxidation equivalent from the higher oxidized state $[(PPIX^{\cdot+})Fe^{IV}O]$ to arachidonic acid. Support for the special role of the tyrosyl residues in the cyclooxygenase reaction comes from the finding that the mutation $Tyr^{385} \rightarrow Phe$ leads to a complete loss of the cyclooxygenase activity whereas the peroxidase activity remains [58]. Furthermore, the chemical modification of tyrosyl residues with tetranitromethane inactivates the cyclooxygenase stronger than the peroxidase (H.-J. Scherer and H.H. Ruf,

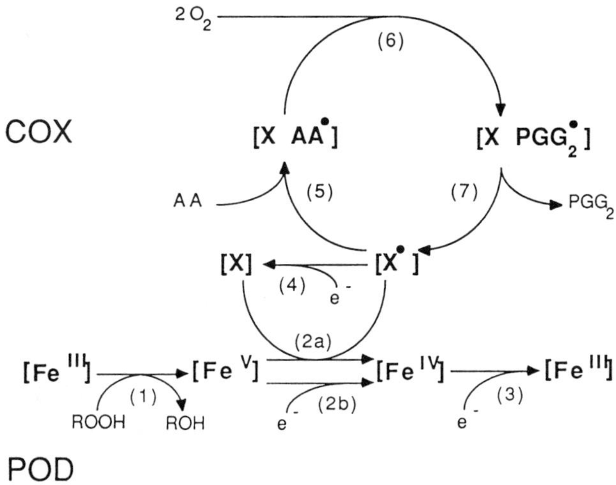

Figure 4 Hypothetical reaction scheme for PGH synthase. The peroxidase reaction (POD; reactions 1, 2b, and 3), the cyclooxygenase cycle (COX; reactions 5, 6, and 7), and the activation (reaction 2a) and deactivation (reaction 4) are depicted. The iron states are formal oxidation states. X represents an amino acid side chain residue, which is oxidized by the Fe(V) state, $[(PPIX^{\cdot +})Fe^{IV}O]$ to a radical. This transfer of oxidation aquivalent is proposed to lead to the initiation of cyclooxygenase reaction by hydrogen abstraction from arachidonic acid. For further explanation, see text.

unpublished data), and Tyr^{385} is modified by tetranitromethane [58]. From the report by Schöneich et al. [68] that a thiyl radical from cysteine can abstract hydrogen at the allylic C-13 of arachidonic acid, a similar reactivity can be assumed for a tyrosyl radical.

A critical evaluation of the Ruf model was published recently [65]. As a main objection it was noted that the spectroscopic observation of a tyrosyl radical is not concomitant with the cyclooxygenase activity, i.e., at low concentrations of electron donors, where cyclooxygenase is even activated, the EPR signal of the tyrosyl radical is not detected. Mn-substituted PGH synthase does not exhibit an EPR doublet signal which corresponds to its cyclooxygenase activity [45]. The EPR-observable tyrosyl radical has been rather attributed to the autoinactivation of the enzyme [45,65], which presumably occurs by oxidative damage to the enzyme. Since the participation of the EPR-detectable tyrosyl radical in the cyclooxygenase reaction is still open, in the reaction scheme (Figure 4) a radical X′ from an as-yet unidentified amino acid side chain is introduced. Tyrosine is still the first candidate for X and Tyr· can have catalytic competence even at low steady-state concentrations which are not detected by EPR. (See also the Note Added in Proof, p. 185.)

The depicted reaction scheme (Figure 4) consists of a peroxidase pathway (reactions 1, 2b, and 3), a cyclooxygenase cycle (reactions 5, 6, and 7), and the reaction for the activation (2a) and the deactivation (4) of the cyclooxygenase cycle. Iron and manganese PGH synthase follow the same reaction scheme with the main difference that the reaction of Mn(III)-PGH synthase with hydroperoxide is lower by three orders of magnitude and thus is the rate-limiting step both of the peroxidase reaction and cyclooxygenase activation in the Mn enzyme. The details of this reaction model for the Mn-PGH synthase are described in Ref. 64.

III. PROSTACYCLIN SYNTHASE

As outlined above, our working hypothesis for the isolation of prostacyclin synthase started out from the analogy of hydroperoxide cleavage and oxygen atom transfer by heme thiolate-containing enzymes. Since the spectral properties of these cytochrome P450 proteins were well established we investigated known tissues with PGI synthase activity for spectral indications for cytochrome P450. Additional help was provided by the fact that *trans*-2-phenylcyclopropylamine (tranylcypromine), a monoamine oxidase inhibitor, had also been reported as an inhibitor for prostacyclin synthase [69]. This strengthened our hypothesis of a heme-catalyzed reaction since basic nitrogen atoms are well-known ligands for hemoproteins and heme thiolate-catalyzed monooxygenations are inhibited by a large number of nitrogen-containing compounds that can bind as ligands and then are detectable by the spectral perturbations of their optical and EPR spectra.

A. Isolation and Characterization

We selected the microsomal fraction of pig aorta as a starting material and recorded difference spectra with tranylcypromine, which appeared with a maximum at 435 nm and a minimum at 415 nm [9]. In the same preparation a peak resulted at 450 nm in the presence of dithionite and carbon monoxide (CO). Essential for our isolation procedure was the solubilization of the pig aortic microsomes for which a mixture of cholic acid and a sulfobetaine was found suitable [1]. The resulting 100,000*g* supernatant was chromatographed on DEAE-Sephacel and at 70 mM Na-phosphate PGI synthase activity was eluted. Thanks to a generous gift of the PGH_2 analog U 46619 by Dr. J. Pike (Upjohn), an affinity column could be prepared which yielded PGI synthase as a 49.2-kDa protein with a specific activity of 1.97 U/mg. The heme thiolate protein copurified with the enzymatic activity and from the spectrum of the homogeneous protein 18 nmol heme/mg protein was calculated, which corresponded to one heme per polypeptide chain (Figure 5). Addition of dithionite only slowly and partially reduced the hemoprotein even in the presence of CO, which was different from cytochrome P450 monooxygenases. Final proof for the heme thiolate nature of PGI synthase could be obtained by EPR spectra [70] (Figure 6).

The g tensor with values at g = 1.90, 2.25, and 2.46 was identical with that of cytochrome P450 monooxygenases [71]. The spectrum also proved that the enzyme as isolated is in its ferric state. Another approach to purify PGI synthase from bovine aorta was used by DeWitt and Smith [72] using monoclonal antibodies in an immunoaffinity column. They could isolate PGI synthase as a 52-kDa protein with a maximum specific activity of 1 U/mg protein and a major heme peak between 418 and 420 nm. The difference spectrum treated with sodium dithionite and CO exhibited a broad absorbance maximum centered around 440 nm. They also reported a K_M value of 5–10 μM and a loss of enzyme activity during catalysis. After determination of the

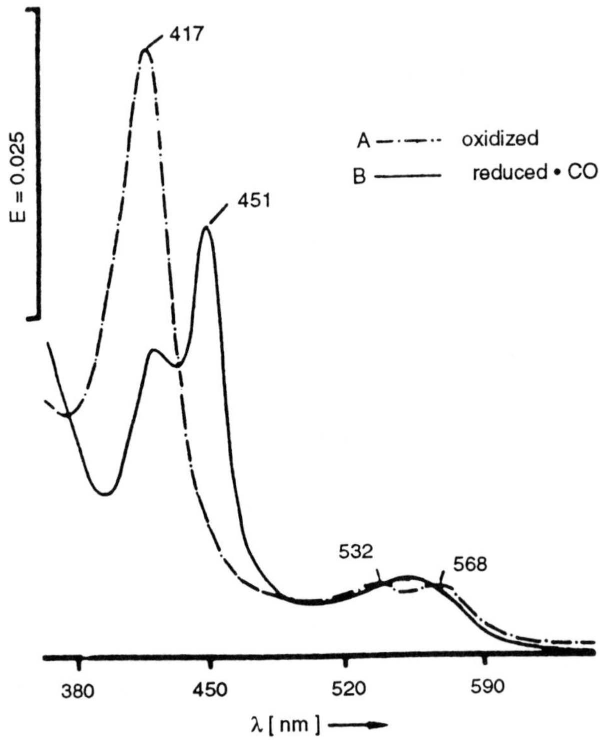

Figure 5 Optical absorption spectra of PGI synthase [1,74].

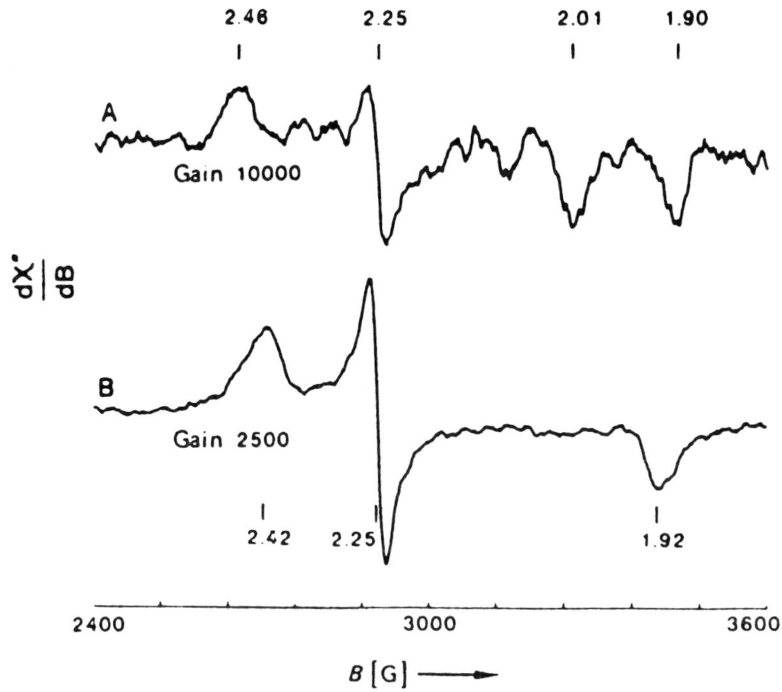

Figure 6 EPR spectra of PGI synthase (A) and of cytochrome P450 in liver microsomes (B) [74].

amino acid composition, the NH_2 terminal amino acid sequence, and the sequence of three TPCK trypsin-derived peptides of the enzyme, and a search for homologies between these peptides and various P450 proteins, no clear answer for significant sequence homologies was obtained [73]. The total sequence of PGI synthase was recently obtained and indicated a moderate (max. 39%) homology to other P450 proteins (Tanabe, personal communication).

It was interesting to note that the various TxA synthase inhibitors based on pyridine or imidazole were completely inactive toward PGI synthase [74, 75]. This already suggested a different stereochemistry at the active site although both enzymes have PGH_2 as a common substrate. The substrate specificity of PGI synthase was studied with 1-^{14}C-labeled PGH_1, PGG_2, PGH_2, 8-iso-PGH_2, 15-keto-PGH_2, and PGH_3 [74]. PGG_2 and 15-keto-PGH_2 were converted to their corresponding 6-keto $PGF_{1\alpha}$ metabolites. 8-*cis*-HHT was predominantly formed in the presence of 8-iso-PGH_2. PGH_1 was converted to 12(*S*)-hydroxy-8(*E*),10(*E*)-heptadecadienoic acid (HHD) and malondialdehyde (MDA). This fragmentation could be measured photometrically by the formation of HHD at 233 nm [76] and proceeded with a molecular activity of 45 s^{-1}.

B. Chemical Mechanism

The initial step of PGI_2 and TxA_2 formation is supposed to be the interaction of the PGH_2 endoperoxide group with the heme iron and the following cleavage of the endoperoxide bond. For PGI_2 biosynthesis the oxygen atom on C-9 should attack the Δ^5 double bond and as discussed in literature this could occur by a polar mechanism after heterolytic cleavage of the O-O bond [77, 78].

From optical difference spectra with the PGH_2 analogs 15(*S*)-hydroxy-9α,11α-expoxymethano-5(*Z*),13(*E*)-prostadienoic acid (U44069) and the 9α,11α-methanoepoxy derivative (U46619) with PGI and TxA synthase, we could conclude that PGI synthase obtains its specifity through an interaction of the ferric heme with the endoperoxide oxygen at C-11, whereas the TxA synthase interacts with the C-9 endoperoxide oxygen [74] (see following section).

This finding contradicted our first proposal of a heme tiolate–stabilized oxonium structure at C-9 [70] and suggested a stabilization of a C-11 alkoxy radical by the heme thiolate group and a subsequent attack of the C-9 oxygen radical at the Δ^5 double bond (Scheme 4). A heterolytic cleavage leading to a complexed negatively charged oxygen atom at C-11 and a free oxonium at C-9 did not seem to be compatible with the properties of a thiolate-Fe^{III} structure and the thermodynamics of such a process.

After the cyclization step a carbon radical at C-5 is formed which had to be oxidized by the formally Fe^{IV} iron. Loss of the C-6 proton yields the vinyl ether PGI_2. It should be pointed out that according to this mechanism the electron transfer step from the C-5 radical to the Fe^{IV} species represents a crucial event in the process.

The postulated sequence also explains the formation of malondialdehyde and the corresponding C17 acid with PGH_1 and 8-iso-PGH_2 by fragmentation of the carbon radical intermediate. Reasons must be the lack of the Δ^5 double bond in PGH_1 and obviously the wrong localization of the double bond in the case of 8-iso-PGH_2 [78]. The absolute geometry of the Δ^5 double bond, however, does not seem to be important [79]. The homolytic cleavage of the endoperoxide does not seem to be the rate-limiting step of PGI_2 biosynthesis if one compares the molecular activity of the PGH_1 fragmentation (45 s^{-1}) with the low turnover number of PGI_2 formation (3.5–4.3 s^{-1}) [74]. Probable rate-limiting steps could then be the cyclization of the alkoxy radical or the oxidation of the carbon radical. In summary,

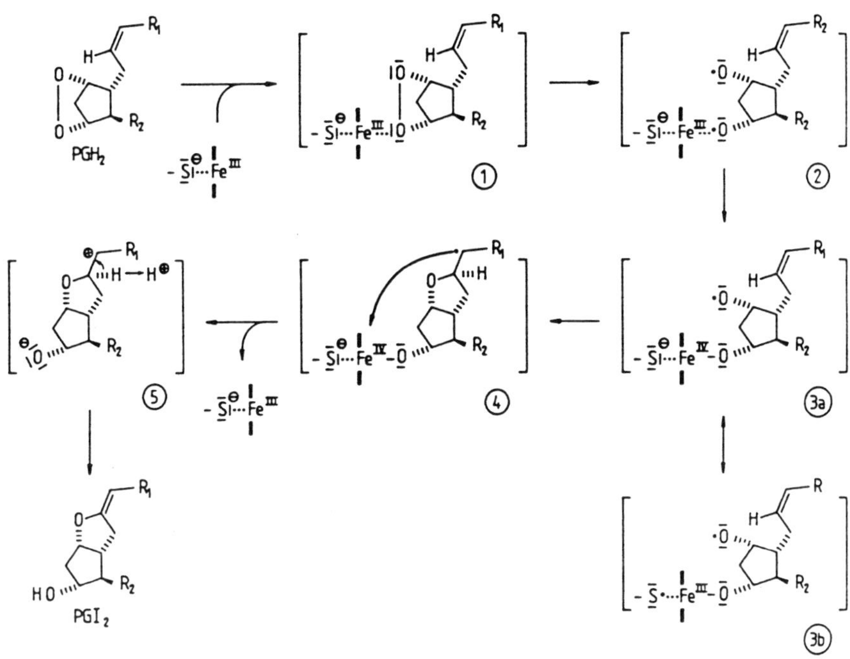

Scheme 4

it is beyond doubt that the isomerization of PGH_2 is catalyzed by a heme thiolate protein which does not require reducing equivalents and hence is different from the oxygen-activating monooxygenases.

IV. THROMBOXANE SYNTHASE

Encouraged by the first evidence for the heme thiolate nature of PGI synthase, our laboratory then concentrated on the isolation of TxA_2 synthase. This enzyme occurs in the microsomal fraction of blood platelets [80] but also of other white cells [81,82] and converts PGH_2 to a mixture of TxA_2 and 12(*S*)-hydroxy-5(*Z*),8(*E*),10(*E*)-heptadecatrienoic acid (HHT) plus malondialdehyde (MDA) [83]. All three products are formed in a close to 1:1:1 ratio. TxA_2 itself is unstable ($\pi_{1/2}$ ~ 30 s) and can only be detected as the stable hydration product TxB_2 [84,85].

A. Isolation and Characterization

Hammarström and Falardeau [86] in 1977 first employed nonionic detergents to solubilize a TxB_2 synthesizing activity from bovine lung but failed to purify the enzyme. Our approach consisted of the spectral investigation of microsomal membranes from human platelets [87], and in agreement with earlier results [88], a cytochrome P450–like CO difference spectrum was obtained with sodium dithionite as a chemical reductant. Proof for a link of this hemoprotein with TxA_2 synthase was obtained by the spectral interaction of specific inhibitors of this enzyme like UK-37.248, OKY-1581, or CGS-13080 in platelet microsomal suspensions [89]. Such inhibitors are now abundantly available and all consist of a pyridine or imidazole ring with a carboxylic side chain [74]. All inhibitors tested showed a difference spectrum with platelet microsomes with a peak around 427 nm and a through around 415 nm. These ligand binding spectra allowed the purification of a platelet heme thiolate protein through seven steps resulting in a protein with an apparent molecular weight of 58,800 on SDS gels [87] (Figure 7). The homogeneous protein exhibited a Soret band at 418 nm which upon reduction by dithionite in the presence of CO shifted to 450 nm. For a final identification as a heme thiolate protein EPR spectra were recorded which showed a g tensor of 1.9, 2.25, and 2.4, typical for P450 proteins [71] (Figure 8).

Most importantly, the purified heme thiolate protein converted PGH_2 to TxB_2 and had a specific activity 24.1 μmol TxB_2 mg prot.$^{-1}$ min^{-1} (20°C) corresponding to a molecular activity of about 24 s^{-1}. Since the protein simultaneously formed HHT and MDA in a reaction sensitive to active site inhibitors, the overall molecular activity was about 50 s^{-1}.

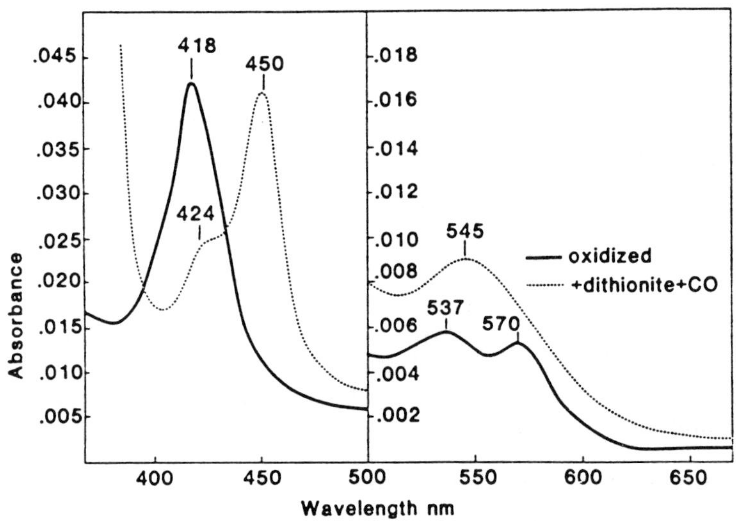

Figure 7 Optical absorption spectra of TxA synthase [92].

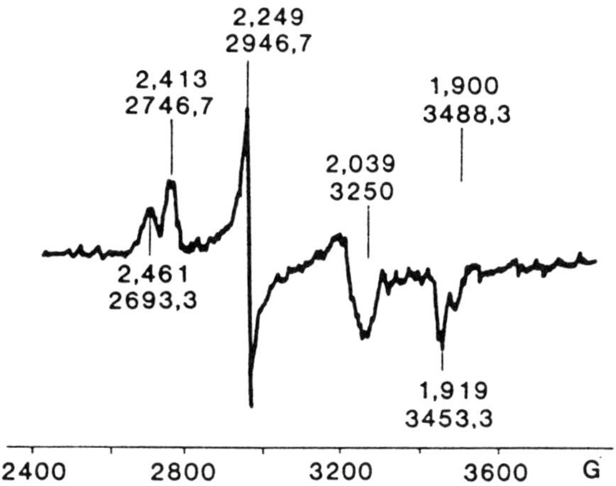

Figure 8 EPR spectra of TxA synthase [92].

Subsequently, Shen and Tai [90] reported on an immunoaffinity preparation of TxA synthase from porcine lung. However, this enzyme only had about 0.1% of the activity of the platelet preparation and possibly was largely inactivated by the elution conditions. This assumption could be supported by our own immunoaffinity purifications [91].

After partial amino acid sequence data became available [91], the total sequence was unraveled by using the PCR method [92,93]. A comparison of the 533 amino acids (60.6 kDa) with known cytochrome P450 monooxygenase sequences confirmed a moderate relationship between these enzymes (sequence homology maximal 36%) and permitted the conclusion that the cysteine in position 479 interacted with the heme moiety. Since a P450 gene family arbitrarily was defined as having less than 40% homology with other genes [94], TxA synthase must be considered as a member of a separate and new heme thiolate family. Surprisingly, the homology with PGI synthase does not exceed 15%, which separates both enzymes into two heme thiolate families. A more detailed comparison of monooxygenase P450 proteins with PGI and TxA synthase in terms of mechanistic implications will be given below.

B. Spectral Probes for the Active Site

Special interest in TxA synthase is concerned with the binding of PGH_2 at the active site, which should be largely different from that of PGH_2 binding to PGI synthase not only because different products arise but because the inhibitors for both heme thiolate enzymes are structurally entirely different

Table 1 Inhibitors of Thromboxane Synthase

Structure	Inhibitors	IC_{50}[a] nmol/L	K_s[b] μmol/L	SC_{50}[c] nmol/L
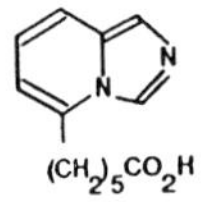	CGS-13080	2.6	0.242	4.4
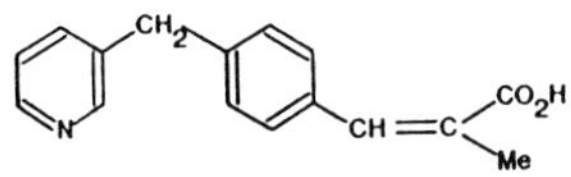	OKY-1581	3.9	0.200	3.6
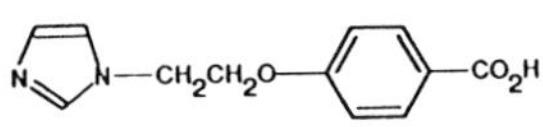	UK-37248	11	0.275	5
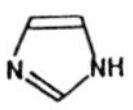	Imidazol	48000	9.8	180
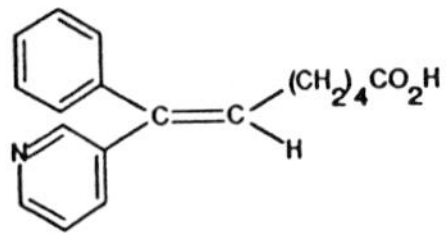	CV-4151	2.1	0.128	2.3
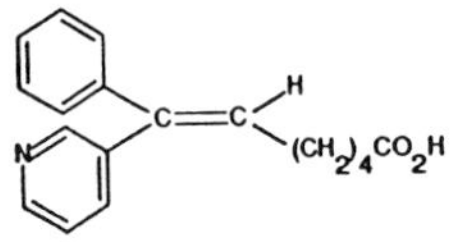	CV-4193	5.2	0.128	2.3
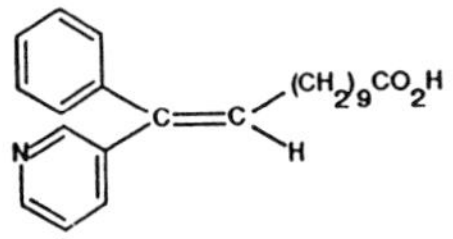	CV-5356	5.4	0.128	2.3
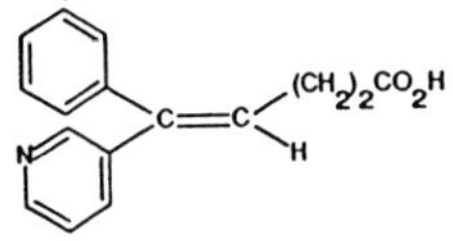	CV-5236	11	0.308	5.6
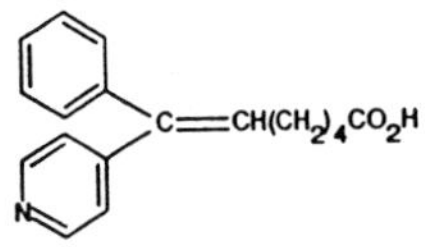	CV-4388	65	3	55
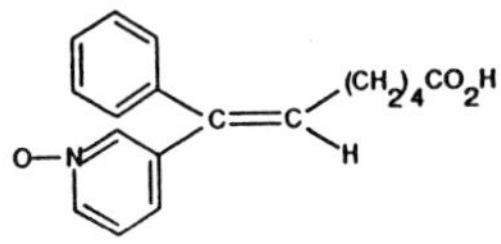	CV-5334	3800	132	2400
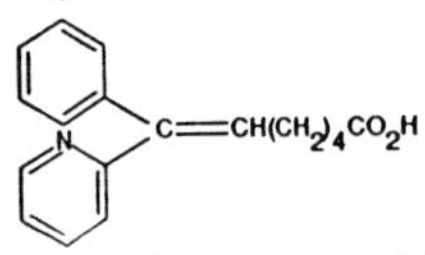	CV-4387	250000	56	1000

[a]Half inhibition of TxA_2 formation.
[b]Spectral dissociation constant.
[c]Corrected spectral dissociation constant for stoichiometric binding.

except that both bind through a basic nitrogen atom to the ferric iron at the catalytic site.

From a series of inhibitor analogs we were able to determine structure–activity relationships that allowed a probing of the geometry of the active site [75].

Table 1 contains commercially available inhibitors and inhibitors from the CV series together with their IC_{50} values and the concentrations for the half-maximal spectral change in the difference spectra of solubilized platelet microsomes. Except the last two derivatives all pyridine compounds caused the same red shift of the Soret band and showed affinities that for the first three compounds corresponded to almost stoichiometric binding. The *N*-oxide CV-5334 exhibited only a weak binding with a peak at 434 nm but most interestingly the *o*-substituted pyridine CV-4387 produced a difference spectrum with a peak of 388 nm and a trough at 419 nm.

Such spectra are known as "substrate" binding spectra in case of the monooxygenases and are believed to arise from removal of a coordinating water molecule at the sixth ligand position of the heme generating a five-coordinated high-spin complex [95]. This would indicate that CV-4387 can bind to the active site replacing the liganded water molecule but being unable to coordinate with its sterically hindered nitrogen to the Fe^{III}. The styrene side chain was most potent in the range of six to eight carbon atoms so that from the inhibitor studies the following picture of the active site could be derived.

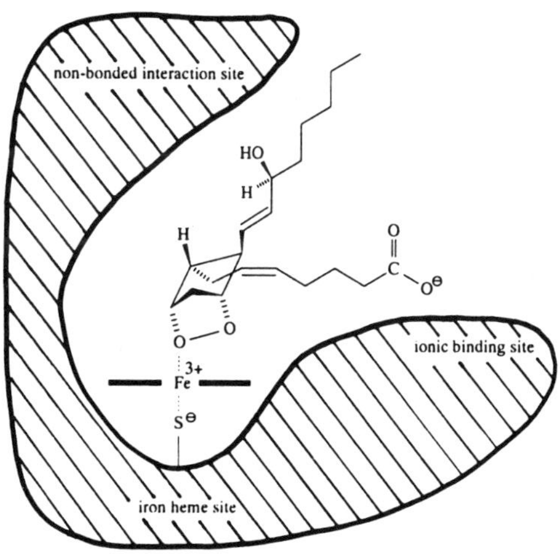

Figure 9 Schematic drawing of substrate binding at the active site of TxA_2 synthase.

It is obvious that the endoperoxide group is in close proximity to the heme iron, which also must be a postulate of any chemical model for TxA synthase. But it is beyond the resolution of this method to decide which oxygen coordinates in the enzyme–substrate complex. This became possible by using the PGH_2 analogs U-44069 and U-46619. Both compounds are stable substrate analogs and can be expected to spectrally interact with TxA synthase if the enzyme could discriminate between the 9 and 11 position. Indeed, 9,11-epoxymethano-PGH_2 formed a difference spectrum with a peak at 409 nm and a trough at 426 nm, whereas the regioisomer produced a weaker change with a maximum at 391 nm and a minimum at 422 nm [74]. The latter closely resembles the situation when the coordinated water molecule is replaced by a noncoordinating substrate analog like CV-4387 whereas the first

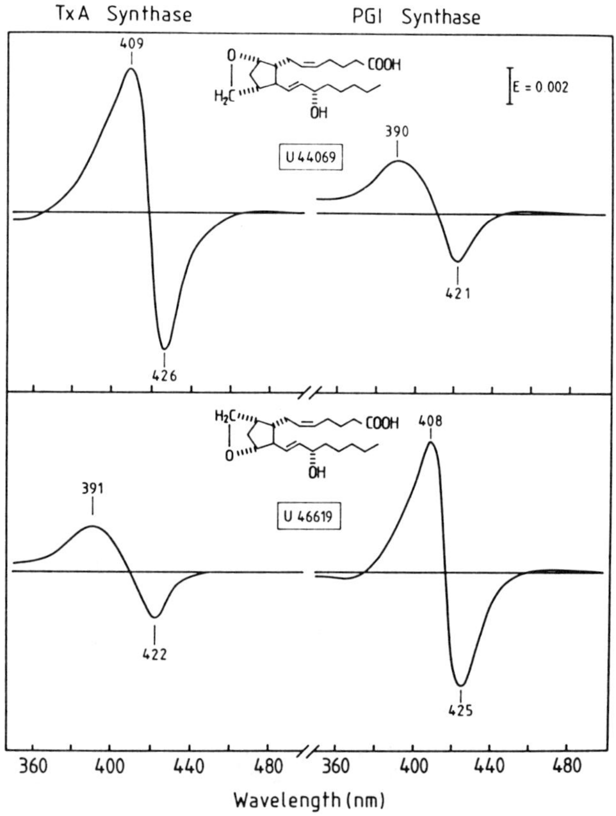

Figure 10 Optical difference spectra of solubilized TxA_2 and PGI synthase with the PGH_2 analog U44069 and U46619 [79].

spectrum could be derived from the replacement of water by a ligated ether oxygen. Such complexes absorb somewhat lower than the aquo complex [96].

Such interpretations would be somewhat speculative without the corresponding results on PGI synthase. With this enzyme the difference spectra are just opposite and hence the binding of the endoperoxide group in PGI synthase should occur through the C-11 endoperoxide oxygen to the heme whereas TxA synthase has a preference for the 9-oxygen atom [74].

C. Chemical Mechanism

The different interactions of the endoperoxide oxygen atoms with both isomerases allowed the conclusion that in PGI synthase a reactive oxygen atom at C-9 should be formed since otherwise no attack at the Δ^5 double bond of PGH_2 would occur. Vice versa, TxA synthase should leave the C-11 oxygen free for the rearrangement process. From the two possible mechanisms discussed for PGI synthase only the homolytic cleavage by the heme thiolate catalytic site seems to be compatible with the above experiment. For TxA synthase a homolytic cleavage would leave an oxy radical at C-11. Its tendency to form an oxo group by radical β scission would be dependent on the possibility for stabilization of a carbon radical at C-12. In the case of a double bond between C-13 and C-14 as in PGH_2 the radical would be allylic and hence stabilized (Scheme 5). The carbon radical at C-12 could undergo fragmentation to MDA and HHT. If, however, a rapid electron transfer to the heme group would occur the resulting carbo cation at C-12 could allow a rearrangement reaction postulated by several authors. Kishi [97] could show experimentally that the analogous 11,12-*seco*-prostaglandin rearranges with 80% yield, in an electrophilic solvolytic reaction, to the corresponding TxB_2 derivative.

Hence it appears that the most critical step in the reaction sequence consists of the rapid or even concerted electron transfer from the C-12 radical to the heme. Formally this heme entity bears an oxidation equivalent shared between the iron and the sulfur (Scheme 6).

It may indeed be the quasi-radical structure of the sulfur which allows an almost instantaneous transfer of the electron from C-12. If this does not occur fragmentation will be the immediate consequence and all conditions leading to a disturbance of the interaction between the heme thiolate and the C-12 radical will end in a fragmentation. This can be observed with PGH_1 and 8-iso-PGH_2 as substrates which form heptadecadienoic acid/MDA and HHT/MDA, respectively.

It should be mentioned that an electron transfer–based mechanism for TxA_2 formation was already postulated by Turner and Herz [98] in 1977 with Fe^{II} being the CO-cleaving species and Fe^{III} as the radical oxidizing agent. We can exclude, however, that the heme in TxA synthase ever becomes ferrous

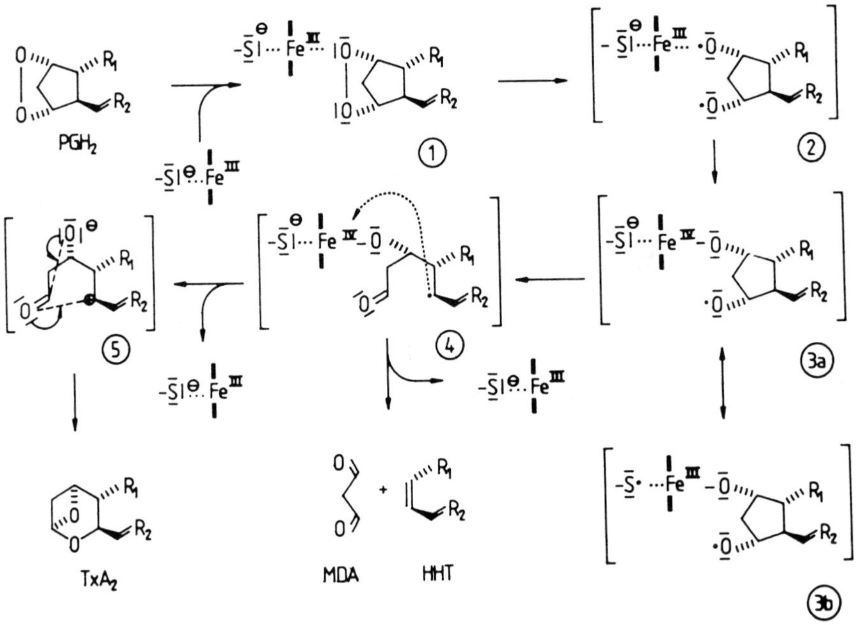

Scheme 5

since no reducing system is involved and CO is never inhibitory as in the monooxygenating heme thiolate catalysis.

Several questions still remain open. One is concerned with the almost 1:1 ratio between the ring closure products and the fragmentation. Does this reflect an imperfect enzyme or have HHT or MDA additional functions as mediators in the regulatory pathways? Our efforts with regard to HHT were negative but 12-keto-5(*Z*),8(*E*),10(*E*)-heptadecatrienoic acid (KHT), an enzymatic oxidation product of HHT formed mainly in the kidney, showed antiaggregatory properties and may therefore act as an antagonist of TxA_2 during platelet aggregation [99].

Scheme 6

V. COMMON PRINCIPLES OF HEME THIOLATE CATALYSIS

At a first sight hydroxylation and epoxidation reactions catalyzed by the heme thiolate–dependent monooxygenases bear little or no resemblance to the isomerization of the prostaglandin endoperoxides to the oxygen heterocycles PGI_2 and TxA_2. In fact, only now that the sequence data became available both isomerases are accepted as new families within the superfamily of heme thiolate proteins. Let us therefore consider in closer detail the possible mechanistic implications of heme thiolate catalysis in monooxygenases and the two isomerases.

The reaction cycle of cytochrome P450 clearly contains two phases: (a) the reduction of dioxygen to the active oxygen and (b) the transfer of the oxygen atom into the substrate. This two-phase sequence became clearly apparent from our first observation that iodosylbenzene can be used as an artificial oxene donor for most heme thiolate monooxygenases [100], which then do not require an external electron source. The same was shown for hydroperoxides and most recently even a physiological enzymic rearrangement of a hydroperoxide to an allene oxide by a heme thiolate protein was reported by Song and Brash [101]. Therefore, electron transfer is an additional and auxiliary function of monooxygenases for which the thiolate ligand is not of primary importance.

If then the heme thiolate structure was critical for the oxene transfer it should be possible to use iodosylbenzene in PGI synthase or TxA synthase for a transfer of the oxygen atom to a suitable substrate. This was indeed achieved with $9\alpha,11\alpha$-methanoepoxy-$PGF_{2\alpha}$ (U-46619) as a substrate [102] which in the presence of platelet microsomes and coincubation with iodosylbenzene yielded three metabolites, which according to their mass spectra were monohydroxylated derivatives. Their formation was blocked by TxA synthase inhibitors and the same metabolites were also generated in liver microsomal incubations in the presence of $NADPH_2$. The oxygen atoms introduced were not found in the side chains but most likely were introduced in the cyclopentane ring and at the methanocarbon. Thus TxA synthase showed the same oxene transfer properties as heme thiolate monooxygenases. With PGI synthase unspecific oxidations by iodosylbenzene at the enzyme did not allow the same experiments.

We then asked whether in a reversed situation heme thiolate monooxygenases could catalyze TxA_2 formation from PGH_2. This was indeed observed, but only with yields around 1–2%. In contrast, HHT amounted to 20–50% from the amount of PGH_2, which confirmed that fragmentation is an unspecific process following the homolytic O-O bond cleavage [74].

Such experiments support the hypothesis that the heme thiolate structure allows two very different catalytically important reactions: both an

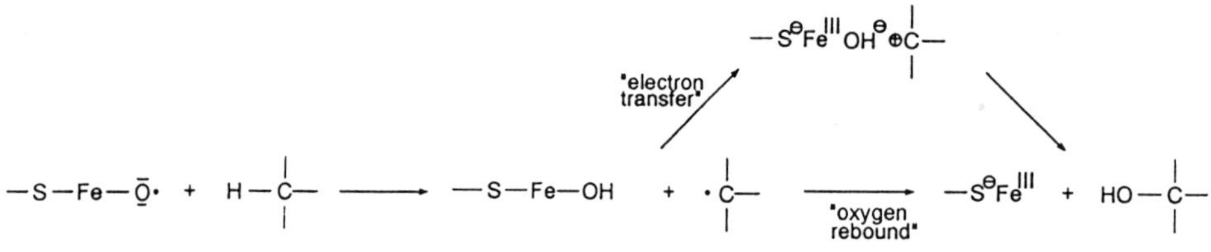

Scheme 7

oxygen atom transfer from iodosylbenzene or organic hydroperoxides intra- or intermolecularly and a homolytic cleavage of endoperoxides with subsequent electron abstraction from carbon radicals.

The latter reaction seems to be essential and crucial for TxA_2 and PGI_2 formation from PGH_2 by the two isomerases, but hydroxylations by monooxygenases may also contain an electron transfer step alternately to a proposed "oxygen rebound" mechanism (Scheme 7). The upper pathway of electron transfer to the ferryl species would be the exact equivalent to the corresponding steps in PGI_2 and TxA_2 biosynthesis. Since an oxygen rebound mechanism is not applicable for the cyclization pathways we would like to suggest electron transfer from a carbon radical to the ferryl heme as a common step in the catalysis of all heme thiolate proteins.

In essence all heme thiolate-catalyzed mechanisms consist of an initial radical step with its advantage of a low activation energy, which then by electron transfer is converted to a subsequent ionic step with its high stereochemical specificity. Also the risk for radical chain reactions of the carbon radical intermediates is minimized by the rapid or even concerted electron transfer to the ferryl thiolate (or ferric thiyl) entity.

It remains to be shown whether allyl oxide synthase [101] and chloroperoxidase, which also contains a thiolate ligand [103], can also be fitted in their mechanisms to the postulated carbon radical–carbo cation transition by the heme thiolate group. The electron transfer may even not be restricted to carbon radicals since a heme thiolate protein has also been identified as a nitrite reductase. So eventually heme thiolate structures may turn out as versatile catalysts in caged radical reactions in order to quench intermediate radical steps through rapid electron transfer.

ACKNOWLEDGMENTS

Essential parts of own results were supported by the Deutsche Forschungsgemeinschaft. We appreciate the help of R. Brugger and G. Naschwitz in preparing the manuscript.

NOTE ADDED IN PROOF

The Three-Dimensional Structure of Prostaglandin Endoperoxide Synthase

Since this chapter was written, the structure of PGH synthase (isoenzyme 1) has been published [104]. The structure of the enzyme, complexed with the nonsteroidal antiinflammatory drug flurbiprofen, was resolved to 3.5 Å, using crystals grown in the presence of detergents. The large catalytic domain is a globular structure with a helical content of about 40%. It contains the clearly separated cyclooxygenase and peroxidase active sites.

The Peroxidase Active Site

The heme is located in a shallow cleft near the surface of the protein, and His-388 coordinates the iron at the proximal side of the heme. AT the distal side of the heme, Gln-203 and His-207 approach the iron to within 5 Å but do not coordinate the metal. This result corroborates a five-coordinate iron as proposed from EPR data. A striking topological similarity (in secondary and tertiary structure) to myeloperoxidase has been noted which parallels the spectroscopic similarities between the two heme proteins. The heme-binding cleft is rather open to the solvent and may provide access for bulky peroxidase substrates such as PGG_2.

The Cyclooxygenase Active Site

The binding site for the cyclooxygenase substrate arachidonic acid has been identified and modelled from the localization of the competitive inhibitor flurbiprofen and the vicinity of Ser-530 which is acetylated by aspirine under inhibition of cyclooxygenase activity. This binding site consists of a long narrow channel extending from the outer surface to the center of the protein. Tyr-385, the residue proposed for the hydrogen abstraction at C-13 of arachidonic acid, is located at the apex of the channel, exactly at a position suitable for the proposed function, and at a distance of 10 Å from the heme edge. Tyr-385 is surrounded by aromatic residues and connected by a short helical backbone to His-388, the distal heme ligand. These features can make up an efficient electron transfer path between Tyr-385 and the heme, which is required for the oxidation of this tyrosine by the porphyrin π cation radical to a catalytically competent tyrosyl radical.

In summary, it is surprising how many of the features predicted from spectroscopic and site-directed mutagenesis work are now verified in the resolved structure. Especially, the reaction mechanism as suggested by the Ruf model fits nicely into the structure of the enzyme. The structure provides a rational basis for the future elucidation of the reaction mechanism of the enzyme.

REFERENCES

1. H. Graf and V. Ullrich, Prostacyclin synthase as a cytochrome P450 enzyme, *Cytochrome P-450, Biochemistry, Biophysics and Environmental Implications* (E. Hietanen, M. Laitinen, and O. Hänninen, eds.), Elsevier, Amsterdam, 1982, p. 103.
2. G. Palmer and J. Reedijk, Nomenclature committee of the international union of biochemistry (NC-IVB) nomenclature of electron transfer proteins: recommendations, *Biochim. Biophys. Acta, 1060*: 599 (1989).
3. V. Ullrich, Cytochrome P450 and biological hydroxylation reactions, *Top. Curr. Chem., 83*: 68 (1979).
4. W.L. Smith and L.J. Marnett, Prostaglandin endoperoxide synthase: structure and catalysis, *Biochim. Biophys. Acta, 1083*: 1 (1991).
5. S. Yamamoto, "Enzymatic" lipid peroxidation: reactions of mammalian lipoxygenases, *Free Rad. Biol. Med., 10*: 149 (1991).
6. P. Needleman, J. Turk, B. A. Jakschik, A.R. Morrison, and J.B. Lefkowith, Arachidonic acid metabolism, *Annu. Rev. Biochem., 55*: 69 (1986).
7. W.L. Smith, L.J. Marnett, and D.L. DeWitt, Prostaglandin and thromboxane biosynthesis, *Pharmacol. Ther., 49*: 153 (1991).
8. J. Capdevila, L.J. Marnett, N. Chacos, R. Prough, and R. Estabrook, Cytochrome P-450 dependent oxygenation of arachidonic acid to hydroxyeicosatetraenoic acids, *Proc. Natl. Acad. Sci. USA, 79*: 767 (1982).
9. V. Ullrich, L. Castle, and P. Weber, Spectral evidence for the cytochrome P450 nature of prostacyclin synthase, *Biochem. Pharmacol., 30*: 2033 (1981).
10. B. Samuelsson, On the incorporation of oxygen in the conversion of 8,11,14-eicosatrienoic acid to prostaglandin E_1, *J. Am. Chem. Soc., 87*: 3011 (1965).
11. D.L. DeWitt, Prostaglandin endoperoxide synthase: regulation of enzyme expression, *Biochim. Biophys. Acta, 1083*: 121 (1991).
12. R.J. Kulmacz and W.E.M. Lands, Cyclooxygenase: measurement, purification and properties, in *Prostaglandins and Related Substances: A Practical Approach* (C. Benedetto, R.G. McDonald-Gibson, S. Nigam, and T.F. Slater, eds.), IRL Press, Washington, DC, 1987, p. 209.
13. L.J. Marnett, P.H. Siedlik, R.C. Ochs, W.D. Pagels, M. Das, K.V. Honn, R.H. Warnock, B.E. Tainer, and T.E., Mechanism of the stimulation of prostaglandin H synthase and prostacyclin synthase by the antithrombotic and antimetastatic agent, nafazatrom, *Mol. Pharmacol., 26*: 328 (1984).
14. W. Nastainczyk, D. Schuhn, and V. Ullrich, Spectral intermediates of prostaglandin hydroperoxidase, *Eur. J. Biochem., 144*: 381 (1984).
15. H.J. Scherer, R. Karthein, S. Strieder, and H.H. Ruf, Chemical modification of prostaglandin endoperoxide synthase by N-acetylimidazol, *Eur. J. Biochem., 205*: 751 (1992).
16. G.J. Roth, E.T. Machuga, and P. Strittmatter, The heme-binding properties of prostaglandin synthetase from sheep vesicular gland, *J. Biol. Chem., 256*: 10018 (1981).
17. T. Miyamoto, N. Ogino, S. Yamamoto, and O. Hayaishi, Purification of prostaglandin endoperoxide synthetase from bovine vesicular gland microsomes, *J. Biol. Chem., 251*: 2629 (1976).

18. F.J. van der Ouderaa, M. Buytenhek, F.J. Slikkerveer, and D.A. van Dorp, On the haemoprotein character of prostaglandin endoperoxide synthetase, *Biochim., Biophys. Acta, 572*: 29 (1979).
19. R.J. Kulmacz and W.E.M. Lands, Prostaglandin H synthase, stoichiometry of heme cofactor, *J. Biol. Chem., 259*: 6358 (1984).
20. R. Karthein, W. Nastainczyk, and H.H. Ruf, EPR study of ferric native prostaglandin H synthase and its ferrous NO derivative, *Eur. J. Biochem., 166*: 173 (1987).
21. H.H. Ruf, D. Schuhn, R. Dietz, W. Nastainczyk, and M. Nielsen, Target size analysis of prostaglandin endoperoxide synthase, *Eur. J. Biochem., 204*: 1069 (1992).
22. D.L. DeWitt and W.L. Smith, Primary structure of prostaglandin G/H synthase from sheep vesicular gland determined from the complementary DNA sequence, *Proc. Natl. Acad. Sci. USA, 85*: 1412 (1988).
23. J.P. Merlie, D. Fagan, J. Mudd, and P. Needleman, Isolation and characterization of the complementary DNA for sheep seminal vesicle prostaglandin endoperoxide synthase (cyclooxygenase), *J. Biol. Chem., 263*: 3550 (1988).
24. C. Yokoyama, T. Takai, and T. Tanabe, Primary structure of sheep prostaglandin endoperoxide synthase deduced from cDNA sequence, *FEBS Lett., 231*: 347 (1988).
25. D.L. DeWitt, E.A. El-Harith, S.A. Kraemer, M.J. Andrews, E.F. Yao, R.L. Armstrong, and W.L. Smith, The aspirin and heme-binding sites of ovine and murine prostaglandin endoperoxide synthases, *J. Biol. Chem., 265*: 5192 (1990).
26. C. Yokoyama and T. Tanabe, On the haemoprotein character of prostaglandin endoperoxide synthetase, *Biochem. Res. Commun., 165*: 888 (1989).
27. M.K. O'Banion, H.B. Sadowski, V. Winn, and D.A. Young, A serum- and glucocorticoid-regulated 4-kilobase mRNA encodes a cyclooxygenase-related protein, *J. Biol. Chem., 266*: 23261 (1991).
28. T. Hla and K. Neilson, Human cyclooxygenase-2 cDNA, *Proc. Natl. Acad. Sci. USA, 89*: 7384 (1992).
29. S.H. Lee, E. Soyoola, P. Chanmugam, S. Hart, W.Q. Sun, H. Zhong, S. Liou, D. Simmons, and D. Hwang, Selective expression of mitogen-inducible cyclooxygenase in macrophages stimulated with lipopolysaccharide, *J. Biol. Chem., 267*: 25934 (1992).
30. D. Picot and R.M. Garavito, Crystallization of a membrane protein: prostaglandin H synthase, in *Cytochrome P450: Biochemistry and Biophysics* (I. Schultz, ed.), Taylor and Francis, London, 1989, p. 29.
31. K. Jahnke, G.H. Degen, and M. Buehner, Crystallization of prostaglandin-H synthase for X-ray structure analysis, *Environ. Health Persp., 88*: 33 (1990).
32. R.M. Garavito and D. Picot, Crystallization of membrane proteins: a minireview, *J. Crystal Growth, 110*: 89 (1991).
33. R.J. Kulmacz, A.L. Tsai, and G. Palmer, Heme spin states and peroxide-induced radical species in prostaglandin H synthase, *J. Biol. Chem., 262*: 10524 (1987).
34. R.J. Kulmacz, Y. Ren, A.L. Tsai, and G. Palmer, Prostaglandin H synthase: spectroscopic studies of the interaction with hydroperoxides and with indomethacin, *Biochemistry, 29*: 8760 (1990).

35. T. Yonetani and H. Anni, Yeast cytochrome c peroxidase, coordination and spin states of heme prosthetic group, *J. Biol. Chem., 262*: 9547 (1987).
36. S. Dasgupta, D.L. Rousseau, H. Anni, and T. Yonetani, Structural characterization of cytochrome c peroxidase by resonance Raman scattering, *J. Biol. Chem., 264*: 654 (1989).
37. T. Shimokawa, and W.L. Smith, Essential histidines of prostaglandin endoperoxide synthase, *J. Biol. Chem., 266*: 6168 (1991).
38. L.B. Dugad, G.N. La Mar, H.C. Lee, M. Ikeda-Saito, K.S. Booth, and W.S. Caughey, A nuclear Overhauser effect study of the active site of myeloperoxidase, *J. Biol. Chem., 265*: 7173 (1990).
39. M. Ikeda-Saito, Spectroscopic ligand binding, and enzymatic properties of the spleen green heme protein, *J. Biol. Chem., 260*: 11688 (1985).
40. R. Wever, W.A. Oertling, H. Hoogland, B.G.J.M. Bolscher, Y. Kim, and G.T. Babcock, Denaturation and renaturation of myeloperoxidase, *J. Biol. Chem., 266*: 24308 (1991).
41. R. Dietz, W. Nastainczyk, and H.H. Ruf, Higher oxidation states of prostaglandin H synthase, *Eur. J. Biochem., 171*: 321 (1988).
42. A.-M. Lambeir, C.M. Markey, H.B. Dunford, and L.J. Marnett, Spectral properties of the higher oxidation states of prostaglandin H synthase, *J. Biol. Chem., 260*: 14894 (1985).
43. R.J. Kulmacz, Prostaglandin H synthase and hydroperoxides: peroxidase reaction and inactivation kinetics, *Arch. Biochem. Biophys., 249*: 273 (1986).
44. R. Karthein, R. Dietz, W. Nastainczyk, and H.H. Ruf, Higher oxidation states of prostaglandin H synthase, *Eur. J. Biochem., 171*: 313 (1988).
45. G. Lassmann, R. Odenwaller, J.F. Curtis, J.A. DeGray, R.P. Mason, L.J. Marnett, and T.E. Eling, Electron spin resonance investigation of tyrosyl radicals of prostaglandin H synthase, *J. Biol. Chem., 266*: 20045 (1991).
46. A. Tsai, G. Palmer, and R.J. Kulmacz, Prostaglandin H synthase, *J. Biol. Chem., 267*: 17753 (1992).
47. J.A. Degray, G. Lassmann, J.F. Curtis, T.A. Kennedy, L.J. Marnett, T.E. Eling, and R.P. Mason, Spectral analysis of the protein-derived tyrosyl radicals from prostaglandin H synthase, *J. Biol. Chem., 267*: 23583 (1992).
48. I.D. Macdonald, G. Graff, L.A. Anderson, and H.B. Dunford, Optical spectra and kinetics of reactions of prostaglandin H synthase, *Arch. Biochem. Biophys., 272*: 194 (1989).
49. Y. Hsuanyu and H.B. Dunford, Kinetics of the reaction of prostaglandin H synthase compound II with ascorbic acid, *Arch. Biochem. Biophys., 281*: 282 (1990).
50. Y. Hsuanyu and H.B. Dunford, Reduction of prostaglandin H synthase compound II by phenol and hydroquinone, and the effect of indomethacin, *Arch. Biochem. Biophys., 292*: 213 (1992).
51. Y. Hsuanyu and H.B. Dunford, Prostaglandin H synthase kinetics, *J. Biol. Chem., 267*: 17649 (1992).
52. M. Bakovic and H.B. Dunford, Kinetics of the oxidation of p-coumaric acid by prostaglandin H synthase and hydrogen peroxide, *Biochemistry, 32*: 833 (1993).

53. T.E. Eling, D.C. Thompson, G.L. Foureman, J.F. Curtis, and M.F. Hughes, Prostaglandin H synthase and xenobiotic oxidation, *Annu. Rev. Pharmacol. Toxicol., 30*: 1 (1990).
54. R.W. Egan, P.H. Gale, W.J. Vandenheuvel, E.M. Baptista, and F.A. Kuehl, Mechanism of oxygen transfer by prostaglandin hydroperoxidase, *J. Biol. Chem., 255*: 323 (1980).
55. P. Plé and L.J. Marnett, Alkylaryl sulfides as peroxidase reducing substrates for prostaglandin H synthase, *J. Biol. Chem., 264*: 13983 (1989).
56. M. Hamberg and R. Samuelsson, On the mechanism of the biosynthesis of prostaglandins E1 and F1α, *J. Biol. Chem., 242*: 5336 (1967).
57. P.-Y. Kwok, F.W. Muellner, and J. Fried, Enzymatic conversions of 10,10-difluoroarachidonic acid with PGH synthase and soybean lipoxygenase, *J. Am. Chem. Soc., 109*: 3692 (1987).
58. T. Shimokawa, R.J. Kulmacz, D.L. DeWitt, and W.L. Smith, Tyrosine 385 of prostaglandin endoperoxide synthase is required for cyclooxygenase catalysis, *J. Biol. Chem., 265*: 20073 (1990).
59. M. Hecker, V. Ullrich, C. Fischer, and C.O. Meese, Identification of novel arachidonic acid metabolites formed by prostaglandin H synthase, *Eur. J. Biochem., 169*: 113 (1987).
60. R.P. Mason, B. Kalayanaraman, B.E. Tainer, and T.E. Eling, A carbon-centered free radical intermediate in the prostaglandin synthetase oxidation of arachidonic acid, *J. Biol. Chem., 255*: 5019 (1980).
61. J. Schreiber, T.E. Eling, and R.P. Mason, The oxidation of arachidonic acid by the cyclooxygenase activity of purified prostaglandin H synthase, *Arch. Biochem. Biophys., 249*: 126 (1986).
62. M.E. Hemler and W.E.M. Lands, Evidence for a peroxide-initiated free radical mechanism of prostaglandin biosynthesis, *J. Biol. Chem., 255*: 6253 (1980).
63. R. Odenwaller, K.R. Maddipati, and L.J. Marnett, Detection of a higher oxidation state of manganese-prostaglandin endoperoxide synthase, *J. Biol. Chem., 267*: 13863 (1992).
64. S. Strieder, K. Schaible, H.-J. Scherer, R. Dietz, and H.H. Ruf, Prostaglandin endoperoxide synthase substituted with manganese protoporphyrin IX, *J. Biol. Chem., 267*: 13870 (1992).
65. W.L. Smith, T.E. Eling, R.J. Kulmacz, L.J. Marnett, and A. Tsai, Tyrosyl radicals and their role in hydroperoxide-dependent activation and inactivation of prostaglandin endoperoxide synthase, *Biochemistry, 31*: 3 (1992).
66. P.J. Marschall and R.J. Kulmacz, Prostaglandin H synthase: distinct binding sites for cyclooxygenase and peroxides substrates, *Arch. Biochem. Biophys., 266*: 162 (1988).
67. Y.-N.P. Chen, M.J. Bienkowski, and L.J. Marnett, Controlled tryptic digestion of prostaglandin H synthase, *J. Biol. Chem., 262*: 16892 (1987).
68. C. Schöneich, K.-D. Asmus, U. Dillinger, and F. von Bruchhausen, Thiyl radical attack on polyunsaturated fatty acids, *Biochem. Biophys. Res. Commun., 16*: 113 (1989).
69. R.J. Gryglewski, S. Bunting, S. Moncada, R.J. Flower, and J.R. Vane, Arterial walls are protected against deposition of platelet thrombin by a substance (pros-

taglandin X) which they make from prostaglandin endoperoxide, *Prostaglandins, 12*: 685 (1976).

70. H. Graf, H.H. Ruf, and V. Ullrich, Die Prostacyclin-Synthase ein Cytochrom-P450-Enzym, *Angew. Chemie, 95*: 497 (1983).
71. R. Tsai, C.A. Yu, I.C. Gunsolus, J. Peisach, W. Blumberg, W.H. Orme-Johnson, H. Beinert, Spin-state changes in cytochrome P-450_{cam} on binding of specific substrates, *Proc. Natl. Acad. Sci. USA, 66*: 1157 (1970).
72. D.L. DeWitt and W.L. Smith, Purification of prostacyclin synthase from bovine aorta by immunoaffinity chromatography, *J. Biol. Chem., 258* (1983).
73. M. Inoue, W.L. Smith, and D.L. DeWitt, Molecular characterization of the prostacyclin synthase, in *Advances in Prostaglandin, Thromboxane, and Leukotriene Research*, Vol. 17 (B. Samuelsson, R. Paoletti, and P.W. Ramwell, eds.), Raven Press, New York, 1987, p. 29.
74. M. Hecker and V. Ullrich, On the mechanism of prostacyclin and thromboxane A_2 biosynthesis, *J. Biol. Chem., 264*: 141 (1989).
75. M. Hecker, M. Haurand, V. Ullrich, and S. Terao, Spectral studies on structure–activity relationships of thromboxane synthase inhibitors, *Eur. J. Biochem., 157*, 217 (1986).
76. M. Hecker, M. Haurand, V. Ullrich, U. Diczfalusy, and S. Hammarström, Products, kinetics, and substrate specifity of homogenous thromboxane synthase from human platelets: development of a novel enzyme assay, *Arch. Biochem. Biophys., 254*: 124 (1987).
77. J. Fried and J. Barton, Synthesis of 13,14-dehydroprostacyclin methyl ester: a potent inhibitor of platelet aggregation, *Proc. Natl. Acad. Sci. USA, 74*: 2199 (1977).
78. U. Diczfalusy and S. Hammarström, Biosynthesis of thromboxanes, in *Advances in Prostaglandin and Thromboxane Research*, Vol. 6 (B. Samuelsson, P.W. Ramwell, and R. Paoletti, eds.), Raven Press, New York, 1980, p. 267.
79. R.A. Johnson, Synthesis of thromboxanes, prostacyclin, and the endoperoxides, in *Advances in Prostaglandin, Thromboxane and Leukotriene Research*, Vol. 14 (J.E. Pike and D.R. Morton, eds.), Raven Press, New York, 1985, p. 131.
80. E.R. Hall and H.H. Tai, Purification of thromboxane synthetase and evidence of two distinct mechanisms for the formation of 12-L-hydroxy-5,8,10-heptadecatrienoic acid by porcine lung microsomes, *Biochim. Biophys. Acta, 665*: 498 (1981).
81. R. Nüsing, R. Lesch, and V. Ullrich, Immunohistochemical localization of thromboxane synthase in human tissues, *Eicosanoids, 3*: 53 (1990).
82. R. Nüsing and V. Ullrich, Immunoquantitation of thromboxane synthase in human tissues, *Eicosanoids, 3*: 175 (1990).
83. A. Raz, D. Aharony, and R. Kenig-Wakshal, Biosynthesis of thromboxane B_2 and 12-L-hydroxy-5,8,10-heptadecatrienoic acid in human platelets, *Eur. J. Biochem., 86*: 447 (1978).
84. M. Hamberg, J. Svensson, and B. Samuelsson, Thromboxanes: a new group of biologically active compounds derived from prostaglandins endoperoxides, *Proc. Natl. Acad. Sci. USA, 72*: 2994 (1975).

85. J. Svensson, M. Hamberg, and B. Samuelsson, On the formation and effects of thromboxane A_2 in human platelets, *Acta Physiol. Scand., 98*: 285 (1976).
86. S. Hammarström and P. Falardeau, Resolution of prostaglandin endoperoxide synthase and thromboxane synthase of human platelets, *Proc. Natl. Acad. Sci. USA, 74*: 3691 (1977).
87. M. Haurand and V. Ullrich, Isolation and characterization of thromboxane synthase from human platelets as a cytochrome P-450 enzyme, *J. Biol. Chem., 260*: 15059 (1985).
88. D.L. Cinti and M.B. Feinstein, Platelets cytochrome P-450: a possible role in arachidonate-induced aggregation, *Biochem. Biophys. Res. Commun., 73*: 171 (1976).
89. V. Ullrich and M. Haurand, Thromboxane synthase as a cytochrome P450 enzyme, in *Advances in Prostaglandins, Thromboxane, and Leukotriene Research*, Vol. 11 (B. Samuelsson, R. Paoletti, and P.W. Ramwell, eds.), Raven Press, New York, 1983, p. 105.
90. R.F. Shen and H.H. Tai, Immunoaffinity purification and characterization of thromboxane synthase from porcine lung, *J. Biol. Chem., 261*: 11592 (1986).
91. R. Nüsing, S. Schneider-Voss, and V. Ullrich, Immunoaffinity purification of human thromboxane synthase, *Arch. Biochem. Biophys., 280*: 325 (1990).
92. C. Yokoyama, A. Miyata, H. Ihara, V. Ullrich, and T. Tanabe, Molecular cloning of human platelet thromboxane A2 synthase, *Biochem. Biophys. Res. Commun., 178*: 1479 (1991).
93. K. Ohashi, K. Ruan, R.J. Kulmacz, K.K. Wu, and L. Wang, Primary structure of human thromboxane synthase determined from the cDNA sequence, *J. Biol. Chem., 267*: 789 (1992).
94. D.R. Nelson, T. Kamataki, D.J. Waxman, F.P. Guengerich, R.W. Estabrook, R. Feyereisen, F.J. Gonzalez, M.J. Coon, I.C. Gunsalus, O. Gotoh, K. Okuda, and D.W. Nebert, The P450 superfamily: update on new sequences, gene mapping, accession numbers, early trivial names of enzymes, and nomenclature, *DNA Cell Biol., 12*: 1 (1993).
95. P.R. Ortiz de Montellano, *Cytochrome P-450: Structure, Mechanism, and Biochemistry*, Plenum Press, New York, 1986.
96. S.C. Tang, S. Koch, G.C. Papfthymiuo, S. Foner, R. Frankel, J.A. Ibers, and R.H. Holm, Axial ligation modes in iron(III)porphyrins, models for the oxidized reaction states of cytochrome P-450 enzymes and the molecular structure of iron(III)protoporphyrin IX dimethylester p-nitrobenzenethiolate, *J. Am. Chem. Soc., 98*: 2414 (1976).
97. M. Kishi, 11,12-Secoprostaglandin cyclization model in thromboxane A_2 biosynthesis, *J. Chem. Soc. Chem. Commun.,* 885 (1986).
98. J.A. Turner and W. Herz, Fe(II)-induced decomposition of epidioxides, a chemical model for prostaglandin E, prostacyclin and thromboxane biosynthesis, *Experientia (Basel), 33*: 1133 (1977).
99. M. Hecker and V. Ullrich, 12(S)-Hydroxy-5,8,10(Z,E,E)-heptadecatrienoic acid (HHT) is preferentially metabolized to its 12-keto derivative by human erythrocytes in vitro, *Eicosanoids, 1*: 19 (1988).

100. F. Lichtenberger, W. Nastainczyk, and V. Ullrich, Cytochrome P450 as an oxene transferase, *Biochem. Biophys. Res. Commun., 70*: 939 (1976).
101. W.C. Song and A.R. Brash, Purification of an allene oxide synthase and identification of the enzyme as a cytochrome P450, *Science, 253*: 427 (1991).
102. M. Hecker, W.J. Baader, P. Weber, and V. Ullrich, Thromboxane synthase catalyses hydroxylations of prostaglandin H_2 analogs in the presence of iodosylbenzene, *Eur. J. Biochem., 169*: 563 (1987).
103. J.H. Dawson, J.R. Trudell, G. Barth, R.E. Linder, E. Bunnenberg, D. Djerassi, R. Chiang, and L.P. Hager, Chloroperoxidase: evidence for P-450 type heme environment from magnetic circular dichroism spectroscopy, *J. Am. Chem. Soc., 98*: 3709 (1976).
104. D. Picot, P.J. Noll, and R.M. Garavito, The structure of the membrane protein prostaglandin H_2 synthase-1 determined by X-ray crystallography, *Nature, 367*: 243–249 (1994).

7

Synthetic Aspects of Porphyrin and Metalloporphyrin Chemistry

Tilak P. Wijesekera
Sun Company, Inc., Marcus Hook, Pennsylvania
David Dolphin
University of British Columbia, Vancouver, British Columbia, Canada

I. INTRODUCTION

Nature has found a prominent place for porphyrins and related macrocycles such as chlorins, bacteriochlorins, and corrins (Figure 1), particularly in their metallated forms. Iron protoporphyrin IX is the prosthetic group of heme proteins that serve diversified biological roles (Figure 2) including oxygen and electron transport and storage (hemoglobin and myoglobin), electron transfer (cytochromes), and biocatalysis (catalase, peroxidase, lignin peroxidase, cytochrome P450). The nonmetallated ("free base") porphyrins are generally present in organisms as precursors of metalloporphyrins and are accumulated and/or excreted in certain physiological disorders such as porphyrias [1]. The preferential localization of certain porphyrins in tumor cells, coupled with the ability of porphyrins to act as photosensitizers in the production of highly reactive singlet oxygen (which in turn exerts a cytotoxic action), has been used extensively in the novel treatment of malignant tissues and pathogens, a technique referred to as photodynamic therapy (PDT) [2,3].

Catalytic activity of heme enzymes is known to be controlled by changes associated with the macrocycle and by modifications of the protein matrix, which in addition to providing a suitable environment for substrate binding may also provide axial ligands to the metal center. As such, synthetic metalloporphyrins have been used extensively as biomimetic agents and thereby in

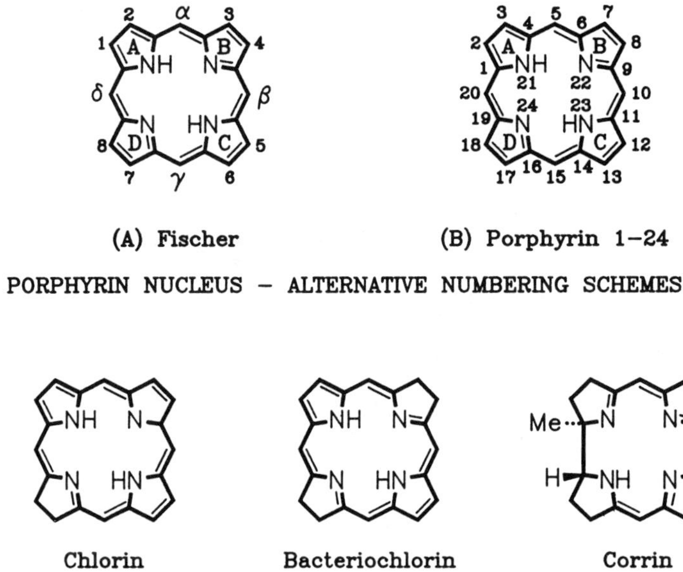

Figure 1 Tetrapyrrolic macrocycles and their numbering systems.

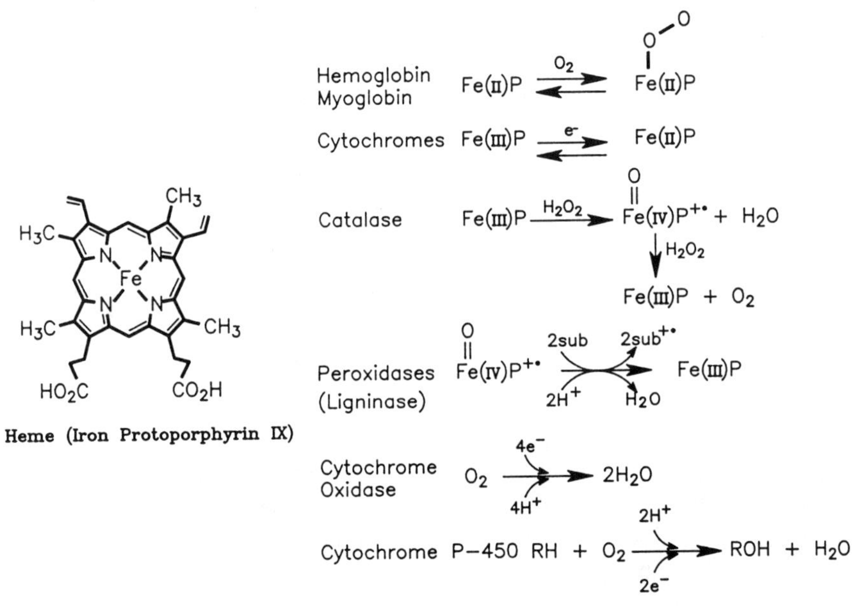

Figure 2 Biological roles of selected heme proteins.

attempts to understand the complex biological processes involved in enzymatic reactions. Furthermore such studies help in the design and development of protein-free in vitro catalysts which could be used to catalyze chemical reactions of industrial importance. In this context, cytochrome P450, the enzyme responsible for catalyzing the hydroxylation of nonpolar substrates (including drugs and pollutants) to yield partially water-soluble products, is probably the enzyme that has received the most attention of both chemists and biochemists [4]. Because of its role in dioxygen activation, understanding and successfully mimicking the catalytic cycle of cytochrome P450 has become extremely important in providing new and less expensive routes to hydroxylation of alkanes and epoxidation of olefins, both of which are industrially important reactions. Another enzyme that has received much attention, especially recently, is lignin peroxidase [5], produced by the lignolytic fungus *Phanerochaete crysosporium*, which is responsible for the degradation of the biopolymer lignin. This is primarily due to the requirement for an environmentally acceptable delignification process as a substitute for the chlorine-based bleaching currently used in the pulp and paper industry.

With the increase in interest in porphyrins and metalloporphyrins from chemists of diverse backgrounds the necessity to access variously substituted porphyrins has been greater than in the past. Apart from the simplest symmetric porphyrins, most useful derivatives either are not commercially available or are priced beyond reach of an average researcher. In this chapter we will review the currently available major synthetic routes to porphyrins and suggest approaches to the construction of the macrocycle based on the substituents required at the periphery. It is appropriate that some of the salient features of porphyrins and metalloporphyrins be discussed first.

II. STRUCTURE AND PROPERTIES OF PORPHYRINS AND METALLOPORPHYRINS

The porphyrin nucleus (Figure 1) is a cyclic tetrapyrrolic system consisting of a 20-carbon skeleton, the four pyrroles being linked by single carbon atom bridges. The nomenclature commonly used in porphyrin chemistry [6] was developed by Hans Fischer and is based on the numbering system shown in Figure 1A, where the pyrrolic (β) positions are numbered from 1 to 8 and the bridging (meso) positions named α, β, γ, and δ. Fischer nomenclature involves a large number of trivial names based on the substituents, which, however, do not convey any structural information. In addition, it involves a "type-isomer" system to distinguish between the different positional isomers. The inability of the Fischer system to name the large number of synthetic and newly isolated porphyrins led to the adoption of a systematic nomenclature based on the 1–24 numbering system (Figure 1B), developed by a joint commission on biochemical nomenclature, consisting of the International Union of Pure and Applied Chemistry (IUPAC) and the International Union of Biochemistry (IUB) [7].

The porphyrin macrocycle contains a total of 22 π electrons with 18 π electrons in direct conjugation, conforming to Huckel's $4n+2$ rule for aromaticity. The two peripheral double bonds Δ^7 and Δ^{17} of rings B and D, respectively (Figure 1B) are cross-conjugated and can therefore be reduced while retaining the aromaticity of the molecule. As such, chlorin (the parent macrocycle of a variety of chlorophylls) with one peripheral double bond reduced and bacteriochlorin with both double bonds reduced still retain the aromaticity of the macrocycle. Of the 12 peripheral positions available for substitution, the reactivity of the eight pyrrolic positions (β) differ from that of the four bridging (meso) positions. It is also important to note that although porphyrins react primarily by substitution (electrophilic or radical) due to the aromatic nature of the macrocycle, the Δ^7 and Δ^{17} double bonds, which are not part of the direct conjugation, are known to react differently under certain reaction conditions [8].

Porphyrins exhibit characteristic absorption properties in the UV-visible region. The metal-free, their diprotonated forms, as well as metallated porphyrins have an intense absorption near 400 nm ($\epsilon \sim 10^5$), referred to as the Soret band, which is characteristic of the macrocyclic system. In addition, the metal-free systems exhibit four less intense bands designated IV, III, II, and I, between 450 and 700 nm. The intensity and the exact positions of these bands are dependent on the solvent as well as the concentration. More importantly, correlations have been shown to exist between the relative intensities of the four bands and the nature and positions of the porphyrin side chains [9].

Four basic types of spectra have been observed based on the relative intensities of the four visible bands (Figure 3). Gouterman [9] proposed an interpretation of these spectral differences based on the perturbations of the π-electron levels by the electronic nature of the peripheral substituents.

A. Etio Type

This type of spectrum is characterized by a IV > III > II > I order of intensity. Porphyrins in which six or more β positions have alkyl substituents with the other two being unsubstituted will exhibit an etio-type spectrum irrespective of the relative orientations of the substituents. In addition to etioporphyrin isomers (from which the name is derived), most naturally occurring porphyrins such as copro-, uro-, hemato-, proto-, and deuteroporphyrins exhibit this type of spectrum.

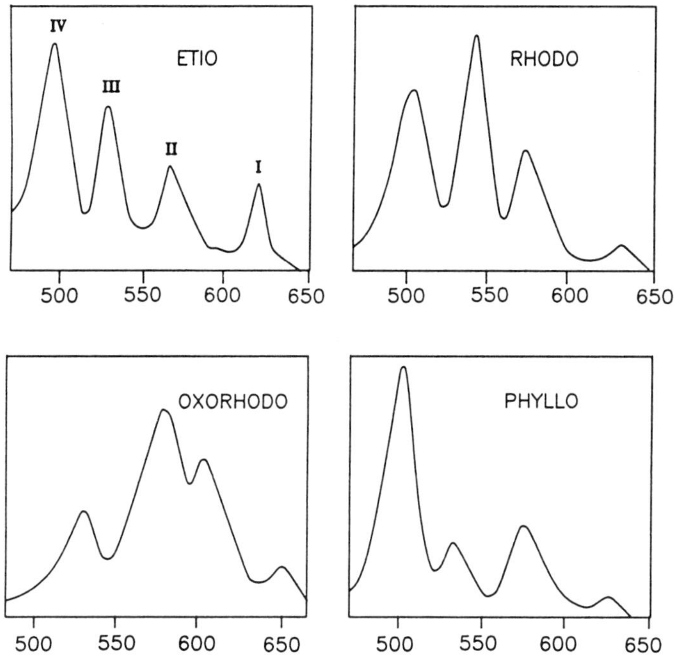

Figure 3 Typical absorption spectra of nonmetallated porphyrins.

B. Rhodo Type

One strongly electron-withdrawing group (e.g., formyl, acetyl, or carboxyl) conjugated with the porphyrin ring causes a change in the intensity pattern (III > IV > II > I) resulting in the rhodo-type spectrum (named after rhodoporphyrin). In addition, this produces a bathochromic shift of all bands in the spectrum. The rhodofying effect of one electron-withdrawing group is canceled by the presence of a similar group on an adjacent pyrrolic group, resulting in an etio-type spectrum. However, the effect on the red shift of absorption maxima is additive. We have shown that a rhodofying effect is also produced when the macrocycle is distorted by short-strap bridging, even without the introduction of any electron-withdrawing groups [10].

C. Oxorhodo Type

This spectral pattern, in which the intensities of the absorption maxima follow the order III > II > IV > I, is characteristic of porphyrins having two electron-withdrawing groups on diagonally opposite pyrrole rings. This can be viewed as a further enhancement of the rhodofying effect.

D. Phyllo Type

This spectral pattern (IV > II > III > I) named after phylloporphyrin is distinguished from the etio type by less intense bands III and I. Two substitution patterns on the periphery are known to produce the phyllo-type spectrum: (a) a single *meso*-alkyl substitution and (b) four or more unsubstituted β positions.

As shown in Figure 4a, porphin (the unsubstituted parent porphyrin) exhibits a phyllotype spectrum [490, 520, 560, 568(sh), 614; Soret: 394 nm] while β-octaethylporphyrin ($OEPH_2$) exhibits a typical etio-type spectrum (Figure 4b; 498, 534, 566, 620; Soret: 400 nm). However, most of the synthetic porphyrins, particularly those used in catalytic studies, exhibit spectral patterns that are different from any of the four types given above. The unsubstituted *meso*-tetraphenylporphyrin ($TPPH_2$) shows a modified etio-type four-band spectrum (Figure 4c; 514, 550, 588, 644; Soret: 418 nm) with an intense band IV. Tetraphenylporphyrins carrying ortho substituents on the phenyl groups show a significant reduction in the intensities of bands III and I relative to the corresponding para-substituted systems [11,12]. This is particularly so for the *ortho*-halogen (F, Cl, Br)–substituted porphyrins in which relative band intensities change to IV > II > III > I. This may be attributed to the restricted rotation of the phenyl groups resulting from steric interaction of the *ortho*-halogen substituent and the β-pyrrole hydrogen. *meso*-Tetrakis(pentafluorophenyl)porphyrin ($TPFPPH_2$) with limited rotation exhibits a spectrum (506, 582; Soret: 412 nm) in which bands III (536 sh) and I (636 wk) are suppressed, while both *meso*-tetrakis(2,6-dichlorophenyl)porphyrin ($TDCPPH_2$; 512, 588; Soret: 418 nm) and *meso*-tetrakis(pentachlorophenyl)porphyrin ($TPCPPH_2$; 514, 590; Soret: 420 nm), whose bulky *ortho*-chloro substituents prevent free rotation of the phenyl groups, exhibit essentially two-band (corresponding to bands IV and II) spectra.

At low pH, the inner nitrogens of porphyrins are protonated to produce dicationic species whose visible spectra differ from those of the corresponding free base. As shown in Figure 4a, protonation changes the phyllo-type four-band spectrum of porphin to a single strong absorption at 544 nm with shoulders at 516 and 570 nm while the etio-type four-band spectrum of octaethylporphyrin changes to a two-band spectrum (548 and 590 nm) with a shoulder at 568 nm (Figure 4b). In acid, unsubstituted tetraphenylporphyrin shows a single broad band at 654 nm with a shoulder at 600 nm (Figure 4c) while the benzene ring–substituted tetraphenylporphyrins show a red shift

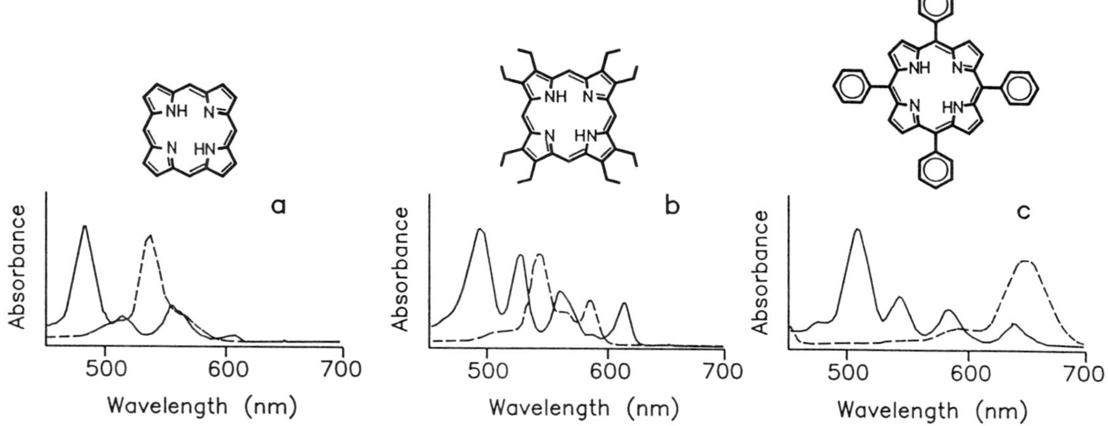

Figure 4 Absorption spectra of selected porphyrins: (a) porphine; (b) octaethylporphyrin; (c) tetraphenylporphyrin. ______, free base; ----, protonated dication.

of the two-band spectra [e.g., 576 and 624 nm for both *meso*-tetrakis(2,6-dichlorophenyl)porphyrin and *meso*-tetrakis(pentafluorophenyl)porphyrin].

The porphyrin dianion, formed by the removal of the inner NH protons, acts as a tetradentate ligand capable of binding a wide variety of metal ions. Divalent metals such as Zn^{2+}, Mg^{2+}, Cu^{2+}, Fe^{2+}, Co^{2+}, and Ni^{2+} form electrically neutral complexes which may either have a square-planar or octahedral geometry depending on the presence of axial ligation. Trivalent metals such as Fe^{3+} and Mn^{3+} carry anionic axial ligands to form electrically neutral pentacoordinate square-pyramidal complexes or hexacoordinate complexes with a neutral sixth ligand. Metallation leads to a simplification of the visible

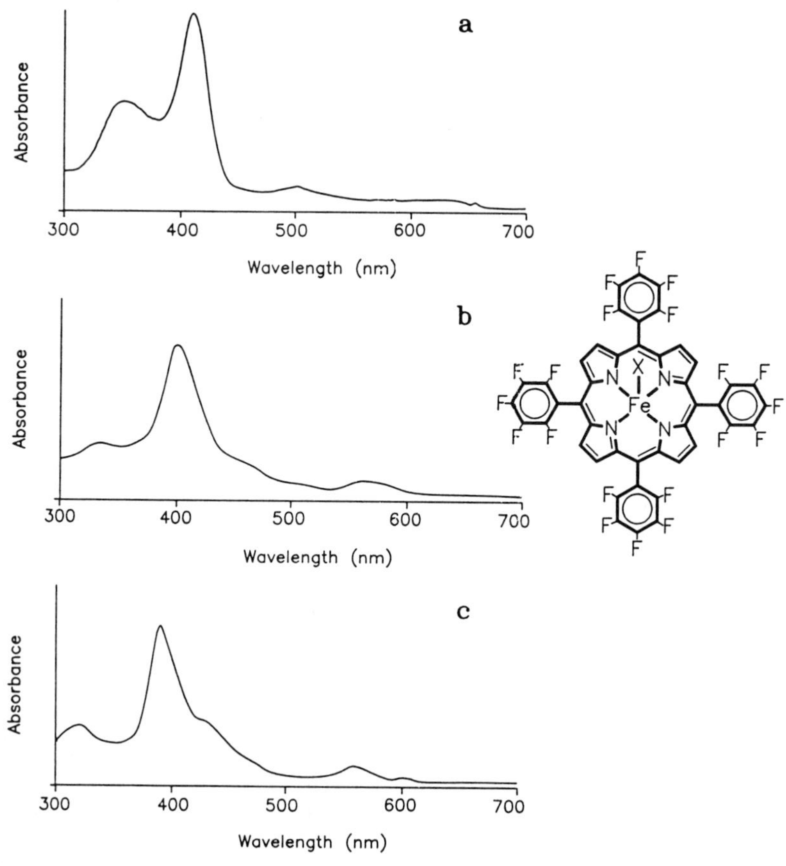

Figure 5 Absorption spectra of *meso*-tetrakis(pentafluorophenyl)porphyrinato-iron(III): (a) chloride; (b) hydroxide; (c) μ-oxo dimer.

region of the absorption spectrum while retaining the Soret absorption near 400 nm, primarily due to the change in the conjugated ring symmetry. For octaethylporphyrin and other porphyrins exhibiting etio-type spectra, coordination of a divalent metal results in two bands designated β and α (between 500 and 600 nm), whose peak positions and relative intensities depend on the metal as well as the nature of the porphyrin ligand (e.g., OEPCu: 524, 560; Soret: 398 nm). In the tetraphenylporphyrin series, the visible region consists essentially of a single strong band with two shoulders [TPPCu: 502(sh), 538, 572(sh); Soret: 414 nm]. In the case of metalloporphyrins of trivalent metal ions, the UV-visible spectrum is significantly changed by the nature of the axial anionic ligand. *meso*-Tetrakis(pentafluorophenyl)porphyrinatoiron(III), (TPFPPFeX; Figure 5a) shows a characteristic split Soret (350 and 410 nm) with two weak bands at 502 and 628 nm when the axial ligand X = Cl^-. When X = OH^- (Figure 5b), a strong Soret band is observed at 406 nm with a weak band at 563 nm [13]. When steric factors permit, as in the case of TPFPPFeX, Fe(III)porphyrins are capable of forming oxo-bridged dimers (μ-oxo) whose UV-visible spectrum (e.g., Figure 5c) differs from that of the hydroxo complex in that the Soret and the β bands are blue-shifted. Chromatographic purification of the hemin (following metal insertion) generally leads to the exchange of the axial ligand but the hemin can be regenerated by treating the resulting hydroxide (or the μ-oxo dimer) with hydrochloric acid.

Changes in the conjugation path affect the optical spectra of the porphyrin macrocycle significantly. Reduction of the one peripheral double bond, although not affecting the aromaticity of the molecule, produces a visible spectrum characterized by an intense long-wavelength absorption at $\lambda = 650$–680 nm (chlorins) while reduction of both Δ^7 and Δ^{17} double bonds produces a strong absorption at $\lambda \sim 740$ nm (bacteriochlorins).

III. SYNTHESIS OF PORPHYRINS

A. Construction of the Macrocycle

Since the landmark synthesis of protohemin (iron protoporphyrin IX) by Hans Fischer in 1929 (which earned him the Nobel Prize), significant progress has been made in the field of porphyrin synthesis [14,15] especially through the efforts of A. W. Johnson, G. W. Kenner, S. F. MacDonald, and R. B. Woodward. During the first half of the 20th century, synthesis of porphyrins, like other areas of organic synthesis, was driven by the requirement for the establishment of structures of natural products. Each synthesis was based on the availability of structurally similar starting materials and the possibility of the introduction and modification of substituents by known chemical reactions. Since the mid-1960s, a new and more systematic approach to synthesis has been adopted that focuses on the structural features of the target molecule rather than those of the starting material. Popularly known as *retrosynthetic analysis*, this approach transforms a synthetic target, in a stepwise manner, into progressively simpler systems that eventually lead to a readily available starting material [16]. The necessity to synthesize variously substituted porphyrins via multistep syntheses has resulted in proper planning by retrosynthetic analysis.

B. Symmetry and Substitution of the Target Molecule

The substitution pattern on the 12 peripheral positions of the target porphyrin molecule should form the basis of the synthetic strategy employed. Several elegant synthetic methods are available for the construction of the porphyrin macrocycle and a retrosynthetic analysis of the target system will narrow down the choices available even though certain routes are more convenient and/or produce higher yields than others. It is important to realize at the very outset that while certain substituents can be introduced after macrocyclization, others have to be already present in the pyrrolic precursor fragments.

The three most important factors to be taken into account in designing the synthesis are (a) the nature of the substituent; carbon- or heteroatom-bonded group, (b) the position of the substituent; β or meso, and (c) the overall symmetry of the molecule. Although most heteroatom substituents such as halogens, nitro, and sulfonic acid groups can be readily introduced onto the preformed porphyrin macrocycle because of the electrophilic (or radical) nature of their activity, carbon-bonded substituents such as alkyl and aryl as well as their derivatives usually need to be incorporated before or during macrocyclization. Furthermore, if such a substituent is required on a β position, that group or an interconvertible functional group has to be incorporated onto the pyrrole subunit during its synthesis from acyclic precursors. The most important factor, however, is the overall symmetry of the macrocycle. Symmetrically placed peripheral substituents in the target molecule allow a synthetic chemist to use a simple one- or two-step method for the construction of the macrocycle while highly unsymmetric systems such as those found in nature require multistep manipulations that require the skills of an experienced porphyrin chemist.

C. Retrosynthetic Analysis

Based on the currently available routes for the construction of the porphyrin macrocycle, retrosynthetic analysis of the target porphyrin would lead to several simple pyrrolic systems (Scheme 1). The choice of the route explored would depend on the nature and position of the substituent as well as how

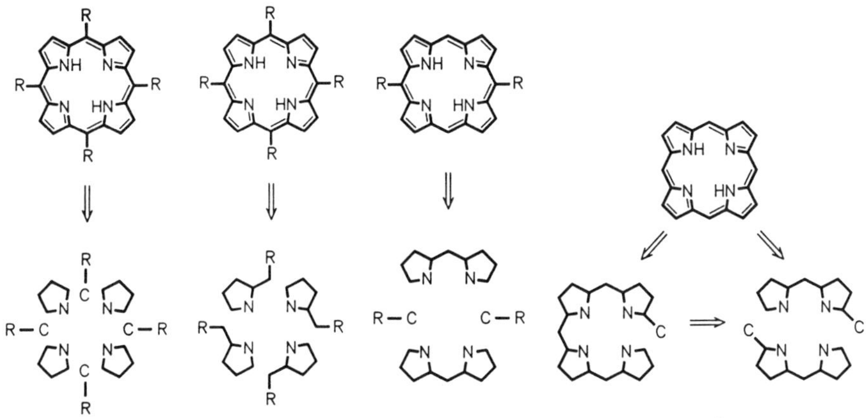

a b c d

Scheme 1

that group would affect the reactivity of the pyrrolic precursor. This is particularly important in the case of electron-withdrawing β substituents since such groups reduce the electron density of the pyrrole, thereby affecting the electrophilic reactivity at the α position. The following skeletal disconnections lead to several simple synthetic precursors for porphyrins.

Type 1 (*Scheme 1a*): This approach disconnects the macrocycle into four pyrrole units and the meso carbons in a single step. This is possible only in highly symmetric porphyrins where the meso substituents as well as the β substituents are identical. In the synthetic direction, the most suitable functional group for the meso carbon is the aldehyde group with the substituent R attached to it. The pyrrole should be unsubstituted at the α position while the β positions may either be unsubstituted or may carry substituents that do not affect its reactivity. Although it is possible to prepare porphyrins having either different R groups or two different β substituents on each pyrrole, such approaches lead to a mixture of porphyrin products and/or isomers that will require tedious separation procedures.

Type 2 (*Scheme 1b*): This approach differs from type 1 in that the meso carbon is already attached to the 2 position of the pyrrole, which in turn carries the R group. In the synthetic direction this involves a self-condensation of a 2-substituted pyrrole via a pyrryl carbo cation that undergoes a head-to-tail coupling of four units to give the porphyrin.

Type 3 (*Scheme 1c*): This disconnection leads to two identical dipyrrolic units and two meso carbons carrying substituent (R) groups. The dipyrrolic system is a 5,5′-unsubstituted dipyrromethane which can be coupled with an appropriate aldehyde. This approach can be used where the target molecule carries different β substituents (symmetrically placed) and different meso groups in the opposite positions.

Type 4 (*Scheme 1d*): This approach leads to dipyrrolic starting materials either directly by a simultaneous two-carbon disconnection or via linear tetrapyrroles by *two* one-carbon disconnections. This is undoubtedly the best method if the target molecule is unsymmetrically substituted. The linear tetrapyrroles can be built in a stepwise manner starting from monopyrroles that carry the required β substituents and cyclized in the final step to give the porphyrin.

D. Synthetic Precursors

The above-mentioned disconnections lead to three types of pyrrolic precursors to porphyrins: monopyrroles, dipyrrolic systems, and linear tetrapyrroles. In order to transform these intermediates to porphyrins, they should be appropriately substituted at the α positions and the β substituents should be so positioned that the reactivity of the fragment is not affected. The common precursors one encounters during porphyrin synthesis are given in Figure 6.

1. Dipyrromethanes

Dipyrromethanes contain two pyrrolic units linked by a methylene (CH_2) bridge (Figure 6a). They resemble monopyrroles in their reactivity especially in that the α position is more reactive than the β position. They are prepared from monopyrroles by acid-catalyzed coupling methods but may undergo rearrangement to give isomeric mixtures under certain acidic conditions. In the coupling of dipyrromethanes to yield porphyrins, the meso carbons are introduced as formyl groups, either directly linked to it (Scheme 1b) or coupled as an aldehyde (Scheme 1c).

2. Dipyrromethenes

Dipyrromethenes contain two pyrrolic units linked by a methine (CH) bridge and hence are oxidized forms of dipyrromethanes. They are more stable in their protonated form. Figure 6b gives the four common dipyrromethene pairs that are used for coupling where the meso carbons are provided by methyl or bromomethyl groups. A dipyrrole of lesser importance is dipyrroketone where two pyrroles are linked by a carbonyl group.

3. Linear Tetrapyrroles

The nomenclature of linear tetrapyrroles is related to that of bile pigments hence the prefix "bila") and the numbering is in accordance with the IUPAC

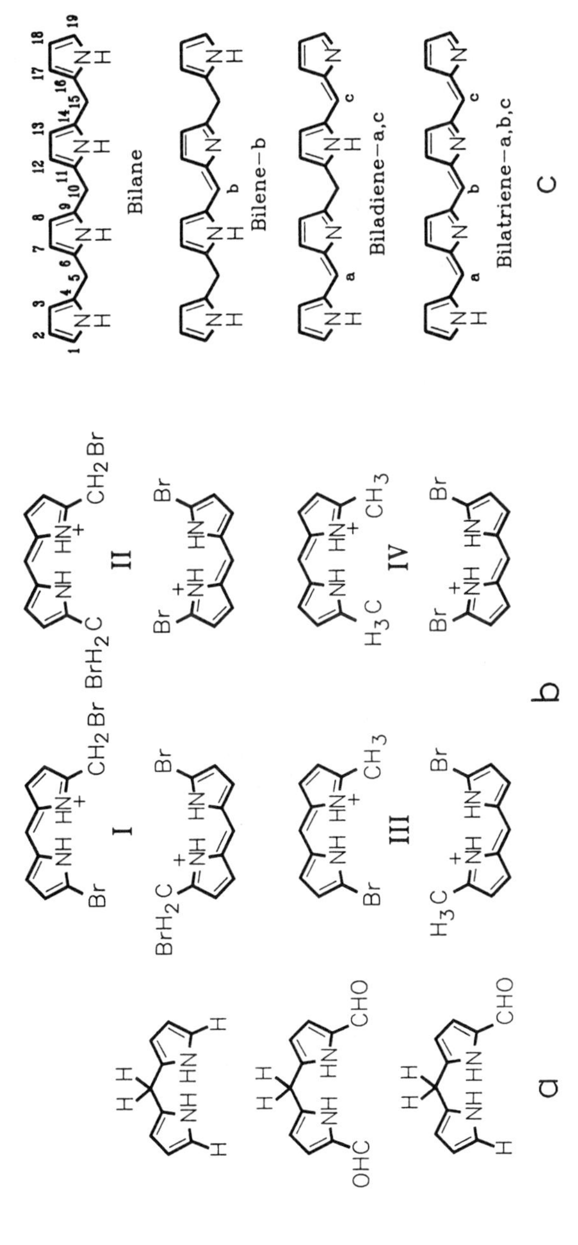

Figure 6 Common pyrrolic precursors to porphyrins: (a) dipyrromethanes; (b) dipyrromethenes; (c) linear tetrapyrroles.

recommendations [7]. The stability of the tetrapyrrole increases with the introduction of unsaturation at the bridging carbons 5, 10, and 15 (a, b, c; Figure 6c) but the stability of saturated systems can be increased by the introduction of electron-withdrawing substituents. Bilenes b and biladienes a,c are the most commonly used linear tetrapyrroles in porphyrin synthesis.

IV. GENERAL SYNTHETIC METHODS

As stated earlier, the best approach to the synthesis of a desired porphyrin system is to analyze the target molecule retrosynthetically based on the symmetry of the system and the type and relative position of the substituents. This will lead to the most suitable precursors which can be used according to the established synthetic procedures. The synthetic methods currently available are grouped according to the precursors used.

A. Cocondensation of a Pyrrole and an Aldehyde

One of the most commonly used methods for the synthesis of porphyrins has been the reaction of pyrrole (or a 3,4-disubstituted pyrrole) with an aliphatic or aromatic aldehyde. This method, which produces meso-tetrasubstituted porphyrins, has limited synthetic value due to the symmetric nature of the porphyrin produced and the restrictions on the type (and position) of the substituent that can be introduced. However, due to the ease of carrying out the reaction, this method has been used extensively to prepare symmetric porphyrins, especially for catalytic studies.

meso-Tetrasubstituted porphyrins (**4**; A = H; Scheme 2) were first prepared in 1936 by Rothmund who heated pyrrole (**1**; A = H) and an aldehyde (**2**) in pyridine in a sealed tube at 150°C for 24 h [17]. This reaction, which was later extended to include more than 25 aliphatic and aromatic aldehydes [18,19], had earlier been used to prepare the unsubstituted system porphin (**4**; A = R = H) in 0.012% yield by reacting pyrrole and formaldehyde in methanol and pyridine [20]. The yields of porphyrins from this reaction were in general less than 3% and the product was contaminated with the corresponding chlorin, the 2,3-dihydroporphyrin. Ball et al. [21] obtained improved yields for $TPPH_2$ by adding zinc acetate to the reaction mixture, a modification that was later used by Badger et al. [11] to prepare several ortho-substituted tetraphenylporphyrins. However, the major drawbacks in this synthetic procedure continued to be the low yields and the difficulty of introducing sensitive substituents (using the appropriate benzaldehydes) due to the severe reaction conditions employed.

A major improvement in this area of cocondensation of pyrrole and aldehydes came from the work of Adler, Longo, and coworkers [22,23] when they showed that the yields of porphyrins can be greatly increased by carrying

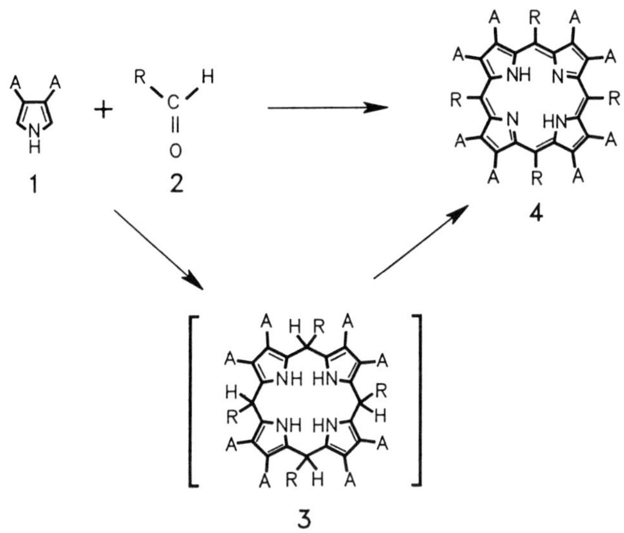

Scheme 2

out the reaction in acidic media, open to air. In refluxing propionic acid, this condensation gave yields of greater than 20%, where the product crystallized out on cooling, thus simplifying the purification procedure. Although higher yields (30–40%) were obtained by using acetic acid as the solvent, the product did not crystallize out due to the higher pK_a of acetic acid, making the isolation and purification more difficult. As in the Rothmund synthesis, the isolated porphyrin was contaminated by the corresponding chlorin, which is indicated by the presence of a strong band ~650 nm in the absorption spectrum. Although this can be separated from the porphyrin by chromatography, significant loss of material occurs due to adsorption. The recommended method of purification is to treat the porphyrin/chlorin mixture with 2,3-dichloro-5,6-dicyanobenzoquinone (DDQ) in order to oxidize the chlorin to the porphyrin [24]. The progress of the reaction can be monitored by the disappearance of the band near 650 nm and subsequent chromatography on alumina removes the oxidant to yield a 95+% recovery of the pure material. Several *ortho-*, *meta-*, and *para*-phenyl–substituted porphyrins have been prepared using this method [12].

Another general synthetic method for *meso*-tetraphenylporphyrins was developed recently by Lindsey et al. [25]. This takes into account the fact that under equilibrium conditions, the acid catalyzed condensation proceeds via the porphyrinogen **3**. The intermediacy of the porphyrinogen in the Adler–Longo synthesis had been shown by Dolphin [26], who was able to isolate β-octamethyl-*meso*-tetraphenylporphyrinogen (**3**; A = CH_3; R = C_6H_5) when 3,4-dimethylpyrrole was reacted with benzaldehyde under mild conditions (acetic acid, 50°C, 5 min). The porphyrin was obtained by treating it with an oxidizing agent (6 equivalents of iodine) since the oxidation of this sterically crowded porphyrinogen in air was slow. The Lindsey modification [25] is a two-step one-pot procedure where pyrrole and the aldehyde are initially condensed with a trace of acid catalysis under anaerobic conditions to produce the porphyrinogen, which is oxidized to the porphyrin by subsequently added oxidant. Extensive reaction optimization studies have shown that maximum yields of the porphyrin (30–40%) are obtained when equimolar concentrations (10^{-2} M) of pyrrole and aldehyde are condensed in the presence of boron trifluoride etherate (10^{-3} M). The reaction is generally carried out in a chlorocarbon solvent at ambient temperature for 1 h before the oxidant (*p*-chloranil or DDQ) is added. In most instances, evaporation of the solvent followed by flash chromatography affords the pure porphyrin.

Rothmund, Alder-Longo, and Lindsey methods (and modifications thereof) have all been used for the synthesis of *meso*-tetraphenylporphyrins with a wide variety of substituents on the phenyl groups. These include several biomimetic systems for heme oxygenation [27] as well as catalytic oxygen activation [28,29]. It is possible to prepare porphyrins carrying different groups at the meso positions by using mixed aldehydes in the cocondensation with pyrrole. However, this results in the formation of isomeric mixtures which have to be separated in order to obtain the desired porphyrin. The mild reaction conditions of the Lindsey method allow porphyrins to be prepared in good yields from sensitive aldehydes, but the optimum reaction concentration of 10^{-2} M poses a problem for any large-scale synthesis. The Adler–Longo method can therefore be considered complementary to the Lindsey method since it is amenable to scale-up for the synthesis of porphyrins in multigram quantities. However, the reaction of ortho-disubstituted benzaldehydes under the Adler–Longo conditions to produce the corresponding sterically hindered porphyrins has not met with much success. ortho-Disubstituted phenylporphyrins have been used extensively as catalysts since the bulky ortho substituents resist inactive μ-oxo dimer formation and oxidative destruction during catalysis (Section VI). Although Longo et al. [30] reported the synthesis of both $TPFPPH_2$ (Adler–Longo conditions; 18% spectrophotometric yield, 11% crude product yield) and $TPCPPH_2$ (monochloroacetic acid/benzene; 7% spectrophotometric yield), no yields have been reported for the pure product. Furthermore, strong absorption bands (659 and 663 nm, respectively) are reported for the two compounds indicating that the porphyrin is contaminated with the corresponding chlorin. Kim et al. [12] have reported the synthesis of $TDCPPH_2$ in 0.7% yield by refluxing a mixture of 2,6-dichlorobenzaldehyde and pyrrole in benzene in the presence of acetic acid. The yields have been improved using a modified Rothmund procedure (refluxing collidine instead of pyridine/sealed tube) whereby the zinc complexes TPCPPZn and TDCPPZn have been synthesized in 4–10% yield [31,32]. The zinc complexes are demetallated with acid and treated with DDQ to obtain the pure porphyrin. However, under Lindsey conditions $TDCPPH_2$ and $TPFPPH_2$ have been synthesized in 40–50% yield [33]. This method allows the synthesis of the free base porphyrin directly with no necessity for removal of chlorin.

meso-Tetramesitylporphyrin [$TMPH_2$, **4**; A = H; R = $(C_6H_2(CH_3)_3)$] is another sterically hindered porphyrin that has been used extensively for catalytic studies. Mesitaldehyde (2,4,6-trimethylbenzaldehyde) does not give the porphyrin product under either Adler–Longo or original Lindsey conditions, but has been converted to $TMPH_2$ in 4.5% yield under modified Rothmund conditions [34,35]. By a modification of the original Lindsey procedure (using ethanol for cocatalysis) it has been shown [36,37] that $TMPH_2$ can be synthesized in up to 29% yield under mild reaction conditions. Lindsey has extended this work to show that cocatalysis gives improved yields with 2-alkyl-, 2-alkoxy-, 2,6-dialkoxybenzaldehydes (in addition to 2,4,6-trialkylbenzaldehyde) but *ortho*-halogen–substituted benzaldehydes (e.g., 2,6-dichloro-, 2,6-difluoro-, and pentafluorobenzaldehydes) show no increase in yields [33].

β-Octasubstituted *meso*-tetraphenylporphyrins have also been synthesized by cocondensation of the 3,4-disubstituted pyrrole and the appropriate benzaldehyde. Dolphin [26] prepared β-octamethyl-*meso*-tetraphenylporphyrin (**4**, A = CH_3; R = C_6H_5) in 43% yield by reacting 3,4-dimethylpyrrole with benzaldehyde in refluxing acetic acid, whereas Evans and Smith [38] prepared the analogous octaethyl-*meso*-tetraphenylporphyrin in 10% yield starting from 3,4-diethylpyrrole. Using 3,4-diphenylpyrrole and benzaldehyde, Medforth and Smith [39] prepared dodecaphenylporphyrin (**4**; A = R = C_6H_5) in 5.7% yield under Lindsey conditions, whereas Takeda et al. [40] recently obtained a 55% yield of this porphyrin under modified Adler–Longo conditions (refluxing in acetic acid for 20 h followed by oxidation with DDQ).

A useful extension of the acid-catalyzed cocondensations of pyrroles and aldehydes is the method developed by Cheng and LeGoff [41] to synthesize meso-unsubstituted β-substituted porphyrins. The reactions were carried out by refluxing ethanolic solutions of 3,4-disubstituted pyrroles with excess of formaldehyde in the presence of hydrochloric or hydrobromic acids. After several hours of reflux, the reaction mixtures were allowed to stand exposed to air to complete oxidation. Several meso-unsubstituted porphyrins carrying the same alkyl group or different substituents (alkyl/acetyl, alkyl/ester, and phenyl/ester) in the two β positions of each pyrrole have been prepared in 64–96% yield. When the two β substituents are different, isomeric mixtures are produced as expected. Changing the solvent (to methanol or tetrahydrofuran) or the aldehyde (to aliphatic or aromatic aldehydes) resulted in little or no porphyrin formation. Treibs and Haberle [42] obtained 77% β-octamethylporphyrin from the reaction of 3,4-dimethylpyrrole with formaldehyde in acetic acid and pyridine (Cheng and LeGoff report a 76% yield), while under similar reaction conditions 3,4-diphenylpyrrole gave β-octaphenylporphyrin (OPPH$_2$; **4**, A = C_6H_5; R = H) in 26% yield [43]. Recently, Takeda and coworkers [44] prepared OPPH$_2$ in 72% yield by condensing 3,4-diphenylpyrrole with formaldehyde in ethanol/aq. HBr [cf. 41] and oxidizing the initially formed porphyrinogen with DDQ.

B. Self-Condensation of 2-Substituted Pyrroles

As shown in Scheme 3, this method uses the head-to-tail condensation of a monopyrrole **5** under acid conditions. The efficiency of this condensation reaction depends on the generation, stability and reactivity of the pyrryl-2-carbo cation **6**, produced by the removal of the leaving group (L) of the monopyrrole **5**. The 5 position of the pyrrole should be unsubstituted or should carry a carboxyl group (X = H or CO_2H), which would undergo facile elimination (decarboxylation) under the reaction conditions. As is not the case in the cocondensation described earlier, the nature of this reaction allows the β substituents of the pyrrole to be different (A $\neq$ B) and yet yield a single porphyrin product, although under certain acid concentrations the linear polypyrrolic intermediates are known to undergo fragmentation/recombination at the bridging carbons leading to isomer formation. Nature has chosen this method to construct the common tetrapyrrolic macrocycle uroporphyrinogen using the monopyrrole porphobilinogen (**5**, A = CH_2CO_2H; B = $CH_2CH_2CO_2H$; R = H; X = H; L = NH_2).

Siedel and Winkler [45] prepared several β-substituted porphyrins by heating 5-carboxy-3,4-dialkyl-2-hydroxymethylpyrroles either dry or in sol-

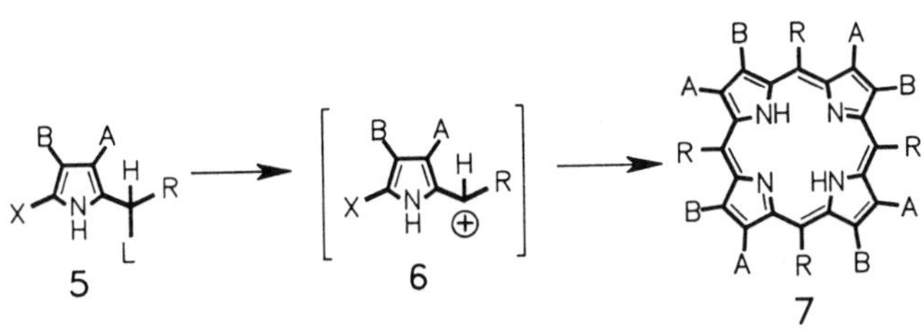

Scheme 3

ution; the 3,4-dimethylpyrrole analog (**5**; A = B = CH_3; R = H; X = CO_2H; L = OH) produced a 46% yield of octamethylporphyrin (OMP) on heating at 160–170°C. However, if one or both β positions were unsubstituted, no porphyrin product was obtained. Octamethylporphyrin has also been prepared (as the copper complex) by reacting 2-aminomethyl-5-carboxy-3,4-dimethylpyrrole (**5**; A = B = CH_3; R = H; X = CO_2H; L = NH_2) with copper(II) acetate in refluxing methanol [46]. A template effect by the metal has been suggested because only traces of porphyrin were produced in the absence of metal. However, octaethylporphyrin (OEP), one of the most widely used porphyrin model systems, has also been made by a similar head-to-tail condensation, without any added metal salts. The precursor pyrrole used is 5-carboxy-3,4-diethyl-2-*N,N*-dimethylaminomethylpyrrole [**5**, A = B = C_2H_5; R = H; X = CO_2H; L = $N(CH_3)_2$]. This is refluxed in near-neutral aqueous solution to give initially the porphyrinogen, which is oxidized to the porphyrin in air [47,48]. The precursor pyrroles for these syntheses are prepared from 3,4-dialkyl-5-methylpyrrole-2-carboxylates, which in turn are obtained from acyclic precursors by standard Knorr reactions [49]. A recent report describes the synthesis of several 2-(substituted methyl) pyrroles and their transformations to porphyrins under near-neutral or basic conditions [50].

Porphyrins carrying β-fluoro substituents have also been synthesized by the cyclocondensation of the appropriately substituted pyrroles. 2,7,12,17-Tetrakis(trifluoromethyl)-3,8,13,18-tetraethylporphyrin (**7**, A = C_2H_5; B = CF_3) has been prepared in 12% yield (as its copper complex) by refluxing 2-acetoxymethyl-5-carboxy-3-ethyl-4-trifluoromethylpyrrole (**5**; A = C_2H_5; B = CF_3; R = H; X = CO_2H; L = $OCOCH_3$) in acetic acid in the presence of copper acetate used as a template [51]. Once again, the starting pyrrole has been prepared via a modified Knorr reaction [52]. The analogous 2-hydroxymethylpyrrole (**5**, A = C_2H_5; B = CF_3; R = H; X = CO_2H; L = OH), however, has been converted to the same porphyrin product in 30% yield by a room temperature reaction in ethanol catalyzed by hydrobromic acid without added metal salts [53]. β-Fluoro-substituted 2-hydroxymethylpyrrole (**5**, A = CH_3; B = F; R = H; X = CO_2H; L = OH) has also been converted to the corresponding porphyrin, 2,7,12,17-tetrafluoro-3,8,13,18-tetramethylporphyrin by refluxing in acetic acid in the presence of potassium ferricyanide although in low 2% yield [54]. In the latter two cases, the starting pyrroles carrying the appropriate β substituents were prepared by special reactions.

Another porphyrin synthesis that involves a self-condensation of a 2-hydroxymethylpyrrole uses as the starting material 3,4-disubstituted pyrrole-2-carboxylate esters. The ester at the 2 position is reduced in situ with lithium aluminum hydride to give the corresponding 2-hydroxymethylpyrrole precursor (**5**, R = X = H; L = OH), which is cyclocondensed in dichloromethane (without isolation) using *p*-toluenesulfonic acid as catalyst [55]. This method has been used to synthesize, in good yields, a series of porphyrins carrying a variety of β substituents including alkyl groups, substituted (2-, 3-, 4- as well as 2,6 and 2,4,6-) phenyl groups and ester groups. It has been shown that the ester group in the α position of the starting pyrrole is selectively reduced by lithium aluminum hydride leaving any ester groups at the β position unaffected [56]. The appropriately β-substituted 5-unsubstituted pyrrole-2-carboxylate esters required for this synthesis are prepared by the reaction of nitroolefins or β-acetoxynitro compounds with α-isocyanoesters in the presence of an organic base [57].

Self-condensation of β-unsubstituted 2-hydroxymethylpyrrole (**5**, A = B = R = X = H; L = OH) has been used to prepare porphin itself. Krol [58] obtained yields of up to 5% by treating dilute solutions of the above pyrrole with potassium persulfate in glacial acetic acid at 60°C. Longo et al. [59] reported yields as high as 18% for this synthesis by carrying out the cyclocondensation in dilute ethylbenzene solutions at 100°C over a period of 11 days. Yalman [60] emphasized the importance of the metal template effect in his reported synthesis of copper porphin (1–20% yield) by adding a precooled solution of 2-hydroxymethylpyrrole in ether/dimethylformamide/acetic acid to a preheated (145°C) mixture of excess copper acetate and dimethylformamide.

Recently, a new synthesis was been reported [61] for the meso-tetrasubstituted porphyrins via the self-condensation of β-unsubstituted 2-hydroxymethylpyrroles **5** (A = B = H; X = CO_2H; L = OH) where the R group is alkyl or aryl. The reactions have been carried out in refluxing propionic acid with yields ranging from 9% to 34%. Addition of zinc acetate to the reaction mixture improves the yield by 30–200%, which once again suggests a template effect of the metal. The R group was introduced into the pyrrole precursor via a Vilsmeier-type acylation followed by reduction with lithium aluminum hydride.

C. Direct Coupling of Dipyrrolic Intermediates

This method, commonly referred to as the "2+2 synthesis," has been used extensively in porphyrin synthesis due to the variety of unsymmetric β-substituent patterns that can be accessed. It is important to note that only one of the two components needs to be symmetric to ensure a single porphyrin product. Dipyrromethanes and dipyrromethenes are the widely used intermediates in the 2 + 2 syntheses and must carry suitable functional groups at the α positions in order to couple to each other. These can be prepared starting from monopyrroles, which are easily obtained by the classical Knorr reactions followed by functional group modifications [49].

1. Porphyrins from Dipyrromethenes

The 2 + 2 synthesis using dipyrromethenes as intermediates formed the backbone of Fischer's classical porphyrin syntheses [62]. The coupling reaction involved the fusion, in succinic acid, of one of four pairs of dipyrromethene intermediates shown in Figure 6b. The yields were generally low since the sensitive β substituents did not survive the harsh reaction conditions. Self-condensation of dipyrromethenes has been used in the synthesis of porphyrins under milder reaction conditions [63]. The appropriately substituted 5′-bromo-5-methyldipyrromethenes (III) in refluxing formic acid (one equivalent of added bromine) produced etioporphyrin I and coproporphyrin I in up to 60% yield. This synthesis can be generalized to prepare porphyrins of C_{2h} symmetry using the more readily available 5′-carboxy-5-methyldipyrromethene, which is brominatively decarboxylated under the reaction conditions. It has also been shown that the dipyrromethene pair IV (Figure 6b) can be used to synthesize porphyrins under similar reaction conditions [64].

2. Porphyrins from Dipyrromethanes

MacDonald [65] developed an alternative 2 + 2 synthesis that uses dipyrromethanes, a procedure later modified by Kenner [66]. This method (Scheme 4) involves the condensation of 5,5′-diformyldipyrromethanes (**8**) with 5,5′-diunsubstituted dipyrromethanes (**9**) under mild acid catalysis. The intermediate porphodimethene (**11**) is rapidly oxidized to the porphyrin in air. Under the same conditions, 5′-unsubstituted 5-formyldipyrromethanes (**10**) can undergo self-condensation to give porphyrins **12a** or **12b** of twofold axial symmetry [67]. This method has been used successfully in the synthesis of highly deformed porphyrins [68], hindered capped porphyrins [69], and functionalized capped porphyrins [70]. Ogoshi and coworkers recently developed a modified 2 + 2 coupling of dipyrromethanes where instead of the 5,5′-diformyldipyrromethane (**8**), a 5,5′-bis(hydroxybenzyl)dipyrromethane was coupled with a 5,5′-diunsubstituted dipyrromethane (**9**) to give unsymmetric 5,15-diphenylporphyrins [71].

A useful extension of this 2 + 2 coupling was reported by Ogoshi et al. [72] who cocondensed β-alkyl-5,5′-unsubstituted dipyrromethanes (**13**) and aromatic aldehydes (**14**) to prepare 5,15-diaryl-β-octaalkylporphyrins (**16**) according to Scheme 5. By carrying out the reactions in refluxing propionic acid in the presence of zinc acetate, several porphyrins with substituted phenyl groups in the meso positions were obtained (as the zinc complex) in 15–25% yield, while the reaction in benzene with catalytic quantities of trifluoroacetic acid gave higher (30–40%) yields of the desired porphyrins after air oxidation. As in the acid-catalyzed cocondensation of pyrrole and benzaldehydes, this reaction was shown by Gunter and Mander [73] to proceed via a porphyrinogen intermediate (**15**). In the reaction of β-tetramethyldi-

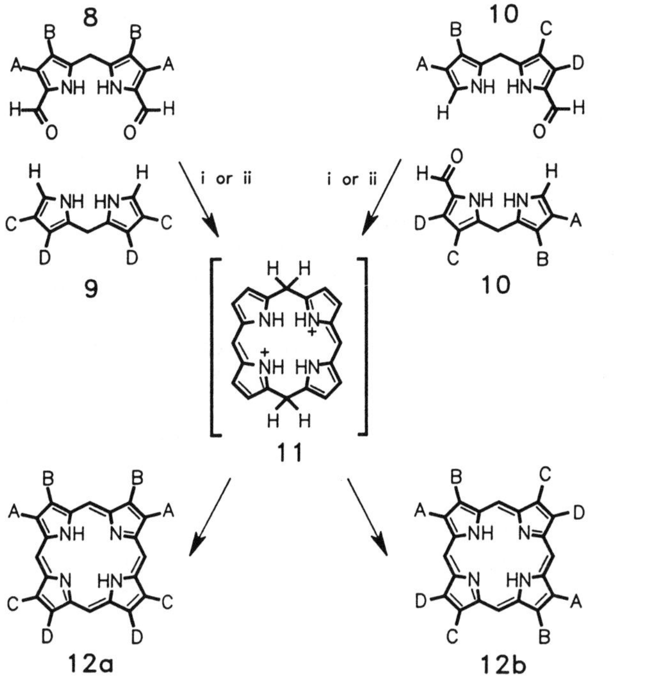

i) HI in CH_3CO_2H, then CH_3CO_2Na and O_2
ii) TsOH in CH_3OH/CH_2Cl_2, then CH_3CO_2Na and O_2

Scheme 4

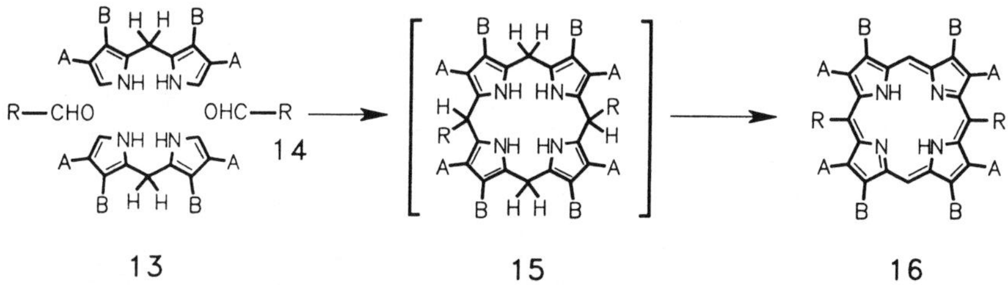

Scheme 5

pyrromethane (**13**, A = B = CH_3) with *ortho*-nitrobenzaldehyde carried out in methanol with catalytic amounts of *p*-toluenesulfonic acid, they isolated the porphyrinogen in >60% yield and subsequently oxidized it quantitatively to the porphyrin using a quinone oxidant.

Several useful modifications of this synthesis have been reported, primarily to improve yields and increase the stability of the sensitive substituents. Manka and Lawrence [74] reported the synthesis of a series of β-unsubstituted 5,15-diarylporphyrins (73–92% yield) by carrying out the cocondensation of 3,3′,4,4′-unsubstituted dipyrromethane (**13**; A = B = H) and substituted benzaldehydes in dichlormethane at room temperature in the presence of catalytic amounts of trifluoroacetic acid, the intermediate being oxidized with *p*-chloranil. Maruyama and coworkers used milder conditions (trichloroacetic acid/acetonitrile) to prepare a series of 5,15-diarylporphyrins (52–87% yield) carrying acid-labile substituents on the phenyl groups [75]. This system appears to be attractive for model building purposes, i.e., for sterically hindered porphyrins [76], strapped porphyrins, and face-to-face dimers [77]; for porphyrin oligomers for photosynthetic charge separation studies [78]; as well as for carotenoid-linked porphyrins used in intramolecular energy transfer studies [79].

D. Cyclization of Linear Tetrapyrroles

The need to synthesize porphyrins with totally unsymmetric β substitution led to the development of procedures involving the coupling of individual pyrroles in an unambiguous manner to give a linear tetrapyrrole, which is cyclized in the final step to give the desired porphyrin. Of the four types of linear tetrapyrroles shown in Figure 6c, bilenes b and biladienes ac are the most commonly used precursors to porphyrins [80].

1. Porphyrins from Bilenes b

Bilenes b (**19**) are prepared [80,81] (Scheme 6) by the acid-catalyzed decarboxylative coupling of 5-carboxydipyrromethane (**17**) with a 5-formyldipyrromethane (**18**). The most commonly prepared bilene is the 1,19-dimethyl derivative **19**, which is readily cyclized to the copper complex of the porphyrin (**20**) using cupric salts. The free base porphyrin (**21**) is obtained by the removal of copper with strong acids.

2. Porphyrins from Biladienes ac

Biladienes ac are usually prepared and handled in their protonated form, commonly as the crystalline dihydrobromides. Their inherent stability makes them the most commonly used linear tetrapyrrole in porphyrin synthesis. Biladienes ac (**24**) were originally prepared by Johnson and Kay [82] by the

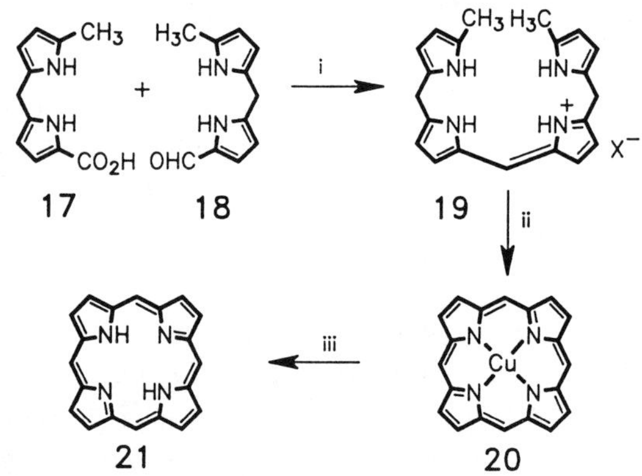

i) CF_3CO_2H, then CH_3OH and HBr

ii) $Cu(OAc)_2$ in CH_3CO_2H/CH_3OH, 100°C;

iii) H_2SO_4/CF_3CO_2H

Scheme 6

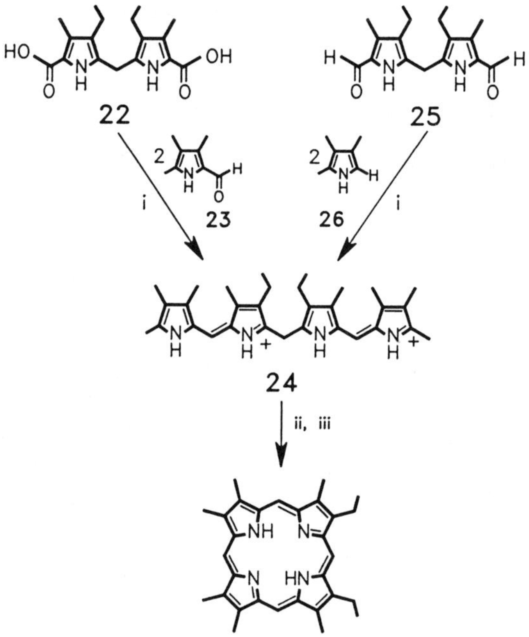

i) HCl in CH_3OH, r.t.;

ii) $Cu(OAc)_2$ in CH_3OH, reflux 48 h;

iii) H_2SO_4

Scheme 7

acid-catalyzed condensation of two equivalents of 2-formyl-5-methylpyrrole (**23**) with a dipyrromethane-5,5′-dicarboxylic acid (**22**) or, alternatively, of 5-unsubstituted 2-methylpyrrole (**26**) with 5,5′-diformyldipyrromethane (**25**) (Scheme 7). Cyclization of the biladiene was effected with cupric salts to give the porphyrin copper complex in 20–30% yield.

The symmetry imposed on this system, i.e., the terminal rings having a symmetric arrangement of substituents, has since been removed by a completely stepwise approach. Differentially protected benzyl *tert*-butyldipyrromethane-5,5′-dicarboxylates can be selectively deesterified (by hydrogenolysis or acid cleavage) and coupled with a 2-formylpyrrole to give initially a tripyrrin. Removal of the remaining ester allows the coupling of a different formylpyrrole to give the biladiene ac whose β-substituent arrangement can be totally unsymmetric [83–85].

Johnson and coworkers [86] developed a regioselective synthesis of octaalkylporphyrins using 1-bromo-19-methylbiladienes ac as the tetrapyrrolic precursor. As shown in Scheme 8, the biladiene is prepared from dipyrrometh*enes* rather than dipyrrometh*anes*. A 5′-bromo-5-bromomethyldipyrromethene **29** is reacted with a 5′-unsubstituted 5-methyldipyrromethene **28** in the presence of tin(IV) chloride to give the tin complex of the biladiene, which on treatment with methanolic hydrobromic acid produces the required biladiene **30** (as its dihydrobromide salt), in 80–95% yield. This is cyclized to the porphyrin either in refluxing *o*-dichlorobenzene (15 min) [86] or by allowing a solution in dimethylsulfoxide/pyridine to stand at room tempera-

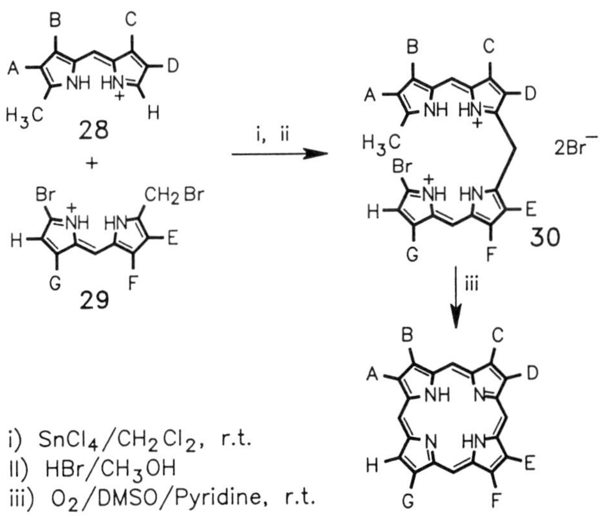

Scheme 8

ture for several days in the dark [87]. The yields reported for the cyclization step vary from 30% to 78%. Although this two-step procedure is by far the mildest for the synthesis of unsymmetric porphyrins in high yield, it was not exploited to the fullest extent for some time due to the limited accessibility of the obligatory dipyrromethene components **28** and **29**. Significant improvements and modifications of some of the existing methodology for the synthesis of these dipyrromethenes (hence the biladienes and porphyrins) have been developed, the details of which were reviewed recently [88]. This two-step procedure has been employed to synthesize several porphyrins in high yield; these include unsymmetric porphyrins, e.g., etioporphyrin III [89], singly and doubly linked porphyrin dimers [90,91], β-unsubstituted porphyrins [92] as well as porphyrins with sensitive functional groups, e.g., mono- and divinylporphyrins [92,93] and uroporphyrins I and III [94].

V. METALLATIONS AND DEMETALLATIONS OF PORPHYRINS

The tetradentate anionic porphyrinato ligand generated by the loss of the two inner NH protons is very versatile in its coordination chemistry, and almost all metals and some metalloids have been combined with it. Five essential stages can be identified in the synthesis of a metalloporphyrin [95]:

A. Protonation–Deprotonation Equilibria

$$H_4P^{2+} \rightleftharpoons H_3P^{+} \rightleftharpoons H_2P \rightleftharpoons HP^{-} \rightleftharpoons P^{2-}$$

Deprotonation of the free base porphyrin H_2P occurs so as to produce the dianion P^{2-} which coordinates to the metal ion. The presence of strong acids therefore retards metallation reactions and any reagent (or solvent) that absorbs the protons generated (decreases the acidity or buffers the system) will favor it.

B. Release of Metal Ion from the "Metal Carrier"

The metal carrier dissociates in the reaction medium producing a coordinatively unsaturated species which can combine with the dianion P^{2-}. The choice of the carrier depends on its activity, its availability and its solubility in the solvent used to dissolve the porphyrin.

C. Formation of the MN_4 System

The initial metallation reaction involves the coordination of the four pyrrole nitrogen atoms to the metal. For divalent metal ions which prefer square-planar arrangements (e.g., Ni, Cu), this produces a stable metalloporphyrin complex.

D. Charge Neutralization

For metal ions with a charge greater than 2^+, the above step produces a charged metalloporphyrin. In order to attain electroneutrality, axial coordinadination is built up with anionic ligands. For Fe^{3+} and Cl^-, the square-pyramidal PFeCl that is formed has the metal center coordinatively saturated, thus producing a stable metalloporphyrin complex. It is important to note that chromatographic purification using added polar solvents (e.g., methanol) generally results in axial ligand exchange which can be distinguished by absorption spectroscopy (see Section II). Treatment with dilute aqueous hydrochloric acid usually regenerates the chloride form.

E. Completion of the Coordination Sphere

For metal ions that prefer octahedral geometry, the coordination sphere is completed by further complexation with neutral donar ligands such as water, pyridine, etc.

As discussed in Section II, metal insertion leads to a change in the absorption spectrum, particularly in the 500- to 700-nm region. Absorption spectroscopy can therefore be used to monitor metal insertion reactions. Buchler [95] discussed in detail the different metallating systems classified by solvent as well as the metal carrier. No single metal insertion procedure is applicable to all metals or porphyrins and special precaution should be taken with porphyrins carrying sensitive functional groups. Metallation is best carried out in solutions as concentrated as possible since dilute solutions tend to slow the bimolecular reaction significantly. This would result in longer reaction times, which for high-boiling solvent systems could lead to decomposition of starting material or product. For the catalytically useful metals such as iron and manganese, insertion is carried out with the M(II) salts, but subsequent autoxidation under aerobic conditions produces the stable higher oxidation states M(III). Some common metallating methods are briefly discussed below [95,96].

F. Chloroform/Methanol Method

A refluxing concentrated chloroform solution of the porphyrin is treated with a saturated solution of the metal(II) acetate and the insertion is followed by absorption spectroscopy. Upon completion, the reaction mixture is concentrated and diluted with methanol to yield the metalloporphyrin in near-quantitative yield. For acid-sensitive systems (e.g., zinc), it is recommended that some sodium acetate be added to buffer the medium. This method has been used to introduce divalent metal ions such as Zn(II), Co(II), Cu(II), and Ni(II), the last of which requires longer reaction times.

Another organic solvent that we have found useful, particularly for the insertion of iron to β-alkylporphyrins, is tetrahydrofuran (THF) because of the good solubility properties of porphyrins in this solvent. Furthermore, both chloroform and THF are low-boiling solvents which reduce the thermal decomposition of sensitive substituents.

G. Acetic Acid/Acetate Method [97–99]

The metal salt, usually the acetate, is heated with the porphyrin in acetic acid at 100°C. The product is crystallized either directly by cooling or by the addition of water or methanol. This method is applicable to the synthesis of metalloporphyrins of most divalent metal ions; Mn(II) and Fe(II) are autoxidized to the more stable high oxidation state. Once again, the use of sodium acetate is recommended for the synthesis of acid-sensitive systems.

H. Dimethylformamide Method [100]

Adler et al. showed that several metals can be conveniently introduced into porphyrins by refluxing a metal salt (usually metal chlorides) in dimethylformamide (DMF). DMF is a uniformly good solvent for both porphyrins and metal salts. The HCl produced in the reaction escapes at the reflux temperature of the solvent. This is the method of choice for phenyl- and/or pyrrole-substituted *meso*-tetraphenylporphyrins primarily due to their insolubility in low-boiling solvents and also due to their higher thermal stability. The product is crystallized directly by dilution with water or with dilute HCl (3–6 N) to dissolve the excess metal salts in the reaction mixture.

A major disadvantage of this solvent is that at reflux it decomposes into dimethylamine which, though neutralizing the HCl produced in the reaction, can lead to complicating side reactions, if long reaction times are required. This is particularly so in the case of highly electron-deficient halogenated porphyrin systems where dimethylamine may act as a nucleophile leading either to halogen substitution [101] or to porphyrin destruction. We have observed that β-octachloro-*meso*-tetrakis(4-sulfonatophenyl)porphyrin undergoes rapid decomposition if iron insertion is attempted in DMF [102]. This is probably due to the lack of steric protection at the meso positions (*ortho*-phenyl positions being unsubstituted), allowing attack by a nucleophile leading to ring opening.

I. Pyridine Method [103]

This is a useful method for metalloporphyrins labile toward acetic acid since pyridine is a good solvent for porphyrins as well as metal salts. The metalloporphyrins are usually isolated as the pyridinates. Pyridine forms complexes with metals of high charge and consequently retards the metallation process.

J. Metal Carbonyl Method [104,105]

This method has been proven to be especially useful for the preparation of metalloporphyrins of groups VI–VIII metals. The metal carbonyl or the carbonyl chloride is heated with the porphyrin in an inert solvent such as benzene, toluene, or decalin.

K. Acetylacetonate Method [106,107]

Metal acetylacetonates make very useful "carriers" due to their ready availability, solubility in organic solvents, and the low pK_a of the acid liberated during the reaction. With metal ions of high charge and small size, a weakly acidic solvent (e.g., phenol) is required to liberate the active metallating species. This method has been used with metals of groups IIIa and IIIb.

The process of demetallation is favored by the presence of acid since protonation of the free base, once formed, drives the equilibrium in the forward direction. Metalloporphyrin stability is often defined in terms of the degree of resistance to the displacement of the metal atom by acid. Several stability classes have been identified based on complete demetallation in (a) acetic acid, (b) aqueous HCl-dichloromethane, and (c) sulfuric acid. While some divalent metals (e.g., zinc) are readily removed by trifluoroacetic acid in dichloromethane, others (e.g., Cu) require 15–25% sulfuric in trifluoroacetic acid or sometimes even stronger conditions.

In order to remove iron from porphyrins, the stable ferric state is best reduced to the ferrous state. Several reagents can be used for this purpose [96] and the choice is usually based on the sensitivity of the substituents. Reduction is commonly carried out by a saturated solution of ferrous sulfate whereas demetallation is effected by hydrochloric acid or gaseous HCl [108].

VI. BIOMIMETIC METALLOPORPHYRIN CATALYSTS

During the 1970s and early 1980s, the synthesis of biomimetic metalloporphyrins was directed towards generating systems that could mimic the reversible binding of dioxygen to the Fe(II) center of hemoglobin and myoglobin [27]. The primary structural requirement therefore was the presence of steric protection to prevent a second Fe(II) system from approaching and reacting to produce an oxidized μ-oxo Fe(III) species. However, since the late 1970s when much of the attention was diverted to understanding the mechanism of action of hemoprotein catalysts, the structural requirements for the metalloporphyrin models also changed. Although cytochrome P450 continues to be the focus of attention in biomimetic studies, lignin peroxidase has also been studied extensively since its discovery in 1984. In order to design efficient model catalysts, it is necessary to understand as best as possible the mechanism by which the enzyme functions.

A. Catalytic Cycle of Lignin Peroxidase

Lignin peroxidase is an extracellular enzyme that catalyzes oxidations in lignin-related aromatic molecules. Much of the attention that this enzyme has received recently is not only due to its oxidative depolymerization of lignin but to its ability to degrade a variety of environmental pollutants including polycyclic aromatic hydrocarbons [109,110], dibenzo-*p*-dioxins [110], and chlorinated phenols [111]. The catalytic cycle of lignin peroxidase (Figure 7) is initiated by a two-electron oxidation of the Fe(III) state of the resting enzyme, by hydrogen peroxide. The active high-valent iron–oxo species (compound I) so produced initiates lignin degradation by one-electron oxidations of methoxylated aromatic substructures of lignin, forming cation-radical species. These undergo bond cleavage and further nonenzymatic reactions with water and/or oxygen leading to lignin degradation. The enzyme reverts to its resting Fe(III) state via compound II, which is a one-electron oxidized form.

B. Catalytic Cycle of Cytochrome P450

Cytochrome P450 catalyzes the insertion of one oxygen atom from dioxygen into the substrate, the other being lost as water; hence its designation as a monooxygenase. Hydroxylation of alkanes and epoxidation of alkenes

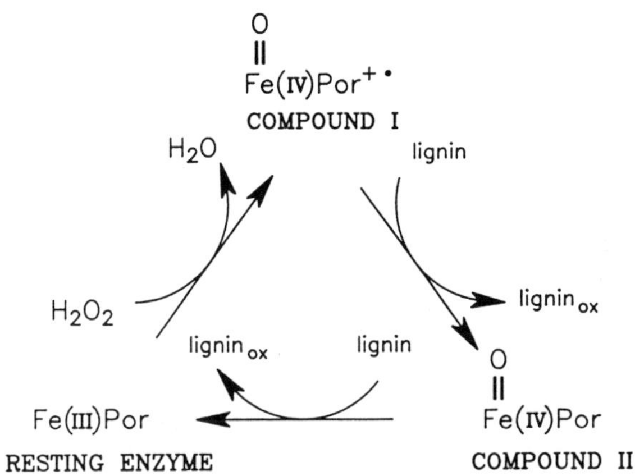

Figure 7 Proposed catalytic cycle of lignin peroxidase.

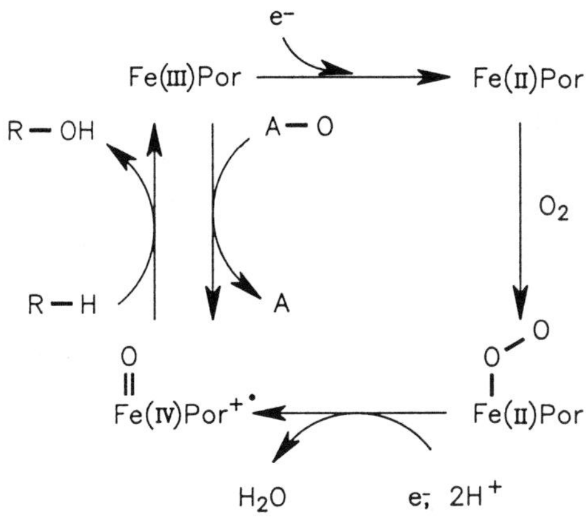

Figure 8 Proposed catalytic cycle of cytochrome P450.

are the two most important reactions catalyzed by this enzyme. As shown in Figure 8, the resting enzyme Fe(III) undergoes a one-electron reduction to the Fe(II) state, which binds dioxygen. Following another one-electron reduction, the rate-limiting enzymatic step, and subsequent protonation of the terminal oxygen atom, a heterolytic cleavage of the O-O bond leads to the active oxygen complex, which is believed to be a high-valent iron–oxo complex [4], the other oxygen atom being released as water. Transfer of the oxygen atom to the substrate results in C-H bond hydroxylation or alkene epoxidation regenerates the resting Fe(III) state of the enzyme.

C. Design of Biomimetic Metalloporphyrin Catalysts

The postulation of a high-valent iron–oxo system as the active species in the proposed catalytic cycles of both lignin peroxidase and cytochrome P450 have resulted in similar structural requirements for the synthetic metalloporphyrin systems used in biomimetic studies. Understanding its nature, stability, and reactivity is important in the design and synthesis of a suitable catalyst. The properties of compounds I of peroxidases are consistent with an oxoiron(IV)porphyrin π-cation radical [112,113] while a similar structure has been postulated for cytochrome P450 as well [4]. However, an oxoiron(IV) species with a protein-centered radical (compound I of cytochrome c peroxidase) [114] has not been ruled out.

It has been shown that the oxidation of simple metalloporphyrins can give rise to similar highly oxidized species corresponding to the catalytic intermediate. Using TMPFeCl and *m*-chloroperbenzoic acid, Groves et al.

[115] prepared and characterized the oxoiron(IV)porphyrin intermediate [FeIV = O(TMP)]$^+$ which has since been studied extensively. More recently, similar species have been generated and spectroscopically characterized for two other iron porphyrins, TDCPPFeCl and *meso*-tetrakis(2,4,6-trimethoxyphenyl)porphyrinatoiron(III) [116]. However, unlike in the enzyme system where the active species is bound to the protein moiety and hence protected, in homogeneous reaction media this highly reactive intermediate of simple metalloporphyrins is capable of self-destruction. In order to minimize such loss of activity, catalyst design has been directed primarily to increasing the electron deficiency, while providing steric bulk around the metalloporphyrin system. This has been shown to improve catalyst stability by resisting oxidative destruction.

It is important to note, however, that the generation of an active oxoiron(IV) species from the resting Fe(III) state of the two enzymes differ, lignin peroxidase using hydrogen peroxide (a two-electron oxidant) directly, while cytochrome P450 uses molecular oxygen and two reducing equivalents provided by NADPH. Thus, mimicking the lignin peroxidase system involves the direct use of a metalloporphyrin catalyst in the presence of hydrogen peroxide or a similar single-atom oxygen transfer reagent. However, the catalytic cycle of cytochrome P450 is very difficult to mimic using simple metalloporphyrin systems with molecular oxygen and a reducing agent. Several model systems using Mn(III)porphyrins as catalysts have been reported [117], but with low rates and yields. Groves et al. [118] were the first to demonstrate that a system consisting of TPPFeCl as the catalyst and iodosylbenzene as an oxygen atom donor was able to perform the reactions catalyzed by cytochrome P450, suggesting the generation of the high-valent intermediate. It is this shunting that has been studied extensively for epoxidation of alkenes, hydroxylation of alkanes, as well as the oxidation of other organic substrates, although various combinations of metalloporphyrins and single-atom donor reagents (AO in Figure 8) have been used [119].

D. Synthesis of Electron-Deficient Porphyrins

Due to the ease of synthesis, the metalloporphyrin catalysts used in biomimetic studies have been of the *meso*-tetraphenylporphyrin type. Since Groves et al. demonstrated that a high-valent oxoiron porphyrin system could be observed with the TMPFe system [115], it was realized that bulky ortho substituents on the phenyl groups could be important in preventing oxidative self-destruction of the catalyst. By using TPFPPFe, Chang and Ebina [120] showed that electron deficiency can enhance the stability of the catalyst and also increase reactivity to the substrate by generating a more electrophilic oxoiron intermediate. Traylor et al. [31] showed that TDCPPFeCl is a very efficient catalyst since both electron deficiency and steric bulk are incorporated into the molecule with *ortho*-chloro groups. In order to adapt this catalyst for use in aqueous media as a model for lignin peroxidase, Dolphin et al. [121] prepared a water-soluble form by introducing sulfonic acid groups to the phenyl moieties, which also helped in increasing the electron deficiency of the catalyst. The synthetic methods available for TMPFe, TPFPPFe, and TDCPPFe catalysts have already been discussed in Section IV.A. Sulfonation of $TDCPPH_2$ has been carried out using fuming sulfuric acid at 165°C for 7 h. Due to the directing influence of the *ortho*-chloro groups, sulfonation takes place at the meta (3-) positions of the phenyl groups.

The development of the next generation of catalysts saw the introduction of more electron-withdrawing substituents into the catalyst molecule. Considering the electrophilic and radical nature of reactivity of the porphyrin macrocycle, β-perhalogenation was the method of choice for creating greater electron deficiency in the system. Chlorination of TDCPP has been achieved by reacting its iron complex (**31**, M = FeCl, X = H), with chlorine gas at 140°C in the presence of anhydrous ferric chloride [122]. The 24-nm red shift of the Soret band indicates the completion of the reaction (20–30 min) and the purified β-octachloroporphyrinatoiron (**32**, M = FeCl) has been isolated in >85% yield. The zinc complex (**31**; M = Zn, X = H) has also

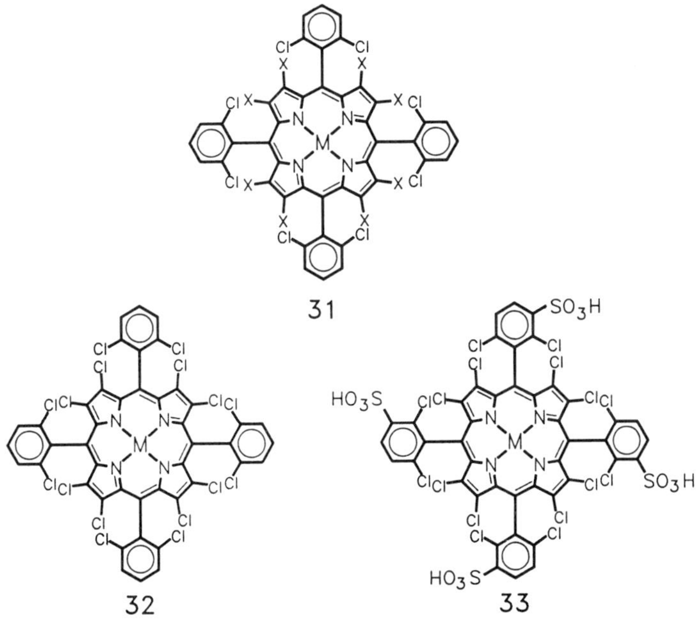

Scheme 9

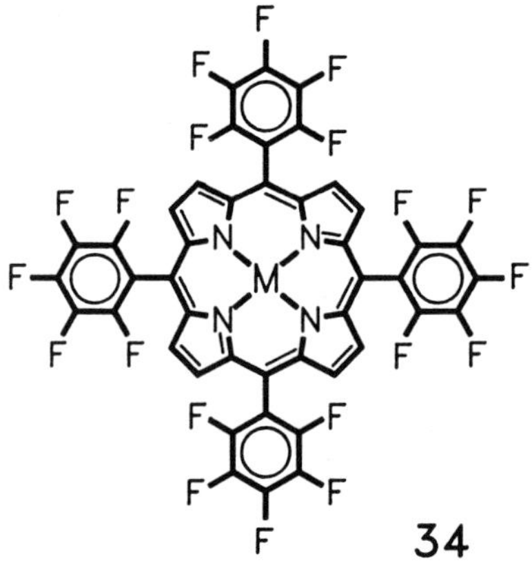

34

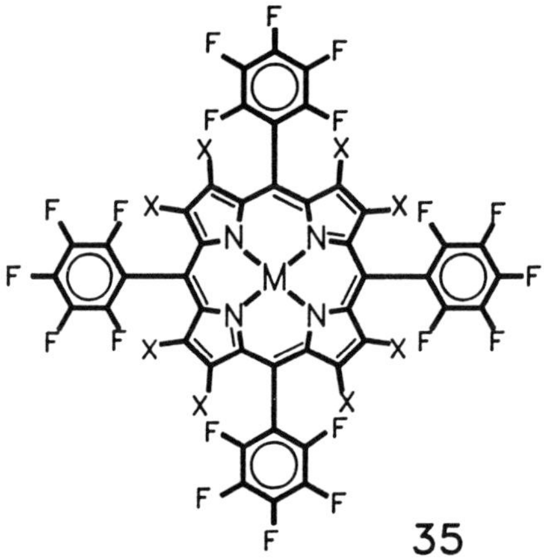

35

Scheme 10

been chlorinated with *N*-chlorosuccinimide [123] in refluxing chlorocarbon solvents to give the analogous zinc complex (**32**; M = Zn) in 78% yield, but had required extended reaction times (22 h). β Perchlorination of TPFPP (**34**) has been achieved [123,124] via both methods while the truly perchloroporphyrin **37** has been prepared by the chlorination of TPCPPFe (**36**, M = FeCl) using the $Cl_2/FeCl_3$ procedure [122]. Chlorination of TMPZn (**38**, M = Zn, X = H) by *N*-chlorosuccinimide, in refluxing methanol, produced **39** (M = Zn, X = Y = Cl) in which one meta position of each phenyl ring is chlorinated in addition to the eight pyrrolic positions [125]. Demetallation with trifluoroacetic acid produced the free base porphyrin in 45–60% overall yield. β Chlorination of unsubstituted TPPFe (**40**, M = FeCl, X = H)

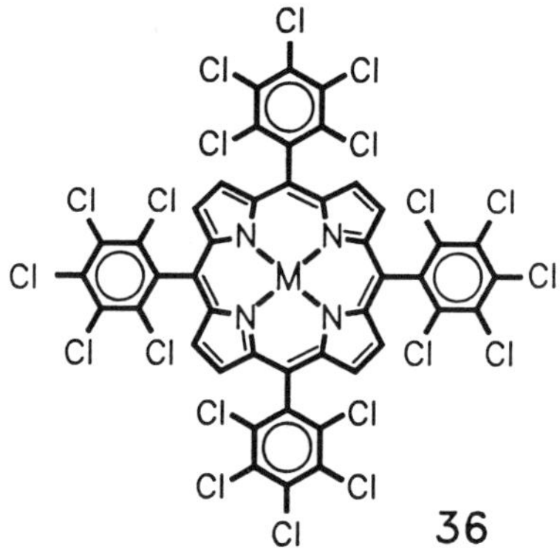

36

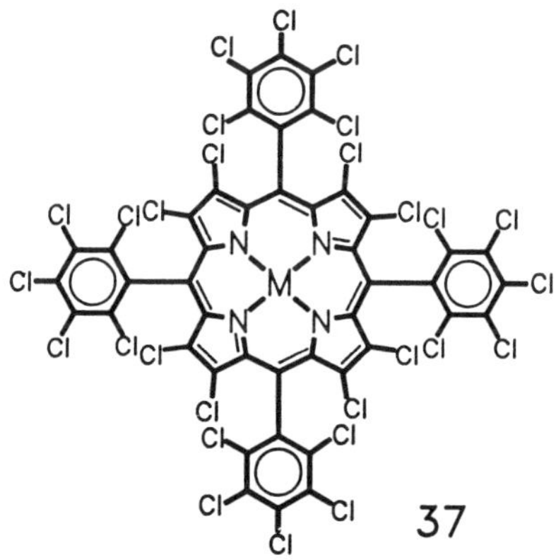

37

Scheme 11

could not be achieved using the $Cl_2/FeCl_3$ system due to destruction of the macrocycle, while attempted chlorination of TPPFe, TPPZn, or TPPCu with *N*-chlorosuccinimide produced the desired β-octachlorinated porphyrin (**41**, M = FeCl, Zn, or Cu) in low yield. However, TPPNi reacted with *N*-chlorosuccinimide at 140°C to give **41** (M = Ni) in 75% yield which was readily demetallated and converted to the ferric complex in >75% overall yield [122]. This confirms the observation that the centrally coordinated metal exerts a directive influence on peripheral substitution [126]. Heating the Ni complex of **41** (M = Ni) in concentrated sulfuric acid for 5 h followed by stirring at room temperature for a further 12 h produced, after iron insertion, the sulfonated hemin **42** (M = FeCl), while the sulfonation of **32** to give **33**

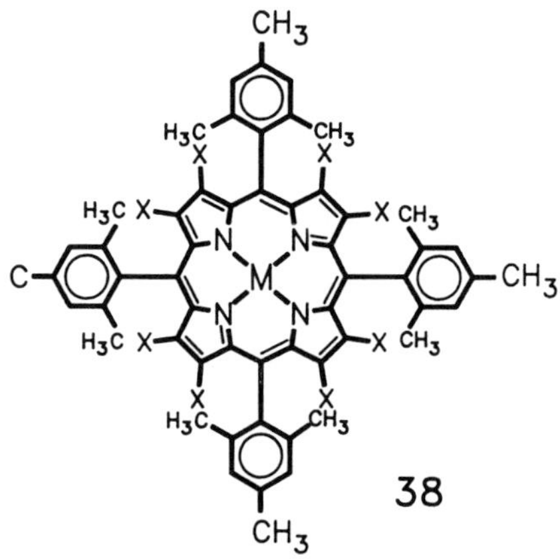

38

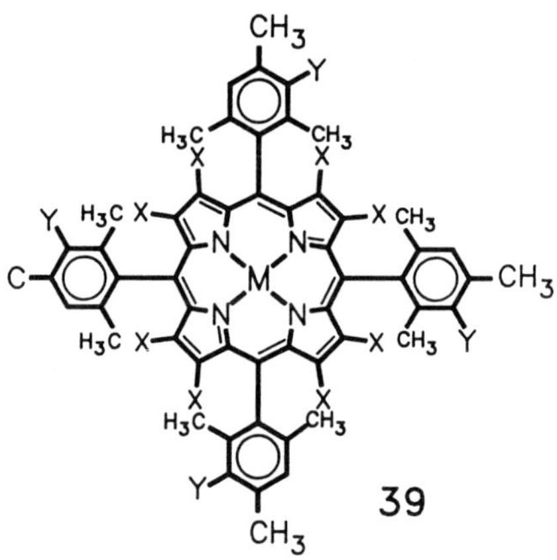

39

Scheme 12

required the use of fuming sulfuric acid [121,122]. These compounds have been used as models for lignin peroxidase.

Perbromination of the porphyrin periphery has also been achieved with either bromine or *N*-bromosuccinimide. Octabromination of TDCPPZn (**31**, M = Zn, X = H) was first reported in refluxing carbon tetrachloride (1 g/100 ml) using 10 equivalents of *N*-bromosuccinimide to give the desired product **31** (M = Zn, X = Br) in 71% yield [127]. However, low solubility of the starting material in this solvent leads to extended reaction times resulting in partial demetallation and incomplete bromination. The use of chloroform or tetrachloroethene as the solvent has been recommended in order to improve the yields [123]. TMP, as its zinc complex (**38**, M = Zn,

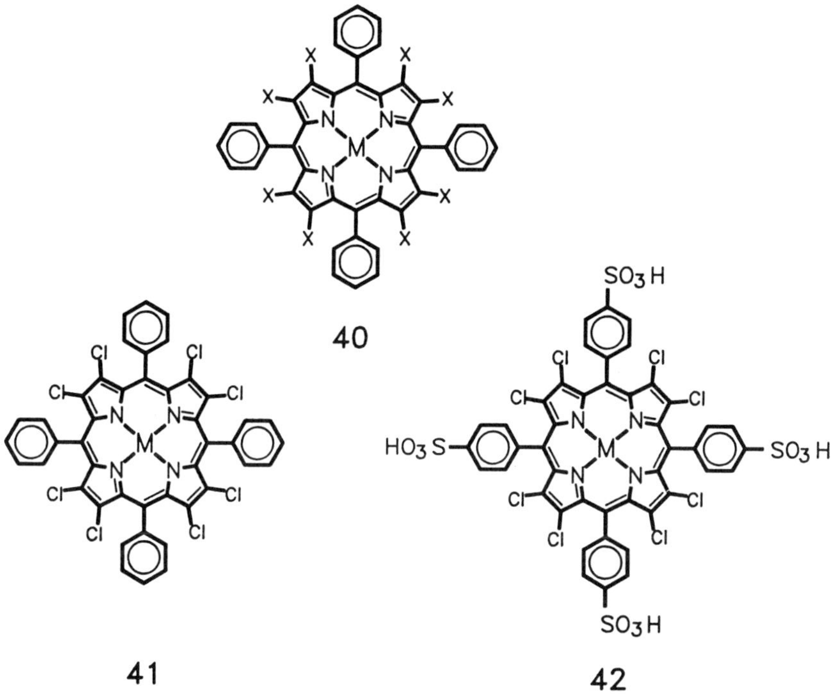

Scheme 13

X = H), has also been octabrominated using 10 equivalents of *N*-bromosuccinimide in refluxing methanol solution [125]. The product **38** (M = Zn, X = Br) has been isolated in 50–65% yield. TPFPP, as its zinc complex (**34**, M = Zn), has been brominated with *N*-bromosuccinimide in refluxing chloroform/tetrachloroethane (1:1) over a period of 48 h while trifluoroacetic acid was slowly added [128]. The octabrominated product (**35**, M = Zn) has been isolated in 70% yield after zinc reinsertion and purification. The chloroiron complex **34** (M = FeCl), however, has been brominated with a 6 M solution of bromine [129], in dry degassed carbon tetrachloride by refluxing for 18 h, to give a 74% yield of **35** (M = FeCl). Unsubstituted TPP, as its copper complex (**40**, M = Cu, X = H), has been octabrominated [130], by stirring a solution in chloroform/carbon tetrachloride (1:1) with liquid bromine at room temperature (4 h) followed by the addition of pyridine and further stirring (12 h). The product **40** (M = Cu, X = Br), isolated in 75% yield, was demetallated with perchloric acid.

The effect of halogen substitution of the porphyrin ligand on the metal center has been demonstrated by cyclic voltammetry of the chloroiron(III) complexes [131]. In general, electron-withdrawing chloro substituents produce anodic shifts of reduction potentials, but the effect was shown to be more pronounced for pyrrolic substitution. This reflects the ease of reduction of the metal center of β-halogenated porphyrins. While the positive shift of $E_{1/2}$ for Fe^{3+}/Fe^{2+} reduction of **31** (M = Fe, X = H) compared to **40** (M = Fe, X = H) was marginal, the corresponding shifts for **32** (M = Fe) compared with **31** (M = Fe, X = H) as well as that for **41** (M = Fe) compared to **40** (M = Fe, X = H) were approximately 0.4 V. Perchlorination of TPPFeCl (**40**, X = H) to TPCPPFeCl (**37**) produced a shift of 0.64 V.

The X-ray crystallographic structures of β-octabromo-TMPNi (**38**, M = Ni, X = Br), TPFPPNi (**35**, M = Ni, X = Br), as well as octachloro-TPFPPCu (**35**, M = Cu, X = Cl) exhibit saddle-shape conformations, created by the buckling of the macrocycle [128,132]. This distortion from planarity minimizes the nonbonded interaction between β-halogen atoms and the *ortho*-phenyl substituents.

E. Synthetic Metalloporphyrins for Commercial Applications

It is not the objective of this chapter to review the catalytic studies carried out with the different metalloporphyrin model systems prepared to date, since that aspect will be discussed in several chapters of this volume (for a recent review, see Ref. 119). However, it should be noted that some of the metalloporphyrin systems developed for biomimetic studies have shown considerable promise in commercial applications.

The water-soluble (sulfonated) polychlorinated metalloporphyrins prepared by Dolphin and coworkers as models for lignin peroxidase have also been examined for their ability to bleach kraft pulp [122,133]. This possible commercial application is particularly important due to the necessity to limit the current use of chlorine in pulp bleaching. *meso*-Tetrakis(2,6-dichloro-3-sulfonatophenyl)-β-octachloroporphyrinatoiron(III) [**33**, M = FeCl], in the presence of *tert*-butylhydroperoxide, has been shown to effect greater than 45% delignification of pulp when incubated for 18 h at 60°C in pH = 4.8 buffer. A second possible application that has been investigated is the decolorization of highly colored effluent from bleach plants, which is an environmental problem. Once again, the above catalyst/oxidant combination was capable of decolorizing (greater than 50%) undiluted effluent in less than 13 min under appropriate conditions. Further studies are underway to determine the optimal reaction conditions as well as the use of cheaper oxidants.

The hydroxylation of light alkanes to alcohols is an important reaction in the commercial production of fuel additives and other oxygenated chemicals. However, the major problem with biomimetic systems is that the use of single-atom oxygen transfer reagents to generate the active intermediate would not necessarily be economically viable. The work of Ellis and Lyons [132,134] is therefore unique in that oxidations of light alkanes to alcohols are carried out using activated metalloporphyrins and molecular oxygen rather than single-atom donor reagents. With β-octachloro- and β-octabromo-*meso*-tetrakis(pentafluorophenyl)porphyrinatoiron(III) as catalysts, it has been possible to oxidize isobutane with air even at room temperature to give *tert*-butyl alcohol in >90% selectivity with >13,000 turnovers. The fact that catalysis is effected without the use of an added coreductant suggests that the mechanism is different from that proposed for cytochrome P450 itself. According to a postulated mechanism, the electron deficiency in the porphyrin ligand facilitates the reduction of the metal center to the Fe(II) state, which binds dioxygen and subsequently produces the catalytically active oxoiron(IV) species by homolytic cleavage of the μ-peroxo dimer. The Fe(II) state and the oxoiron(IV) state are thought to be in equilibrium with a μ-oxo dimer species, the position of this equilibrium being dependent on steric and electronic factors. Although no crystallographic data are yet available for the catalytically active hemins, it is likely that the enhanced catalytic activity observed in β-halogenated systems is related to the saddle shape of the ligand (observed for Ni and Cu complexes [132]), which restricts its ability to form stable μ-oxo dimer species. According to this mechanism, the catalyst essentially serves as a dioxygenase rather than a monooxygenase (cytochrome P450) since both atoms of molecular oxygen are transferred to the substrate via the oxoiron(IV) intermediate.

ACKNOWLEDGMENTS

This work was supported by the U.S. Department of Energy (Morgantown Energy Technology Center), the Gas Research Institute and Sun Company (TPW), and by the Canadian Natural Sciences and Engineering Research Council (DD).

REFERENCES

1. T.K. With, A short history of porphyrins and the porphyrias, *Int. J. Biochem., 11*: 189 (1980).
2. D. Dolphin and E. Sternberg, Medical applications of dyes: a review of photodynamic therapy, in *Infrared Absorbing Dyes* (M. Matsuoka, ed.), Plenum Press, New York, 1990, p. 193.

3. B.W. Henderson and T.J. Dougherty, How does photodynamic therapy work? *Photochem. Photobiol., 55*: 145 (1992).
4. P.R. Ortiz de Montellano, Oxygen activation and transfer, *Cytochrome P-450: Structure, Mechanism and Biochemistry* (P.R. Ortiz de Montellano, ed.), Plenum Press, New York, 1986.
5. M. Tien and T.K. Kirk, Lignin-degrading enzyme from *Phanerochaete chrysosporium*: purification, characterization and catalytic properties of a unique H_2O_2-requiring oxygenase, *Proc. Natl. Acad. Sci. USA, 81*: 2280 (1984).
6. R. Bonnett, Nomenclature, *The Porphyrins*, Vol. 1 (D. Dolphin, ed.), Academic Press, New York, 1978, p. 1.
7. Nomenclature of tetrapyrroles; Recommendations of IUPAC-IUB Joint Commission on Biochemical Nomenclature, *Eur. J. Biochem., 178*: 277 (1988).
8. M.J. Crossley, P.L. Burn, S.S. Chew, F.B. Cuttance, and I.A. Newsom, Regiospecific introduction of four substituents to porphyrin systems at antipodal pyrrolenic positions, *J. Chem. Soc. Chem. Commun.*, 1564 (1991).
9. M. Gouterman, Effects of substitution on the absorption spectra of porphin, *J. Chem. Phys., 30*: 1139 (1959).
10. T.P. Wijesekera, J.B. Paine III, D. Dolphin, F.W.B. Einstein, and T. Jones, Enforced deformation of porphyrins by short-strap bridging, *J. Am. Chem. Soc., 105*: 6747 (1983).
11. G.M. Badger, R.A. Jones, and R.L. Laslett, The synthesis of porphyrins by the Rothmund reaction, *Aust. J. Chem., 17*: 1028 (1964).
12. J.B. Kim, J.J. Leonard, and F.R. Longo, A mechanistic study of the synthesis and spectral properties of meso-tetraarylporphyrins, *J. Am. Chem. Soc., 94*: 3986 (1972).
13. K. Jeyarajah, A. Gold, G.E. Toney, and J.H. Helms, Preparation and characterization of the μ-oxo dimer and hydroxo complexes of [tetrakis(pentafluorophenyl)porphinato]iron(III), *Inorg. Chem., 25*: 3516 (1986).
14. D. Dolphin (ed.), *The Porphyrins*, Vols. 1 and 2, Academic Press, New York, 1978.
15. A.H. Jackson and K.M. Smith, The total synthesis of pyrrole pigments: 1973–1980, in *Total Synthesis of Natural Products*, Vol. 6 (J. ApSimon, ed.), Wiley-Interscience, New York, 1984, p. 237.
16. E.J. Corey and H.-M. Cheng, *The Logic of Chemical Synthesis*, Wiley, New York, 1989.
17. P. Rothmund, A new porphyrin synthesis. The synthesis of porphin, *J. Am. Chem. Soc., 58*: 625 (1936).
18. P. Rothmund, Porphyrin studies. III. The structure of the porphine ring system, *J. Am. Chem. Soc., 61*: 2912 (1939).
19. P. Rothmund and A.R. Menotti, Porphyrin studies. IV. The synthesis of α, β, γ, δ-tetraphenylporphine, *J. Am. Chem. Soc., 63*: 267 (1941).
20. P. Rothmund, Formation of porphyrins from pyrrole and aldehydes, *J. Am. Chem. Soc., 57*: 2010 (1935).
21. R.H. Ball, G.D. Dorough, and M. Calvin, A further study of the porphine-like products of the reaction of benzaldehyde and pyrrole, *J. Am. Chem. Soc., 68*: 2278 (1946).

22. A.D. Adler, F.R. Longo, and W. Shergalis, Mechanistic investigations of porphyrin synthesis. Preliminary studies on ms-tetraphenylporphin, *J. Am. Chem. Soc., 86*: 3145 (1964).
23. A.D. Adler, F.R. Longo, J. Finarelli, J. Goldmacher, J. Assour, and L. Korsakoff, A simplified synthesis for meso-tetraphenylporphin, *J. Org. Chem., 32*: 476 (1967).
24. K. Rousseau and D. Dolphin, A purification of meso-tetraphenylporphyrin, *Tetrahedron Lett.*, 4251 (1974).
25. J.S. Lindsey, I.C. Schreiman, H.C. Hsu, P.C. Kearney, and A. M. Marguerettaz, Rothmund and Adler-Longo reactions revisisted: synthesis of tetraphenylporphyrins under equilibrium conditions, *J. Org. Chem., 52*: 827 (1987).
26. D. Dolphin, Porphyrinogens and porphodimethenes. Intermediates in the synthesis of meso-tetraphenylporphyrins from pyrroles and benzaldehyde, *J. Heterocyclic Chem., 7*: 275 (1970).
27. B. Morgan and D. Dolphin, Synthesis and structure of biomimetic porphyrins, in *Structure and Bonding*, Springer-Verlag, New York, 1987, p. 116.
28. D. Mansuy, P. Battioni, and J.-P. Battioni, Chemical model systems for drug-metabolizing cytochrome P-450 dependent monooxygenases, *Eur. J. Biochem., 184*: 267 (1989).
29. D. Mansuy, Biomimetic catalysts for selective oxidation in organic chemistry, *Pure Appl. Chem., 62*: 741 (1990).
30. F.R. Longo, M.G. Finarelli, and J.B. Kim, The Synthesis and Some Physical Properties of ms-Tetra(pentachlorophenyl)porphin, *J. Heterocyclic Chem., 6*: 927 (1969).
31. P.S. Traylor, D. Dolphin, and T.G. Traylor, Sterically protected hemins with electronegative substituents: efficient catalysts for hydroxylation and epoxidation, *J. Chem. Soc. Chem. Commun.,* 279 (1984).
32. M.M. Williamson, C.M. Prosser-McCartha, S. Mukundan, Jr., and C.L. Hill, Isolation and characterization of the principal kinetic products in the Rothmund synthesis of sterically hindered tetraarylporphyrins. Crystal and molecular structures of [tetrakis(2,6-dichlorophenyl)porphinato]zinc(II) and bis[(meso-2,6-dichlorophenyl)-5-(o,o′-dichlorobenzyl)dipyrromethene]zinc(II) complexes, *Inorg. Chem., 27*: 1061 (1988).
33. J.S. Lindsey and R.W. Wagner, Investigation of the synthesis of ortho-substituted tetraphenylporphyrins, *J. Org. Chem., 54*: 828 (1989).
34. S.S. Eaton and G.R. Eaton, Rotation of phenyl rings in metal complexes of substituted tetraphenylporphyrins, *J. Am. Chem. Soc., 97*: 3660 (1975).
35. J.T. Groves and T.E. Nemo, Aliphatic hydroxylation catalyzed by iron porphyrin complexes, *J. Am. Chem. Soc., 105*: 6243 (1983).
36. R.W. Wagner, D.S. Lawrence, and J.S. Lindsey, An improved synthesis of tetramesitylporphyrin, *Tetrahedron Lett., 28*: 3069 (1987).
37. M. Kihn-Botulinski and B. Meunier, Synthesis of tetramesitylporphyrin, *Inorg. Chem., 27*: 209 (1988).
38. B. Evans and K.M. Smith, Sterically crowded porphyrins: meso-tetraphenyloctaethylporphyrin, *Tetrahedron Lett., 18*: 443 (1977).
39. C.J. Medforth and K.M. Smith, The synthesis and solution conformation of dodecaphenylporphyrin, *Tetrahedron Lett., 31*: 5583 (1990).

40. J. Takeda, T. Ohya, and M. Sato, A ferrochelatase transition-state model. Rapid incorporation of copper(II) into nonplanar dodecaphenylporphyrin, *Inorg. Chem., 31*: 2877 (1992).
41. D.O. Cheng and E. LeGoff, Synthesis of substituted porphyrins, *Tetrahedron Lett., 17*: 1469 (1977).
42. A. Treibs and N. Haberle, Synthesis and electronic spectra of ms-substituted porphines, *Justus Liebigs Ann. Chem., 718*: 183 (1968).
43. M. Friedman, Octaarylporphyrins, *J. Org. Chem., 30*: 859 (1965).
44. J. Takeda, T. Ohya, and M. Sato, A new synthesis of octaarylporphyrins: naturally occuring porphyrin mimics, *Chem. Pharm. Bull., 38*: 264 (1990).
45. W. Siedel and F. Winkler, Oxidation of pyrrole derivatives with Pb tetraacetate. A new kind of porphyrin synthesis, *Justus Liebigs Ann. Chem., 554*: 162 (1943).
46. G.M. Badger and A.D. Ward, The synthesis of porphyrins from 2-aminomethylpyrroles, *Aust. J. Chem., 17*: 649 (1964).
47. J.B. Paine III, W.B. Kirshner, D.W. Moskowitz, and D. Dolphin, An improved synthesis of octaethylporphyrin, *J. Org. Chem., 41*: 3857 (1976).
48. C.-B. Wang and C.K. Chang, A convenient synthesis of pyrrole precursors for octaalkylporphyrins, *Synthesis*, 548 (1979).
49. J.B. Paine III, Synthesis of pyrroles and of porphyrins via single-step coupling of dipyrrolic intermediate, in *The Porphyrins* (D. Dolphin, ed.), Academic Press, New York, 1978, p. 101.
50. H. Kinoshita, S. Tanaka, N. Nishimori, H. Dejima, and K. Inomata, Synthesis of 2-(substituted methyl)-3,4-disubstituted pyrroles and their conversion into the corresponding porphyrins, *Bull. Chem. Soc. Jpn., 65*: 2660 (1992).
51. M. Homma, K. Aoyagi, Y. Aoyama, and H. Ogoshi, Electron deficient porphyrins I. Tetrakis(trifluoromethyl)porphyrin and its metal complexes, *Tetrahedron Lett., 24*: 4343 (1983).
52. H. Ogoshi, M. Homma, K. Yakota, H. Toi, and Y. Aoyama, Synthesis of β-trifluoromethylpyrroles, *Tetrahedron Lett., 24*: 929 (1983).
53. K. Aoyagi, H. Toi, Y. Aoyama, and H. Ogoshi, Facile syntheses of perfluoroalkylporphyrins. Electron deficient porphyrins II, *Chemistry Lett.*, 1891 (1988).
54. H. Onda, H. Toi, Y. Aoyama, and H. Ogoshi, Fluoropyrroles and tetrafluoroporphyrins, *Tetrahedron Lett., 26*: 4221 (1985).
55. N. Ono, H. Kawamura, M. Bougauchi, and K. Maruyama, Porphyrin synthesis from nitrocompounds, *Tetrahedron, 46*: 7483 (1990).
56. N. Ono, K. Sugi, and T. Ogawa, Preparation of porphyrins having ester or nitro functions, *Chem. Express, 6*: 869 (1991).
57. D.H.R. Barton, J. Kervagoret, and S.Z. Zard, A useful synthesis of pyrroles from nitroolefins, *Tetrahedron, 46*: 7587 (1990).
58. S. Krol, A new synthesis of porphin, *J. Org. Chem., 24*: 2065 (1959).
59. F.R. Longo, E.J. Thorne, A.D. Adler, and S. Dym, Notes on the synthesis of porphin, *J. Heterocyclic Chem., 12*: 1305 (1975).
60. R.G. Yalman, Preparation of porphin, substituted porphin and metal chelates thereof, U.S. Patent 3,579,533 (1971).
61. Y. Kuroda, H. Murase, Y. Suzuki, and H. Ogoshi, A new route for Meso-substituted porphyrin, *Tetrahedron Lett., 30*: 2411 (1989).

62. H. Fischer and H. Orth, *Die Chemie des Pyrrols*, Vol. 2, Part 1, Johnson Reprint, New York, 1968.
63. K. Smith, Porphyrins and bile pigments from brominated pyrromethenes, *J. Chem. Soc. Perkin Trans., 1*: 1471 (1972).
64. J.B. Paine III, C.K. Chang, and D. Dolphin, The synthesis of porphyrins via dipyrromethenes, *Heterocycles, 7*: 831 (1977).
65. G.P. Arsenault, E. Bullock, and S.F. MacDonald, Pyrromethanes and porphyrins therefrom, *Can. J. Chem., 82*: 4384 (1960).
66. J.A.S. Caveleiro, A.M.D'A Rocha Gonsalves, G.W. Kenner, and K.M. Smith, Pyrroles and related compounds. 33. Total synthesis of deuterated derivatives of protoporphyrin IX for nuclear magnetic resonance studies of hemoproteins, *J. Chem. Soc. Perkin Trans., 1*: 1771 (1974).
67. A. Markovac and S.F. Macdonald, Syntheses with 5-dibromomethyl- and 5-formylpyrromethenes, *Can. J. Chem., 43*: 3364 (1965).
68. T.P. Wijesekera, J.B. Paine III, and D. Dolphin, Improved synthesis of covalently strapped porphyrins. Application to highly deformed porphyrin synthesis, *J. Org. Chem., 53*: 1345 (1988).
69. T.P. Wijesekera, S. David, J.B. Paine III, B.R. James, and D. Dolphin, Durene-capped porphyrins: synthesis and characterization, *Can. J. Chem., 66*: 2063 (1988).
70. H. Tang, T.P. Wijesekera, and D. Dolphin, Synthesis and characterization of derivatized capped porphyrins, *Can. J. Chem., 70*: 1366 (1992).
71. T. Ema, Y. Kuroda, and H. Ogoshi, Selective syntheses of unsymmetrical meso-arylporphyrins, *Tetrahedron Lett., 32*: 4529 (1991).
72. H. Ogoshi, H. Sugimoto, T. Nishiguchi, T. Watanabe, Y. Matsuda, and Z.-I. Yoshida, Synthesis of 5-aryl- and 5,15-diaryl-2,3,7,8,12,13,17,18-octaethylporphyrins, *Chemistry Lett.*, p. 29 (1978).
73. M.J. Gunter and L.N. Mander, Synthesis and atropisomer separation of porphyrins containing functionalization at the 5,15-meso positions: application to the synthesis of binuclear ligand systems, *J. Org. Chem., 46*: 4792 (1981).
74. J.S. Manka and D.S. Lawrence, High yield synthesis of 5,15-diarylporphyrins, *Tetrahedron Lett., 30*: 6989 (1989).
75. A. Osuka, T. Nagata, F. Kobayashi, and K. Maruyama, An improved synthesis of 5,15-diaryloctaalkylporphyrins, *J. Heterocyclic Chem., 27*: 1657 (1990).
76. R. Young and C.K. Chang, Synthesis and characterization of blocked and ligand-appended hemes derived from atropisomeric meso-diphenylporphyrins, *J. Am. Chem. Soc., 107*: 898 (1985).
77. A. Osuka, F. Kobayashi, T. Nagata, and K. Maruyama, One-pot synthesis of strapped porphyrins and face-to-face dimeric porphyrins, *Chem. Lett.,* 287 (1990).
78. T. Nagata, A. Osuka, and K. Maruyama, Synthesis and optical properties of conformationally constrained trimeric and pentameric porphyrin arrays, *J. Am. Chem. Soc., 112*: 3054 (1990).
79. A. Osuka, H. Yamada, and K. Maruyama, Synthesis of conformationally restricted carotenoid-linked porphyrins, *Chem. Lett.*, p. 1905 (1990).
80. A.W. Johnson, Synthesis of porphyrins from 1,19-dideoxy biladienes-ac and 1,19-dideoxybilenes-b, in *The Porphyrins*, Vol. 1 (D. Dolphin, ed.), Academic Press, New York, 1978, p. 235.

81. P.S. Clezy and L. van Thuc, The chemistry of pyrrolic compounds, The oxidative cyclization of derivatives of 1,19-dideoxybilenes-b, *Aust. J. Chem., 37*: 2085 (1984).
82. A.W. Johnson and I.T. Kay, The formation of porphyrins by the cyclization of bilenes, *J. Chem. Soc.,* 2418 (1961).
83. J.A.P.B. de Almeida, G.W. Kenner, J. Rimmer, and K.M. Smith, Pyrrole and related compounds. 35. A stepwise general synthesis of unsymmetrically substituted porphyrins, *Tetrahedron, 32*: 1793 (1976).
84. J. Engel and A. Gossauer, Synthesis of new biladiene-ac derivatives and their cyclization to uroporphyrin III, 12-decarboxyuroporphyrin III and phyriaporphyrin III methyl esters, *Liebigs Ann. Chem.,* 1637 (1976).
85. K. M. Smith and G.W. Craig, Porphyrin synthesis through tripyrrins: An alternative approach, *J. Org. Chem., 48*: 4302 (1983).
86. R.L.N. Harris, A.W. Johnson, and I.T. Kay, A stepwise synthesis of unsymmetrical porphyrins, *J. Chem. Soc. C,* 22 (1966).
87. P. Bamfield, R.L.N. Harris, A.W. Johnson, and I.T. Kay, Synthesis of rhodoporphyrin XV diethyl ester and related porphines, *J. Chem. Soc. C,* 1436 (1966).
88. T.P. Wijesekera and D. Dolphin, 1-Bromo-19-methylbiladienes-ac; useful precursors to porphyrins, *Synlett*, 235 (1990).
89. J.B. Paine III, J. Hiom, and D. Dolphin, Bromination of dipyrromethenes for porphyrin synthesis, *J. Org. Chem., 53*: 2796 (1988).
90. J.B. Paine III and D. Dolphin, Synthesis of covalently linked dimeric porphyrins, *Can. J. Chem., 56*: 1710 (1978).
91. J. Hiom, J.B. Paine III, U. Zapf, and D. Dolphin, The synthesis of cofacial porphyrin dimers, *Can. J. Chem., 61*: 2220 (1983).
92. P. Yon-Hin, T.P. Wijesekera, and D. Dolphin, Transformation of a monovinylporphyrin to benzoporphyrins via Diels–Adler adducts, *Tetrahedron Lett., 30*: 6135 (1989).
93. P. Yon-Hin, T.P. Wijesekera, and D. Dolphin, An efficient route to vinylporphyrins, *Can. J. Chem., 68*: 1867 (1990).
94. T.P. Wijesekera and D. Dolphin, unpublished results.
95. J.W. Buchler, Synthesis and properties of metalloporphyrins, in *The Porphyrins*, Vol. 1 (D. Dolphin, ed.), Academic Press, New York, 1978, p. 389.
96. J.-H. Fuhrhop and K.M. Smith, Laboratory methods, in *Porphyrins and Metalloporphyrins* (K.M. Smith, ed.), Elsevier, Amsterdam, 1975, p. 757.
97. J.O. Alben, W.H. Fuchsman, C.A. Beaudreau, and W.S. Caughey, Substituted deuteroporphyrins. III. Iron(II) derivatives. Reactions with oxygen and preparations from chloro and methoxo hemins, *Biochemistry, 7*: 624 (1968).
98. P.J. Crook, A.H. Jackson, and G.W. Kenner, Pyrroles and related compounds. 28. Synthesis of alpha-oxymesoporphyrin-IX, *Justus Liebigs Ann. Chem., 748*: 26 (1971).
99. J.W. Buchler and L. Puppe, Metal complexes with tetrapyrrole ligands. II. Metal chelates of alpha, gamma-dimethyl, gamma-dihydrooctaethylporphine via reductive methylation of (octaethylporphinato)zinc, *Justus Liebigs Ann. Chem., 740*: 142 (1970).
100. A.D. Adler, F.R. Longo, F. Kampas, and J. Kim, On the preparation of metalloporphyrins, *J. Inorg. Nucl. Chem., 32*: 2443 (1970).

101. K.M. Kadish, C. Araullo-McAdams, B.C. Han, and M.M. Franzen, Syntheses and spectroscopic characterization of $[T(p\text{-}Me_2N)F_4PP]H_2$ and $[T(p\text{-}Me_2N)F_4\text{-}PP]M$ where $T(p\text{-}Me_2N)F_4PP$ is the dianion of meso-tetrakis[o,o,m,m-tetrafluoro-p-(dimethylamino)phenyl]porphyrin and M = Co(II), Cu(II), or Ni(II). Structures of $[T(p\text{-}Me_2N)F_4PP)Co$ and [meso-tetrakis(pentafluorophenyl)porphinato]cobalt(II), $(TF_5PP)Co$, *J. Am. Chem. Soc., 112*: 8364 (1990).
102. T.P. Wijesekera and D. Dolphin, unpublished observations.
103. J.-H. Fuhrhop and D. Mauzerall, One-electron oxidation of metalloporphyrins, *J. Am. Chem. Soc., 91*: 4174 (1969).
104. M. Tsutsui, M. Ichakawa, F. Vohwinkel, and K. Suzuki, Unusual metalloporphyrins. I. Preparation of chromium mesoporphyrin IX dimethyl ester, *J. Am. Chem. Soc., 88*: 854 (1966).
105. M. Tsutsui, C.P. Hrung, D. Ostfeld, T.S. Srivastava, D.L. Cullen, and E.F. Meyer Jr., Unusual metalloporphyrins. 23. Unusual metalloporphyrin complexes of rhenium and technicium, *J. Am. Chem. Soc., 75*: 3952 (1975).
106. J.W. Buchler, M. Foltz, H. Habets, J. van Kaam, and K. Rohbock, Metal complexes of tetrapyrrole ligands. 17. Octacoordinate zirconium and hafnium porphines containing axial 1,3-diketonate ligands, *Chem. Ber., 109*: 1477 (1976).
107. J.W. Buchler, L. Puppe, K. Rohbock, and H. Schneehage, Metal complexes of octaethylporphine. Preparation, axial ligand substitution and reduction, *Ann. N.Y. Acad. Sci., 206*: 116 (1973).
108. D.B. Morell, J. Barrett, and P.S. Clezy, Prosthetic group of cytochrome oxidase. I. Purification as porphyrin a and conversion into hemin a, *Biochem. J., 78*: 793 (1961).
109. J.A. Bumpas, M. Tien, D. Wright, and S.D. Aust, Oxidation of persistent environmental pollutants by a white rot fungus, *Science, 228*: 1434 (1985).
110. K.E. Hammel, B. Kalyanaraman, and T.K. Kirk, Oxidation of polycyclic aromatic hydrocarbons and dibenzo[p]dioxins by *Phanerochaete chrysosporium* ligninase, *J. Biol. Chem., 261*: 16948 (1986).
111. V.-B. Huynh, H.-M. Chang, T.W. Joyce, and T.K. Kirk, Dechlorination of chloroorganics by a white-rot fungus, *Tappi J., 68*: 98 (1985).
112. D. Dolphin, A. Forman, D.C. Borg, J. Fajer, and R.H. Felton, Compounds I of catalase and horse radish peroxidase: π-cation radicals, *Proc. Natl. Acad. Sci. USA, 68*: 614 (1971).
113. J.E. Roberts, B.M. Hoffman, R. Rutter, and L.P. Hager, Electron-nuclear double resonance of horseradish peroxidase compound I. Detection of the porphyrin π-cation radical, *J. Biol. Chem., 256*: 2118 (1981).
114. B.M. Hoffman, J.E. Roberts, C.H. Kang, and E. Margoliash, Electron paramagnetic and electron nuclear double resonance of the hydrogen peroxide compound of cytochrome c peroxidase. *J. Biol. Chem., 256*: 6556 (1981).
115. J.T. Groves, R.C. Haushalter, M. Nakamura, T.E. Nemo, and B.J. Evans, High-valent iron–porphyrin complexes related to peroxidase and cytochrome P-450, *J. Am. Chem. Soc., 103*: 2884 (1981).
116. D. Mandon, R. Weiss, K. Jayarajah, A. Gold, J. Terner, E. Bill, and A.X. Trautwein, Models for peroxidase compound I: generation and spectroscopic

characterization of new oxoferryl porphyrin π-cation radical species, *Inorg. Chem., 31*: 4404 (1992).

117. I. Tabushi, Reductive dioxygen activation by use of artificial P-450 systems, *Coord. Chem. Rev., 86*: 1 (1988).
118. J.T. Groves, T.E. Nemo, and R.S. Meyers, Hydroxylation and epoxidation catalyzed by iron–porphine complexes. Oxygen transfer from iodosylbenzene, *J. Am. Chem. Soc., 101*: 1032 (1979).
119. B. Meunier, Metalloporphyrins as versatile catalysts for oxidation reactions and oxidative DNA cleavage, *Chem. Rev., 92*: 1411 (1992).
120. C.K. Chang and F. Ebina, NIH shift in haemin-iodosylbenzene mediated hydroxylations, *J. Chem. Soc. Chem. Commun.,* 778 (1981).
121. D. Dolphin, T. Nakano, T.E. Maione, T.K. Kirk, and R.K. Farrell, Synthetic model ligninases, in *Lignin Enzymic and Microbial Degradation* (E. Odier, ed.), INRA, Paris, 1987, p. 157.
122. D. Dolphin, T. Nakano, T.K. Kirk, T.P. Wijesekera, R.L. Farrell, and T.E. Maione, Porphyrins and uses thereof, U.S. Patent 4,892,941 (1990).
123. A.M. d'A, Rocha Gonsalves, R.A.W. Johnstone, M.M. Pereira, J. Shaw, and A.J.F. do N. Sobral, Metal-assisted reactions. 22. Synthesis of perhalogenated porphyrins and their use as oxidation catalysts, *Tetrahedron Lett., 32*: 1355 (1991).
124. J.F. Bartoli, O. Brigaud, P. Battioni, and D. Mansuy, Hydroxylation of linear alkanes catalyzed by iron porphyrins: particular efficacy and regioselectivity of perhalogenated porphyrins, *J. Chem. Soc. Chem. Commun.,* 440 (1991).
125. P. Hoffmann, A. Robert, and B. Meunier, Preparation and catalytic activities of the manganese and iron derivatives of Br_8TMP and $Cl_{12}TMP$, two robust porphyrin ligands obtained by halogenation of tetramesitylporphyrin, *Bull. Chem. Soc. Fr., 129*: 85 (1992).
126. M.M. Catalano, M.J. Crossley, M.M. Harding, and L.G. King, Control of reactivity at the porphyrin periphery by metal–ion coordination: a general method for specific nitration at the beta-pyrrolic position of 5,10,15,20-tetraaryporphyrins, *J. Chem. Soc. Chem. Commun.,* 1535 (1984).
127. T.G. Traylor and S. Tsuchiya, Perhalogenated tetraphenylhemins: stable catalysts of high turnover catalytic hydroxylations, *Inorg. Chem., 26*: 1338 (1987).
128. D. Mandon, P. Ochsenbein, J. Fischer, R. Weiss, K. Jayarajah, R.N. Austin, A. Gold, P.S. White, O. Brigaud, P. Battioni, and D. Mansuy, β-Halogenated-pyrrole porphyrins. Molecular structures of 2,3,7,8,12,13,17,18-octabromo-5,10,15,20-tetramesitylporphyrin, nickel(II) 2,3,7,8,12,13,17,18-octabromo-5,10,15,20-tetramesitylporphyrin and nickel(II) 2,3,7,8,12,13,17,18-octabromo-5,10,15,20-tetrakis(pentafluorophenyl)porphyrin, *Inorg. Chem., 31*: 2044 (1992).
129. J.E. Lyons and P.E. Ellis, Jr., Selective low temperature hydroxylation of isobutane by molecular oxygen catalyzed by an iron perhaloporphyrin complex, *Catalysis Lett., 8*: 45 (1991).

130. P. Bhyrappa and V. Krishnan, Octabromotetraphenyporphyrin and its metal derivatives: electronic structure and electrochemical properties, *Inorg. Chem., 30*: 239 (1991).
131. T. Wijesekera, A. Matsumoto, D. Dolphin, and D. Lexa, Perchlorinated and highly chlorinated meso-tetraphenylporphyrins, *Angew. Chem. Int. Ed. Engl., 29*: 1028 (1990).
132. J.E. Lyons, P.E. Ellis Jr., R.W. Wagner, P.B. Thompson, H.B. Gray, M.E. Hughes, and J.A. Hodge, Air-oxidation of light alkanes by first-row transition metals in macrocyclic ligand environments, in *Proceedings of American Chemical Society*, Petroleum Chemistry Division, San Francisco, 1992, p. 307.
133. P.S. Skerker, R.L. Farrell, D. Dolphin, F. Cui, and T. Wijesekera, Biomimetic bleaching of kraft pulps, in *Applications of Biotechnology in Pulp and Paper Manufacture* (T.K. Kirk and H.-M. Chang, eds.), Butterworth-Heinemann, Boston, 1990, p. 203.
134. P.E. Ellis Jr. and J.E. Lyons, Selective air oxidation of light alkanes catalyzed by activated metalloporphyrins—the search for a suprabiotic system, *Coord. Chem. Rev., 105*: 181 (1990).

8

Asymmetric Oxidation with Chiral Porphyrin Catalysts

Yoshinori Naruta *

Kyoto University, Kyoto, Japan

I. INTRODUCTION

Asymmetric reactions have not yet been fully exploited in organic synthesis [1]. Recently, there have been many reports of enantioselective reactions using stoichiometric amount of chiral sources. However, the most elegant and practical solution is to use catalytic amounts of the chirality-inducing agent. Such catalytic asymmetric syntheses have been successful in enantioselective C-C bond formation [2] and hydrogenation [3]. Catalytic and asymmetric oxidation is a more difficult reaction to accomplish because chiral ligands are often not stable under the oxidation conditions. This requirement severely limits the type of ligands that can be applied in asymmetric oxidations. Currently, three catalytic asymmetric oxidation systems have been proposed: asymmetric osmium-catalyzed hydroxylation, Sharpless epoxidation of allylic alcohols, and Mn–Schiff base complex–catalyzed oxidation of unfunctionalized olefins [4]. Among them, Sharpless allylic epoxidation with *t*-BuOOH as an oxidant is the most popular one and is widely applied in many synthetic reactions [5]. To expand the variety of substrate olefins, metal complexes with various Schiff base ligands have been applied to the epoxidatiion of simple unfunctionalized olefins with use of NaOCl or PhIO as oxidants [6a–j]. In a recent report [6k], O_2-*t*-BuCHO system was applied to the reaction of Mn–Schiff base–catalyzed epoxidation to give reasonable ee's. This system is very interesting in terms of the modeling reaction of cytochrome P450 monooxygenases. These catalysts gave high ee's only with cyclic olefins containing sterically different groups attached to

**Present affiliation*: Institute for Molecular Science, Okazaki, Japan

Table 1 Selected Results of Enantioselective Epoxidation with Chiral Mn–Schiff Base Catalysts

Olefin	Catalyst	Oxidant	Yield (%)	ee (%)	Ref.
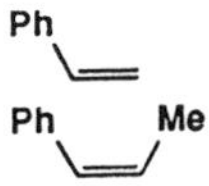	A	$Me_3C_6H_2IO^a$	75	57	6a
	A	$Me_3C_6H_2IO$	73	84	6a
	B	PhIO	19	46	6j
	B/Py-N-Ox[b]	PhIO	49	65	6j
	C	NaOCl	88	84	6g
	D	NaOCl	84	92	6g
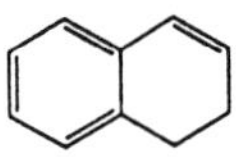	C	$Me_3C_6H_2IO$	72	78	6a
	B	PhIO	93	49	6i,j
	E/N-MeIm[b]	O_2/Me_3CCHO	78	63	6k
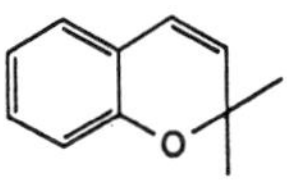	D	NaOCl	87	98	6h

[a]$Me_3C_6H_2IO$ = 2,4,6-trimethylphenyliodosylbenzene.
[b]Py-N-Ox = pyridine *N*-oxide, *N*-MeIm = *N*-methylimidazole.

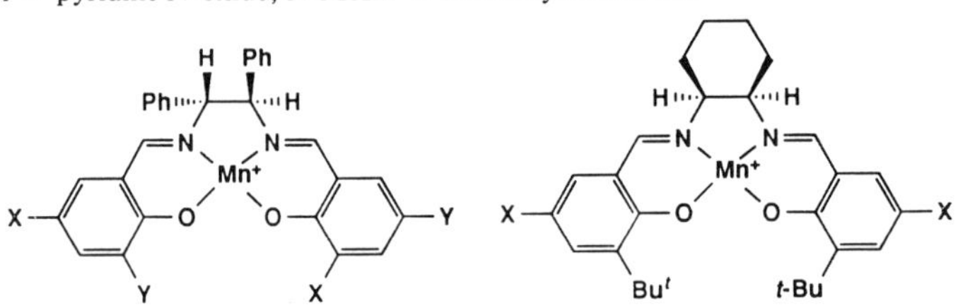

A X = H, Y = *t*-Bu

B X = H, Y = C*H(Et)Ph

C X = Me, Y = *t*-Bu

D X = *t*-Bu

E X = Me

the ends of the double bond (Table 1). From the relationship between catalyst structure and enantioselectivity, bulky substituents on the aromatic rings of the Schiff base ligand would control the direction of substrate approach to the metal center. These substituents also exert electronic effects on the resultant ee, and electron-donating groups showed higher ee's [6g].

Among natural enzymatic systems including a heme as a prosthetic group, the three-dimensional structure of cytochrome $P450_{cam}$ has been fully elucidated by X-ray crystallographic analysis [7]. The substrate binding pocket

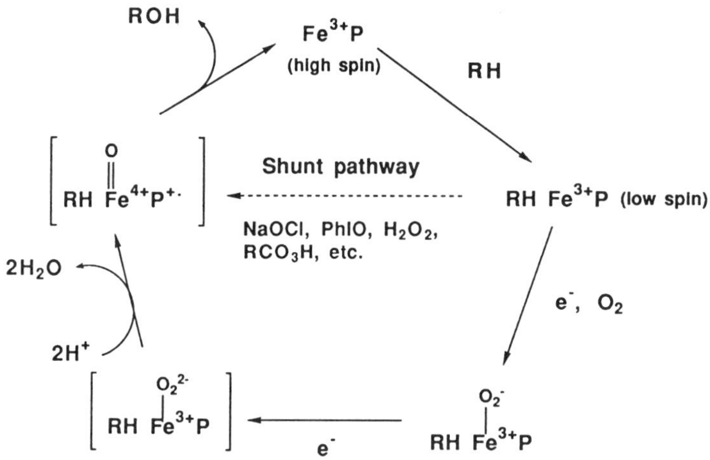

Scheme 1 Cytochrome P450 catalytic reaction cycle. Substrate-bound Fe^{3+} protoporphyrin IX (low spin) is reduced to the corresponding Fe^{2+}, and the resultant O_2 complex is further reduced to the Fe^{3+}-O_2^{2-} complex, which is converted to the high-valent $P^{+}\cdot Fe^{4+}=O$ complex by the protonation of the terminal oxygen. This is assumed to be the active species of the monooxygenation. RH and ROH represent substrate and oxidized product, respectively. Those intermediates shown in brackets are hypothetical and have not directly been observed.

on the heme molecule is chiral. This enables the enantioselective oxidation of olefins [8–12] and sulfides [13] to the corresponding epoxides and sulfoxides, respectively, in fair to good ee's. On the other hand, model oxidation reactions mimicking these enzymatic reactions have been developed with a simple metalloporphyrin catalyst and appropriate single-oxygen donors (shunt pathway) (Scheme 1) [14]. Modification of porphyrins with chiral auxiliaries creates a chiral reaction cavity at the reaction center and those porphyrins should provide catalysts for the asymmetric oxidation of organic substrates by the application of the artificial oxidation systems. To this end, various chiral porphyrins have been designed and applied to the oxidation of simple substrates.

II. CATALYST DESIGN AND SYNTHESIS

There are four problems to be considered in the design of a chiral oxidation catalyst. First, the catalyst symmetry is the most important to give rise to enantioselectivity. Asymmetric oxidations, such as olefin epoxidation,

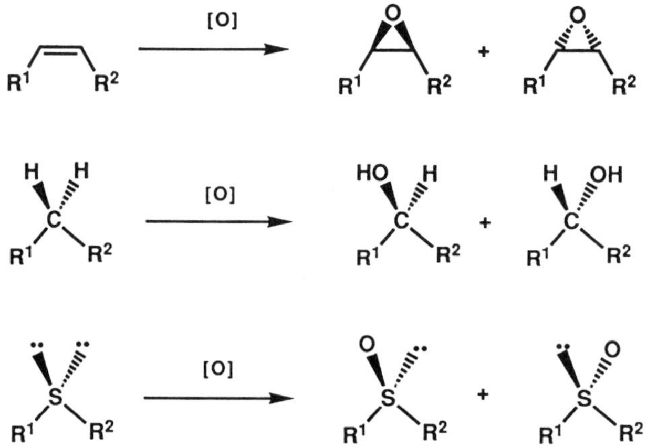

Figure 1 Asymmetric oxidation of olefins, saturated hydrocarbon, and sulfides based on prochiral face recognition.

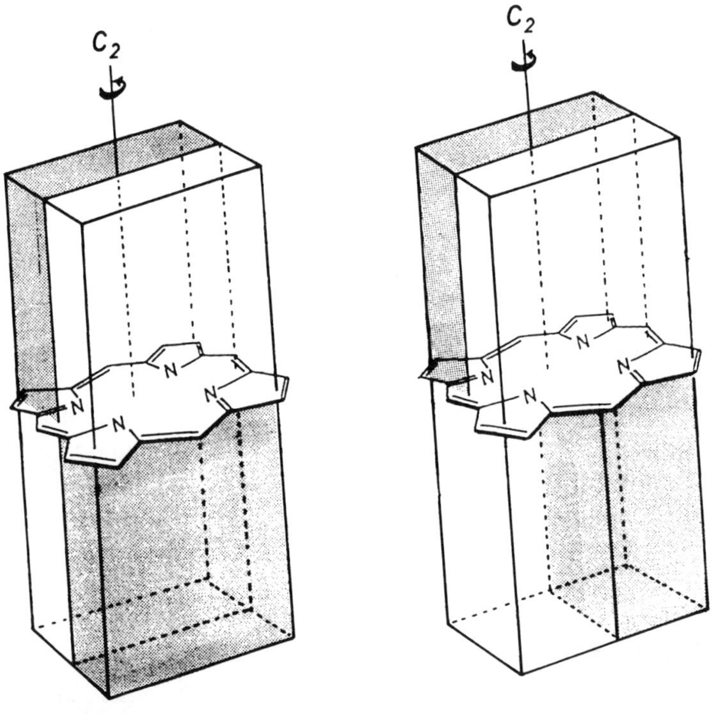

Figure 2 Schematic drawing of chiral porphyrins applicable for asymmetric reaction. All four reaction cavities (shaded and unshaded) on each porphyrin molecule should be equivalent and the molecular symmetry is D_2 or D_4.

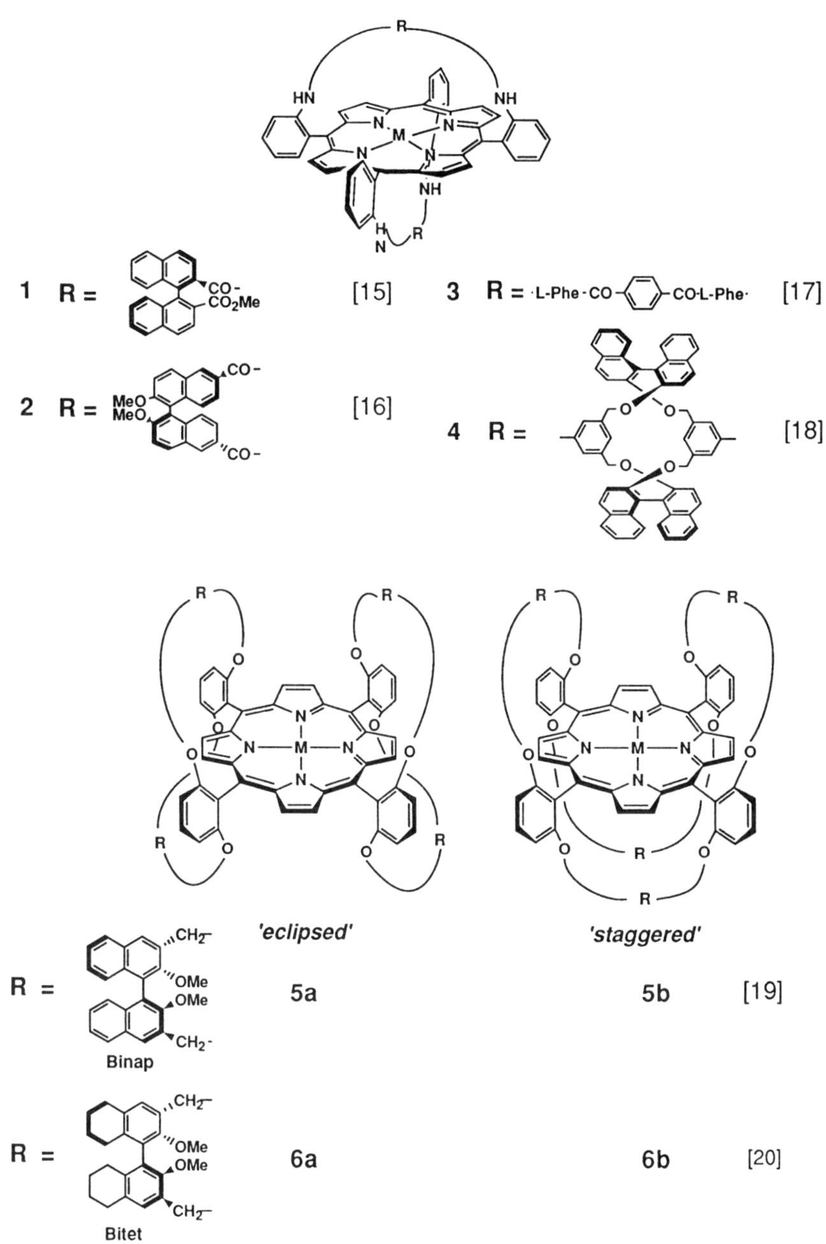

Figure 3 Porphyrins linked by a chiral appendage.

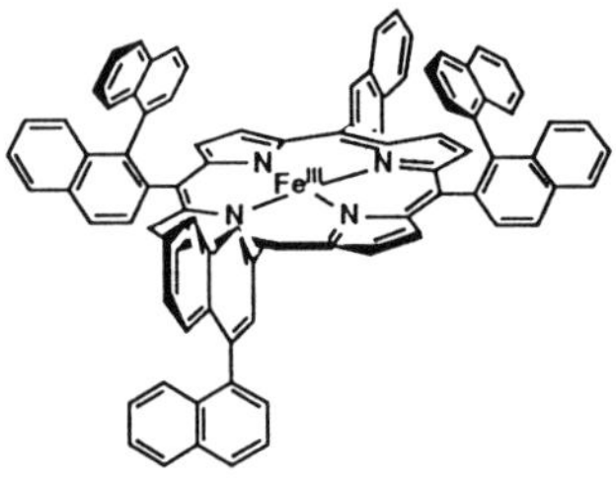

7 [21]

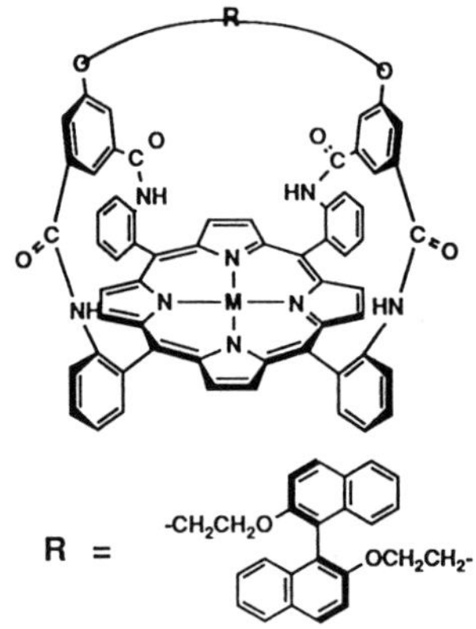

8 [22]

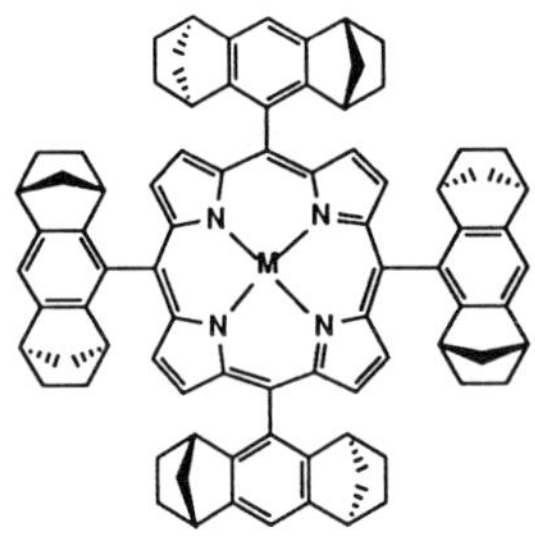

9 [23]

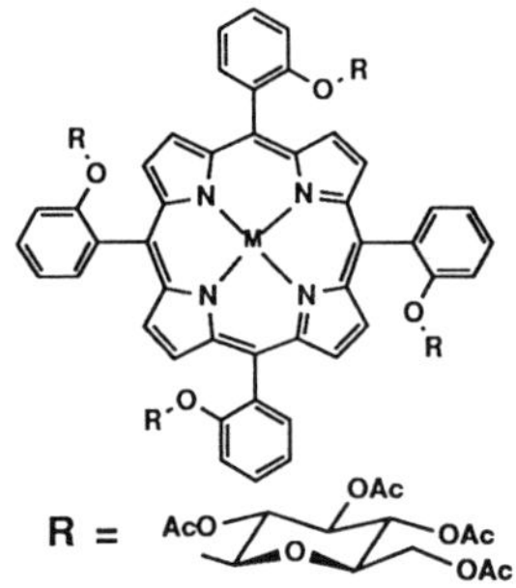

10 [24]

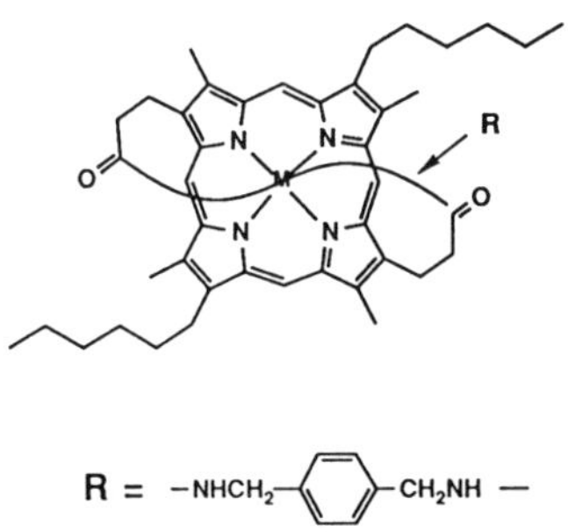

11 [25]

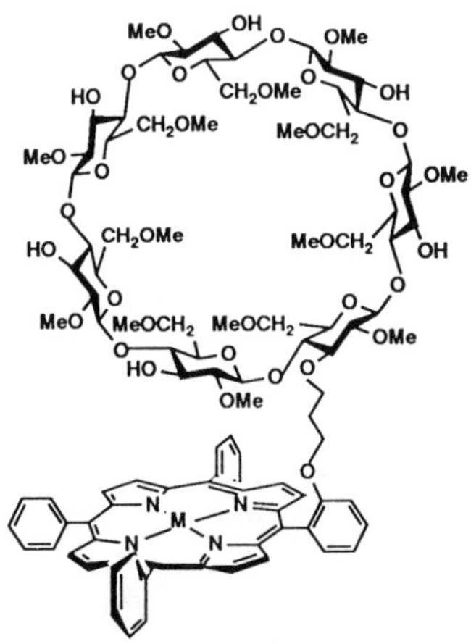

12 [26]

Figure 3 (Continued)

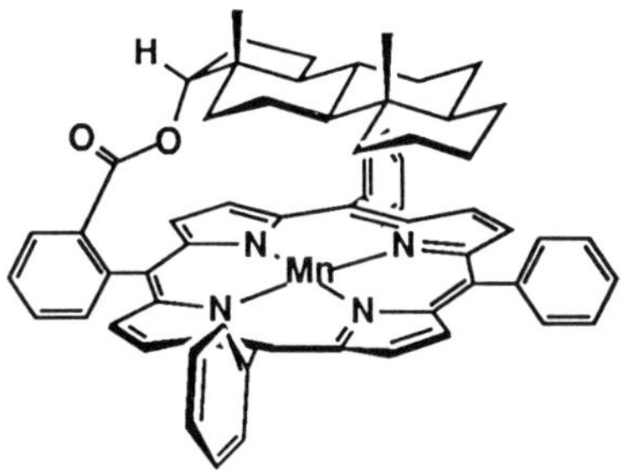

13 [27]

Figure 3 (Continued)

sulfoxidation, and hydroxylation of saturated hydrocarbons, involve prochiral face recognition (Figure 1). Thus the reaction sites of the applied catalysts should have C_2-symmetry. Consequently, the chiral porphyrin itself should belong to a D_2 (or D_4) symmetry group. To meet these criteria, four reaction cavities (shaded and unshaded in Figure 2) above and below a porphyrin plane should be equivalent after modification by chiral groups. Until now, versatile porphyrins (**1–11**) bearing chiral auxiliaries have been synthesized according to these criteria (Figure 3) [15–27].

The second problem to be considered is how one can construct a chiral environment close enough to the metal center. However, even if the stereogenic center on the catalyst is somewhat far from the metal, bulky groups on the auxiliary can contribute to create a suitable environment for efficient prochiral recognition on the catalyst.

Third, one has to keep in mind the mode of the prochiral face recognition of a substrate, i.e., how one can fix a substrate in the chiral cavity at the oxo transfer stage (Scheme 2). This is especially important for substrates which are not able to coordinate strongly to the catalyst.

The major synthetic strategy that has been followed is modification of the porphyrin ligand with an appropriate chiral auxiliary. Ligands based on *meso*-tetra(*o*-aminophenyl)porphyrin connected by bridging chiral amide linkages have been used most frequently because the synthesis and separation of its atropisomers has been established. However, since porphyrins bearing carboxamido groups are somewhat fragile under oxidation conditions [28], one would not expect high catalytic turnovers with these ligands. From the viewpoint of catalyst degradation, ligands containing auxiliaries linked by a C-C bond or an ether bond are more robust than those with the carboxamido groups.

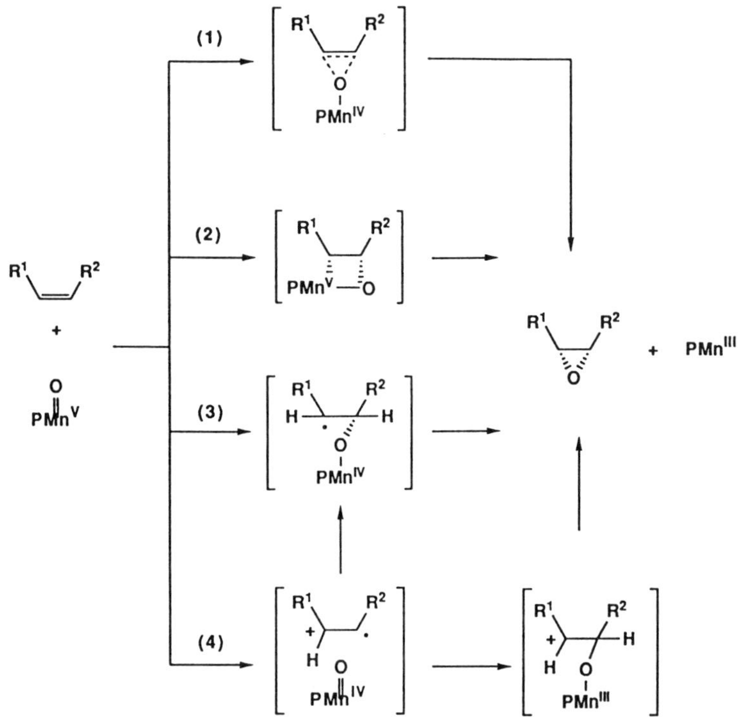

Scheme 2 Possible mechanisms of oxotransfer from a high-valent oxo-Mn porphyrin in the epoxidation of olefins: (1) direct oxygen atom transfer, (2) metallaoxetane formation, (3) free radical addition and successive fast ring closure, and (4) single-electron transfer followed by a collapse to a radical or to a carbo cation. Similar scheme has been considered for the corresponding iron complex $P^{+}\cdot Fe^{IV}=O$. Since the mechanism seems to be depend on substrate, metal, and ligand, currently it cannot be explained by universal mechanism (see Ref. 14).

III. ASYMMETRIC OXIDATION OF OLEFINS

Since olefin epoxidation is one of the most intensively studied of metalloporphyrin-catalyzed reactions, much information is available. Table 2 compares the enantioselective epoxidation of styrene with various chiral metalloporphyrin catalysts. As the reaction conditions vary with the different examples, one cannot directly compare their turnover numbers and ee's. However, they do give some idea about the catalyst efficiency on chiral induction. When catalysts have sufficient room at the reaction site or the chiral appendages are not tightly fixed to the catalyst, lower ee's have been obtained.

Table 2 Comparison of Enantioselective Epoxidation of Styrene with Various Chiral Porphyrin Catalysts

Catalyst	Turnover number[a]	ee (%)	Ref.
1	14	48	15
2	23	30	16
3	2.5[b]	50[b]	17
4	68	48	18
5a	49	22	19
6a	28 (485[c])	58[d]	20
7	240	20	21
8	–	20	21
9	180	52	23
10	50[b]	27.5[b]	24
11	68	50	25

[a]Turnover numbers were calculated based on the amount of charged oxidant or applied styrene.
[b]Values obtained in the reaction of *p*-chlorostyrene.
[c]After repeated addition of iodosylbenzene.
[d]After many turnovers, this value was almost the same as that obtained at the initial stage.

In most examples, one cannot distinguish clear steric or electronic effects with various substituted styrenes. To understand how the olefins can be discriminated at their prochiral faces, systematic studies with many substituted styrene derivatives have been carried out in epoxidations with the catalysts **5a** (M = Fe) and **6a** (M = Fe) and PhIO as an oxidant [19,20]. These catalysts have rather tight pockets on both sides of the porphyrin ring and two MeO groups are at positions very close to the central metal (Figure 4). Except for the steric bulkiness of the two auxiliaries, these catalysts have similar reaction cavities. In these systems, the major configuration of the obtained epoxides can be rationally explained by the interaction of a substrate with the chiral auxiliaries (Figure 5) [19,20]. The most impressive results were obtained when the substituted styrene had electron-withdrawing group(s): with catalyst **5a** (M = Fe) 80% ee for 2-nitrostyrene and with catalyst **6a** (M = Fe), 96% ee for 3,5-dinitrostyrene. These are the highest ee values ever recorded with metalloporphyrin-based catalysts.

During analysis of the substituent electronic effects on ee, a very interesting relationship between $\Delta\Delta G^{\neq}$ (free energy difference between the formation of the two enantiomeric epoxides) and Hammett σ values of sub-

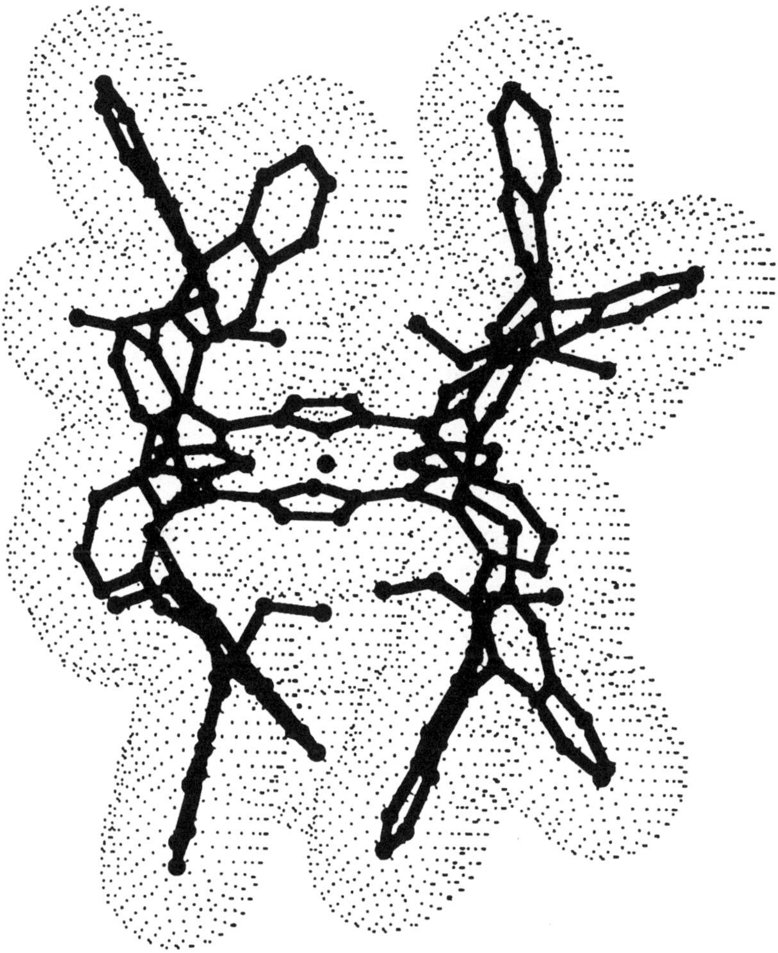

Figure 4 Computer-generated model of twin-coronet porphyrin **5a** (M = Fe). The interactive modeling was performed on a Silicon Graphics IRIS 4D/35 computer using MAXIMIN2 package inside SYBYL 5.5.

stituent(s) was found (Figure 6). Some nitro-substituted styrenes showed unexpectedly low ee's in comparison with related styrenes, e.g., 2,4-$(NO_2)_2$ (76% ee) compared with 2-NO_2 (89% ee) and 3.5-$(NO_2)_2$ (96% ee). This difference could not be rationally explained on the basis of steric interactions. Inspection of the configuration at the oxo transfer stage (Figure 5) shows that the substrate aromatic ring can overlap with the π system of the chiral

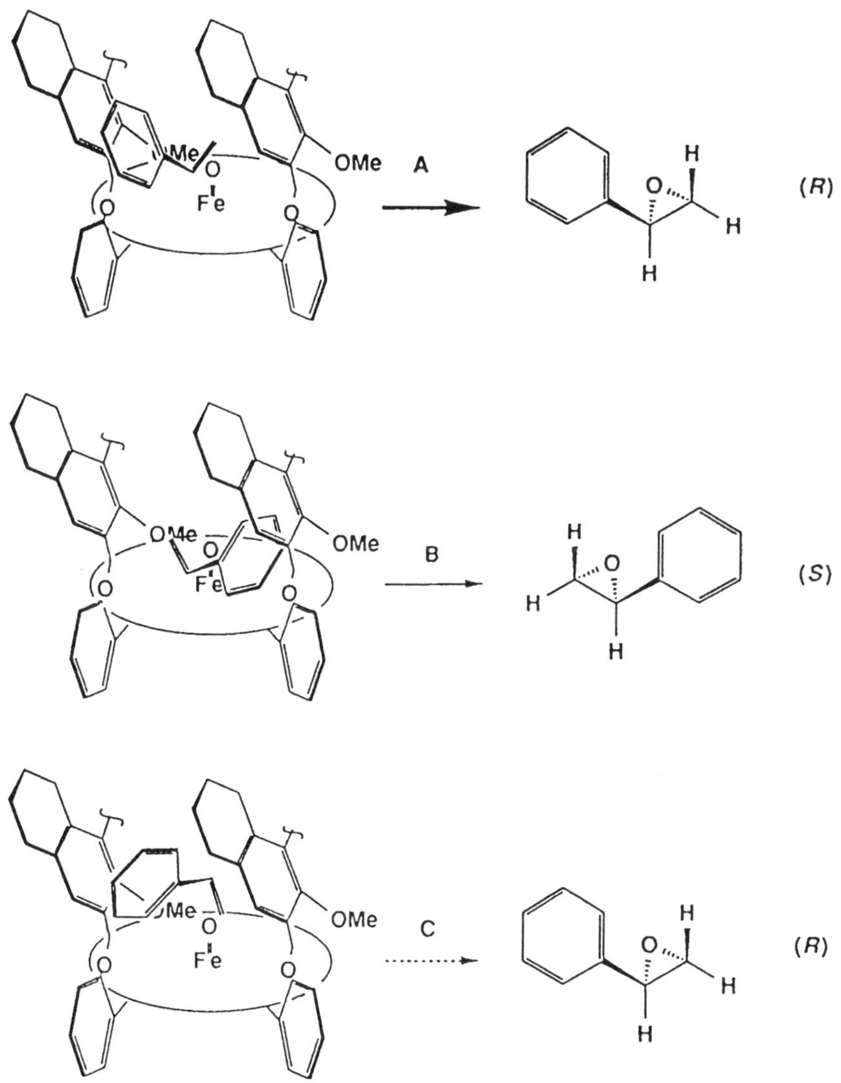

Figure 5 Possible geometries of styrene in the cavity over the catalyst **6a** in the transition state of oxygen transfer [20]. Partial structure of the catalyst is shown for clarity.

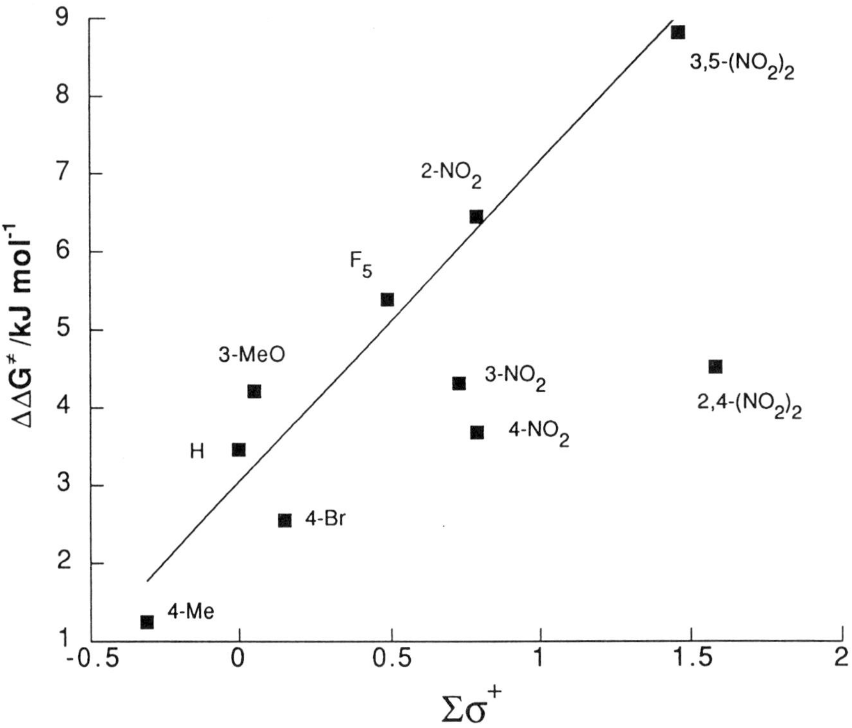

Figure 6 Plot of $\Delta\Delta G^{\neq}$ between two enantiomers of epoxides vs. $\Sigma\sigma^+$ of the substrate styrene. The values were obtained in the catalytic epoxidation with **6a** [20].

appendage. This can lead to electronic interactions between the two π systems. From the detailed analysis of substrate LUMO (lowest unoccupied molecular orbital) and the appendage HOMO (highest occupied molecular orbital), it is proposed that the enantioselectivity is controlled by the interaction of their frontier orbitals at the oxo transfer arrangement to an olefin. These results suggest a new role of the chiral appendage, which can discriminate prochiral faces of the substrate not only by steric effects but by CT-type interaction as well.

Finally, the effect of the central metal ion on epoxide ee should be mentioned. In most of the catalytic reactions studied, either the Fe or the Mn derivative of the porphyrin was used. Recently, the two metals were compared in epoxidations of various olefins [29]. For styrene, no difference in ee was observed between Mn and Fe. In contrast, cis-1,2-disubstituted olefins, e.g., *cis*-β-methylstyrene, were oxidized in remarkably high ee with the Mn derivative **5a** (M = Mn). The reason is not clear, but the transition state of the Mn-catalyzed oxidation will be later in the reaction and the activation energy will be high because the Mn catalyst is less reactive than the corresponding Fe catalyst and the reaction time is longer for Mn than for Fe. Thus, the energy difference between the diastereomeric transition states is expected to be larger for the Mn catalyst than Fe.

IV. ASYMMETRIC OXIDATION OF SATURATED HYDROCARBONS

Asymmetric hydroxylation of saturated hydrocarbons is a completely undeveloped field in organic synthesis. It is one of the most difficult reactions to achieve because of the requirement of a high potential oxidizing agent, which is usually incompatible with high chemo- and stereoselectivity. Thus, a very limited number of examples are known. Using the iron derivative of the catalyst **2** (M = Fe) and PhIO, the hydroxylation of benzylic hydrocarbons has been reported to give the corresponding benzylic alcohol in 40–72% ee [16]. In these instances, the iron derivative gave both higher chemical yield and higher ee than the corresponding Mn one. A mechanistic study using deuterium-labeled ethylbenzene indicated that abstraction of one prochiral hydrogen (*pro-R*) and succeeding rebound of ·OH proceeded with retention of configuration (>90%) in spite of the intermediate formation of carbon radicals. On the other hand, if the other hydrogen (*pro-S*) was removed, up to 40% racemization was observed. These results suggest that the substrate arrangement is fixed by nonbonded interaction between the intermediate radical and the catalyst.

Oxidation of α-pinene under irradiation of visible light in the presence of moelcular oxygen was reported to proceed enantioselectively in the presence of the cyclodextrin-linked metalloporphyrin catalyst **12** (M = Mn, Fe) [26]. In this instance, the corresponding chiral epoxide and homoallylic alcohols were obtained in up to 67% ee (Figure 7). Though the stereochemical course is unclear, direction of attack of active oxygen molecule ($^1O_2$ and $^3O_2$)

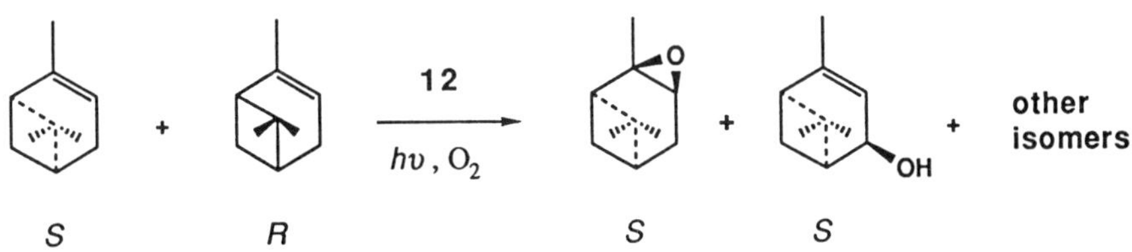

Figure 7 Photochemical oxidation of α-pinene with cyclodextrin-linked porphyrin **12** [26].

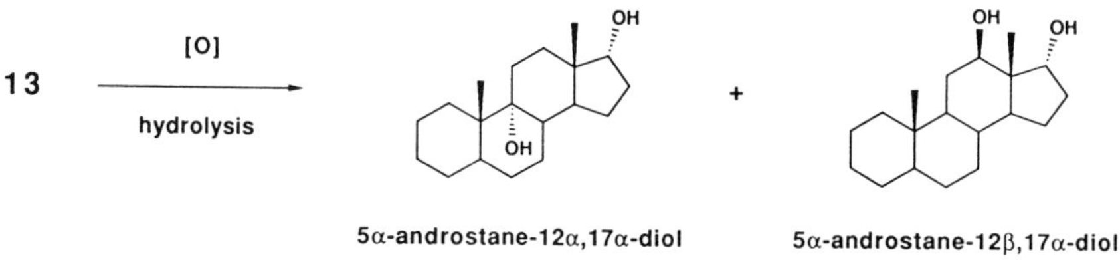

Figure 8 Regio- and stereoselective hydroxylation of androsteron-linked porphyrin **13** [27].

to the substrate seems to be regulated by the inclusion of the substrate to the cylodexitrin appendage.

Cytochrome P450–dependent enzymes also catalyze versatile regio- and stereoselective hydroxylation in the metabolism of steroids. These reactions are very interesting from a synthetic standpoint. In order to mimic this reaction, the steroid-linked porphyrin **13** was synthesized and stoichiometric oxidation of the attached steroid moiety was examined as a kind of remote oxidation [27]. Since the relative arrangement between the steroid moiety and the metal center is restricted, very high regio- and stereoselectivities were observed. When the carboxylate ester of the porphyrin linked by 5α-androstane-17β-ol was oxidized, the corresponding 9α,17α-diol and 12β,17α-diol were selectively obtained (Figure 8).

In a related example, a steroid-linked porphyrin was used as a catalyst for the regioselective oxidation of polyene or substrate steroid itself in the bilayer membrane of a phospholipid [30].

Since this field is one of the most challenging frontiers spanning biology and chemistry, further progress is strongly expected.

V. ASYMMETRIC OXIDATION OF SULFIDES

Sulfides are the most easily oxidizable substrates among those mentioned above. Since enantiomerically pure sulfoxides are useful chiral synthons in organic synthesis [31], catalytic asymmetric sulfoxidation is an important synthetic goal. The reaction seems superficially to be much easier than olefin epoxidation or alkane hydroxylation, and one can selectively obtain sulfoxide over sulfone if one applies suitable oxidation conditions. However, the observed ee's range from 14% to 48% ee [16] and from 17% to 73% ee [32] (Table 3). In comparison with other substrates, these values are modest even if one uses sulfides substituted by electron-withdrawing groups. In contrast to enzymatic [13] or stoichiometric asymmetric sulfoxidations [33], catalytic oxidations of sulfides with a Mn–Schiff base catalyst show moderate enan-

Table 3 Selected Examples of Monooxidation of Sulfides with Various Oxidation Conditions

Substrate	Catalyst	Oxidant	Yield (%)	ee (%)	Ref.
$PhSCH_3$	Mn–Schiff base complex[a]	H_2O_2	90	47	34
o-$BrC_6H_4SCH_3$	Mn–Schiff base complex[a]	H_2O_2	80	68	34
p-$MeC_6H_4SCH_3$	Mn–Schiff base complex[a]	H_2O_2	95	42	34
p-$NO_2C_6H_4SCH_3$	Mn–Schiff base complex[a]	H_2O_2	86	66	34
$PhSCH_3$	Horseradish peroxidase	H_2O_2	95	47	35
p-$CH_3C_6H_4SC_2H_5$	Rat liver Cyt. P-450s	O_2	—	58–68	36
$PhSCH_3$	$Ti(OiPr)_4$, (+)-diethyl tartrate[b]	t-BuOOH	80	89	33b
p-$MeC_6H_4SCH_3$	$Ti(OiPr)_4$, (+)-diethyl tartrate[b]	t-BuOOH	90	88	33b
$PhSCH_3$	**5a** (M = Fe)/N-MeIm[c]	PhIO	70	46	32
p-$MeC_6H_4SCH_3$	**5a** (M = Fe)/N-MeIm[c]	PhIO	72	54	32
$C_6F_5SCH_3$	**5a** (M = Fe)/N-MeIm[c]	PhIO	28	73	32
$PhSCH_3$	**2** (M = Fe)	PhIO	84	24	16
o-$BrC_6H_4SCH_3$	**2** (M = Fe)	PhIO	74	48	16

[a]Mn–Schiff base complex (X = OMe) with chiral cyclohexyldiamine moiety in Table 1.
[b]Stoichiometric amount of the Ti complex and the ligand was used.
[c]N-MeIm = N-methylimidazole.

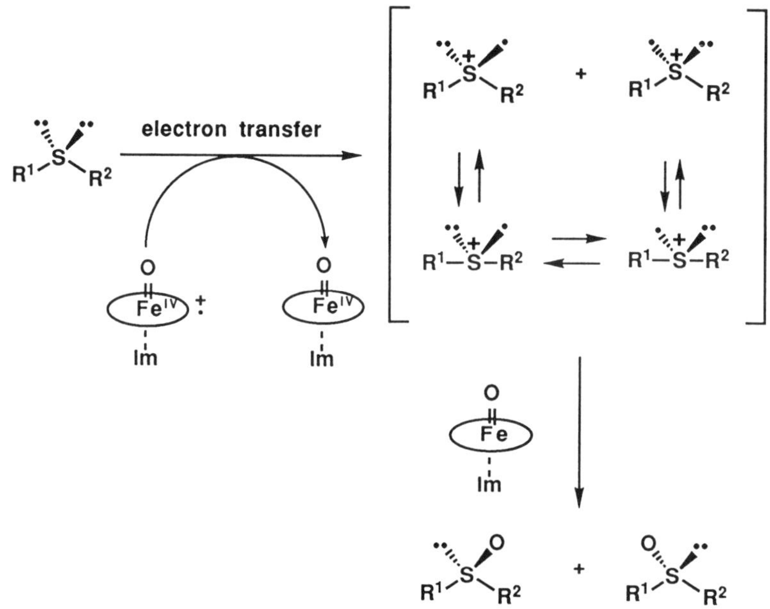

Figure 9 Reaction pathway of sulfide oxidation with a high-valent iron porphyrin [32].

tioselectivity (34–68% ee) [34]. It is worth mentioning that microbial as well as flavin-dependent oxygenases gave much higher ee's (20–100% ee) than cytochromes P450 [31].

Mechanistic analysis of this reaction has been performed using the catalyst **5a** (M = Fe). A Hammett plot of this reaction ($\rho^+ \approx 0$) suggests that the rate-determining step of the oxidation is the formation of the putative high-valent oxoiron species rather than oxygen transfer from the oxoiron intermediate to the substrate. It is assumed that the reaction proceeds via a single-electron transfer process analogous to that occurring in the enzymatic system (Figure 9). Consequently, the intermediate cation radical would lose its stereochemical integrity, thus decreasing the efficiency of prochiral face recognition. The same mechanism was also proposed for the racemic oxidation of sulfide with simple iron *meso*-tetraphenylporphyrin catalyst [37].

ACKNOWLEDGMENT

This work was supported by a grant-in-aid for scientific research from the Ministry of Education, Science, and Culture. The author thanks T. Nishimura, Sumitomo Pharmaceutical Co., for his preparation of the computer-generated model.

REFERENCES

1. S.G. Davis, J.M. Brown, A.J. Pratt, and G. Fleet, Asymmetric synthesis—meeting the challenge, *Chem. Br.*, 259 (1987).
2. R. Noyori and M. Kimura, Enantioselective catalysis with metal complexes: an overview, in *Modern Synthetic Chemistry* (R. Scheffold, ed.), Springer-Verlag, Berlin, 1989, p. 155.
3. J. Halpern, Asymmetric catalytic hydrogenation: mechanism and origin of enantioselection, in *Asymmetric Synthesis*, Vol. 5 (J.D. Morrison, ed.), Academic Press, New York, 1985, p. 41.
4. (a) J.S.M. Wai, I.E. Markó, J.S. Svendson, M.G. Finn, E.N. Jacobsen, and K.B. Sharpless, A mechanistic insight leads to a greatly improved osmium-catalyzed asymmetric dihydroxylation, *J. Am. Chem. Soc., 111*: 1123 (1989). (b) E.J. Corey, P.D. Jardine, S. Virgil, P.W. Yuen, and R.D. Connell, Enantioselective vicinal hydroxylation of terminal and E-1,2-disubstituted olefins by a chiral complex of osmium tetroxide, *J. Am. Chem. Soc., 111*: 9243 (1989). (c) K. Tomioka, M. Nakajima, Y. Iitaka, and K. Koga, Mechanistic aspects of asymmetric cis-dihydroxylation of olefins with osmium tetroxide employing a C_2-symmetric chiral diamine, *Tetrahedron Lett., 29*: 573 (1988). (d) T. Yamada and K. Narasaka, Asymmetric oxidation of olefins with osmium tetroxide coordinated by chiral diamines derived from L-tertaric acid, *Chem. Lett.*, 131 (1986). (e) M. Hirama, T. Oishi, and S. Ito, Asymmetric dihydroxylation of alkenes with osmium tetroxide: chiral *N,N′*-dialkyl-2,2′-bipyrrolidine complex, *J. Chem. Soc. Chem. Commun.*, 665 (1989).

5. M.G. Finn and K.B. Sharpless, On the mechanism of asymmetric epoxidation with titanium-tartrate catalysts, in *Asymmetric Synthesis*, Vol. 5 (J.D. Morrison, ed.), Academic Press, New York, 1985, p. 247.
6 (a) W. Zhang, J.L. Leobach, S.R. Wilson, and E.N. Jacobsen, Enantioselective epoxidation of unfunctionalized olefins catalyzed by (salen)manganese complexes, *J. Am. Chem. Soc., 112*: 2801 (1990). (b) R. Irie, K. Naota, Y. Ito, N. Matsumoto, and T. Katsuki, Catalytic asymmetric epoxidation of unfunctionalized olefins, *Tetrahedron Lett., 31*: 7345 (1990). (c) R. Irie, K. Naota, Y. Ito, and T. Katsuki, Enantioselective epoxidation of unfunctionalized olefins using chiral (salen)manganese(III) complexes, *Tetrahedron Lett., 32*: 1055 (1991). (d) W. Zhang and E.N. Jacobsen, Asymmetric olefin epoxidation with sodium hypochlorite catalyzed by easily prepared chiral Mn(III) salen complexes, *J. Org. Chem., 56*: 2296 (1991). (e) H. Fu, G.C. Look, W. Zhang, E.N. Jacobsen, and C.-H. Wong, Mechanistic study of a synthetically useful monooxygenase model using the hypersensitive probe *trans*-2-phenyl-1-vinylcyclopropane, *J. Org. Chem., 56*: 6497 (1991). (f) E.N. Jacobsen, W. Zhang, A.R. Muci, J.R. Ecker, and L. Deng, High enantioselective epoxidation catalysts derived from diaminocyclohexane, *J. Am. Chem. Soc., 113*: 7063 (1991). (g) E.N. Jacobsen, W. Zhang, and M.L. Güler, Electronic tuning of asymmetric catalysts, *J. Am. Chem. Soc., 113*: 6703 (1991). (h) N.H. Lee, E.N. Jacobsen, Enantiomerically pure epoxychromanes via asymmetric catalysis, *Tetrahedron Lett., 32*: 5055 (1991). (i) N.H. Lee and E.N. Jacobsen, Enantioselective epoxidation of conjugated dienes and enynes. trans-Epoxides from cis-olefins, *Tetrahedron Lett., 32*: 6533 (1991). (j) R. Irie, K. Noda, Y. Ito, N. Matsumoto, and T. Katsuki, Catalytic asymmetric epoxidation of unfunctionalized olefins using chiral (salen)manganese(III) complexes, *Tetrahedron: Asymmetry, 2*: 481 (1991). (k) Y. Yamda, K. Imagawa, T. Nagata, and T. Mukaiyama, Enantioselective epoxidation of unfunctionalized olefins with molecular oxygen and aldehyde catalyzed by optical active manganese(III) complexes, *Chem. Lett.*, 2231 (1992).
7. (a) T.L. Poulos, B.C. Finzel, and A.J. Howard, The high resolution crystal structure of cytochrome $P450_{cam}$, *J. Mol. Biol., 192*: 687 (1987). (b) T.L. Poulos, B.C. Finzel, and A.J. Howard, Crystal structure of substrate-free *P. putida* cytochrome P-450, *Biochemistry, 25*: 5314 (1986). (c) R. Raag and T.L. Poulos, Crystal structure of the carbon monoxide–substrate–cytochrome $P\text{-}450_{cam}$ ternary complex, *Biochemistry, 28*: 7586 (1989).
8. V. Schurig and D. Wistuba, Asymmetric microsomal epoxidation of simple prochiral olefins, *Angew. Chem. Int. Ed. Engl., 23*: 796 (1984).
9. C. Harris, R.M. Philpot, O. Hernandez, and J.R. Bend, Rabbit pulmonary cytochrome P-450 monooxygenase system: isozyme differences in the rate and stereoselectivity of styrene oxidation, *J. Pharm. Exp. Ther. 236*: 144 (1986).
10. G.L. Fourenass, C. Harris, F.P. Guengerich, and J.R. Bend, Stereoselectivity of styrene oxidation in microsomes and in purified cytochrome P-450 enzymes from rat liver, *J. Pharm. Exp. Ther., 248*: 429 (1989).
11. P.R. Ortiz de Montellano, J.A. Fruetel, J.R. Collins, D.L. Camper, and G.H. Loew, Theoretical and experimental analysis of the absolute stereochemistry of *cis*-β-methylstyrene epoxidation by cytochrome $P450_{cam}$, *J. Am. Chem. Soc., 113*: 3195 (1991).

12. H. Fu, M. Newcomb, and C.-H. Wong, *Pseudomonas oleovorans* monooxygenase catalyzed asymmetric epoxidation of allyl alcohol derivatives and hydroxylation of allyl alcohol derivatives and hydroxylation of a hypersensitive radical probe with the radical ring opening rate exceeding the oxygen rebound rate, *J. Am. Chem. Soc., 113*: 5878 (1991).
13. (a) B.J. Auret, D.R. Boyd, H.B. Henbest, and C.G. Watson, Species-strain dependence of stereoselectivity in microbial oxidation of thioethers, *Phytochemistry, 13*: 65 (1974). (b) T. Takata, M. Yamazaki, K. Fujimori, Y.H. Kim, T. Iyanagi, and S. Oae, Enzymatic oxygenation of sulfides with cytochrome P-450 from rabbit liver. Stereochemistry of sulfoxide formation, *Bull. Chem. Soc. Jpn., 56*: 2300 (1981). (c) D.J. Waxman, D.R. Light, and C. Walsh, Chiral sulfoxidation catalyzed by rat liver cytochrome P-450, *Biochemistry, 21*: 2499 (1982). (d) S. Colonna, N. Gaggero, G. Carrea, and P. Pasta, Horseradish peroxidase catalyzed sulfoxidation is enantioselective, *J. Chem. Soc. Chem. Commun.,* 357 (1992).
14. (a) K.A. Jørgensen and B. Schiøtt, Metallaoxetanes as intermediate in oxygen-transfer reaction—reality or fiction? *Chem. Rev., 90*: 1483 (1990). (b) B. Meunier, Metalloporphyrins as versatile catalysts for oxidation reactions and oxidative DNA cleavage, *Chem. Rev., 92*: 1411 (1992).
15. J.T. Groves and R.S. Myers, Catalytic asymmetric epoxidation with chiral iron porphyrins, *J. Am. Chem. Soc., 105*: 6243 (1983).
16. J.T. Groves and P. Viski, Asymmetric hydroxylation by a chiral iron porphyrin, *J. Am. Chem. Soc., 111*: 8537 (1989); J.T. Groves and P. Viski, *J. Org. Chem., 55*: 368 (1990).
17. D. Mansuy, P. Battioni, J.-P. Renaud, and P. Guerin, Asymmetric epoxidation of alkenes catalyzed by a "basket-handle" iron-porphyrin bearing amino acid, *J. Chem. Soc. Chem. Commun.,* 155 (1985).
18. J.P. Collman, X. Zhang, V.J. Lee, and J.I. Brauman, Asymmetric epoxidation catalyzed by an iron "binap capped" porphyrin, *J. Chem. Soc. Chem. Commun.,* 1647 (1992).
19. Y. Naruta, F. Tani, and K. Maruyama, *Chem. Lett.,* 1269 (1989); Y. Naruta, F. Tani, N. Ishihara, and K. Maruyama, *J. Am. Chem. Soc., 113*: 6865 (1991).
20. Y. Naruta, N. Ishihara, F. Tani, and K. Maruyama, Synthesis of chiral bitetralin-strapped "twin coronet" porphyrins. Catalytic and asymmetric epoxidation of styrene derivatives, *Chem. Lett.,* 1933 (1991); Y. Naruta, N. Ishihara, F. Tani, and K. Maruyama, Asymmetric epoxidation of simple olefins by chiral bitetralin-linked "twin-coronet" porphyrin catalysts, *Bull. Chem. Soc. Jpn., 66*: 158 (1993).
21. S. O'Malley and T. Kodadek, Synthesis and characterization of the "chiral wall" porphyrin: a chemically robust ligand for metal-catalyzed asymmetric epoxidation, *J. Am. Chem. Soc., 111*: 9116 (1989).
22. J.P. Collman, X. Zhang, R.H. Hembre, and J.I. Brauman, Shape-selective olefin epoxidation catalyzed by manganese picnic basket porphyrins, *J. Am. Chem. Soc., 112*: 5356 (1990).
23. R.L. Halterman and S.-T. Jan, Catalytic asymmetric epoxidation of unfunctionalized alkenes using the first D_4-symmetric metallotetraphenylporphyrin, *J. Org. Chem., 56*: 5253 (1991).

24. P. Maillard, J.L. Guerquin-Kern, and M. Momenteau, Catalytic properties of iron and manganese glycoconjugated porphyrins, *Tetrahedron Lett., 32*: 4901 (1991).
25. K. Konishi, K. Oda, K. Nishida, T. Aida, and S. Inoue, Asymmetric epoxidation of olefins catalyzed by manganese complexes of chiral "strapped" porphyrins with diastereotopic faces. A novel strategy for stereochemical modeling of the active site of cytochrome P-450, *J. Am. Chem. Soc., 114*: 1313 (1992).
26. L. Weber, I. Imiolczyk, G. Haufe, D. Rehorek, and H. Henning, Photocatalytic enantiodiscriminating oxygenation with cyclodextrin-linked porphyrins and molecular oxygen, *J. Chem. Soc. Chem. Commun.*, 301 (1992).
27. T.L. Stuk, P.A. Grieco, and M.M. Marsh, Site-selective hydroxylation of steroids via oxometalloporphinates covalently linked to ring D: introduction of hydroxyl groups into the C(9) and C(12) position of 5α-androstanes, *J. Org. Chem., 56*: 2957 (1991).
28. A. El-Kamsi, D. Lexa, P. Maillad, M. Momenteau, and J.M. Savéant, Reversible two-electron–one-proton systems in the ring-centered oxidation of metalloporphyrins bearing secondary amide-linked superstructures, *J. Am. Chem. Soc., 113*: 1586 (1991).
29. Y. Naruta, F. Tani, and K. Maruyama, Remarkable effects of metal ions and axial bases on catalytic and asymmetric oxidation of simple olefins with a "twin-coronet" porphyrin, *Tetrahedron Lett., 33*: 6323 (1992).
30. J.T. Groves and R. Neuman, Regioselective oxidation catalysts in synthetic phospholipid vesicles. Membrane-spanning steroidal metalloporphyrins, *J. Am. Chem. Soc., 111*: 2900 (1989).
31. For reviews, see M. Madesclaire, Synthesis of sulfoxides by oxidation of thioethers, *Tetrahedron, 42*: 5459 (1986); H.L. Holland, Chiral sulfoxidation by biotransformation of organic sulfides, *Chem. Rev., 88*: 473 (1988).
32. Y. Naruta, F. Tani, and K. Maruyama, Catalytic and asymmetric oxidation of sulphides with iron complexes of chiral "twin coronet" porphyrins, *J. Chem. Soc. Chem. Commun.,* 1378 (1990); Y. Naruta, F. Tani, and K. Maruyama, Enantioselective oxidation of sulfides catalyzed by iron complexes of chiral "twin coronet" porphyrins, *Tetrahedron: Asymmetry, 2*: 533 (1991).
33. (a) F.A. Davis and A.C. Sheppard, Application of oxaziridines in organic synthesis, *Tetrahedron, 45*: 5703 (1989). (b) H.B. Kagan, E. Duñach, C. Nemecek, P. Pitchen, O. Sammuel, and S.-H. Zhao, A short route to chiral sulfoxides using titanium-mediated asymmetric oxidation, *Pure Appl. Chem., 57*: 1911 (1985).
34. M. Paulucki, P. Hanson, and E.N. Jacobsen, Asymmetric oxidation of sulfides with H_2O_2 catalyzed by (salen)Mn(III) complexes, *Tetrahedron Lett., 33*: 7111 (1992).
35. S. Colonna, N. Gaggero, G. Carrea, and P. Pasta, Horseradish peroxidase catalyzed sulfoxidation is enantioselective, *J. Chem. Soc. Chem. Commun.* 357 (1992).
36. D.J. Waxman, D.R. Light, and C. Walsh, Chiral sulfoxidations catalyzed by rat liver cytochrome P-450, *Biochemistry, 21*: 2499 (1982).
37. S. Oae, Y. Watanabe, and K. Fujimori, Biomimetic oxidation of organic sulfides with TPPFe(III)Cl/imidazole/hydrogen peroxide, *Tetrahedron Lett., 23*: 1189 (1982).

9

Metalloporphyrin-Catalyzed Oxidation of Hydrocarbons with Dioxygen

Teresa Mlodnicka
Institute of Catalysis and Surface Chemistry, Polish Academy of Sciences, Kraków, Poland

I. INTRODUCTION

Oxidation of organic compounds with molecular oxygen in which a radical chain mechanism is involved is called *autoxidation.* Oxygen is an abundant and inexpensive reagent which can effect a variety of useful oxidation reactions, thus being an attractive oxidizing agent. In its ground state the dioxygen molecule is a triplet. Consequently, reactions between dioxygen and organic molecules which are in a singlet state are symmetry-forbidden and require high activation energy to overcome the symmetry barrier.

In principle, both molecular oxygen and organic molecules may be activated by coordination to a transition metal. The molecule of dioxygen behaves generally as an electron acceptor and undergoes partial reduction. Coordination of hydrocarbon molecules, on the other hand, can facilitate the generation of organic radicals which react further with dioxygen. Therefore, transition metal complexes may act as catalysts for autoxidation reactions.

The interactions between molecular oxygen and transition metal complexes have been studied by many authors. In particular, oxygen-containing complexes of metalloporphyrins have become the subject of intensive investigations. This interest was stimulated by the well-known efficiency of biological oxidation processes catalyzed by a variety of enzymes that contain iron porphyrin complexes as the prosthetic group, e.g., the cytochrome P450–dependent monooxygenases which mediate a broad range of biological oxygenations.

Simple metalloporphyrins can serve as suitable model compounds for studying the formation and reactions of the active intermediates which play a fundamental role in these biocatalytic reactions. The rate of formation and decomposition of these intermediates is governed by the electronic effects imposed by such factors as the character of the metal center, the nature of substituents on the porphyrin macrocycle, and the electron-donating or accepting properties of solvents and additives. The electronic structure of metalloporphyrins creates the possibility of controlling the transfer of electron density from the periphery to the metal center by introducing appropriate peripheral substituents and/or axial ligands. Thus the formation of the metal–oxygen bond and its character may be controlled by appropriate selection of both the substituents and the axial ligands.

II. AUTOXIDATION OF HYDROCARBONS

A. Fundamentals of Autoxidation

The liquid phase autoxidation of hydrocarbons has been studied extensively and is the subject of numerous reviews and monographs [1–15]. Autoxidations proceed in three stages, which may be described by the following sequence of reactions [12]:

Initiation

$$In_2 \xrightarrow{R_i} 2\ In\bullet \quad (1)$$

$$In\bullet + RH \longrightarrow InH + R\bullet \quad (2)$$

Propagation:

$$R\bullet + O_2 \longrightarrow RO_2\bullet \quad (3)$$

$$RO_2\bullet + RH \xrightarrow{k_p} RO_2H + R\bullet \quad (4)$$

Termination:

$$R\bullet + RO_2\bullet \longrightarrow ROOR \quad (5)$$

$$2RO_2\bullet \xrightarrow{k_t} RO_4R \longrightarrow O_2 + \text{nonradical products} \quad (6)$$

Such compounds as hydrogen peroxide, dialkyl peroxides, *tert*-butylhydroperoxide, and peroxy esters, which yield free radicals on thermal decomposition, are common initiators of autoxidation reactions. Once the radicals are generated the reaction proceeds via a chain mechanism (Eqs. 3 and 4) and alkylperoxy radicals play the most important role in the propagation step. Alkylhydroperoxides are the main intermediates. The reaction of the alkyl radical (R·) with dioxygen is very rapid and the rate-determining step of the autoxidation reaction is hydrogen transfer from the hydrocarbon to the alkylperoxy radical (Eq. 4) which depends on the C-H bond energy of the hydrocarbon as well as on the structure and reactivity of the alkylperoxy radical. This latter is determined by polar and steric effects imposed by its substituents. Termination occurs mainly via reaction 6, i.e., self-reaction of two alkylperoxy radicals. The energy of the allylic and aldehydic C-H bonds is 355 and 360 kJ mol^{-1}, respectively, and is comparable to the tertiary C-H bond energy in saturated hydrocarbons. The latter is lower than the energies of the secondary and primary C-H bonds. Alkylperoxy radicals preferentially abstract the most weakly bound hydrogen. The overall rate of autoxidation depends on the ratio of the propagation and termination rate constants:

$$-dO_2/d_t = k_p[RH](R_i/2k_t)^{1/2} \qquad (7)$$

where R_i = initiation rate , k_p = propagation rate constant
and k_t = termination rate constant

where R_i = initiation rate, k_p = propagation rate constant, and k_t = termination rate constant.

In the course of the autoxidation of olefins allylic radicals are generated as a result of hydrogen abstraction by alkylperoxy radicals:

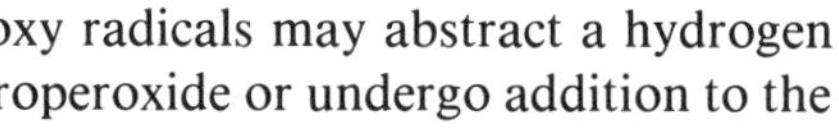 (8)

The allylic radicals subsequently react with dioxygen to give allylperoxy radicals (Eq. 9):

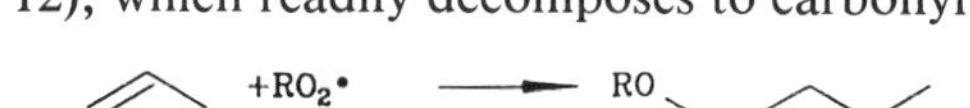 (9)

Peroxy radicals may abstract a hydrogen atom from the olefin to give the hydroperoxide or undergo addition to the double bond (Eq. 10). The resulting compound undergoes decomposition to produce epoxide and alkoxy radical (Eq. 11), and its further reaction with oxygen yields a polyperoxide (Eq. 12), which readily decomposes to carbonyl compounds (Eq. 13).

 (10)

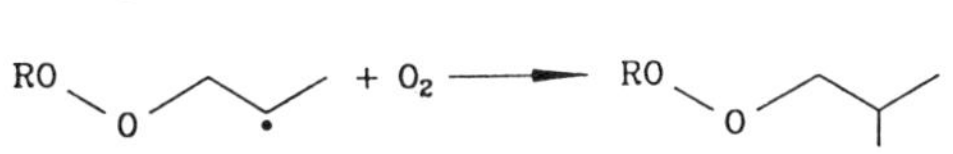 (11)

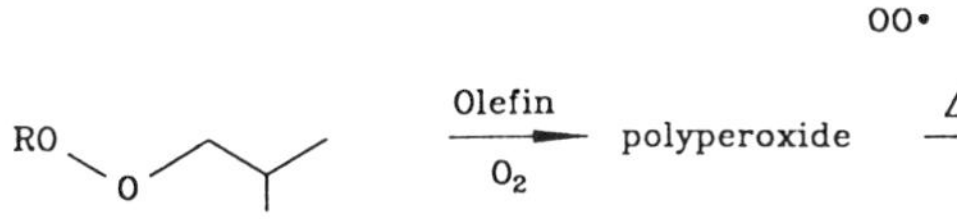 (12)

 (13)

Abstraction and addition are competing processes and their ratio depends on the structure of the olefin and the oxygen pressure. The variety of reaction pathways occurring during the autoxidation process is responsible for the lack of selectivity.

Aldehydes are autoxidized via the formation of acylperoxy radicals and peroxy acids are the primary products.

$$RCHO + In\cdot \longrightarrow R\dot{C}O + InH \tag{14}$$

$$R\dot{C}O + O_2 \longrightarrow RCO_3\cdot \tag{15}$$

$$RCO_3\cdot + RCHO \longrightarrow RCO_3H + R\dot{C}O \tag{16}$$

In so-called cooxidations a mixture of two substrates is autoxidized and the added substrate exerts an important effect on the rate of autoxidation of the main compound. These processes are of significant importance from both a theoretical and practical point of view.

B. Catalytic Autoxidation of Hydrocarbons

1. Modes of Initiation

Transition metal compounds are often used as initiators of autoxidation reactions. The autoxidation of organic substrates catalyzed by metalloporphyrins has been dealt with in numerous papers [16–28] and reviews [29–33]. Table 1 lists metalloporphyrins which have been applied as catalysts in the autoxidation of hydrocarbons and aldehydes. The listed data demonstrate that the metalloporphyrin complexes capable of one-electron oxidation–reduction transformations and those binding dioxygen are active catalysts of the autoxidation reactions. In principle, several different pathways are possible for catalysis of autoxidations by metalloporphyrins. The first one involves the metal-catalyzed decomposition of trace amounts of hydroperoxides into chain-initiating radicals (Reactions 17 and 18). This is known as the Haber–Weiss mechanism [13]. Reactive hydrocarbons, e.g., cyclohexene, always contain trace amounts of hydroperoxides unless they are rigorously purified by passing over a column of basic alumina prior to use.

$$PM^{n-1} + RO_2H \longrightarrow PM^{n}OH + RO\cdot \tag{17}$$

$$PM^{n} + RO_2H \longrightarrow PM^{n-1} + RO_2\cdot + H^{+} \tag{18}$$

A second possibility is radical formation via direct reaction of the metalloporphyrin with the substrate, probably via initial electron transfer followed by loss of a proton (Reactions 19 and 20).

Table 1 Metalloporphyrin-Catalyzed Autoxidations

Metalloporphyrin[a]	Substrate	Product(s)	Catalyst/substrate ratio	Solvent[b]	Temp. (°C)	Ref.
Co(TPP)	CH_3CHO	CH_3CO_3H	1×10^{-2}	EA, EB	10	16
Co(TPP)	CH_3CHO	CH_3CO_3H	8×10^{-4}	EA, EB	10	17,18
	PhCHO	$PhCO_3H$	4×10^{-4} to 5×10^{-3}	EA, EB	10	17
	CH_3CH_2CHO	$CH_3CH_2CO_3H$	4.3×10^{-5} 1.8×10^{-1}	EA, AA, DMK DCM, DMF, DMSO, THF	30	33
	Cyclohexene	Cyclohexen-3-one	5×10^{-3}	EA	30	20
Co(p-ClTPP)	CH_3CHO	CH_3CO_3H	1×10^{-2}	EA	10	16
Co(p-NO_2TPP)	CH_3CHO	CH_3CO_3H	1×10^{-2}	EA	10	16
Co(OEP) and Co(OEP)OH	Cyclohexene	Cyclohexen-3-one,	2×10^{-4}	B	25	25
	squalene	α,β-unsaturated aldehydes, epoxides	1×10^{-3}	B	60	26
Cu(TPP)	CH_3CHO	CH_3CO_3H	1×10^{-2}	EA	10	16
Fe(TPP)Cl	Cyclohexene	Cyclohexen-3-one	9×10^{-5}	EA	25	19
	Cycloheptene	Cyclohepten-3-one	9×10^{-5}	EA	25	19
	Tetralin	α-Tetralone	9×10^{-5}	EA	25	19
	α-Methylstyrene	Acetophenone	9×10^{-5}	EA	25	19
Fe(III)protoporphyrin IX dimethyl ester μ-oxodimer	Cyclohexene	Cyclohexen-3-one	2×10^{-4}	B	25	25

(continued)

Table 1 (Continued)

Metalloporphyrin[a]	Substrate	Product(s)	Catalyst/substrate ratio	Solvent[b]	Temp. (°C)	Ref.
$[FeTPP]_2O$	Cyclohexene	Cyclohexen-3-one	2×10^{-4}	B	30	22
$[FeTPPBr_4]_2O$	Cyclohexene	Cyclohexen-3-one	2×10^{-4}	B	30	22
$[FeTPPCN_4]_2O$	Cyclohexene	Cyclohexen-3-one	2×10^{-4}	B	30	22
Mn(TPP)Cl	Cyclohexene	Cyclohexen-3-one	2×10^{-4}	B	30	24
Mn(TPP)Cl	CH_3CH_2CHO	$CH_3CH_2CO_2H$	1.25×10^{-3}	EA	30	33
Ni(TPP)	CH_3CHO	CH_3CO_3H	1×10^{-2}	EA	10	16
Rh(TPP)Cl	Cyclohexene	Cyclohexene-3-hydroperoxide	9×10^{-5}	None	25	19
Rh(CO)(TPP)Cl	Cyclohexene	Cyclohexene-3-hydroperoxide	9×10^{-5}	None	25	19

[a]TPP = tetraphenylporphyrin; TTP = tetratolylporphyrin; OEP = octaethylporphyrin; TMP = tetramesitylporphyrin.
[b]EA = ethyl acetate; EB = ethyl isobutyrate; AA = acetic acid; B = benzene; DMK = acetone; DCM = dichloromethane; DMF = dimethylformamide; DMSO = dimethylsulfoxide; THF = tetrahydrofuran.

$$PM^n + RH \longrightarrow [PM^{n-1} + RH^{+\bullet}] \tag{19}$$

$$[PM^{n-1} + RH^{+\bullet}] \longrightarrow PM^{n-1} + R\bullet + H^+ \tag{20}$$

The ease of radical formation via this pathway will depend on the oxidation potentials of the metalloporphyrin and the substrate and the strength of the C-H bond. A third possibility involves the abstraction of a hydrogen atom from the substrate by coordinated dioxygen, i.e., a superoxo-metal complex according to Eqs. 21 and 22.

$$PM^{n-1} + O_2 \longrightarrow PM^n\text{-O-O}\bullet \tag{21}$$

$$PM^n\text{-O-O} + RH \longrightarrow PM^n\text{-O-OH} + R\bullet \tag{22}$$

However, it is difficult to distinguish between initiation via reactions 21 and 22, or 17 and 18, and it is doubtful whether the former play a role in hydrocarbon autoxidations. Nevertheless the reaction between metalloporphyrins and dioxygen can play an important role in the regeneration of the oxidized form of the catalyst in metalloporphyrin-catalyzed oxidations.

A fourth possibility is that reactive, high-valent oxometal species formed by reaction of dioxygen (Eq. 23) or alkylhydroperoxides (Eq. 24) with the metalloporphyrin can effect the direct oxidation of the substrate. Obviously, the role of reaction 24 could become more important once an autoxidation is underway and alkylhydroperoxide is accumulating.

$$PM^n\text{-O-O}\bullet \xrightarrow{PM^{n-1}} PM^n\text{-O-O-}M^nP \longrightarrow 2\ PM^{n+1}{=}O \tag{23}$$

$$PM^n + RO_2H \longrightarrow PM^n \leftarrow \underset{H}{OOR} \longrightarrow \underset{\substack{| \\ OR}}{PM^{n+2}{=}O} \tag{24}$$

2. Metalloporphyrin Complexes with Dioxygen

From among the metals applied as centers of the porphyrins listed in Table 1, Mn(II), Fe(II), Co(II), Cu(I), and Rh(II) can bind dioxygen. The character of bonding and the degree of oxygen activation as well as the properties and reactivity of the coordinated oxygen species depend on the electronic structure and distribution of the charge density at the metal center which transfers electrons to the coordinated dioxygen. The electron transfer may be additionally controlled by axial ligands and/or peripheral substituents.

Extensive studies have been devoted to the oxygen-binding ability of manganese [35–43], iron [44–76], and cobalt [80–97] prophyrins. Some data relevant to the configuration and structure of the oxygenated complexes of metalloporphyrins obtained in a direct reaction with dioxygen and with other oxidizing agents are given in Table 2.

Because of their low oxidation potentials Mn(II) and Fe(II) porphyrins are very unstable in solution and in the presence of oxygen undergo rapid oxidation. However, the resulting species have different structures. At room temperature, the existence of Mn(II)tetraphenylporphyrin (TPP) species have been demonstrated spectrometrically only in carefully purified benzene solution and in the presence of the reducing agent, $NaBH_4$. On introduction of oxygen immediate conversion to an Mn(III) complex followed [24]. Nevertheless at $-79°C$ Mn(II)(TPP)–pyridine complex in toluene solution appeared to bind dioxygen reversibly and electron spin resonance (ESR) studies revealed very little electron transfer to the dioxygen molecule. It is generally accepted that the oxygenated complex has side-on geometry [36,38, 39]. Irreversible oxidation to a dimeric species is also possible [35,41]. The ESR investigations carried out on the oxidation of Mn(II)(TPP) with O_2 in chlorobenzene solution at 10°K provided evidence for the following transformations:

Table 2 Oxygenated Complexes of Some Metalloporphyrins

Metalloporphyrin	Reaction	Oxygenated complex	Type of the oxygen species	Ref.
Mn(TPP)Cl	O_2	Mn^{IV}-O_2^{2-}	Peroxo	36,38,39
Mn(TPP)Cl	O_2^-	$[Mn^{III}$-$O_2]^-$	Peroxo	79
Mn(TMP)Cl	CPBA[a]	Mn^{IV}-O^{2-}	Oxo	43
		Mn^{V}-O^{2-}	Oxo	78
Fe(TPP)Cl	O_2	Fe^{II}-O_2	Superoxo	44,45,51, 63,65
		Fe^{III}-O_2-Fe^{III}	μ-Peroxo	30,48,66
		Fe^{IV}-O^{2-}	Oxo	72,73,75
		Fe^{III}-O-Fe^{III}	μ-Oxo	47,74
BFeT(piV)PP	O_2	BFe^{II}-O_2	Superoxo	61
Fe(TPP)Cl	O_2^-	Fe^{III}-O_2^{2-}	Peroxo	68
Fe(TMP)Cl	CPBA[a]	Fe^{IV}-$O^{2-}(TMP)^{+\cdot}$	Oxo cation radical	77
Co(TPP)	O_2	Co^{III}-O_2^-	Superoxo	30,80–83,85 88,96
Rh(TPP)	O_2	Rh^{III}-O_2^-	Superoxo	98

[a]*m*-Chloroperbenzoic acid.

$$Mn^{(II)}(TPP) + O_2 \longrightarrow [Mn^{(III)} TPP]^+ + O_2^{\bullet -} \quad (25)$$

$$Mn^{(II)}(TPP) + O_2^{\bullet -} \longrightarrow [Mn^{(III)}(TPP)O_2]^- \quad (26)$$

leading further to the mixed valence binuclear manganese complexes [(TPP)-Mn(II)-O-Mn(III)(TPP)]$^-$, (TPP)Mn(III)-O-Mn(III)TPP, and a (TPP)Mn(IV) = O species [42]. Special attention has been attached to the oxygen-binding ability of iron porphyrins because of the important role which they play in oxidation processes both in vivo and in vitro. Hemoglobin and myoglobin, which combine dioxygen in blood and tissues of numerous living organisms, contain iron(II) porphyrins as prosthetic groups [71]. Their natural environment provides favorable conditions for reversible binding of oxygen in end-on configuration. In contrast, oxygenation of simple iron porphyrins in solution at ambient temperature results in the formation of μ-oxo dimers. The oxidative transformations proceed according to the following sequence of reactions [74,75]:

$$PFe^{(II)} \xrightarrow{O_2} PFeO_2 \xrightarrow{PFe^{(II)}} PFe^{(III)}{-}O{-}O{-}Fe^{(III)}P$$
$$\longrightarrow PFe^{(IV)}{=}O \xrightarrow{PFe^{(II)}} PFe^{(III)}{-}O{-}Fe^{(III)}P \quad (27)$$

The resulting μ-oxo complex has a linear five-coordinate geometry [76].

Cobalt(II)tetraphenylporphyrins maintain their identity in noncoordinating solvents even in the presence of dioxygen and at ambient temperature. However, when stimulated by donor axial ligands they form stable 1:1 oxygen complexes which are paramagnetic and therefore the interaction of dioxygen with cobalt porphyrins is easily detectable by ESR spectroscopy [85,86,89,90,95,96]. From among cobalt porphyrins Co(II)tetraphenylporphyrin and its para-substituted derivatives have been most commonly used in ESR investigations of dioxygen binding in the presence of a variety of coordinating agents. These porphyrins exhibit square-planar geometry with phenyl rings situated out of plane from the porphyrin macrocycle [97]. Two free coordination sites in the trans position are equally accessible for substrates and/or molecules of solvent. The single unpaired electron in the d_z^2 orbital is perfectly located for axial and peripheral electronic stimulations. The ESR studies proved that Co(TPP) can activate molecular oxygen to its superoxo, end-on bonded form and the electron from a metal d_z^2 orbital is transferred to the oxygen antibonding orbitals where it spends 80% of its time [96].

$$Co^{(II)} TPP + O_2 \longrightarrow Co^{(III)}(TPP)\ldots\ldots O_2^{\bullet -} \quad (28)$$

The process of dimerization to μ-peroxo or μ-oxo complexes, which is a common phenomenon in alkyl-substituted porphyrins, is strongly inhibited by steric hindrance of the freely rotating phenyl rings in Co(TPP).

Similarly, the d^7 configuration of Rh(II) porphyrins results in their ability to bind dioxygen, and comparatively stable adducts with Rh(OEP) (OEP = octaethylporphyrin) have been obtained at $-80°C$. Their g values indicate an end-on configuration [98]. However, on heating to 20°C the ESR signal slowly diminishes, which coincides with the formation of a diamagnetic μ-peroxo complex of rhodium(III).

As will be shown in the following section, the reactivities of the various metalloporphyrin–oxygen complexes are reflected in their catalytic behavior.

3. Course of the Catalytic Reactions

It may be seen from the data quoted in Table 1 that cyclohexene was the most commonly investigated olefin and cyclohexen-3-one was the main product of the autoxidation. It indicates that the dominant pathway of the reaction is the abstraction of the allylic hydrogen followed by reaction of the resulting cyclohexenyl radical with dioxygen to produce hydroperoxide, which undergoes metal-mediated decomposition:

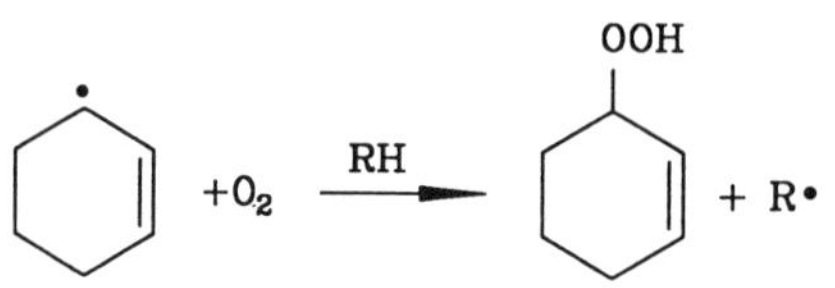

(29)

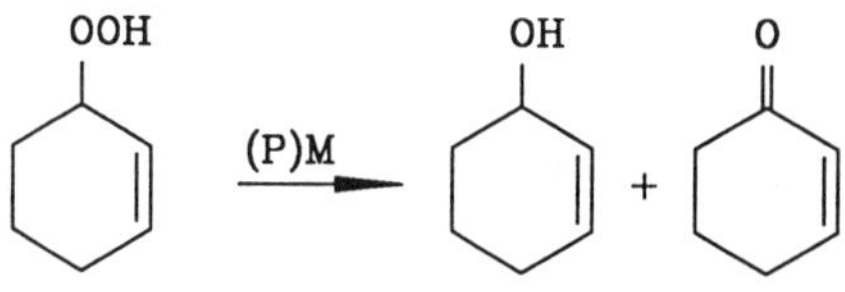

(30)

The initial formation of a radical from the olefin substrate may involve direct reaction with the metal center (Eq. 19) or with the oxygenated adduct (Eq. 22). Detailed investigations of the autoxidation reactions carried out with various metalloporphyrins indicate that initiation proceeds mainly by the electron transfer (Eq. 19). Thus, the induction period observed in the oxidation of cyclohexene depends on whether Co(II)OEP or Co(III)OEP is used as the initial catalyst. For the former, an induction period of 1 h is necessary, while the latter initiates the reaction immediately. ESR investigations of the system containing diamagnetic Co(III)(OEP)OH under argon showed the rapid appearance of the signal characteristic of Co(II)OEP on the introduction of cyclohexene, indicating an electron transfer from the substrate to the catalyst [21,25]. Thus the observation of an induction period with Co(II)OEP is consistent with the generation of Co(III) species by oxidation of Co(II)OEP with dioxygen. These data suggest that the Co(III) porphyrin induces the generation of the hydrocarbon radicals which can react directly with dioxygen. It has also been reported that Co(III)(TPP)X solutions contain small amounts of π-cation radicals, resulting from one-electron oxidation of the porphyrin ligand (Eq. 31), which would be very effective initiators of the autoxidation reactions in the electron transfer mechanism.

$$Co^{(III)}\ (TPP) \xrightarrow{-e} Co^{(III)}\ (TPP)^{+\bullet} \tag{31}$$

The (P)Co(III)O_2^- adduct is a stable and detectable species in the reaction medium and it has been shown that it plays an important rôle in the autoxidation of aldehydes [16–18,33] by abstracting a hydrogen atom from the aldehyde group. It also participates in the insertion of an oxygen atom in the phenol molecule to produce quinone. In these compounds, however, the C-H and O-H bond energies are lower than the energy of the C-H bond in hydrocarbons. It is not likely that this complex would be able to effect the abstraction of hydrogen from molecules for which the bond dissociation energy exceeds 370 kJ mol^{-1}. It has been suggested that dioxygen activated on Co(TPP) owing to its electrophilic nature, as shown by the ESR g value, may undergo addition to the double bond. It seems, however, that the contribution from this reaction pathway is not very important.

Co(III)$(TPP)^+$ is not able to initiate the autoxidation of cumene [20]. This means that it is inert toward relatively reactive tertiary C-H bonds. Similar results were obtained for tetralin.

The investigations of cyclohexene autoxidation using Mn(II)(TPP) as the initial catalyst have revealed that the reaction requires a 1-h induction period [24]. Taking into account the high reactivity of Mn(II) porphyrins toward dioxygen it may be concluded that its rapid oxidation takes place at the beginning of the reaction, resulting in the formation of some species in which the access of the hydrocarbon molecules to the metal center is strongly hindered. Thus, for instance, in five-coordinate Mn(III) porphyrins the metal center is situated out of the plane of the porphyrin ligand and the access of the coordinating molecule to the remaining coordination site is difficult and may be realized only by strongly coordinating molecules. The access to the metal center involved in a dimeric structure would also be difficult.

Similarly, Fe(III) porphyrins such as Fe(TPP)Cl and Fe(OEP)Cl, in which the metal center is also situated out of plane, initiate the autoxidation after an induction period of 1–3 h [19]. However, the introduction of the catalyst in the form of iron(II)porphyrin, which in the presence of oxygen is immediately transformed into the oxygen-bridged dimer, eliminates the induction period. Taking into account Eq. 27, which describes the behavior of iron(II)porphyrins in the presence of oxygen, one can speculate that some more reactive species than the μ-oxo dimer, such as a high-valent oxoiron species is responsible for the initiation step. It should be noted that the solutions of Fe(II) porphyrin and olefins were stable under argon atmosphere, and the introduction of dioxygen initiated a competitive oxidation of both the substrate and the catalyst.

On the basis of the described properties of the oxygenated adducts and the data provided by the autoxidation reactions, it may be concluded that the metalloporphyrin complexes with dioxygen, generated in the course of the reaction, are not very effective for the generation of the hydrocarbon radicals or addition to the double bond. Moreover, since hydrocarbon autoxidations may also be initiated by metal-assisted decomposition of adventitious hydroperoxides present as impurities or generated by exposure of hydrocarbons to light, one should be very careful in drawing conclusions concerning the mechanism of initiation for reactions with long induction times.

Oxidation of cyclohexene with dioxygen catalyzed by rhodium porphyrins such as Rh(III)(TPP)(CO)Cl and Rh(III)(TPP)Cl led to the formation of cyclohexene hydroperoxide as the predominant product [19]. It has been suggested that incorporation of the dioxygen may occur when both the hydrocarbon and the dioxygen are bound to the same metal center [99]. However, it should always be borne in mind that the yield of hydroperoxide in a free radical autoxidation is dependent on the chain length of the process, e.g., a chain length of 100 equates with a hydroperoxide yield of 99%.

C. Autoxidation of Aldehydes

Aldehydes are often used as additives in cooxidations with O_2 [100,101]. Due to the low energy of the C-H bond the abstraction of the aldehydic hydrogen is a facile process and is followed by reaction of the acyl radicals with dioxygen to afford peroxy acids as the primary product (Eqs. 14–16). Both steps of the reaction—formation of the acyl radicals and decomposition of the peroxy acid—are catalyzed by transition metal complexes.

Detailed investigations of the autoxidation of aldehydes were carried out for cobalt, nickel [16–18], and manganese [33] porphyrins. ESR investigations carried out in the course of the autoxidation reaction of propionaldehyde catalyzed by Co(TTP) indicated that the oxygenated adduct formed at the beginning of the reaction interacts with a molecule of aldehyde (Eq. 32). The resulting complex decomposes in a rate-determining step to yield acyl radicals and $Co(III)TTP^+$ species [33].

$$PCo^{(III)}\ O_2{\bullet} + RCHO \longrightarrow PCo^{(III)}\ O_2{\bullet}H(CO)R \qquad (32)$$

Kinetic investigations of the autoxidation of acetaldehyde catalyzed by Co(TPP) led to the following equation for the rate of initiation [18].

$$R_i = k_1 [RCHO] [O_2] [CoTPP] \tag{33}$$

This equation is consistent with an initiation step involving interactions among aldehyde, dioxygen, and catalyst.

Although copper(II)tetraphenylporphyrin does not bind molecular oxygen readily, it was an active catalyst in the autoxidation of acetaldehyde. In this case, however, the reducing properties of the substrate favor generation of either $[Cu(I)(TPP)]^-$ or a porphyrin anion radical which could possibly initiate the aldehyde autoxidation. It is noteworthy that a Cu(I)–dioxygen species has been frequently postulated as a kinetic intermediate in enzymatic systems [76].

The initiation step of aldehyde autoxidation can also be realized in a direct reaction with a metal center with an appropriate oxidation potential:

$$M^n(TPP) + RCHO \longrightarrow M^{n-1}(TPP) + R\dot{C}O + H^+ \tag{34}$$

Aldehydes are known to be able to release an electron easily and when coordinated to the metalloporphyrin molecule are able to transfer this electron to the ligand bound at the trans position. Single-electron donation to the metalloporphyrin is easier than electron pair addition because it does not involve antibonding interactions [103]. Simultaneously, coordination through the electron-rich carbonyl group prevents coordination in the axial trans position by a second donor molecule, leaving it accessible for dioxygen.

D. Reactions of Hydroperoxides

Hydroperoxides generated in the course of the autoxidation reactions (Eq. 4) are readily decomposed into free radicals by transition metal complexes according to the Haber–Weiss mechanism (Eqs. 17 and 18; see above). Similariy, peroxy acids which are intermediate products in the autoxidation reactions of aldehydes react with transition metal complexes:

$$RCO_3H + M^{n-1} \longrightarrow RCO_2\bullet + M^n + HO^- \tag{35}$$

$$RCO_3H + M^n \longrightarrow RCO_3\bullet + M^{n-1} + H^+ \tag{36}$$

The acyloxy radicals generated in Reaction 35 may decompose to produce carbon dioxide:

$$RCO_2\bullet \longrightarrow R\bullet + CO_2 \tag{37}$$

Transition metal complexes differ in their ability to decompose hydroperoxides and peroxy acids and metalloporphyrins have been found to react with hydroperoxides and peroxy acids via both homolytic and heterolytic, two-electron mechanisms [104–115].

$$PM^{n} + RO_2H \longrightarrow PM^{n+2}{=}O + ROH \tag{38}$$

$$PM^{n} + RCO_3H \longrightarrow PM^{n+2}{=}O + RCO_2H \tag{39}$$

Which pathway is favored depends on such factors as the electronic structure of the metal center and its charge density, the structure of the hydroperoxide, and the character of the leaving group (ROH or RCO_2H) and of the solvent used. Heterolytic pathways are strongly favored in protic solvents [112,113] and with peracids as the oxidant. The high-valent oxometal species generated in Reactions 38 [115] and 39 [43,77,78] have been found to be effective agents for the epoxidation of olefins and hydroxylation of alkanes with high chemo-, regio-, and stereoselectivity. This subject is dealt with in other chapters of this book.

Coordination of hydroperoxides to Co(p-X)TPP (X = –H, –Cl, $-OCH_3$) prior to their decomposition has been detected by following spectral changes at 235–245 K [116]. It has been found that the rate constant of complex formation depends on the electronegativity of the para substituents and increases with their electron-donating properties. Also, the nature of the hydroperoxide used influences the rate constant. In contrast, no spectral evidence suggesting complex formation has been observed for Ni(II) and Cu(II) porphyrins. This is ascribed to the difficulties in coordination due to too high an oxidation potential in the case of Ni(TPP) and difficulty in assuming a five-coordinate structure for Cu(TPP). At 283 K the decomposition reaction was already too fast to be measured. Investigations carried out on the decomposition of 2,3-dimethylindolenylhydroperoxide catalyzed by Mn(TPP)Br, Co(TPP), Cu(TPP), and Ni(TPP) have shown that with manganese and cobalt porphyrins the decomposition is as fast, or even faster, than the overall oxidation process [27]. In the case of Cu(TPP) the process is much less efficient, and Ni(TPP) is inactive toward hydroperoxide decomposition. This is in agreement with the earlier statement that these porphyrins are not efficient in the reactions requiring a valence change mechanism.

Investigations carried out on the decomposition of cumylhydroperoxide in benzene-cyclohexane solution at 20°C in the presence of various metalloporphyrins have shown that Fe(TPP)Cl is the best catalyst for the decomposition of the hydroperoxide with simultaneous formation of cyclohexanol (40%) and cyclohexanone (20%). Moreover, the catalyst appeared unchanged at the end of the reaction [117]. It is not clear as to whether a homolytic or heterolytic mechanism is involved. Similar behavior was observed with Mn(TPP)Cl.

Coordination of peroxy acid to the metalloporphyrin complex is also possible and has been observed prior to the oxidation of Fe(III) porphyrin into an oxoiron(IV)porphyrin cation radical [118]. It should be emphasized that the porphyrin ligand is sensitive to oxidizing conditions in the reaction medium, and therefore metalloporphyrins may operate as catalysts only at the initial stages of the reactions, before the excess of the oxidizing agent causes their degradation. The resistance of metalloporphyrins to degradation may be improved by introducing substituents at the periphery of the macrocycle. Halogen substituents present on the phenyl and pyrrole rings increase considerably the stability of these porphyrins under the oxidizing conditions (see Chapter 10).

III. REDUCTIVE OXYGEN ACTIVATION

A comparison of the products obtained in the Mn(TPP)Cl-catalyzed cyclohexene oxidation with dioxygen in the presence and absence of reducing agents is made in Figure 1. Upon addition of reducing agents such as $NaBH_4$ or H_2–colloidal Pt the product distribution changes dramatically. In the ab-

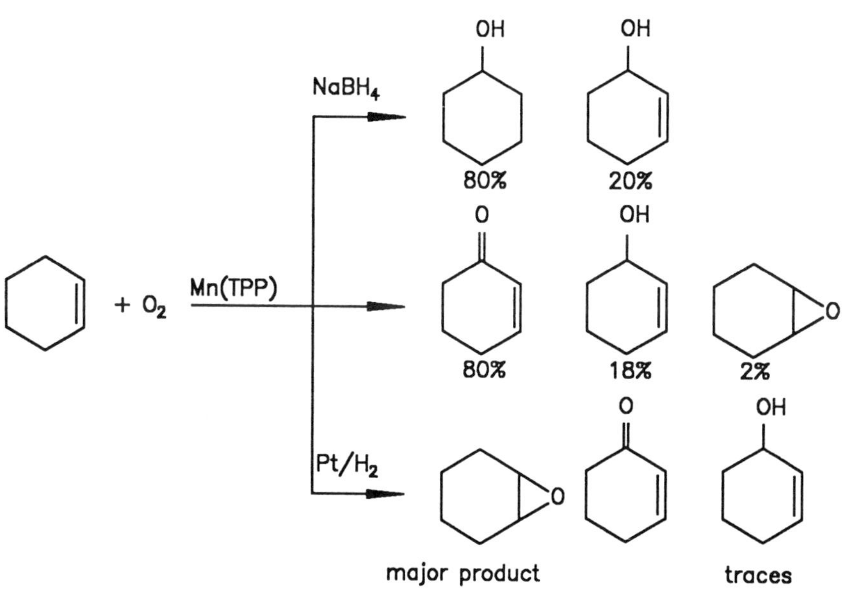

Figure 1 Product distribution in oxidation of cyclohexene. (From Ref. 33.)

sence of the reducing agent an induction period of 1 h is observed, and cyclohexen-3-one is the main product. The reaction is a typical autoxidation reaction sensitive to radical inhibitors [24]. In the presence of an excess amount of $NaBH_4$ rapid oxidation of cyclohexene follows, yielding cyclohexanol and cyclohexenol in a 4:1 ratio. It is believed that the epoxide is an intermediate which in the presence of $NaBH_4$ is reduced to cyclohexanol. Similarly, when *tert*-butylammonium borohydride was added to a solution of Mn(TPP)Cl in methylene chloride oxidation of terminal olefins with dioxygen afforded methyl ketones and alcohols [119]. Rh(III) octaethylporphyrin in the presence of $NaBH_4$ also showed a very good catalytic activity in the oxidation of cyclohexene with dioxygen to yield cyclohexanol [120].

With H_2–colloidal Pt in ethanol-benzene solution cyclohexene oxide is obtained as the main product and the addition of imidazole enhances the epoxidation rate [40]. This system is also active in the hydroxylation of adamantane to give the tertiary alcohol as the predominant product. The reactivity of the system was similar to that observed with Mn(TPP)Cl and iodosobenzene, in which high-valent oxomanganese species are generated, which are powerful oxidizing agents. Supplementary experiments showed that the reactions carried out in the presence of the reducing agents were insensitive to radical inhibitors. This observation and the divergence in the reaction products indicates that these reactions follow different pathways to the autoxidation reaction. The presence of reducing agents presumably results in the generation of active oxometal intermediates, such as PMn(V)=O, capable of incorporating an oxygen atom into the organic substrate. The observed phenomenon is called reductive oxygen activation and is completely analogous to the activation mechanism in the cytochrome P-450 system where NADPH is the reductant.

Stereospecific and regioselective monoepoxidation of polyolefins with O_2 in the presence of Mn(III)(TPP)Cl as catalyst and H_2–colloidal Pt was also reported [121]. A more complex system composed of water-soluble manganese(III)tetrasulfonatephenylporphyrin (MnTSPP), dioxygen, flavin mononucleotide (FMN), or methyldihydronicotinamide (MeNAH), which accelerate Mn(III) reduction in water-ethanol solution, was very efficient for the epoxidation of cyclohexene and regioselective epoxidation of nerol [122]. Similar systems containing an Fe center instead of a Mn center displayed much lower activity [123].

A detailed kinetic study aimed at explaining the phenomenon of reductive oxygen activation was carried out for the system consisting of the "picket fence" iron porphyrin: T(piv)PPFe(III)Cl in conjunction with H_2–colloidal Pt as the reducing agent. The pivaloyl groups situated at the ortho positions of the phenyl groups provide steric shielding for the metal center and the bound oxygen, and hence inhibit dimerization [124]. The system contained

T(piv)PPFe(III)Cl, dioxygen, methylimidazole, and small amounts of hydrochloric acid and benzoic anhydride. It was suggested that the formation of a (P)Fe(III)-O-O species and subsequent heterolytic cleavage of the O-O bond to give a high-valent oxometal species are the key steps of the process. The cleavage of the O-O bond is facilitated by the presence of protons and acid anhydride. It was pointed out that the reducing agent, which was present in excess, competes with the molecules of substrate in the reaction with the active, oxygenated intermediates. This is reflected in the low yields of oxygenated products observed.

The system composed of Mn(III)(TPP)Cl and sodium ascorbate as the reducing agent, in H_2O-benzene solution containing a phase transfer agent, was reported to effect the epoxidation of olefins and the oxidation of alkanes with O_2. In the latter case ketones were the main products [125].

Recently, a system using Zn powder as the reducing agent, acetic acid as the source of protons, Mn(TPP)Cl, and methylimidazole as cocatalyst was employed for the epoxidation of alkenes and hydroxylation of alkanes with dioxygen. Cyclooctene is epoxidized with 95% conversion and 90% yield. Cyclooctane is hydroxylated with a yield of 15% based on the reducing agent [126].

All of these systems employing dioxygen and reducing agents are able to reproduce qualitatively the main reactions of cytochrome P450, i.e., epoxidation of olefins and hydroxylation of alkanes. However, the turnover numbers are inferior to those found for cytochrome P450. Comparable activities are obtained only for the systems employing Mn(III)(TSPP) with flavin mononucleotide as reducing agent and that with Mn(TPP)Cl and Zn powder. Similarly, the system based on iron porphyrin Fe(T-piv)PPCl in the presence of zinc amalgam, methylviologen as electron transfer agent, and acetic anhydride in acetonitrile has been found to oxidize cyclohexane and epoxidize cyclohexene with good yield and selectivity [127,128].

When ferrocene was used as the reducing agent the oxidation of cyclohexane with dioxygen catalyzed by metalloporphyrins afforded cyclohexanol and cyclohexanone [129].

An electrochemical system using Mn(TPP)Cl as catalyst in the presence of benzoic anhydride was also employed for reductive oxygen activation in the epoxidation of cyclooctene. Although the yield based on reducing agent was high (56%), the rate remained low [130]. The efficiencies of the various systems are compared in Table 3.

IV. COOXIDATION

A. Cooxidation of Hydrocarbons with Aldehydes

Cooxidation allows for the utilization of peroxidic intermediates generated in the reaction medium for additional oxidation processes [12]. For example,

Table 3 Some Metalloporphyrin-Based Systems for the Reductive Oxygen Activation

Oxidizing system	Substrate	Product	Turnover frequency (cycles/min)	Ref.
Mn^{III}(TSPP)	Cyclohexene	Epoxide (90%)	0.5	122
O_2,MeNAH,FMN,		Cyclohexenone (10%)		
$(PhCO)_2O$,	Nerol 6.5	6,7-Epoxide	2.1	
N-MeIm in H_2O/EtOH,				
Mn(TPP)Cl,	Cyclohexene	Epoxide	1.9	126
O_2, Zn powder,		Cyclohexenone	0.26	
CH_3COOH,	Cyclooctene	Epoxide	3.3	
1-Me-Im,				
	Cyclooctane	Cyclooctanol	0.2	
		Cyclooctanone	0.26	
Fe(T-piv)PPCl,	Cyclohexene	Epoxide	5.6	128
O_2, Zn(Hg),		Cyclohexen-3-ol		
$MV(ClO_4)_2$	Cyclohexane	Cyclohexanol (52%)	1.1	
Tritylimidazole, in CH_3CN.		Cyclohexanone (48%)		

the process of cooxidation of aldehydes and olefins utilizes the active oxygen of acylperoxy radicals (Eq. 40) and peracids (Eq. 41) for the epoxidation of olefins:

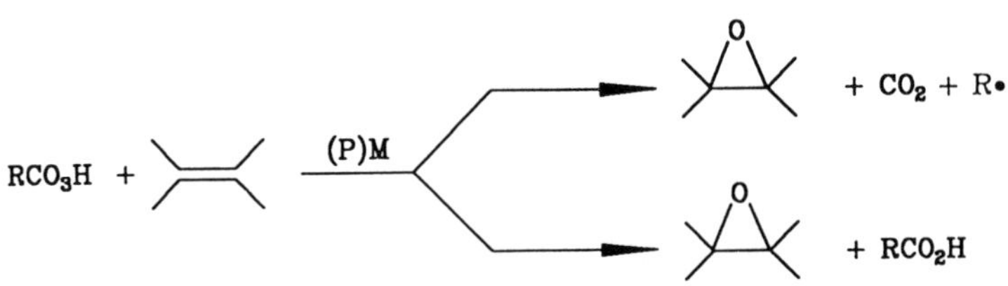

(40)

(41)

In the reactions of olefins with acylperoxy radicals addition to the double bond (Eq. 10) is favored over abstraction of hydrogen. In cooxidations catalyzed by metalloporphyrins, the aldehyde plays a multiple role. It reduces the initial metalloporphyrins to forms which can bind dioxygen, as is the case with Mn(III) and Fe(III) porphyrins. Simultaneously, protons are generated (Eq. 34) which are known to promote the heterolytic cleavage of the peroxidic O-O bond. Last but not least, it serves as a substrate for generation of acylperoxy radicals and peroxy acid.

Detailed investigations have been carried out for systems composed of metalloporphyrins, dioxygen, aldehyde, solvent, and hydrocarbons such as propene (Eq. 42) or cyclohexane (Eq. 43) as the substrate for oxidation [131–140]:

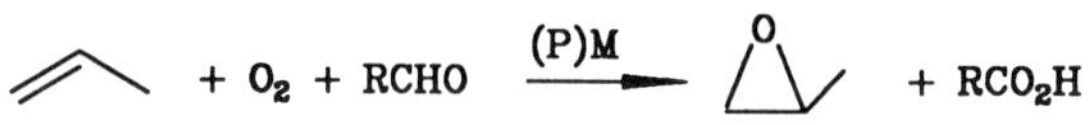

(42)

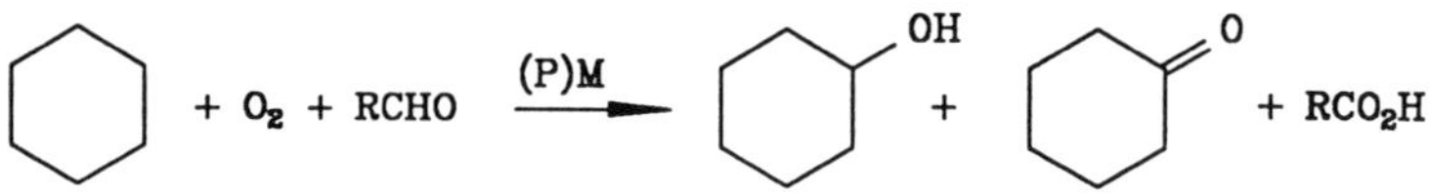

(43)

When first-row transition metal *p*-tetratolylporphyrins (MTTP) were used as catalysts in the oxidation of propene and cyclohexane by dioxygen, in ethyl acetate or benzene in the presence of propionaldehyde, the main product of propene oxidation was propene oxide whereas cyclohexane was oxidized to cyclohexanol and cyclohexanone. In both cases peroxypropionic acid, propionic acid, and carbon dioxide were also found. No products of allylic oxidation were detected and no oxidation took place in the absence of any of the components of the system, i.e., aldehyde, dioxygen, or MTTP. The experimental data are summarized in Tables 4 and 5, where the values of the reduction potentials of the investigated metalloporphyrins are also given.

It should be noted that for porphyrins of manganese, iron, and cobalt the induction period is short. Although the total yields of all products in the propene oxidation reactions are comparable, the product distributions differ considerably. Similarly, the distribution of products obtained in cyclohexane oxidation varies with the metal used. A short induction period is observed when the reduction potential of the metalloporphyrin used is low because the generation of the acyl radicals from aldehyde in the first step of the reaction is facile and takes place at a high rate according to Eq. 34.

In the case of Ni(II), Cu(II), and Zn(II)tetratolylporphyrins the reduction potentials are too high to allow facile electron transfer from the aldehyde to the metal complex. Moreover, with these porphyrins a transferred electron becomes located on the porphyrin ring. Such an anion radical is not expected to activate dioxygen because the latter can bind only to the metal center, where the appropriate conditions for the orbital overlap exist.

Figure 2 illustrates the oxygen uptake during the cooxidation of propene and cyclohexane with RCHO catalyzed by Fe, Co, and Mn porphyrin. A surprisingly high rate of oxygen consumption was observed with cobalt porphyrin compared with Mn(III) and Fe(III) porphyrins. The readiness of Co(II) porphyrin to bind dioxygen to form a monomeric oxygenated adduct

Table 4 Distribution of the Reaction Products in the Epoxidation of Propene

Catalyst	Reduction[a] potential $E_{1/2}$ vs. SCE (V)	Induction time (min)	Conversion of aldehyde (%)	Oxygen consumption (mmol)	Yield of main products (mmol)			Turnover frequency[b] (cycles/min)
					Epoxide	Peroxy acid	Carbon dioxide	
Mn(TPP)Cl	−0.35 III→II	1	100	5.8	0.79 (16.3%)	0.3 (6.2%)	3.76 (77.5%)	6.6
Mn(TPP)Cl + *n*-hexylimidazole		1	100	6.2	0.36 (11.3%)	0.75 (23.9%)	2.02 (64.6%)	
Mn(TDCPP)Cl	−0.12 III→II		86.3	3.8	0.37 (30.3%)	0.05 (4.0%)	0.8 (65.5%)	
Fe(TTP)Cl	−0.33 III→II	5	92	4.5	0.18 (3.9%)	3.5 (76%)	0.93 (20.1%)	0.8
Co(TTP)	−0.85 II→I	1	100	8.4	0.72 (14.8%)	2.0 (41%)	2.15 (44.1%)	12
Ni(TTP)[c]	−1.41	35	91	0.9	Trace	1.7	0.1	
Cu(TTP)[c]	−1.39	No reaction in 24 h						
Zn(TTP)[c]	−1.45	No reaction in 24 h						

[a]P. Leduc and T. Mlodnicka, unpublished results.
[b][epoxide]:[catalyst] × min based on the initial rates.
[c]Ring reduction.
Conditions: [catalyst] = 2×10^{-3} M, [aldehyde] = 1 M, propene:oxygen = 1:2, [imidazole] = 5×10^{-2} M, TDCPP—tetra-*ortho*-dichlorophenylporphyrin.

Table 5 Distribution of the Reaction Products in the Cooxidation of Cyclohexane with Propionaldehyde

Catalyst	Induction time (min)	Oxygen consumption (mmol)	Yield of the main products (mmol)				Turnover frequency[a] (cycles·min^{-1})	
			Alcohol	Ketone	Peroxy acid	CO_2	1	2
Mn(TPP)Cl	10	5.8	0.03	0.05	0.1	2.2	8.7	0.3
Mn(TDCPP)Cl[b]	5	6.5	0.3	0.4	0.02	2.46	12	3
Fe(TTP)Cl	0	7.8	0.03	0.06	5.7	2.4	68	0.7
Co(TTP)	10	8.4	0.075	0.07	0.6	4.0	16	0.5
Ni(TTP)	33	6.9	0.05	0.03	6.5	0.3	8	0.1

[a] 1, Corresponds to the total amount of products; 2, to the amount of cyclohexanol and cyclohexanone. Turnover frequencies are based on initial rates: [products]:[catalyst] × min.

[b] [cyclohexane] = 0.9 M.

Conditions: [cyclohexane] = 1.8×10^{-1} M, [propionaldehyde] = 1 M, [catalyst] = 2×10^{-3} M, react. temp. 30°C.

Source: Refs. 137 and 140.

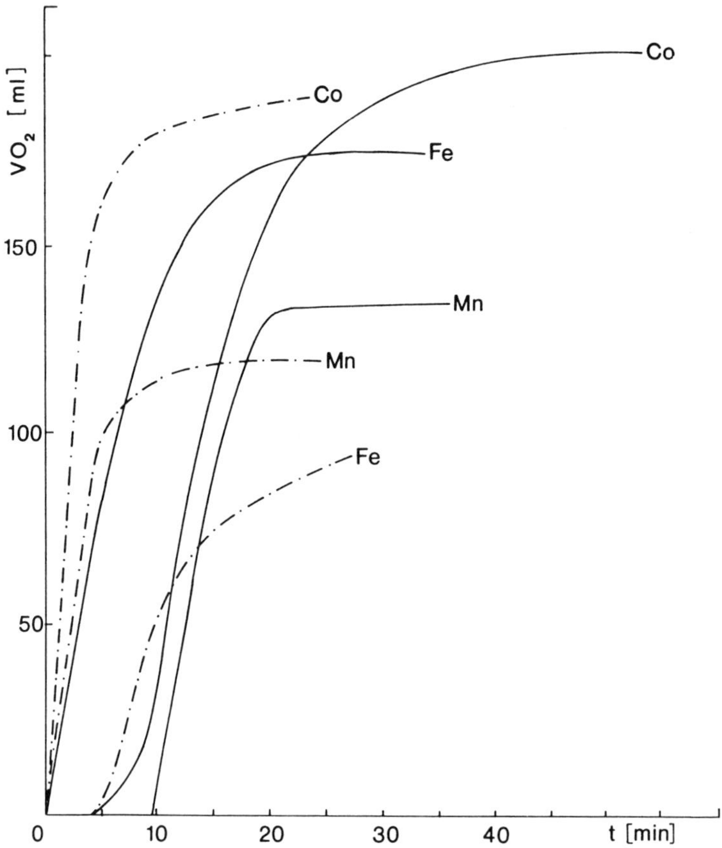

Figure 2 Oxygen uptake in the course of oxidation of cyclohexane (full lines) and propene (dashed lines). Conditions as given in Tables 4 and 5.

and the facile reduction of a Co(III) porphyrin complex due to its low reduction potential render the catalytic cycle of the generation of acyl radicals described by Eq. 34 very efficient, which is reflected in the high turnover frequencies of the process, as shown in Table 4.

Peroxy acid is a more efficient oxidant than molecular oxygen, and it interacts with the porphyrin species to form an intermediate complex responsible for the incorporation of an oxygen atom into the hydrocarbon substrates. Heterolytic cleavage of the peroxy acid metalloporphyrin complex results in the formation of high-valent oxometal species (Eq. 44) which effect the epoxidation of olefins and the hydroxylation of alkanes. The homolytic cleavage of the acylperoxymetalloporphyrin (Eq. 45), in contrast, affords carbon dioxide via acyloxy radical intermediates. Thus the amount of carbon dioxide evolved is indicative of the contribution from the homolytic reaction pathway.

$$(P)M^{n} + RCO_3H \longrightarrow (P)M^{n}O_3CR \longrightarrow (P)M^{n+2}{=}O + RCO_2^{-} \quad (44)$$

$$(P)M^{n}O_3CR \longrightarrow (P)M^{n+1}{=}O + RCO_2^{\bullet} \quad (45)$$

The catalytic activities of tetratolylporphyrins in the epoxidation of propene follow the order [140] Co(TTP) > Mn(TTP)Cl > Fe(TPP)Cl. The highest yield (8%) in epoxide based on aldehyde cosubstrate was found for Mn porphyrin.

In the oxidation of cyclohexane the observed reactivity order was Fe(TTP)Cl > Co(TTP) > Mn(TTP)Cl. Only low yields of cyclohexanol and cyclohexanone were observed (1.5% or less) indicating that most of the aldehyde was converted directly to the carboxylic acid without any cooxidation of cyclohexane.

The different orders of activities and selectivities observed for the investigated porphyrins indicate that the active, oxygen-containing metalloporphyrin species show different reactivities vis-à-vis the olefinic double bond and the C-H bond, i.e., oxygen atoms in these intermediates show different electrophilic properties. They may also reflect different contributions from homolytic and heterolytic pathways with the various metalloporphyrins.

The active oxygen-containing species responsible for the epoxidation of olefins and hydroxylation of alkanes are generated in the reaction medium from the initial metalloporphyrins and the oxidizing agent. In cooxidations, active oxomanganese(IV) and (V) species have been detected spectrometrically. In the case of iron porphyrins only the final species, i.e., the μ-oxo dimer, was observed [135].

It has also been found that the course of the catalytic reaction sometimes depends on the nature of the oxidant used [141]. This suggests that an intermediate complex composed of the metalloporphyrin and the oxidant rather than a discrete oxometal species may be the active oxidant. Thus the yield and rate of the reaction would depend on the structure of the oxidant and the nature of the leaving group. In the case of Co porphyrin its oxidized forms such as $(P^{+\cdot})$Co(III) π-cation radical and dication radical, in which one or two electrons are removed from the porphyrin periphery, were detected in the reaction medium [33,139,142]. This suggests that a cobalt porphyrin species containing both an oxidized porphyrin ligand and a peroxy acid ligand is responsible for the insertion of oxygen atom into the hydrocarbon molecule.

B. Effect of Substituents and Axial Ligands

Due to the high conjugation within the porphyrin macrocycle the electron density at the metal center, which is one of the driving forces of the catalytic reactions, may be controlled by introducing substituents with different electron-donating and electron-withdrawing properties. The effect is more evident when more than one substituent is introduced into the phenyl groups of tetraphenylporphyrins. Thus it has been found that the epoxidation of propene in the dioxygen–propionaldehyde–ethyl acetate system is much more selective when the phenyl groups of manganese tetraphenylporphyrin are chlorinated at all the ortho positions (MnTDCPPCl) [138]. This phenomenon may be explained in terms of an increased electrophilicity of the metal center caused by the presence of eight chlorine substituents, as evidenced by the measurements of the redox potentials (see Table 4). High electrophilicity of the metal center and of the oxygen bound to it favors the reaction with the olefinic double bond. Simultaneously, however, lower conversion of aldehyde and consumption of dioxygen are observed, which may be related to the steric hindrance created by the chlorine atoms that impede the approach of the substrates to the metal center. Thus the stimulating effect derived from the presence of electron-withdrawing substituents may be reduced by unfavorable steric effects.

Metalloporphyrins have one or two coordination sites in a trans position which are equally accessible for substrates as well as for other coordinating molecules. Various additives compete with molecules of substrates for coordination to the metal center and when present in a large excess may nullify the catalytic activity of metalloporphyrins. Nevertheless, catalytic amounts of some additives exert a promoting effect on the course of the oxidation reactions. Coordination of electron-donating molecules such as pyridine and imidazoles results in increased electron density at the metal center, which in the case of cobalt(II)porphyrin promotes binding of dioxygen at the trans position and causes back bonding from the metal orbitals to dioxygen which reinforces σ-bond formation and converts oxygen to the superoxo form [88]. It has been shown that the activity of Co(II)(TTP) in the oxidation of acetaldehyde is increased by such basic additives as pyridine and dimethylformamide (DMF) present in catalytic amounts [16]. The effect exerted by pyridine is larger than that of DMF. This may be related to the stronger donating ability of pyridine as is manifested in the corresponding donor numbers, 33.1 and 26.6. Donor numbers as defined by Gutmann reflect the coordinating ability of molecules [143].

In (P)Mn(III)X and (P)Fe(III)X one of the coordination sites is occupied by the bound counterion and the metal center is situated out of the plane of the porphyrin ring. The coordination of N-donor ligands at the remaining coordination site pulls the metal center into the plane of the ring with simultaneous weakening of the metal–halogen bond. This effect facilitates heterolytic reaction of manganese and iron porphyrins with oxygen atom donors such as peroxy acids, hydrogen peroxide, hydroperoxides, etc., leading to facile generation of high-valent metal–oxo species [144–150].

Recently, it was shown that the presence of the azido ligand in iron and manganese tetraphenylporphyrins renders these complexes active and selective catalysts for the hydroxylation of isobutane with dioxygen at 80°C and 100 psig in the absence of any peroxidic or reductive agents (see Chapter 10). Similar results have been obtained with perhalogenated tetraphenylporphyrinato ligand systems [151,152].

C. Effect of Solvent

The appropriate choice of a solvent for the metalloporphyrin-catalyzed reactions is a very important point when considering the solubility of the complex, coordinating properties of the solvent molecules and their protic nature and electronic parameters. Nonfunctionalized metalloporphyrins are soluble in organic solvents and are insoluble in water. Therefore the reactions carried out in biphasic systems require the presence of a phase transfer agent to transport the water-soluble reactant to the organic phase. Tetraalkylammonium, trialkylsulfonium salts, and crown ethers have been used for transport of oxometal anion reagents such as NaOCl [146], $KHSO_5$ [141], and the magnesium salt of monoperoxyphthalic acid (MMPP) [153]. Molecules of solvent may act as axial ligands, playing an important role in the process of activation of molecular oxygen. On the other hand, they may exert an inhibiting effect on the access of the reagent molecules to the active forms of the catalyst by creating a solvating sphere around them. Also, the ease of decomposition of the active intermediate composed of catalyst and reagent molecules may be dependent on the solvent applied. It is a well-known fact that in the catalytic, liquid phase reaction the active form of the catalyst is generated in the reaction medium during the induction period. Thus solvent can exert a significant influence on the induction period, the rate of the reaction, and its selectivity. Attempts have been made to correlate these effects with such solvent properties as Taft's σ values [17], electron donor–acceptor properties, dipole moments, and dielectric constants [33].

The oxidation of propionaldehyde, with Co(TTP) as catalyst, carried out in several solvents revealed that both the induction period and the rate of oxygen uptake are influenced by the electron donor–acceptor properties of the solvent used. The experimental data are given in Table 6. For strongly coordinating solvents such as DMF or dimethylsulfoxide (DMSO), a long induction period, low oxygen uptake, and low conversion of aldehyde are observed while in weakly coordinating solvents such as benzene, ethyl ace-

Table 6 Dependence of the Induction Time of Propionaldehyde Oxidation on Electronic Parameters of the Solvent[a]

Solvent	Dipole moment (μ) (D)	Dielectric constant (ϵ)	DN[b]	Induction time (h)
Benzene	0	2.3	0.1	0.15
Acetic anhydride	2.8	20.1	10.5	0.25
Ethyl acetate	1.8	6.0	17.1	0.25
DMF	3.8	36.7	26.6	1.2
DMSO	4.3	48.9	29.8	4.9
Pyridine	2.26	12.3	33.1	10

[a][aldehyde] = 1 M, [CoTPP] = 2.5 mM, temperature 30°C.
[b]Donor numbers (DN) are taken from Ref. 143.
Source: Ref. 33.

tate, etc., the reaction starts almost immediately and leads to 100% conversion of aldehyde.

Protic solvents, such as acetic acid (acceptor number 52.9), inhibit metalloporphyrin-catalyzed oxygenations, even though protic and hydroxylic solvents are known [112,113] to facilitate heterolytic O-O bond cleavage. Presumably, they inhibit the formation of the metal–dioxygen complex. They may also undergo nucleophilic addition to the porphyrin ring, leading to deactivation.

The effect of the donor number of the solvent applied has also been observed in the cooxidation of benzaldehyde and propene. As in the case of aldehyde oxidation, the induction period of the reaction increases with the increasing donor number of the solvent. Moreover, the product distribution is also solvent-dependent [33]. In noncoordinating solvents the yield in epoxide is higher than in the coordinating ones.

V. CONCLUDING REMARKS AND FUTURE PROSPECTS

Oxidation of hydrocarbons with molecular oxygen catalyzed by simple metalloporphyrins proceeds predominantly through formation of peroxidic intermediates which undergo subsequent decomposition via radical chain reactions. Metalloporphyrins participate in both the formation of the organic radicals and in the decomposition of the peroxidic species generated from the reaction of these radicals with dioxygen. The ability of metalloporphyrins to bind dioxygen facilitates reoxidation of the metal center and sometimes may also result in formation of the oxygenated metalloporphyrin adducts undergoing transformation to the potent high-valent oxometal species. This latter reaction pathway is favored in the presence of protons and reducing agents. The major disadvantages of these reaction mechanisms are the lack of selectivity and the oxidative degradation of the porphyrin ligand.

Recently, two types of metalloporphyrin-based systems have been discovered in which molecular oxygen effects selective oxidation of hydrocarbons in the absence of any peroxidic or reductive agents. One of them employs manganese and iron complexes of perhalogenated porphyrins. Extensive halogenation increases considerably the stability of these porphyrins and enhances the electrophilicity of the oxygen bound to the metal center rendering it more reactive (see next chapter). The second system, which shows true dioxygenase-like activity, is based on ruthenium porphyrins. Certain ruthenium porphyrins which contain bulky substituents at ortho positions of the phenyl rings show a remarkable and unique reactivity vis-à-vis dioxygen yielding Ru(VI)*trans*-dioxo species which are stable in solution at room temperature. The catalytic oxidation reactions are carried out in this system with high chemo-, regio-, and stereoselectivity. The key factor in the oxidations effected by Ru(VI)–dioxo and Ru(IV)–monooxo species seems to be the nature of the trans axial ligand [154]. In short, we conclude that catalytic studies with model metalloporphyrins will make significant contributions to the discovery of effective, industrially attractive methods for selective oxidation of hydrocarbon substrates.

REFERENCES

1. E.T. Denisov and N.M. Emanuel, Catalysis by salts of metals with variable valence in reactions of liquid-phase oxidation (in Russian), *Russ. Chem. Rev., 29*: 1409 (1960).
2. F.R. Mayo, Free radical autoxidations of hydrocarbons, *Acc. Chem. Res., 1*: 193 (1968).
3. L. Reich and S.S. Stivala, *Autoxidation of Hydrocarbons and Polyolefins*, Marcel Dekker, New York, 1969.
4. N.M. Emanuel, E.T. Denisov, and Z.K. Maizus, *Liquid Phase Oxidation of Hydrocarbons*, Plenum Press, New York, 1967.
5. N.M. Emanual (ed.), *Oxidation of Hydrocarbons in the Liquid Phase*, Pergamon Press, Oxford, 1965.
6. W.O. Lundberg (ed.), *Autoxidation and Antioxidants*, Vols. 1 and 2, Wiley, New York, 1962.
7. I.V. Berezin, E.T. Denisov, and N.M. Emanuel, *The Oxidation of Cyclohexane*, Pergamon Press, Oxford, 1965.
8. E.T. Denisov, N.I. Mitskevich, and V.E. Agabekov, *Liquid Phase Oxidation of Oxygen Containing Compounds*, Plenum Press, New York, 1977.
9. F.R. Mayo, ed., *Oxidation of Organic Compounds*, Adv. Chem. Series 75, 76, 77, Am. Chem. Soc., Washington, DC, 1968.
10. J.K. Kochi (ed.), *Free Radicals*, Wiley, New York, 1973.
11. G. Scott, *Atmospheric Oxidation and Antioxidants*, Elsevier, Amsterdam, 1965.

12. R.A. Sheldon and J.K. Kochi, *Metal-Catalyzed Oxidations of Organic Compounds*, Academic Press, New York, 1981.
13. R.A. Sheldon, *The Chemistry of Functional Groups*: *Peroxides* (S. Patai, ed.), Wiley, New York, 1983.
14. J.K. Kochi, *Organometallic Mechanism and Catalysis*, Academic Press, New York, 1978.
15. A. Bielanski and J. Haber, *Oxygen in Catalysis*, Marcel Dekker, New York, 1991, p. 443.
16. Y. Ohkatsu and T. Osa, The liquid-phase oxidation of aldehydes with metal tetra(p-tolyl) porphyrins, *Bull. Chem. Soc. Jpn., 50*: 2945 (1977).
17. Y. Ohkatsu, O. Sekiguchi, and T. Osa, The liquid-phase oxidation of aldehydes with Fe, Cu-polyphthalocyanines and cobalt tetra-p-tolylporphyrin, *Bull. Chem. Soc. Jpn., 50*: 701 (1977).
18. M. Tezuka, O. Sekiguchi, Y. Ohkatsu, and T. Osa, Oxidation of acetaldehyde catalyzed by cobalt(II) tetraphenylporphyrin, *Bull. Chem. Soc. Jpn., 49*: 2765 (1976).
19. D.R. Paulson, R. Ullman, R.B. Sloane, and G.L. Closs, Catalysis of autoxidation by metalloporphyrins, *J. Chem. Soc. Chem. Commun.*, 186 (1974).
20. Y. Ohkatsu and T. Tsuruta, Autoxidation reactions of hydrocarbons catalyzed by Co(II)tetra(p-tolyl) porphyrin, *Bull. Chem. Soc. Jpn., 51*: 188 (1978).
21. M. Baccouche, J. Ernst, J.H. Fuhrhop, R. Schlozer, and H. Arzoumanian, Metalloporphyrin catalyzed epoxidation with molecular oxygen, *J. Chem. Soc. Chem. Commun.*, 821 (1977).
22. H. Ledon, Oxidation de cyclohexene catalysée par des porphyrines de fer (in French), *C. R. Acad. Sci. Paris, Ser. C, 288*: 29 (1979).
23. H. Ledon, Y. Brigandat, M. Primet, M. Negre, and M. Bartholin, Synthesis and characterization of porphyrins linked to functionalized polystyrenes, *C. R. Acad. Sci. Paris, Ser. C, 288*: 77 (1979).
24. I. Tabushi and N. Koga, P-450 type oxygen activation by porphyrin-manganese complex, *J. Am. Chem. Soc., 101*: 6456 (1979).
25. J.H. Fuhrhop, M. Baccouche, H. Grabow, and H. Arzoumanian, Metalloporphyrin-catalyzed oxidations (1) 2. Inhibition and re-activation of metalloporphyrin catalysts. A simple model for hormone action, *J. Mol. Catal., 7*: 245 (1980).
26. J.H. Fuhrhop, M. Baccouche, and G. Penzlin, Metalloporphyrin-catalyzed oxidations. 3. Epoxidation and aldehyde formation by oxidative carbon–carbon bond cleavage in squalene, *J. Mol. Catal., 7*: 257 (1980).
27. M.N. Dufour, A.L. Crumbliss, G. Johnston, and A. Gaudemer, Reaction of indoles with molecular oxygen catalyzed by metalloporphyrins, *J. Mol. Catal., 7*: 277 (1980).
28. H. Sakurai, Y. Hataya, T. Goromaru, and H. Matsuura, Model system for drug metabolism in isolated hepatocytes; oxidation of cyclohexene by metalloporphyrin complexes, *J. Mol. Catal., 29*: 153 (1985).
29. F. Basolo, B.M. Hoffman, and J. Ibers, Synthetic oxygen carriers of biological interest, *Acc. Chem. Res., 8*: 384 (1975).
30. R.D. Jones, D.A. Summerville, and F. Basolo, Synthetic oxygen carriers related to biological systems, *Chem. Rev., 79*: 139 (1979).

31. G. McLendon and A.E. Martell, Inorganic oxygen carriers as models for biological systems, *Coord. Chem. Rev., 19*: 1 (1976).
32. N.S. Enikolopian, K.A. Bogdanova, and K.A. Askarov, Metalloporphyrin and azaporphyrin complexes as catalysts for oxidations with molecular oxygen (in Russian), *Usp. Khim., 52*: 20 (1983).
33. T. Mlodnicka, Metalloporphyrins as catalysts in autoxidation processes, *J. Mol. Catal., 36*: 205 (1986).
34. T. Szymanska-Buzar and J.J. Ziolkowski, Activation of hydrocarbon molecules via Co(III)-RH interaction in trifluoroacetic acid solution, *J. Mol. Catal., 5*: 341 (1979).
35. C.J. Weschler, B.M. Hoffman, and F. Basolo, Synthetic oxygen carrier. A dioxygen adduct of a manganese porphyrin, *J. Am. Chem. Soc., 97*: 5278 (1975).
36. B.M. Hoffman, C.J. Weschler, and F. Basolo, The dioxygen adduct of meso-tetraphenylporphyrin–manganese(II), a synthetic oxygen carrier, *J. Am. Chem. Soc., 98*: 5473 (1976).
37. R.D. Jones, D.A. Summerville, and F. Basolo, Manganese(II) porphyrin oxygen carriers. Equilibrium constants for the reaction of dioxygen with para-substituted meso-tetraphenylporphyrinatomanganese(II) complexes, *J. Am. Chem. Soc., 100*: 4416 (1978).
38. B.M. Hoffman, T. Szymanski, T.G. Brown, and F. Basolo, The dioxygen adducts of several manganese(II) porphyrins. Electron paramagnetic resonance studies, *J. Am. Chem. Soc., 100*: 7253 (1978).
39. L.K. Hanson and B.M. Hoffman, Griffith model bonding in dioxygen complexes of manganese porphyrins, *J. Am. Chem. Soc., 102*: 4602 (1980).
40. I. Tabushi and A. Jazaki, P-450-type dioxygen activation using H_2/colloidal Pt as an effective electron donor, *J. Am. Chem. Soc., 103*: 7371 (1981).
41. J.A. Smegal and C. Hill, Hydrocarbon functionalization by the (iodosylbenzene)manganese(IV)porphyrin complex from the (tetraphenylporphyrinato)manganese(III)-iodosylbenzene catalytic hydrocarbon oxidation system. Mechanism and reaction chemistry, *J. Am. Chem. Soc., 105*: 3515 (1983).
42. G.C. Dismukes, J.E. Sheats, and J.A. Smegal, Mn^{2+}/Mn^{3+} and Mn^{3+}/Mn^{4+} mixed valence binuclear manganese complex of biological interest, *J. Am. Chem. Soc., 109*: 7202 (1987).
43. J.T. Groves and M.K. Stern, Olefin epoxidation by manganese(IV) porphyrins: evidence for two reaction pathways, *J. Am. Chem. Soc., 109*: 3812 (1987).
44. L. Pauling, Nature of the iron–oxygen bond in oxyhemoglobin, *Nature, 203*: 182 (1964).
45. J.J. Weiss, Nature of the iron–oxygen bond in oxyhemoglobin, *Nature, 203*: 183 (1964).
46. I.A. Cohen and W.S. Caughey, Substituted deuteroporphyrins. IV. On the kinetics and mechanism of reactions of iron(II) porphyrins with oxygen, *Biochemistry, 7*: 636 (1968).
47. E.B. Fleisher and T.S. Srivastava, The structure and properties of μ-oxo-bis (tetraphenylporphyrin)–iron(III), *J. Am. Chem. Soc., 91*: 2403 (1969).
48. N. Sadasivan, H.I. Eberspeacher, W.H. Fuchsman, and W.S. Caughey, Substituted deuteroporphyrins. VI. Ligand exchange and dimerization reactions of deuterohemins, *Biochemistry, 8*: 534 (1969).

49. H. Kobayashi and Y. Yanagawa, Electronic spectra and electronic structure of iron(II) tetraphenylporphyrin, *Bull. Chem. Soc. Jpn., 45*: 450 (1972).
50. C.K. Chang and T.G. Traylor, Proximal base influence on the binding of oxygen and carbon monoxide to heme, *J. Am. Chem. Soc., 95*: 8477 (1973).
51. J.P. Collman, R.R. Gagne, T.R. Halbert, J.C. Marchon, and C.A. Reed, Reversible oxygen adduct formation in ferrous complexes derived from a "picket fence" porphyrin. A model for oxomyoglobin, *J. Am. Chem. Soc., 95*: 7868 (1973).
52. C.K. Chang and T.G. Traylor, Syntheses of the myoglobin active site, *Proc. Natl. Acad. Sci. U.S.A., 70*: 2647 (1973).
53. C.K. Chang and T. G. Traylor, Solution behaviour of a synthetic myoglobin active site, *J. Am. Chem. Soc., 95*: 5810 (1973).
54. D.L. Anderson, C.J. Weschler, and F. Basolo, Reversible reactions of simple ferrous porphyrins with molecular oxygen at low temperatures, *J. Am. Chem. Soc., 96*: 5599 (1974).
55. C.J. Weschler, D.L. Anderson, and F. Basolo, Kinetics of the dissociation of iron(II) porphyrin oxygen adducts. Axial base effects, *J. Chem. Soc. Chem. Commun.,* 757 (1974).
56. C.J. Weschler, D.L. Anderson, and F. Basolo, Kinetics and thermodynamics of oxygen and carbon monoxide binding to simple ferrous porphyrins at low temperatures, *J. Am. Chem. Soc., 97*: 6707 (1975).
57. J. Almog, J.E. Baldwin, R.L. Dyer, J. Huff, and C. Wilkerson, Reversible binding of dioxygen to mesoporphyrin IX derivatives at low temperatures, *J. Am. Chem. Soc., 96*: 5600 (1974).
58. J. Almog, J.E. Baldwin, and J. Huff, Reversible oxygenation and autoxidation of a "capped" porphyrin iron(II) complex, *J. Am. Chem. Soc., 97*: 227 (1975).
59. W.S. Brinigar, C.K. Chang, J. Geibel, and T.G. Traylor, Solvent effects on reversible formation and oxidative stability of heme–oxygen complexes, *J. Am. Chem. Soc., 96*: 5597 (1974).
60. G.C. Wagner and R.J. Kassner, Spectroscopic properties of protoheme complexes undergoing reversible oxygenation, *J. Am. Chem. Soc., 96*: 5593 (1974).
61. J.P. Collman, R.R. Gagne, C.A. Reed, W.T. Robinson, and G.A. Rodley, Structure of an iron(II) dioxygen complex. Model for oxygen carrying hemoproteins, *Proc. Natl. Acad. Sci. USA, 71*: 1326 (1974).
62. J.P. Collman, R.R. Gagne, C.A. Reed, T.R. Halbert, G. Lang, and W.T. Robinson, "Picket fence porphyrins." Synthetic models for oxygen binding hemoproteins, *J. Am. Chem. Soc., 97*: 1427 (1975).
63. J.P. Collman, J.I. Brauman, E. Rose, and K.S. Suslick, Cooperativity in oxygen binding to iron porphyrins, *Proc. Natl. Acad. Sci. USA, 75*: 1052 (1978).
64. A. Dedieu, M.M. Rohmer, M. Benard, and A. Veillard, Oxygen binding to iron porphyrins. An ab initio calculation, *J. Am. Chem. Soc., 98*: 3717 (1976).
65. J.P. Collman, Synthetic models for the oxygen binding hemoproteins, *Acc. Chem. Res., 10*: 265 (1977).
66. D.H. Chin, J. DelGaudio, G.N. LaMar, and A.L. Balch, Detection and characterization of the long-postulated-Fe-O-O-Fe intermediate in the autoxidation of ferrous porphyrins, *J. Am. Chem. Soc., 99*: 5486 (1977).

67. D. Dolphin, B.R. James, and H.C. Welborn, Oxygenation and carbonylation of a reduced P-450 cam enzyme and derivatives reconstituted with meso-deutero-, dibromodeutero-, and diacetyldeuteroheme, *J. Mol. Catal., 7*: 201 (1980).
68. E. McCandlish, A.R. Miksztal, M. Nappa, A.Q. Sprenger, J.S. Valentine, J.D. Stong, and T.G. Spiro, Reactions of superoxide with iron porphyrins in aprotic solvents. A high spin ferric porphyrin peroxo complex, *J. Am. Chem. Soc., 102*: 4268 (1980).
69. B. Ward, C.B. Wang, and C.K. Chang, Nonbonding steric effect on CO and O_2 binding to hemes. Kinetics of ligand binding in iron–copper cofacial diporphyrins and strapped hemes, *J. Am. Chem. Soc., 103*: 5236 (1981).
70. C.E. Castro, Mechanism of reactions of hemoproteins with oxygen and hydrogen peroxide in the oxidation of organic substrates, *Pharm. Ther., 10*: 171 (1980).
71. P.R. Ortiz de Montellano (ed.), *Cytochrome P-450: Structure, Mechanism and Biochemistry*, Plenum Press, New York, 1986.
72. J.O. Alben, W.H. Fuchsman, C.A. Beaudreau, and W.C. Caughey, Substituted deuteroporphyrins. (III). Iron(II) derivatives. Reactions with oxygen and preparation from chloro and methoxohemins, *Biochemistry, 7*: 624 (1968).
73. G.S. Hammond and C.S. Wu, Oxidation of iron(II) chloride in non-aqueous solvents, *Adv. Chem. Ser., 77*: 186 (1968).
74. A.B. Hoffman, D.M. Collins, W.V. Day, E.B. Fleischer, T.S. Strivastava, and J.L. Hoard, The crystal structure and molecular stereochemistry of μ-oxo-bis[$\alpha,\beta,\gamma,\delta$-tetraphenylporphyrinato-Iron(III)], *J. Am. Chem. Soc., 94*: 3620 (1972).
75. A.L. Balch, Y.W. Chan, R.J. Cheng, G.N. LaMar, L. Latos-Grazynski, and M.W. Renner, Oxygenation patterns for iron(II) porphyrins. Peroxo and ferryl ($Fe^{IV}O$) intermediates detected by ^1H nuclear magnetic resonance spectroscopy during the oxygenation of (tetramesitylporphyrin) iron(II), *J. Am. Chem. Soc., 106*: 7779 (1984).
76. B.R. James, *Interaction of Dioxygen with Metalloporphyrins*, in *The Porphyrins*, Vol. 5 (D. Dolphin, ed.), Academic Press, New York, 1978, p. 207.
77. J.T. Groves, R.C. Haushalter, M. Nakamura, T.E. Nemo, and B.J. Evans, High-valent iron–porphyrin complexes related to peroxidase and cytochrome P-450, *J. Am. Chem. Soc., 103*: 2884 (1981).
78. J.T. Groves, Y. Watanabe, and T.J. McMurry, Oxygen activation by metalloporphyrins. Formation and decomposition of an acylperoxymanganese(III) complex, *J. Am. Chem. Soc., 105*: 4489 (1983).
79. R.B. Van Atta, C.E. Strouse, L.K. Hanson, and J.S. Valentine, [Peroxotetraphenylporphinato]manganese(III) and [Chlorotetraphenylporphinato]manganese(II) anions. Syntheses, crystal structure and electronic structure, *J. Am. Chem. Soc., 109*: 1425 (1987).
80. F.A. Walker, An electron spin resonance study of coordination of the fifth and sixth positions of $\alpha,\beta,\gamma,\delta$-tetra(p-methoxyphenyl)porphinatocobalt(II), *J. Am. Chem. Soc., 92*: 4235 (1970).
81. K. Yamamoto and T. Kwan, Oxygenation of cobalt tetraphenylporphine as investigated by ESR spectroscopy, *J. Catal., 18*: 354 (1970).
82. K. Yamamoto and T. Kwan, Interaction of metal porphyrins with molecular oxygen in solution, *Bull. Chem. Soc. Jpn., 45*: 664 (1972).

83. M. Momenteau, M. Fornier, and M. Rougee, Etude spectrometric de la réduction de quelques metalloporphyrines de cobalt et de la vitamine B_{12} milieu organique (in French), *J. Chim. Phys., 67*: 926 (1970).
84. B.B. Wayland and D. Mohajer, An electron spin resonance study of the interaction of sulphur dioxide with tetraphenylporphyrincobalt(II), *J. Chem. Soc. Chem. Commun.,* 776 (1972).
85. H.C. Stynes and J.A. Ibers, A pronounced solvent effect on the reversible oxygenation of a cobalt(II) porphyrin system, *J. Am. Chem. Soc., 94*: 5125 (1972).
86. H.C. Stynes and J.A. Ibers, Thermodynamics of the reversible oxygenation of amine complexes of cobalt(II) protoporphyrin IX dimethyl ester in a nonaqueous medium, *J. Am. Chem. Soc., 94*: 1559 (1972).
87. F.A. Walker, Steric and electronic effects in the coordination of amines to a cobalt(II) porphyrin, *J. Am. Chem. Soc., 95*: 1150 (1973).
88. D.V. Stynes, H.C. Stynes, B.R. James, and J.A. Ibers, Thermodynamics of ligand and oxygen binding to cobalt protoporphyrin IX dimethyl ester in toluene solutions, *J. Am. Chem. Soc., 95*: 1796 (1973).
89. D.V. Stynes, H.C. Stynes, J.A. Ibers, and B.R. James, Kinetics of the reaction of amine complexes of cobalt(II) protoporphyrin IX dimethyl ester with oxygen. Evidence for hydrogen bonding with coordinated oxygen, *J. Am. Chem. Soc., 95*: 1142 (1973).
90. B.B. Wayland, J.W. Minkiewicz, and M.E. Abd-Elmageed, Spectroscopic studies for tetraphenylporphyrincobalt(II) complexes of CO. NO, O_2, RNC and $(RO)_3P$ and a bonding model for complexes of CO, NO, and O_2 with cobalt(II) and iron(II) porphyrins, *J. Am. Chem. Soc., 96*: 2795 (1974).
91. B.B. Wayland and M.E. Abd-Elmageed, Tetraphenylporphyrincobalt(II) complexes as a probe for the nature of phosphine σ-donor orbitals. Electron paramagnetic resonance studies of phosphine and dioxygen complexes, *J. Am. Chem. Soc., 96*: 4809 (1974).
92. B.B. Wayland, M.E. Abd-Elmageed, and L.F. Mehne, Low-spin cobalt(II)–Schiff base and -porphyrin complexes of CO, MeNC, $AsMe_3$ and PX_3. Electron paramagnetic resonance studies, *Inorg. Chem., 14*: 1456 (1975).
93. J.E. Linar, P.E. Ellis, J.R. Budge, R.D. Jones, and F. Basolo, Oxygenation of iron(II) and cobalt(II) "capped porphyrins," *J. Am. Chem. Soc., 102*: 1896 (1980).
94. M. Kozuka and K. Nakamoto, Vibrational studies of tetraphenylporphyrinatocobalt(II) and its adducts with CO, NO, and O_2 in gas matrices, *J. Am. Chem. Soc, 103*: 2162 (1981).
95. P. Doppelt and R. Weiss, Synthesis and spectral properties of five-coordinate mercapto-cobalt(II) porphyrins and their O_2 adducts: synthetic analogs for the active site in cytochromes P-450, *Nouv. J. Chim., 7*: 341 (1983).
96. F.A. Walker, ESR Studies of Co(II) tetraphenylporphyrins and their oxygen adducts: complex formation with aromatic molecules and sterically hindered Lewis bases, *J. Magn. Reson., 15*: 201 (1974).
97. F.A. Walker, D. Beroiz, and K.M. Kadish, Electronic effects in transition metal porphyrins. 2. The sensitivity of redox and ligand addition reactions in para-substituted tetraphenylporphyrin complexes of cobalt(II), *J. Am. Chem. Soc., 98*: 3484 (1976).

98. B.B. Wayland and A.R. Newman, Dioxygen complexes of rhodium porphyrins, *J. Am. Chem. Soc., 101*: 6472 (1979).
99. B.R. James and E. Ochiai, Mechanism of oxygenation reactions catalyzed by rhodium(I) complexes in solution, *Can. J. Chem., 49*: 975 (1971).
100. F. Tsuchija and T. Ikawa, Liquid-phase oxidation of 2-butene in the presence of benzaldehyde, *Can. J. Chem., 47*: 3191 (1969).
101. A.D. Vreugdenhil and H. Reit, Liquid-phase cooxidation of aldehydes and olefins. Radical versus nonradical epoxidation, *Recl. Trav. Chim. Pays-Bas, 91*: 237 (1972).
102. X.Y. Wang, R.J. Motekaitis, and E.A. Martell, Metallotetraphenylporphyrin-catalyzed oxidation of 2,6-di-tert-butylphenol, *Inorg. Chem., 23*: 271 (1984).
103. K.M. Kadish, L.A. Bottomley, and D. Beroiz, Reactions of pyridine with a series of para-substituted tetraphenylporphyrincobalt and -iron complexes, *Inorg. Chem., 17*: 1124 (1978).
104. W.A. Lee and T.C. Bruice, Homolytic and heterolytic oxygen–oxygen bond scission accompanying oxygen transfer to iron(III) porphyrins by percarboxylic acids and hydroperoxides. A mechanistic criterion for peroxidase and cytochrome P-450, *J. Am. Chem. Soc., 107*: 513 (1985).
105. L.C. Yuan and T.C. Bruice, Oxygen donation to manganese(II) tetraphenylporphyrin chloride. Low reactivity of hydroperoxides as oxygen donors to manganese(III) porphyrins, *Inorg. Chem., 24*: 986 (1985).
106. W.A. Lee, L.C. Yuan, and T.C. Bruice, Oxygen transfer from percarboxylic acids and alkyl hydroperoxides to (meso-tetraphenylporphinato) iron(III) and chromium(III), *J. Am. Chem. Soc., 110*: 4277 (1988).
107. G.X. He and T.C. Bruice, Nature of the epoxidizing species generated by reactions of alkyl hydroperoxides with iron(III) porphyrins. Oxidations of cis-stilbene and (Z)-1,2-bis(trans-2,trans-3-diphenylcyclopropyl)ethene by t-BuOOH in the presence of [meso-tetrakis(2,4,6-trimethylphenyl)porphinato]-, [meso-tetrakis(2,6-dichlorophenylphorphinato]-, and [meso-tetrakis(2,6-dibromophenyl)porpinato]iron(III)chloride, *J. Am. Chem. Soc., 113*: 2747 (1991).
108. T.C. Bruice, Reactions of hydroperoxides with metallotetraphenylporphyrins in aqueous solutions, *Acc. Chem. Res., 24*: 243 (1991).
109. T.G. Traylor, W.A. Lee, and D.V. Stynes, Model compound studies related to peroxidases. Mechanism of reactions of hemins with peracids, *J. Am. Chem. Soc., 106*: 755 (1984).
110. T.G. Traylor and F. Xu, A biomimetic model for catalase: the mechanisms of reaction of hydrogen peroxide and hydroperoxides with iron(III) porphyrins, *J. Am. Chem. Soc., 109*: 6201 (1987).
111. T.G. Traylor and R. Popovitz-Biro, Hydrogen bonding to the proximal imidazole in heme protein model compounds: effects upon oxygen binding and peroxidase activity, *J. Am. Chem. Soc., 110*: 239 (1988).
112. T.G. Traylor and J.P. Ciccone, Mechanism of reactions of hydrogen peroxide and hydroperoxides with iron(III) porphyrins. Effects of hydroperoxide structure on kinetics, *J. Am. Chem. Soc., 111*: 8413 (1989).
113. T.G. Traylor, Wen-Pang Fann, and D. Bandyopadhyay, A common heterolytic mechanism for reactions of iodosobenzenes, peracids, hydroperoxides

and hydrogen peroxide with iron(III) porphyrins, *J. Am. Chem. Soc., 111*: 8009 (1989).

114. D. Mansuy, P. Battioni, and J.P. Renaud, In the presence of imidazole, iron- and manganese-porphyrins catalyze the epoxidation of alkenes by alkyl hydroperoxides, *J. Chem. Soc. Chem. Commun.,* 1255 (1984).
115. D. Mansuy, Cytochrome P-450 and synthetic models, *Pure Appl. Chem., 59*: 759 (1987).
116. T. Yamada and Y. Kamiya, Spectrophotometric study on the complex formation of hydroperoxide with tetraphenylporphyrin cobalt(II), *Bull. Chem. Soc. Jpn., 53*: 1077 (1980).
117. D. Mansuy, J.F. Bartoli, J.C. Chottard, and M. Lange, Metalloporphyrin-catalyzed hydroxylation of cyclohexane by alkyl hydroperoxides: special efficiency of iron porphyrins, *Angew. Chem. Int. Ed. Engl., 19*: 909 (1980).
118. J.T. Groves and Y. Watanabe, Oxygen activation by metalloporphyrins related to peroxidase and cytochrome P-450. Direct observation of the O-O bond cleavage step, *J. Am. Chem. Soc., 108*: 7834 (1986).
119. M. Pérree-Fauvet and A. Gaudemer, Manganese porphyrin-catalyzed oxidation of olefins to ketones by molecular oxygen, *J. Chem. Soc. Chem. Commun.,* 874 (1981).
120. Y. Aoyama, T. Watanabe, H. Onda, and H. Ogoshi, Efficient olefin oxygenation with tetrahydroborate and dioxygen catalyzed by a rhodium porphyrin complex, *Tetrahedron Lett., 24*: 1183 (1983).
121. I. Tabushi and K. Morimitsu, Stereospecific, regioselective and catalytic monoepoxidation of polyolefins by the use of a P-450 model, H_2-O_2-TPPMn-colloidal platinum, *J. Am. Chem. Soc., 106*: 6871 (1984).
122. I. Tabushi and M. Kodera, Flavin-catalyzed reductive dioxygen activation with N-methyldihydronicotinamide, *J. Am. Chem. Soc., 108*: 1101 (1986).
123. T. Santa, T. Mori, and M. Hirobe, Chemical studies on drug metabolism. 8. Oxygen activation and olefin oxygenation by iron(III) porphyrin as a model system of cytochrome P-450, *Chem. Pharm. Bull., 33*: 2175 (1985).
124. I. Tabushi, M. Kodera, and M. Yokoyama, Kinetics and mechanism of reductive dioxygen activation catalyzed by the P-450 model system. Iron picket fence as a catalytic center, *J. Am. Chem. Soc., 107*: 4466 (1985).
125. M. Fontecave and D. Mansuy, Monooxygenase-like oxidations of olefins and alkanes catalyzed by manganese porphyrins: comparison of systems involving either O_2 and ascorbate or iodosylbenzene, *Tetrahedron, 40*: 4297 (1984).
126. P. Battioni, J.F. Bartoli, P. Leduc, M. Fontecave, and D. Mansuy, A new and efficient biomimetic system for hydrocarbon oxidation by dioxygen using manganese porphyrins, imidazole and zinc, *J. Chem. Soc. Chem. Commun.,* 791 (1987).
127. F.I. Karasevich, A.M. Khenkin, and A.E. Shilov, A chemical model of cytochrome P-450; monooxygenase-like activation of dioxygen, *J. Chem. Soc. Chem. Commun.,* 731 (1987).
128. A.M. Khenkin and A.E. Shilov, Biomimetic alkane oxidation in the presence of iron complexes, *Nouv. J. Chem., 13*: 659 (1989).
129. G.B. Shulpin and A.N. Druzhinina, Biomimetic activation of the carbon–hydrogen bond, *Izv. Acad. Nauk, SSSR, Ser. Khim., 12*: 2739 (1991).

130. S.E. Creager, S.A. Raybuck, and R.W. Murray, An efficient electrocatalytic model cytochrome P-450 epoxidation cycle, *J. Am. Chem. Soc., 108*: 4225 (1986).
131. I. Apostol, J. Haber, T. Mlodnicka, and J. Poltowicz, Cobalt-tetra(p-tolyl)porphyrin as catalyst in oxidation of propionaldehyde, *J. Mol. Catal., 14*: 197 (1982).
132. I. Apostol, J. Haber, T. Mlodnicka, and J. Poltowicz, Kinetics and mechanism of cooxidation of propylene and benzaldehyde in the presence of some cobalt complexes, *J. Mol. Catal., 26*: 239 (1984).
133. T. Mlodnicka, J. Haber, I. Apostol, and J. Poltowicz, Epoxidation of propylene in the presence of benzaldehyde and Co(II)p-CH_3TPP as catalyst, *Oxidation Commun., 5*: 189 (1983).
134. I. Apostol, J. Haber, T. Mlodnicka, and J. Poltowicz, Selectivity–electronic structure correlation in the liquid phase cooxidation of aldehydes and olefins in the presence of cobalt porphyrins as catalysts, *Proceedings of 8th Int. Congr. Catalysis*, Vol. 4, Berlin, 1984, p. 497.
135. J. Haber, T. Mlodnicka, and J. Poltowicz, Metal-dependent reactivity of some metalloporphyrins in oxidation with dioxygen, *J. Mol. Catal., 54*: 451 (1989).
136. J. Haber, T. Mlodnicka, and M. Witko, Reactivity–structure correlations in oxidation with metalloporphyrins, *J. Mol. Catal., 52*: 85 (1989).
137. R. Iwanejko, T. Mlodnicka, and J. Poltowicz, Metalloporphyrin-catalyzed oxidation of cyclohexane with dioxygen, in *New Developments in Selective Oxidation* (G. Centi and F. Trifiro, eds.), Elsevier, Amsterdam, 1990, p. 195.
138. R. Iwanejko, P. Leduc, T. Mlodnicka, and J. Poltowicz, Metalloporphyrin-catalyzed epoxidation of propylene, in *Dioxygen Activation and Homogeneous Catalytic Oxidation* (L. Simandi, ed.), Elsevier, Amsterdam, 1991, p. 113.
139. R. Iwanejko and T. Mlodnicka, Transition of cobalt tetratolylprophyrin under the influence of peroxy acid in the process of generation of catalytically active porphyrin species, *Bull. Pol. Acad. Sci., 35*: 403 (1987).
140. J. Haber and T. Mlodnicka, Oxidation of hydrocarbons with dioxygen,*J. Mol. Catal., 74*: 131 (1992).
141. A. Robert and B. Meunier, Oxygenation of hydrocarbons by $KHSO_5$ catalyzed by manganese porphyrin complexes, *Nouv. J. Chim., 12*: 885 (1988).
142. W.A. Lee and T.C. Bruice, Transfer of oxygen from percarboxylic acids and alkyl hydroperoxides to (meso-tetraphenylporphyrinato)cobalt(III) chloride, *Inorg. Chem., 25*: 131 (1986).
143. V. Gutmann, *The Donor–Acceptor Approach to Molecular Interactions*, Plenum Press, New York, 1978.
144. J.A.S. Rozenberg, A.W. van der Made, J.W.H. Smeets, and R.J.M. Nolte, Cyclohexene epoxidation by the mono-oxygenase model (tetraphenylporphyrinato)manganese(III) acetate–sodium hypochlorite, *J. Mol. Catal., 31*: 271 (1985).
145. P. Battioni, J.P. Renaud, J.F. Bartoli, M. Momenteau, and D. Mansuy, Biomimetic oxidation of hydrocarbons by H_2O_2 catalyzed by manganese porphyrins: the two roles of imidazole, *Recl. Trav. Chim. Pays-Bas, 106*: 332 (1987).
146. B. Meunier, M.E. de Carvalho, O. Bartolini, and M. Momenteau, Proximal effect of the nitrogen ligands in the catalytic epoxidation of olefins by the NAOCl/manganese(III) porphyrin system, *Inorg. Chem., 27*: 161 (1988).

147. B. Meunier, M.E. de Carvalho, and A. Robert, Aspects of metalloporphyrin-catalyzed oxygenation of hydrocarbons with anionic single oxygen donors, NaOCl and $KHSO_5$, *J. Mol. Catal., 41*: 185 (1987).
148. P. Battioni, J.P. Renaud, J.P. Bartoli, M. Reina-Artiles, M. Fort, and D. Mansuy, Monooxygenase-like oxidation of hydrocarbons by H_2O_2 catalyzed by manganese porphyrins and imidazole: selection of the best catalytic system and nature of the active oxygen species, *J. Am. Chem. Soc., 110*: 8462 (1988).
149. S. Banfi, A. Maiocchi, F. Montanari, and S. Quici, Investigations on factors ruling catalytic efficiency and chemical stability of Mn(III)-porphyrins in HOCl olefin epoxidation: conditions for practical application, *J. Org. Chem., 54*: 1850 (1989).
150. J. Haber, R. Iwanejko, and T. Mlodnicka, Oxidation of cyclohexane with H_2O_2 and chlorotetratolylporphyrinatochromium(III) as catalyst, *J. Mol. Catal., 55*: 268 (1989).
151. P. Ellis, Jr. and J.E. Lyons, Effect of axial azide on the selective, low temperature metalloporphyrin-catalyzed reactions of isobutane with molecular oxygen, *J. Chem. Soc. Chem. Commun.,* 1187 (1989).
152. P.E. Ellis, Jr. and J.E. Lyons, Selective air oxidation of light alkanes catalyzed by activated metalloporphyrins—the search for a suprabiotic system, *Coord. Chem. Rev., 105*: 181 (1990).
153. C. Querci and M. Ricci, A very efficient alkene epoxidation by magnesium monoperoxyphthalate catalyzed by manganese porphyrins, *J. Chem. Soc. Chem. Commun.,* 889 (1989).
154. T. Mlodnicka and B.R. James, Oxidations catalyzed by ruthenium porphyrins, in *Catalysis by Metal Complexes* (F. Montanari, ed.), in press.

10
Reactions of Alkanes with Dioxygen: Toward Suprabiotic Systems

James E. Lyons and Paul E. Ellis, Jr.
Sun Company, Inc., Marcus Hook, Pennsylvania

I. INTRODUCTION

Catalytic alkane conversion science and technology is maturing to the point that one can envision economical processes for upgrading the light alkanes currently found in great abundance in natural gas and light refinery streams. Application of this science to the conversion of these light alkanes directly to alcohols using molecular oxygen would make possible the utilization of inexpensive aliphatic hydrocarbons for producing valuable chemical and fuel products.

Until recently, there have been no effective catalysts for the selective efficient conversion of the light alkanes—methane, ethane, propane, and the butanes—to alcohols using air or oxygen as the *sole* oxidant. Processes based on heterogeneous catalysts use air or oxygen but normally are conducted at elevated temperatures and produce deeply oxidized products such as carboxylic acids and anhydrides. Homogeneous catalysts have also proven to be relatively ineffective for the air oxidation of light alkanes to alcohols and this fact has spawned tremendous interest in biological systems.

Cytochrome P450 and methane monooxygenase have been found to convert alkanes to alcohols in vitro [1–7]. Both heme and nonheme iron centers have been implicated in the oxygen atom transfer step. Growing understanding of the biological systems together with the need for catalysts which can activate aliphatic C-H bonds for selective hydroxylation has stimulated intense activity in attempting to mimic biological systems with synthetic analogs [6–14]. Much important knowledge has been gained which is highly relevant to the design of catalytic systems for selective air oxidation of alkanes to alcohols. In addition, this work has brought into focus the problems with the biological systems for operation in an abiotic medium. An acceptable catalyst will be one which effectively applies relevant aspects of the biological systems while circumventing or solving the problems that accompany the applications of enzymic systems.

II. BIOMIMETIC OXIDATION

A. Cytochrome P450

The monooxygenase cytochrome P450 contains an iron center in a porphyrinic environment which, in the presence of reduced nicotinamide adenine dinucleotide (NADH), can activate molecular oxygen to form a high-oxidation state iron–oxo (Fe = O) species which is capable of hydroxylating the C-H bond of an alkane [1–4]. In vitro experiments have shown that hydrocarbons such as norbornane [15], octane [16,17], and hexane [18] can be catalytically hydroxylated using this system. Conditions are mild (20°C, 1 atm air) and reaction rates are moderately high (>40 mol alcohol/mol enzyme/min) [19]. The function of NADH in this system is to furnish the necessary electrons and protons to the iron center to generate the active species (Eq. 1):

$$\text{“Fe”} + \text{NADH} + H^+ + O_2 \rightarrow \text{“Fe}=\text{O”} + \text{NAD}^+ + H_2O \quad (1)$$

B. Useful Concepts for Catalysis

Among the most useful concepts emanating from the many attempts to mimic enzymatic systems is the idea that high-oxidation-state iron intermediates are the active sites for alkane oxidation [8]. In the case of the heme–iron system of cytochrome P450, an iron–oxo intermediate has been implicated [4a]. Porphyrinatoiron(III) complexes have been used in attempts to model this biocatalyst.

As shown by the dotted line in Figure 1, single-oxygen-atom transfer reagents, XO, have been used to circumvent the need for two electrons and two protons provided in the natural system by NADH.

Groves [9], in 1979, first used iodosylbenzene as an oxygen atom donor to Fe in FeTPPCl and was able to hydroxylate adamantane to adamantanol-1 and adamantanol-2. Groves [20] using Cr TPPCl* and iodosylbenzene was able to isolate an intermediate Cr(V) complex which was characterized as a Cr(V) = O, chromyl complex.

*For definitions, see list of abbreviations preceding the references.

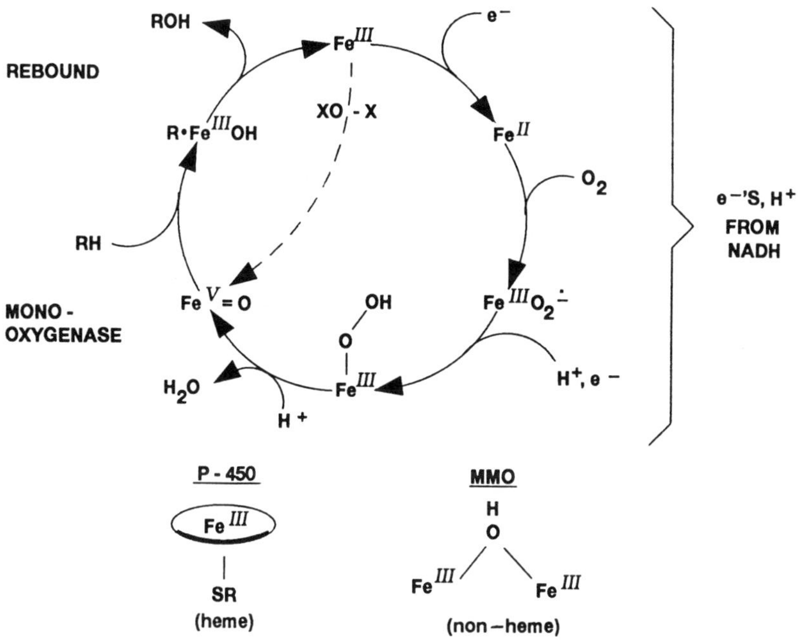

Figure 1 Biological and biomimetic alkane oxidation.

Many other workers have used a variety of single-oxygen donors to act as a “shunt” reagent to form high-oxidation-state metal–oxo species that are catalytically active. These high-oxidation-state metal–oxo complexes apparently convert aliphatic hydrocarbon substrates to alcohols. It appears, therefore, that if one can generate these species in situ, alkanes will be converted to alcohols. In this chapter we will concentrate on metal porphyrin–catalyzed alkane oxidations which use air or oxygen as the oxidant. A recent review of other oxygen sources including a thorough review of single-oxygen-atom donor work was published by Meunier [21].

Another useful concept arising from the studies of biological oxidations carried out to date is the utility of a rigid hydrophobic pocket around the iron which attracts the alkane from an aqueous medium and rejects alcohol product to prevent overoxidation. The shape and size of the pocket orients the hydrocarbon molecule for hydroxylation. Its rigidity also helps to prevent self-oxidation which can result in destruction of the complex. Thus a catalyst which could generate high-oxidation-state iron–oxo units in a hydrophobic pocket could exhibit high oxidation activity without ligand destruction and with high selectivity to the alcohol.

C. O_2-Based Hydroxylations with Sacrificial Reductants

The major problem with using either biological or biomimetic systems investigated to date for practical catalytic oxidations is the stoichiometric consumption of expensive coreductants or oxygen atom transfer reagents. Biological systems use NADH stoichiometrically in the production of the active iron (Fe = O) centers. The NAD which is formed in this process must be reduced by some reagent which is sacrificed in order to return the NAD to NADH. In synthetic systems, it has been possible to replace the single-atom oxygen donors with dioxygen for metalloporphyrin-catalyzed hydroxylation by the use of sacrificial coreductants. The simplest means which has been found to date to furnish the required two electrons and two protons to a biomimetic catalyst systems is to split the hydrogen molecule over a group VIII metal catalyst in situ (Figure 2). Tabushi and Yazaki [22] were able to use MnTPPCl with imidazole and Pt/H_2 to hydroxylate adamantane with air, although the rate was very slow. Although it has been successful only for conversion of very reactive hydrocarbons, such an approach could provide the possible pathway shown in Eq. 2 for the direct conversion of alkanes to alcohols. But even this relatively straightforward approach would probably be too costly to convert natural gas components or other inexpensive alkane feedstocks to alcohols for fuel use since it requires stoichiometric consumption of 1 mol of hydrogen per mol of alkane converted. However, if the catalytic reaction shown in Eq. 2 could be accomplished in good yield at a reasonable rate, it could have utility in producing chemicals valuable enough to bear the cost of hydrogen consumption.

$$RH + O_2 + H_2 \xrightarrow{\text{Cat}} ROH + H_2O \qquad (2)$$

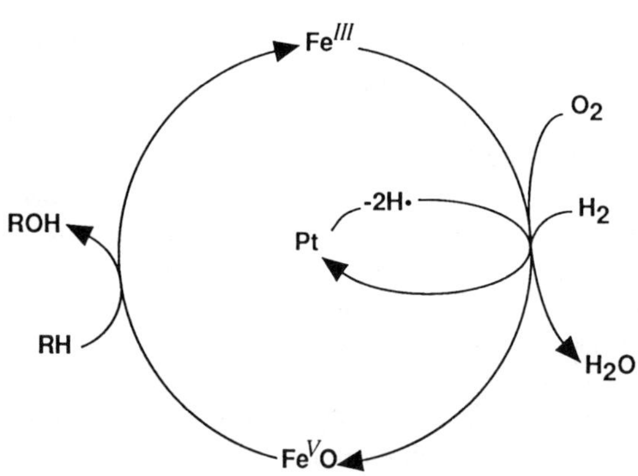

Figure 2 Hydrogen as a coreductant in a biomimetic oxidation.

Not only H_2 but several other sacrificial chemical reductants have enabled metalloporphyrins to catalytically activate dioxygen toward alkane hydroxylation. Tabushi and Koga [23] were the first to utilize chemical reductants in metalloporphyrin-catalyzed oxidations. Using $NaBH_4$ as the reductant together with added imidazole, they found that MnTPPCl catalyzed predominant epoxidation of cyclohexene. Tabushi [24] postulates that when colloidal Pt/H_2 is used as the reductant, Mn(III) is reduced to dioxygen-active Mn(II), eventually leading to a high-valent Mn = O active site.

Mansuy et al. [25,26] developed a hydroxylation system that employs ascorbate ion reduction with MnTPPCl, O_2, and a phase transfer agent to aid the transfer of an electron from the aqueous ascorbate to the organic phase containing the metalloporphyrin. Using this system cyclohexane was converted to a mixture of cyclohexanol and cyclohexanone, and heptane to all the heptanones. FeTPPCl and CoTPP were also used [27] with ascorbate ion to hydroxylate cyclohexane. Other reducing systems which have been studied include ones which use propionaldehyde [28], Zn [29], Zn(Hg) [30], and ferrocene [31].

The key concept that this work demonstrates is that reduction to the oxygen-active state of the metal [e.g., Fe(II) or Mn(II)] is essential for completing a catalytic cycle. Whether a high-oxidation-state metal–oxo complex is obtained from these systems is unknown but the ability to utilize dioxygen is inextricably tied to the ability to reach a *low* "oxygen-active" oxidation state of the metal.

D. O_2-Based Hydroxylations Utilizing Electrode Processes

Electrochemical-based catalytic systems provide the electrons which replace the chemical reductants discussed above. For the metalloporphyrin-catalyzed hydroxylation of alkanes two types of approaches are viable. One utilizes a soluble metalloporphyrin and an external electrode, and in the second the metalloporphyrin is incorporated as part of the electrode itself.

Mansuy et al. [32] were able to oxidize alkanes such as cyclooctane, adamantane, and ethylbenzene to mixtures of alcohols and ketones by using O_2 in an electrochemical cell consisting of a platinum gauze working electrode with MnTPPCl, imidazole, electrolyte, and acetic acid in acetonitrile. The yield based on electrons consumed was low and the rate slow; however, hydroxylation was achieved.

Bedioui et al. [33] constructed a graphite electrode coated with a film of a polypyrrole-substituted manganese porphyrin. With this electrode in the presence of 1-methylimidazole, hydrocarbon substrate, supporting electrolyte, and bubbling O_2, they were able to oxidize tetralin, indane, and cyclooctane. The rates were faster than the soluble system of Mansuy but yield per electron was low and actual productivity was quite poor since only 10^{-8} to 10^{-7} mol of manganese porphyrin could be attached to the electrodes.

E. O_2-Based Hydroxylations Utilizing Light

Irradiation into the ligand to metal change transfer band in FeTPPCl caused the photoreduction of the iron to the ferrous state but only in the presence of hydrocarbons with weak C-H bonds such as cumene and cyclohexene [34]. In the presence of these hydrocarbons and oxygen, photoinitiation of autoxidation occurred giving the usual radical chain oxidation products.

In similar work Fe(TDCPP)OH was irradiated with 350–450 nm light producing Fe(II) and initiating oxidation of cyclohexane by O_2 [35].

III. THE SUPRABIOTIC APPROACH

To mimic simply the biological systems may not be sufficient to generate active stable and cost-effective catalysts for alkane oxidations. An abiotic catalyst which employs an active metal–oxo intermediate must have several features that are above and beyond the capability of biological systems. An important feature would be to generate the metal–oxo species directly from molecular oxygen without the costly requirement of a stoichiometric added coreductant.

A major problem with biomimetic systems is deactivation of the catalyst by oxidative destruction. Two ways in which this comes about are by (a) μ-oxo dimer formation and (b) oxidative degradation of the ligand system. Nature uses a hydrophobic pocket to deter these deactivation processes. By rigidly holding the iron porphyrin in place the hydrophobic pocket prevents two iron atoms from getting close enough to form inactive species. It also keeps weak C-H bonds away from the active center so that the organic system stays intact. Nature has regeneration mechanisms to combat destruction of portions of the enzyme system—a synthetic abiotic system would not possess such possibilities. Any major destruction of the ligand system in a synthetic catalyst would result in loss of both selectivity and activity. Abiotic catalysts must be inherently robust to thermal and oxidative degradation in order to survive the rigors of industrial processing or laboratory synthesis.

Thus a suprabiotic catalyst is one that does not require stoichiometric consumption of a reductant ($2e^- + 2H^+$) for cleaving O_2, uses *both* atoms of the O_2 molecule to form product (exhibits dioxygenase activity), and is robust with respect to reaction conditions. In this chapter, we will describe a rationale for creating just such a suprabiotic system. We will provide initial results which indicate that a good measure of success has already been achieved, and we will propose additional work for further development of new suprabiotic catalyst systems in order to produce efficient processes for converting light alkanes to valuable fuels and chemicals.

When searching for a possible model catalytic system that could meet the requirements of a suprabiotic catalyst, we formulated the catalytic cycle [36] shown in Figure 3. Since nature uses a porphyrinic macrocycle as the ligand system, this seemed to be a reasonable starting point for this research. The cycle shown in the figure uses elementary steps, most of which have precedent in inorganic chemistry. In this hypothetical cycle oxygen reacts with the metal center to form a superoxo complex which combines with another center to form a μ-peroxo complex. Cleavage of the O-O bond in the peroxo complex leads to the metal–oxo complex which oxidizes the alkane. One of the reasons why this doesn't happen readily under normal circumstances is that the highly reactive metal oxo complex combines rapidly with a metal center to form a catalytically inactive μ-oxo complex (Eq. 3):

$$M{=}O + M \rightarrow M\text{-}O\text{-}M \quad (3)$$

One goal of suprabiotic catalyst design, therefore, might be to adjust the electronic characteristics of the ligand system about the metal center in such a way as to encourage O-O bond cleavage, and to stabilize monomeric metal species relative to μ-oxo intermediates.

A. Catalyst Design Concept

As we first envisioned a suprabiotic catalyst, it was one which could operate in the absence of costly stoichiometric coreductants using only air to

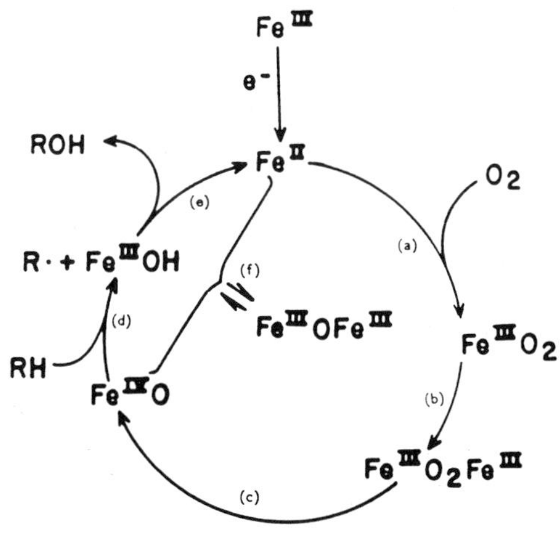

Figure 3 Alkane hydroxylation with O_2 using a suprabiotic catalyst.

oxidize an alkane to an alcohol but in some respects mimic biology in that the active intermediates would be similar. It appears as though the active intermediate which oxidizes alkanes in both cytochrome P450 and methane monooxygenase is the ferryl species Fe = O. We devised a possible conceptual route to active ferryl using only iron, air, and substrate (Figure 3) [36].

Until recently, no success had been achieved in attempting to oxidize alkanes to alcohols with only air or oxygen in this manner. Iron porphyrins have been tried with O_2 and considered as autoxidation catalysts [37] but were able to oxidize only activated benzylic C-H bonds such as those in tetralin. Metallophthalocyanines have been used as autoxidation catalysts [38, 39] but only recently [38] has work with halogenated phthalocyanines appeared to show reactivity that is superior to classical metal-promoted autoxidation.

The requirements for a suprabiotic system for air oxidations of alkanes to alcohols are many. First, although formation of μ-oxo dimer from the iron(II) precursor is detrimental, μ-peroxo dimer formation is crucial. Second, access of the alkane to the active center must be good. Third, the alkyl radical formed by C-H bond homolysis should remain in the coordination sphere until rebound to the alcohol occurs. Fourth, the alcohol should be expelled as quickly from the coordination sphere as possible to prevent overoxidation. Next, μ-oxo dimer formation must be reversed or prevented. Finally, the iron(III)/(II) reduction potential must be high enough to continuously regenerate active Fe(II) from the pool of Fe(III) even in the presence of molecular oxygen. In addition, the ligand system must be robust to oxidation and must not have electrons readily available to reduce the high-oxidation-state ferryl before it can oxidize the alkane. The kinds of porphyrin complexes available in the past did none of these things and consequently were inactive catalysts.

Using perhaloporphyrin complexes it is possible to (a) stabilize active iron–oxygen centers to both of the kinds of oxidative decay referred to above, (b) provide a lipophilic environment to attract the alkane to the active site and thereby provide a hydrophobic environment to expel alcohol from the coordination sphere, and (c) tune the redox potential of the metal center so that the μ-oxo dimer is no longer inactive and so that iron(II) species may be continually regenerated in an oxidizing environment. Figure 4 suggests how perhaloporphyrins may act as a suprabiotic catalyst for converting alkanes to alcohols. We will discuss how the electronic and steric environment about the oxidation-active metal center affects the rate and selectivity of alkane oxidation in homogeneous liquid phase oxidations. Reaction rate seems to be highly dependent on proper tuning of redox potentials. Unprecedented reaction rates have been achieved in the liquid phase and much higher rates may still be possible.

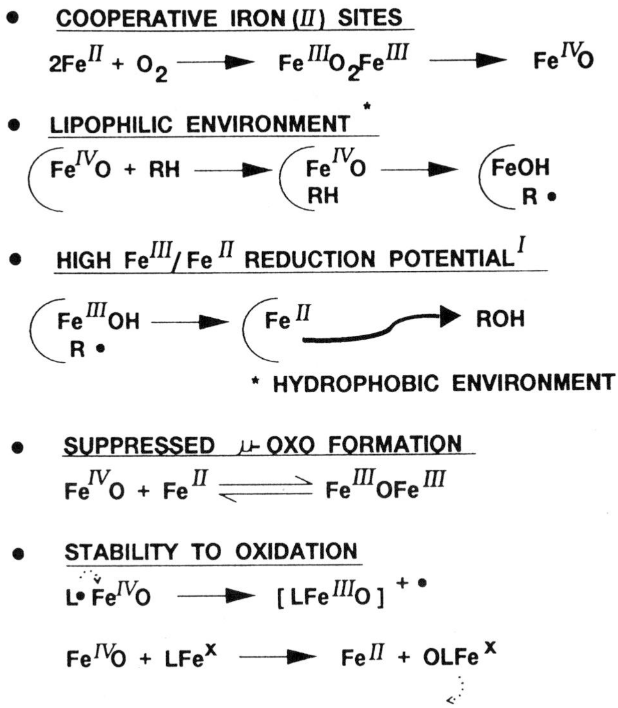

Figure 4 Properties of a suprabiotic catalyst.

While steric effects influence rate as well, we are finding that their prime importance is in reaction selectivity. Of importance here is steric inhibition of μ-oxo dimer formation while still allowing the reductive binding of dioxygen between two metalloporphyrin moieties. Ejecting formed alcohol from the coordination sphere is also of great importance as we have seen in both homogeneous and heterogeneous alkane oxidation. The greatest gains to date have been in tuning the redox potential for high activity, and there is still room for significant improvement here. In the future, more attention must be given to those aspects of catalyst structure that will affect selectivity to alcohol. Of all of these structural factors, one of the most important is hydrophobicity of the ligand environment.

It should be noted at this point that we are proposing a rather novel solution to the μ-oxo dimer problem. Most past efforts sought to prevent formation of this unreactive species solely by steric hindrance, which prevents the iron centers from getting close to one another [40,41]. Our approach is largely an electronic rather than a steric one. An implication of this approach is that even a μ-oxo complex which has been properly tuned electronically might provide a catalytically sufficient standing concentration of active metal–oxo.

An important aspect of catalyst design employed in this work is the utilization of a nonoxidizable ligand. Firstly the ligand cannot be oxidized by the very high oxidation states generated at the metal center. For example, if one attempts to form an iron(V)–oxo species by treatment of FeTPPCl with iodosylbenzene, one forms instead the iron(IV) radical cation by electron transfer from the porphyrin ligand [42]. The ligands used in this work must be resistant to this kind of electron transfer and provide a stable coordination sphere even for metals in very high oxidation states. It goes without saying that the ligands employed must be inert to autoxidation and other potentially destructive processes.

Finally, the alkane must be brought into the coordination sphere of the metal. Biology uses the strategy of the hydrophobic pocket. It should be realized, however, that biological systems operate in an aqueous medium. The complexes used for liquid phase oxidation in our work are hydrocarbon-soluble. Although affinity of the substrate for the catalyst is a desirable kinetic factor, this component is not easily built into a catalyst. Future generations of catalysts may address this opportunity.

IV. CATALYSIS BY ELECTRON-DEFICIENT METALLOPORPHYRINS

A. Effects of Axial Ligation

Initial studies [43] used tetraphenylporphyrinatometal(III) complexes of first-row transition metals including chromium, manganese, and iron as catalysts for the reaction of a series of light alkanes with molecular oxygen (Table 1). This series comprised three hydrocarbons having medium to strong carbon–hydrogen bonds: isobutane (94 kcal/mol), propane (96 kcal/mol), and ethane (98 kcal/mol). Isobutane oxidations were carried out at 80°C, propane was oxidized at 150°C, and ethane at 200°C. When the axial ligand, X, was chloride or acetate, no activity was observed for any of these hydrocarbons under these conditions. When the axial ligands were azide or nitride (Table 1), isobutane and propane were oxidized at an appreciable rate but ethane was unreactive. The macrocyclic ligand system, however, was not oxidatively stable and was rapidly destroyed during the reaction resulting in catalyst deactivation. The azide effect may be related to the ease of autoreduction of the azidometal(III) complex to give metal(II). The reason for the activity of manganese and chromium nitrido complexes is unclear.

Table 1 Oxidation of Isobutane Catalyzed by Metalloporphyrin Azide and Nitride Complexes[a]

Catalyst	Catalyst (mmol)	O_2 Uptake (mmol)[b]	Catalyst turnovers[c]	Products (mmol) Me_2CO	Products (mmol) *t*-BuOH	Selectivity[d] to *t*-BuOH (%)
Fe(TPP)Cl	0.025	0	0	0	0	–
Mn(TPP)Cl	0.025	0	0	0	0	–
Mn(TPP)Cl	0.050	0	0	0	0	–
Cr(TPP)Cl	0.025	0	0	0	0	–
Fe(TPP)N_3	0.013	1.7	130	0.2	2.2	92
Mn(TPP)N_3	0.013	2.3	177	0.4	3.0	88
Mn(TPP)N_3	0.025	5.5	220	0.7	6.2	90
Cr(TPP)N_3	0.013	3.4	262	0.4	3.3	89
Cr(TPP)N_3	0.025	6.6	264	0.8	6.3	89
[Fe(TPP)]$_2$N	0.013	0	0	0	0	–
Mn(TPP)N	0.025	0	0	0	0	–
Cr(TPP)N	0.013	0	0	0	0	–

[a]Isobutane (6–7 g) was added to a solution of the catalyst in benzene (25 ml). The solution was stirred under 100 psig of O_2 at 80°C for 6 h.
[b]Oxygen was added in 10 psig increments as consumed. Gas was analyzed by gas chromatography–mass spectrometry and corrected for small amounts of CO_2 formed. *Caution*: Any experiments run within the explosive limits must be properly barricaded.
[c]Mol oxygen consumed/mol catalyst used.
[d](Mol *t*-BuOH produced/total mol liquid product) × 100. The only products detected were the major product. *t*-BuOH, and the minor oxidative cleavage products, acetone and CO_2.
Source: Ref. 43.

B. Benefits of Substituting Electron-Withdrawing Groups for H on Catalyst Life and Activity

In order to make the macrocycle more oxidatively stable, all of the C-H bonds of the phenyl groups in the tetraphenylporphyrinato ligand were replaced with C-F bonds [44,45]. This had the effect of reducing the oxidizability of the ligand both by molecular oxygen and by the metal itself. Although modest gains in activity were observed for manganese and chromium porphyrin–catalyzed reactions, tetrakispentafluorophenylporphyrinatoiron complexes were more active by an order of magnitude than their unfluorinated counterparts (Table 2). In all cases the half-life of the catalysts improved as well.

It is possible that the superior results observed for the iron complexes were due to the high activity of an iron–oxo intermediate which was stabilized by a ligand with strong electron-withdrawing groups which prevented ligand oxidation and concomitant reduction of the iron. In addition, the extraordinarily high Fe(III)/Fe(II) reduction potential that these complexes

Table 2 Oxidation of Isobutane Catalyzed by Fluorinated Metalloporphyrin Complexes[a]

Complex	Amount used (mmol)	Catalyst turnovers[b]	Selectivity to *t*-BuOH[c]
Fe(TPP)Cl	0.025	0	–
$Fe(TPPF_{20})Cl$	0.016	2040	90
$Fe(TPP)N_3$	0.013	130	93
$Fe(TPPF_{20})N_3$	0.016	2060	89
$Mn(TPP)N_3$	0.013	180	88
$Mn(TPPF_{20})N_3$	0.016	750	87
$Cr(TPP)N_3$	0.025	280	89
$Cr(TPPF_{20})N_3$	0.016	450	97
$[Fe(TPP)]_2O$	0.013	0	–
$[Fe(TPPF_{20})]_2O$	0.007	1730[d]	92

[a]Isobutane (6–7 g) was added to a solution of the catalyst in 25 ml benzene. The solution was stirred under O_2 (100 psig) at 80°C for 6 h (O_2 added as used).
[b]Moles oxygen consumed/mole catalyst used.
[c](Moles *t*-BuOH produced/total moles liquid product) × 100. The only products detected were *t*-BuOH, acetone, and CO_2.
[d]Incomplete solubility of $[Fe(TPPF_{20})]_2O$ makes this a lower limit value.
Source: Ref. 44.

exhibit allows the generation of oxygen binding sites even under oxidizing conditions.

Chang and Ebina [12] first noted the increase in catalytic oxidation by increasing the electron withdrawal in the metalloporphyrin. Fe and $Mn(TPPF_{20})Cl$ were found to catalyze the PhIO-mediated oxidation of cyclohexane. The authors attributed a more electrophilic oxene as the reason for the increase. Nappa and Tolman [46] examined a series of FeTPP complexes and related the Hammett σ constants for phenyl substituents to the reaction rate. They saw enhanced oxidation activity due to electron-withdrawing groups with $Fe(TPPF_{20})$ complexes giving the highest yield.

Although significant oxidative stabilization of the ligand was achieved in these systems by replacing 20 hydrogens with fluorine atoms, there are still 8 more hydrogens attached to the pyrrole rings in the porphyrin. These present additional sites of oxidative instability which should be dealt with. One approach has been to replace these 8 hydrogens by halogenation. The perhaloporphyrin complexes were more robust *and* more active than the partially halogenated catalysts (Tables 3 and 4).

The beneficial effects of further halogenation at the β or pyrrolic positions of tetraphenylporphyrins have been demonstrated and exploited by

Table 3 Isobutane Oxidations Catalyzed by Perhaloporphyrin Complexes[a]

Complex	T (°C)	Turnovers[b]	*t*-BuOH Sel. (%)
$Fe(TPPF_{20}\beta\text{-}Br_8)Cl$	60	1890	91
	40	690	90
	27	130	90
$Fe(TPPF_{20}\beta\text{-}Br_8)N_3$	60	1550	87
	40	670	89
	27	620	NA
$Fe(TPPF_{20}\beta\text{-}Cl_8)Cl$	60	1800	90
	40	760	90
	27	220	91
$Fe(TPPF_{20}\beta\text{-}Cl_8)N_3$	60	1800	90
	40	760	92
$Cr(TPPF_{20}\beta\text{-}Br_8)Cl$	60	0	–
$Cr(TPPF_{20}\beta\text{-}Cl_8)Cl$	60	70	NA
$Cr(TPPF_{20}\beta\text{-}Br_8)N_3$	60	110	92
$Cr(TPPF_{20}\beta\text{-}Cl_8)N_3$	80	450	88
$Mn(TPPF_{20}\beta\text{-}Br_8)Cl$	60	110	87
$Mn(TPPF_{20}\beta\text{-}Cl_8)Cl$	60	0	–
$Mn(TPPF_{20}\beta\text{-}Br_8)N_3$	60	270	87
$Mn(TPPF_{20}\beta\text{-}Cl_8)Cl + NaN_3$[c]	80	1289	86
$Mn(TPPF_{20}\beta\text{-}Cl_8)Cl + NaN_3$[c]	60	156	88
$Co(TPPF_{20}\beta\text{-}Cl_8)$	60	355	91

[a]Isobutane (7 g) was oxidized (6 h) in 25 ml benzene at the designated temperature under 100 psig oxygen using 0.013 mmol of the designated catalyst.
[b]Moles of oxygen consumed/mol catalyst used.
[c]After warming the reaction mixture for 6 h as indicated in note a, the reaction mixture was cooled to room temperature, 10 mg of NaN_3 added, and the procedure repeated according to note a. Turnovers are calculated based on the second 6-h period.

several research groups. Traylor and Tsuchiya [47] brominated the β positions of Zn(TDCPP) ultimately forming $Fe(TDCPP\beta\text{-}Br_8)Cl$ after removal of zinc and insertion of iron. This complex was able to increase the C_6F_5IO-mediated oxidation rate of norbornane over the unbrominated Fe(TDCPP)Cl.

Dolphin and coworkers [48,49] examined chlorinated metalloporphyrins and found that increased chlorination of the ring led to increased rates of cyclohexane oxidation with C_6F_5IO. They were able to relate increased electron withdrawal from ring halogenation to the Fe(III)/Fe(II) reduction potential.

There are some examples in which increased halogenation of the porphyrin ring has not led to increased oxidation rates. Mansuy and coworkers [50] examined a group of electronegative porphyrins and the yields of alkane

Table 4 Propane Oxidations Using First-Row Metal Complexes of $TPPF_{20}\beta$-Br_8 as Catalysts[a]

Complex	Time (h)	T (°C)	Turnovers[b]	IPA/acetone[c]
$Fe(TPPF_{20})N_3$	3	125	330	0.8
	6	125	410	0.7
$Fe(TPPF_{20}\beta$-$Br_8)Cl$	4.5	125	420	1.0
$Fe(TPPF_{20}\beta$-$Br_8)Br$	4.5	125	470	0.8
$Fe(TPPF_{20}\beta$-$Br_8)N_3$	4.5	125	514	0.9
$Cr(TPPF_{20})N_3$	4.5	125	41	<0.1
$Cr(TPPF_{20}\beta$-$Br_8)Cl$	4.5	125	111	0.3
$Cr(TPPF_{20}\beta$-$Br_8)N_3$	4.5	125	87	0.6
$Mn(TPPF_{20})N_3$	3	125	0	—
$Mn(TPPF_{20}\beta$-$Br_8)Cl$	4.5	125	0	—
$Mn(TPPF_{20}\beta$-$Br_8)N_3$	4.5	125	87	1.0
$Mn(TPPF_{20}\beta$-$Br_8)F$	4.5	125	370	0.9

[a]Propane (1.36 mol) was added to benzene (48 ml) containing the catalyst (0.013 mmol). The solution was stirred for 3 h at 125°C under 1000 psig of air in a glass-lined autoclave. Liquids and gases were analyzed by gas chromatography. Production of carbon oxides never amounted to 10% of total products. Isopropyl alcohol and acetone exceeded 90 mol % of carbon-containing reaction products in all cases.
[b]Moles of acetone plus isopropyl alcohol formed per mol of catalyst used.
[c]Molar ratio isopropyl alcohol to acetone formed.

oxidation by PhIO followed the degree of halogenation in most cases. An exception to the trend was the observation that $Fe(TPPF_{20})Cl$ seemed to give higher yields than $Fe(TPPF_{20}\beta$-$Cl_8)Cl$. These workers propose that there is a greater tendency for the active species to behave like PFe(IV)-O· than P^+·Fe(IV)=O with increased electron withdrawal from the porphyrin rings.

Tsuchiya and coworkers [51,52] were able to perfluorinate iron porphyrin with CoF_3 in pyridine leading to $Fe(TPPF_{28})Cl$. They were able to oxidize benzene to phenol with H_2O_2 but surprisingly were unable to oxidize cyclohexane with H_2O_2.

Another approach to achieving catalyst stability is to switch the porphyrinic systems to the similar phthalocyanines. Success was achieved [38] in preparing perfluorophthalocyanato metal azides which have somewhat lower activity than the porphyrins but longer life. Long life is attributable both to the total lack of C-H bonds and to the presence of nitrogens in the bridge positions of the macrocycle. These catalysts do not produce alcohols in selectivity as high as do their porphyrin analogs.

V. ELECTRON-DEFICIENT METALLOPORPHYRINS AS CATALYSTS FOR REACTIONS WITH O_2

It has been determined [36,43–45,53] that by increasing the amount of electron withdrawal from either the meso or β (pyrrolic) position of the porphyrinato macrocycle surrounding an oxidation active first-row metal center, one can greatly increase the M(III)/M(II) reduction potential of the complex. This results in an enormous increase in light alkane oxidation activity and provides us for the first time with a family of metal complexes capable of catalyzing the direct selective air oxidation of these refractory materials. In studies [54,55] encompassing porphyrin complexes of the metals iron, manganese, and chromium, we have shown that the catalytic activity responds positively to electron withdrawal in both the meso and the β positions. It has further been shown that the light alkane air oxidation activity of the iron complexes of electron-deficient porphyrins is superior to that of the other metals.

A. Perhalogenated Metalloporphyrin Catalysts for Combining Alkanes and O_2

Figure 5 shows the porphyrinato macrocycle with the meso positions labeled Z and the β positions labeled Y. Our work has shown that while iron complexes of the macrocyclic ligand wherein Z is phenyl (Y = H) have very little catalytic activity for air oxidations of light alkanes at low temperature, those in which Z is pentafluorophenyl (Y = H) exhibit high catalytic activity. Changing the Y group from H to X, wherein X is a halogen such as chlorine or bromine, causes further increase in activity. At the time of their discovery, perhalogenated iron complexes were more active than any of the then known metal complexes for light alkane oxidation by at least an order of magnitude

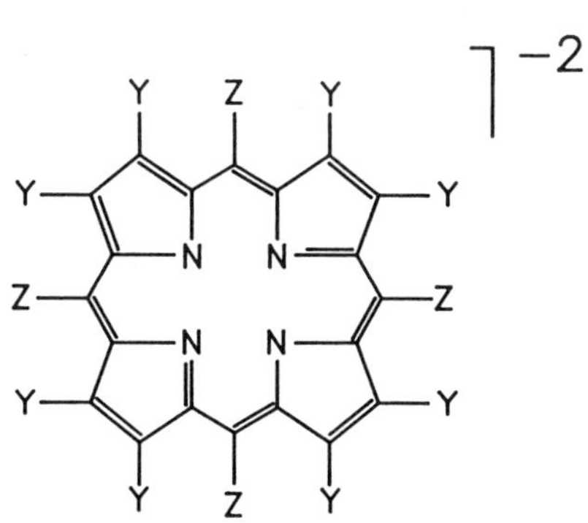

Figure 5 Substituted porphyrin macrocycle.

Table 5 Iron Haloporphyrin–Catalyzed Isobutane Oxidations[a]

Catalyst	T (°C)	t (h)	i-C_4H_{10} (mmol)	O_2 (psi)	Reaction products (mmol) t-BuOH	Acetone	CO	CO_2	Conversion i-C_4H_{10} (%)	Sel. (%) t-BuOH[b]	TON[c]
$Fe(TPPF_{20})OH$	80	3	1869	53	310	46	7	18	17	87	11,330
	24	143	1871	53	382	17	0	18	19	95	11,600
$Fe(TPPF_{20}\beta$-$Cl_8)Cl$	80	3	1865	142	334	38	6	29	20	88	12,400
	21	120	1865	63	366	16	tr	na	20.5	95	12,730
$Fe(TPPF_{20}\beta$-$Br_8)Cl$	80	3	1862	148	414	81	6	28	27	84	16,500
	80	3	1870	53	277	43	8	23	17	87	10,060
	24	71	1862	53	372	35	tr	27	22	92	12,150

[a]Isobutane was oxidized by an oxygen-containing gas mixture (75 atm diluent–N_2) in the liquid phase (180 ml) for 3 h. Oxygen added as consumed.
[b](moles t-BuOH/mol liquid product) × 100.
[c]Moles (t-BuOH and acetone) produced/mol catalyst used.

and made unprecedented selective low-temperature alkane oxidations possible (Table 5). Alcohols are major reaction products and no expensive oxidants or coreductants are required using these catalysts.

Table 6 compares the isobutane and propane oxidation activity of the series of iron porphyrin complexes described in Figure 5. Although these complexes were the most active known, and perhalogenation had created the most stable porphyrin complexes yet prepared, even more stable and active catalysts are still being generated by further enhancing the electron-withdrawing power of the peripheral substituents on the porphyrin ring and by adjusting the geometries of the metal complexes [61].

B. Electron Withdrawal, Reduction Potential, and Catalytic Activity

In the past, the very critical determinations of Fe(III)/Fe(II) reduction potential had been done in different laboratories and not always under identical conditions. Because of the apparent direct relationship between reduction potential and catalytic activity of iron porphyrin complexes we reinvestigated this area. The Fe(III)/Fe(II) redox potentials vs. SCE for the series of iron porphyrins have been determined under a set of consistent conditions.

Table 6 Alkane Oxidations Catalyzed by Iron Porphyrins, $Fe(TPPZ_{20}\beta\text{-}Y_8)X^{a,b}$

Complex								
Z	Y	X	Substrate	t (h)	T (°C)	TON[c]	Selectivity to *t*-BuOH[d]	IPA/acetone
H	H	Cl	Isobutane	6	60	0	—	—
H	Cl	Cl	Isobutane	6	60	0	—	—
H	Br	Cl	Isobutane	6	60	110	88	—
F	H	Cl	Isobutane	6	60	1160	90	—
			Propane	4.5	125	390	—	0.8
F	Cl	Cl	Isobutane	6	60	1800	90	—
F	Br	Cl	Isobutane	6	60	1890	91	—
			Propane	4.5	125	420	—	1.0
F	Br	Br	Propane	4.5	125	470	—	0.8
F	Br	N_3	Propane	4.5	125	514	—	0.9

[a]A solution of 0.013 mmol of the catalyst in 25 ml benzene containing 6 g of isobutane was stirred at the designated temperature under 100 psig of O_2 for the designated time.
[b]Stirred solution of 0.023 mmol of the catalyst in 60 g of propane in 48 ml benzene at 125°C under 1000 psig air.
[c]Moles *tert*-butyl alcohol produced/mol catalyst; or mol (isopropyl alcohol + acetone) formed/mol catalyst used.
[d](Moles *tert*-butyl alcohol produced/total mol liquid product) × 100.

The CV's were all obtained in CH_2Cl_2 and tetrabutylammonium chloride was used as supporting electrolyte. A glassy carbon electrode was used. Table 7 lists the half-wave potentials determined for several of the compounds in the series.

As the C-H bonds around the periphery of the tetraphenylporphyrinato ligand system are changed to C-X bonds (X = F, Cl, Br), the Fe(III)/(II) reduction potential increases. Table 7 shows that the catalytically inactive parent iron porphyrin, Fe(TPP)Cl, exhibits an Fe(III)/(II) half-wave potential at −0.22 V whereas the $E_{1/2}$ exhibited by the highly active perhaloporphyrin complex, Fe(TPPF$_{20}\beta$-Cl$_8$)Cl, is fully 0.5 V higher. It appears that there is a direct relationship between the extent of electron withdrawal from the porphyrin ring, which causes an increase in the Fe(III)/(II) reduction potential, and the catalytic activity of the iron complex (Figure 6).

C. Porphyrin Ring Buckling, Steric Hindrance, and Catalytic Activity

What was surprising about Table 7, however, was that the catalytic activity of Fe(TPPF$_{20}\beta$-Br$_8$)Cl was not lower than that of Fe(TPPF$_{20}\beta$-Cl$_8$)Cl. Structural information currently being gathered indicates that the steric bulk of the β-bromo substituents may prevent access of large molecules to the central iron atom. This has two possible consequences which could affect reaction rate. First, a sterically hindered porphyrin might bind dioxygen to give a μ-peroxo complex which could cleave to form an active ferryl–oxo species. It would be unlikely that a highly hindered diiron-μ-oxo complex, however, would be very stable if it is formed at all. To the extent that ferryl–oxo complexes are active intermediates and that diiron-μ-oxo species are not, a steric environment which prevents μ-oxo formation could cause an oxidation rate enhancement.

Table 7 Relationship Between Catalyst Reduction Potential and Activity

Catalyst	Fe(III)/Fe(II) $E_{1/2}$ (V)[a]	Turnovers[b]
Fe(TPP)Cl	−0.221	0
Fe(TPPF$_{20}$)Cl	+0.07	1160
Fe(TPPF$_{20}\beta$-Br$_8$)Cl	+0.19	1890
Fe(TPPF$_{20}\beta$-Cl$_8$)Cl	+0.28	1800

[a]Cyclic voltammetry in CH_2Cl_2 vs. SCE. Tetrabutylammonium chloride supporting electrolyte, glassy carbon electrode.

[b]Catalyst turnovers for 60°C isobutane oxidations in benzene.

Source: Ref. 69.

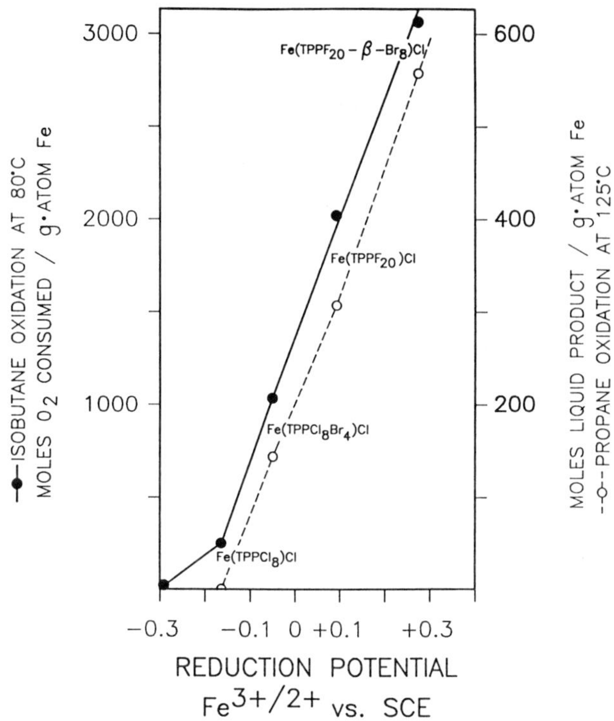

Figure 6 Catalyst activity vs. redox potential.

Structural features of electron-deficient porphyrins also appear to play an important role in catalytic air oxidation reactions. In particular, steric interactions which could limit or prevent μ-oxo dimer formation may be important for high catalytic activity. We have identified two different types of steric interactions which may be important. One way in which μ-oxo dimer formation could be suppressed is through severe buckling of the porphyrin macrocycle due to steric bulk in the β positions. It was seen that as the bulkiness of the β substituent increases from H to Cl to Br, the extent of buckling increases. It would seem difficult to make the μ-oxo species when the β positions bear eight bulky bromines.

Second, large electron-withdrawing groups such as chlorines in the ortho position of the phenyl rings of tetraphenylporphyrinato complexes make μ-oxo dimer formation difficult at best. The Traylors and Dolphin [41] prepared Fe(TDCPP)Cl and Fe($TPPCl_{20}$)Cl, and found them too sterically hindered to form μ-oxo dimers, instead forming monomeric hydroxide complexes when treated with base.

In order to compare steric effects on catalytic activity, we studied two perhaloporphyrin complexes having similar electrochemical properties but differing greatly in steric bulk. The complexes—$Fe(TPPCl_{28})Cl$ and $Fe(TPPF_{20}\beta\text{-}Cl_8)Cl$—have similar electronic properties [$E_{1/2}$ for Fe(III)/Fe(II) = 0.27 and 0.28, respectively (49)] but access to the iron center in $Fe(TPPCl_{28})Cl$ is far more restricted. It is possible that $Fe(TPPF_{20}\beta\text{-}Cl_8)Cl$ could form a μ-oxo dimer. On the other hand, it is not likely that $Fe(TPPCl_{28})Cl$ could form a μ-oxo dimer because of the huge nonbonded interactions resulting from the eight *ortho*-chloro substituents on each macrocycle, which cause significant buckling of the macrocycle.

D. Oxidation of Alkanes Using $Fe(TPPCl_{28})Cl$ as Catalyst

We attempted the reaction of isobutane (20%) in benzene with molecular oxygen under standard test conditions (100 psig O_2, 6 h) at 40, 60, and 80°C using $Fe(TPPCl_{28})Cl$ as catalyst. *No reaction occurred.* Under these conditions nearly 2000 turnovers were observed at 60°C using $Fe(TPPF_{20}\beta\text{-}Cl_8)Cl$ as the catalyst. One interpretation of these results is that not only is the μ-oxo dimer, $[Fe(TPPCl_{28})]_2O$, unable to form, but even the μ-peroxo dimer, $[Fe(TPPCl_{28})]_2O_2$, cannot be made. Inspection of the molecular model of the monomeric Fe(II) complex indicates that the iron is surrounded by a nearly symmetric barrier of chlorine atoms. Thus, it is possible that steric hindrance about the iron centers is so great that this complex may not be able to reductively bind dioxygen.

Small-oxygen-atom transfer reagents, however, should still be able to oxidize the iron center to a ferryl intermediate which will oxidize hydrocarbons. It has been found [48] that $Fe(TPPCl_{28})Cl$ oxidizes cyclohexane to the alcohol and the ketone and oxidizes cyclohexene to the epoxide when iodosylbenzene is the oxidant.

Thus, when a small oxygen donor molecule having oxygen in the reductively bound state is used, oxidations, presumably occurring through an active ferryl intermediate, are possible. When it is necessary to reductively bind a dioxygen molecule in order to get the active catalytic intermediate, steric bulk may prevent this reaction from occurring.

Therefore, it is possible that we may now have spanned the range of steric requirements for both μ-oxo and μ-peroxo complex formation. It would appear that a properly tuned iron complex which can readily form a μ-peroxo dimer but which cannot form a μ-oxo dimer should give the greatest concentration of active ferryl and thus should be the best catalyst.

VI. FUTURE SUPRABIOTIC CATALYSTS

We have shown that catalysts such as $Fe(TPPF_{20}\beta\text{-}Br_8)Cl$ and $Fe(TPPF_{20}\beta\text{-}Cl_8)Cl$ are superior air oxidation catalysts able to selectively transform alkanes containing tertiary C-H bonds such as isobutane and those with secondary C-H bonds such as propane to alcohols. In order to be commercially attractive the rates of these reactions could be faster. Although highly electron-deficient porphyrinato complexes could give highly enhanced autoxidation behavior, our work suggests the possibilities of suprabiotic pathways as well. In order to attack even less reactive alkanes such as ethane and methane, more active catalysts are needed. Can we guess what increasingly active suprabiotic catalysts will look like?

What have we learned at this stage from studies of suprabiotic catalysts? What properties do they possess? We see that several properties are important to the success of a suprabiotic catalyst:

1. Electron withdrawal from the metal center via electron withdrawing groups on the periphery of the macrocycle resulting in high M(III)/M(II) reduction potential
2. Removal of reactive C-H bonds for increased oxidative stability
3. Sufficient solubility and compatibility with alkane substrate and alcohol product
4. Geometric properties which encourage reductive binding of dioxygen to form binuclear μ-peroxo species, and
5. Steric constraints which favor M=O intermediates over binuclear μ-oxo complexes (without preventing reductive binding of O_2).

A. Increasing Electron Withdrawal

Halogenation of groups at the β and meso positions of metalloporphyrins has been successfully used to increase the electron withdrawal from the metal center leading to superior oxidation catalysts. A tabulation of electron-withdrawing capabilities of functional groups such as one of Hammett σ parameters (Table 8), quickly suggests the use of even more potent electron-withdrawing groups than halogens. Porphyrins having C-H bonds replaced with CF_3 groups [56,57], CN groups [58,59], and NO_2 groups [60,61] have been prepared.

We have shown [62] that NO_2 reacts with Fe(OEP)Cl to give Fe(OEP)$(NO_2)_4$Cl and the μ-oxo species $[Fe(OEP)(NO_2)_4]_2O$. This μ-oxo species is able to catalyze hydroxylation of isobutane with 5 bars of O_2 at 60° and 80°C giving 800 and 1600 turnovers respectively in 5 h in benzene. At room temperature 1000 turnovers were achieved in 48 h. The selectivity to *tert*-butyl alcohol is about 90% in each case. Fe(OEP)Cl or its μ-oxo analog is inactive even at 80°C. Clearly nitration at the four meso positions has overcome the effects of the eight electron-donating ethyl groups. Stimulated by recent results in this field, further examples of suprabiotic catalysts enhanced by powerful electron-withdrawing groups will no doubt soon be inundating the literature.

Table 8 Values for Common Functional Groups

Group	σ_p
OMe	−0.28
Me	−0.14
H	0
Ph	0.05
F	0.15
Cl	0.24
Br	0.26
I	0.28
CF_3	0.53
CN	0.70
NO_2	0.81

Source: The σ_p values were excerpted from J. March, *Advanced Organic Chemistry*, 4th Ed., Wiley, New York, 1992, p. 280.

B. Geometric Control Over Catalysis

We have seen that the geometry of the porphyrin around the metal center is important to its catalytic power. Inability to form a μ-peroxo dimer could explain the inactivity of a potentially powerful suprabiotic catalyst such as $Fe(TPPCl_{28})Cl$.

Control over the porphyrin architecture could provide catalysts which position two irons in proximity for formation of μ-peroxo dimers and homolytic cleavage leading to two M=O centers (Eq. 4) (55). Several workers including Chang et al. [63] and Collman et al. [64] developed cofacial diporphyrins, variations of which may someday be able to perform this task.

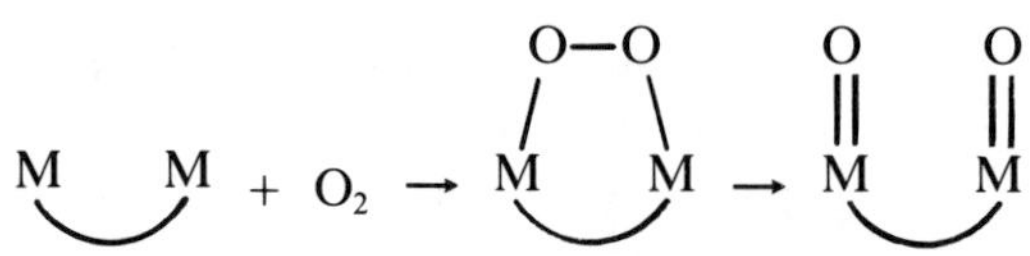

(4)

Other workers such as Herron, Stucky, and Tolman [65] have isolated metal phthalocyanines in the cages of zeolites to prepare oxidation catalysts (although inactive with O_2). Inert supports have also been used to disperse metalloporphyrins [66] and metallophthalocyanines [67]. This approach seems to be a productive one in instances where the active intermediate is a superoxo species, such as in the Merox process for sulfur oxidation [68]. In the case of liquid phase alkane hydroxylation using O_2 as the oxidant, it may be necessary to form a μ-peroxo dimer in order to split dioxygen and form the more highly active ferryl species. Since supported mononuclear metalloporphyrins may have difficulty forming μ-peroxo dimers, it might be preferable either to use soluble monomeric metalloporphyrins which can react with O_2 to form μ-peroxo dimers and ultimately ferryl species, or to use proximate metal centers such as those in the aforementioned cofacial diporphyrins.

VII. CONCLUSION

No doubt the development of suprabiotic catalysts will lead to increasingly active and perhaps commercially important air/oxygen oxidation catalysts. The increased availability of porphyrin synthetic routes will speed this advance. The need for oxygenated fuels and fuel additives as well as the possibilities of new applications for suprabiotic catalysts in fine chemical production and even in pollution abatement should make this an important area of research for many years.

ABBREVIATIONS

TPP	*meso*-tetraphenylporphyrin dianion
TDCPP	*meso*-tetrakis(2,6-dichlorophenyl)porphyrin dianion
$TPPF_{20}$	*meso*-tetrakis(pentafluorophenyl)porphyrin dianion
TDCPPβ-Br_8	*meso*-tetrakis(2,6-dichlorophenyl)-β-octabromoporphyrin dianion
$TPPF_{20}\beta$-Cl_8	*meso*-tetrakis(pentafluorophenyl)-β-octachloroporphyrin dianion
$TPPF_{28}$	*meso*-tetrakis(pentafluorophenyl)-β-octafluoroporphyrin dianion
$TPPF_{20}\beta$-Br_8	*meso*-tetrakis(pentafluorophenyl)-β-octabromoporphyrin dianion
$TPPCl_{20}$	*meso*-tetrakis(pentachlorophenyl)porphyrin dianion
$TPPCl_{28}$	*meso*-tetrakis(pentachlorophenyl)-β-octachloroporphyrin dianion
OEP	β-octaethylporphyrin dianion
$OEP(NO_2)_4$	*meso*-tetranitro-β-octaethylporphyrin dianion

ACKNOWLEDGMENTS

The authors acknowledge the helpful contributions of Prof. Harry B. Gray. The financial assistance of the U.S. Department of Energy (Morgantown Energy Technology Center), the Gas Research Institute, and Sun Company for supporting the work carried out in Sun Laboratories that is reported here is greatly appreciated. We are grateful to Dr. Tilak P. Wijesekera for providing a sample of $Fe(TPPCl_{28})Cl$.

REFERENCES

1. J.T. Groves, Cytochrome P-450 and other heme-containing oxygenases, *Adv. Inorg. Biochem., 1*: 119 (1979).
2. R.E. White and M.J. Coon, Oxygen activation by cytochrome P-450, *Annu. Rev. Biochem., 49*: 315 (1980).
3. F.P. Guengerich and T. MacDonald, Chemical mechanisms of catalysis by cytochromes P-450: a unified view, *Acc. Chem. Res., 17*: 9 (1984).
4. (a) G.A. Hamilton, Chemical models and mechanisms for oxygenases, in *Molecular Mechanisms of Oxygen Activation* (O. Hayaishi, ed.), Academic Press, New York, 1974, p. 405. (b) P.R. Ortiz de Montellano, Oxygen activation and transfer, in *Cytochrome P-450* (P.R. Ortiz de Montellano, ed.), Plenum Press, New York, 1986, pp. 235–239.
5. E.J. McKenna and M.J. Coon, Enzymic ω-oxidation, IV. Purification and properties of the ω-hydroxylase of *Pseudomonas oleovorans, J. Biol. Chem., 245*: 3882 (1970).
6. J. Colby and H. Dalton, Some properties of a soluble methane monooxygenase from *Methylococcus capsulatus* strain Bath, *Biochem. J., 157*: 495 (1976).
7. M.P. Woodland and H.J. Dalton, Purification and characterization of component A of the methane monooxygenase from *Methylococcus capsulatus* (Bath), *J. Biol. Chem., 259*: 53 (1984).
8. J.T. Groves, T.E. Nemo, and R.S. Myers, Aliphatic hydroxylation via oxygen rebound. Oxygen transfer catalyzed by iron, *J. Am. Chem. Soc., 98*: 859 (1976).
9. J.T. Groves, T.E. Nemo, and R.S. Myers, Hydroxylation and expoxidation catalyzed by iron-porphine complexes. Oxygen transfer from iodosylbenzene, *J. Am. Chem. Soc., 101*: 1032 (1979).
10. D. Mansuy, J.F. Bartoli, J.C. Chottard, and M. Lange, Metalloporphyrin-catalyzed hydroxylation of cylohexane by alkyl hydroperoxides: pronounced efficiency of iron-porphyrins, *Angew. Chem. Int. Ed. Eng., 19*: 909 (1980).
11. C.L. Hill and B.C. Schardt, Alkane activation and functionalization under mild conditions by a homogeneous manganese (III) porphyrin-iodosylbenzene oxidizing system, *J. Am. Chem. Soc., 102*: 6374 (1980).
12. C.K. Chang and F. Ebina, NIH shift in haemin-iodosylbenzene-mediated hydroxylations, *J. Chem. Soc. Chem. Commun.,* 778 (1981).
13. J.T. Groves and T.E. Nemo, Aliphatic hydroxylation catalyzed by iron porphyrin complexes, *J. Am. Chem. Soc., 105*: 6243 (1983).
14. B. DePorter, M. Ricci, O. Bortolino, and B. Meunier, Catalytic hydroxylation of saturated hydrocarbons with the sodium hypohalite/manganese porphyrin system, *J. Mol. Cat., 31*: 221 (1985).
15. J.T. Groves, G.A. McClusky, R.E. White, and M.J. Coon, Aliphatic hydroxylation by highly purified liver microsomal cytochrome P-450. Evidence for a carbon radical intermediate, *Biochem. Biophys. Res. Commun., 81*: 154 (1978).
16. G. Cardini and P. Jurtshuk, Cytochrome P-450 involvement in the oxidation of n-octane by cell-free extracts of *Corynebacterium* species strain 7E1C, *J. Biol. Chem., 243*: 6070 (1968).
17. G. Cardini and P. Jurtshuk, Enzymic hydroxylation of n-octane by *Corynebacterium* sp. strain 7E1C, *J. Biol. Chem., 245*: 2789 (1970).

18. K. Morohashi, H. Sadano, Y. Okada, and T. Omura, Position specificity in n-hexane hydroxylation by two forms of cytochrome P-450 in rat liver microsomes, *J. Biochem., 93*: 413 (1983).
19. S.G. Sligar and R.I. Murray, Cytochrome P-450cam and other bacterial P-450 enzymes, in *Cytochrome P-450* (P.R. Ortiz de Montellano, ed.), Plenum Press, New York, 1986, p. 478.
20. J.T. Groves and W.J. Kruper, Jr., Preparation and characterization of an oxoporphinatochromium (V) complex, *J. Am. Chem. Soc., 101*: 7613 (1979).
21. B. Meunier, Metalloporphyrins as versatile catalysts for oxidation reactions and oxidative DNA cleavage, *Chem. Rev., 92*: 1411 (1992).
22. I. Tabushi and A. Yazaki, P-450-type dioxygen activation using H_2/colloidal Pt as an effective electron donor, *J. Am. Chem. Soc., 103*: 7371 (1981).
23. I. Tabushi and N. Koga, P-450 type oxygen activation by porphyrin–manganese complex, *J. Am. Chem. Soc., 101*: 6456 (1979).
24. I. Tabushi, Reductive dioxygen activation by use of artificial P-450 systems, *Coord. Chem. Rev., 86*: 1 (1988).
25. D. Mansuy, M. Fontecave, and J.F. Bartoli, Mono-oxygenase-like dioxygen activation leading to alkane hydroxylation and olefin epoxidation by an Mn(III) (porphyrin)-ascorbate biphasic system, *J. Chem. Soc. Chem. Commun.,* 253 (1983).
26. M. Fontecave and D. Mansuy, Monoxygenase-like oxidation of olefins and alkanes catalyzed by manganese porphyrins: comparison of systems involving either O_2 and ascorbate or iodosylbenzene, *Tetrahedron, 40*: 4297 (1984).
27. L. Ji, M. Liu, A.K. Hsieh, and T.S.A. Hor, Metalloporphyrin-catalyzed hydroxylation of cyclohexane with molecular oxygen, *J. Mol. Cat., 70*: 247 (1991).
28. R. Iwanejko, T. Mlodnicka, and J. Poltowicz, Metalloporphyrin-catalyzed oxidation of cyclohexane with dioxygen, in *New Developments in Selective Oxidation* (G. Centi and F. Trifiro, eds.), Elsevier, Amsterdam, 1990, p. 195.
29. P. Battioni, J.E. Bartoli, P. Leduc, M. Fontecave, and D. Mansuy, A new and efficient biomimetic system for hydrocarbon oxidation by dioxygen using manganese porphyrins, imidazole, and zinc, *J. Chem. Soc. Chem. Commun.,* 791 (1987).
30. E.I. Karasevich, A.M. Khenkin, and A.E. Shilov, A chemical model of cytochrome P-450: mono-oxygenase-like activation of dioxygen, *J. Chem. Soc. Chem. Commun.,* 731 (1987).
31. G.B. Shulpin and A.N. Druzhinina, Biomimetic activation of the C-H bond. Oxygenation of hydrocarbons with O_2 catalyzed by porphyrin complexes of metals in the presence of ferrocene as a reducing agent, *Izv. Akad. Nauk SSSR, Ser. Khim.,* 2739 (1991).
32. P. Leduc, P. Battioni, J.F. Bartoli, and D. Mansuy, A biomimetic electrochemical system for the oxidation of hydrocarbons by dioxygen catalyzed by manganese-porphyrins and imidazole, *Tetrahedron Lett., 29*: 205 (1988).
33. F. Bedioui, S.G. Granados, and J. Devynck, Biomimetic oxidation of hydrocarbons by dioxygen: electrocatalysis using polypyrrole-manganese porphyrin film modified electrodes, *New J. Chem., 15*: 939 (1991).
34. D.N. Hendrickson, M.G. Kinnaird, and K.S. Suslick, Photochemistry of (5, 10, 15, 20-tetraphenylporphyrinato)iron(III) halide complexes, Fe (TPP)(X), *J. Am. Chem. Soc., 109*: 1243 (1987).

35. A. Maldotti, C. Bartocci, R. Amadelli, E. Polo, P. Battioni, and D. Mansuy, Oxidation of alkanes catalyzed by photoactivated iron porphyrins, *J. Chem. Soc. Chem. Commun.,* 1487 (1991).
36. P.E. Ellis, Jr., and J.E. Lyons, Selective air oxidation of light alkanes catalyzed by activated metalloporphyrins—the search for a suprabiotic system, *Coord. Chem. Rev., 105*: 181 (1990).
37. D.R. Paulson, R. Ullman, and R.B. Sloane, Catalysis of autoxidation by metalloporphyrins, *J. Chem. Soc. Chem. Commun.,* 186 (1974).
38. J.E. Lyons and P.E. Ellis, Jr., Azide activation of metallophthalocyanine complexes for the catalytic oxidation of alkanes in the liquid phase, *Appl. Cat. A: Gen., 84*: L1 (1992).
39. L. Weber, M. Grosche, H. Hennig, and G. Haufe, Oxygenation of alkenes with phthalocyaninato manganese (III) and iron (III) complexes and dioxygen, *J. Mol. Cat., 78*: 19 (1993).
40. A.L. Balch, Y.W. Chan, R.J. Cheng, G.N. Lamar, L. Latos-Grazynski, and M.W. Renner, Oxygenation patterns for iron (II) porphyrins. Peroxo and ferryl ($Fe^{IV}O$) intermediates detected by ^1H nuclear magnetic resonance spectroscopy during the oxygenation of (tetramesitylporphyrin)iron(II), *J. Am. Chem. Soc., 106*: 7779 (1984).
41. P.S. Traylor, D. Dolphin, and T.G. Traylor, Sterically protected hemins with electronegative substituents: efficient catalysts for hydroxylation and epoxidation, *J. Chem. Soc. Chem. Commun.,* 279 (1984).
42. A.L. Balch, L. Latos-Grazynski, and M.W. Renner, Oxidation of red ferryl [$(Fe^{IV}O)^{2+}$] porphyrin complexes to green ferryl [$(Fe^{IV}O)^{2+}$] porphyrin radical complexes, *J. Am. Chem. Soc., 107*: 2983 (1985).
43. P.E. Ellis, Jr., and J.E. Lyons, Effect of axial axide on the selective, low-temperature metalloporphyrin-catalysed reactions of isobutane with molecular oxygen, *J. Chem. Soc. Commun.,* 1187 (1989).
44. P.E. Ellis, Jr., and J.E. Lyons, Effect of fluorination of the meso-phenyl groups on selective tetraphenylporphyrinatometal (II)-catalysed reactions of isobutane with molecular oxygen, *J. Chem. Soc. Chem. Commun.,* 1189 (1989).
45. P.E. Ellis, Jr., and J.E. Lyons, Effect of fluorination of the meso-phenyl groups on selective tetraphenylporphyrinatoiron (II)-catalysed reactions of propane with molecular oxygen, *J. Chem. Soc. Chem. Commun.,* 1315 (1989).
46. M.J. Nappa and C.A. Tolman, Steric and electronic control of iron porphyrin catalyzed hydrocarbon oxidations, *Inorg. Chem., 24*: 4711 (1985).
47. T.G. Graylor and S. Tsuchiya, Perhalogenated tetraphenylhemins: stable catalysts of high turnover catalytic hydroxylations, *Inorg. Chem., 26*: 1338 (1987).
48. D.H. Dolphin, T. Nakano, T.K. Kirk, T.E. Maione, R.L. Farrell, and T.P. Wijesekera, Porphyrins, their syntheses and uses thereof, World Int. Patent WO88/07988, October 20, 1988.
49. T. Wijesekera, A. Matsumoto, D. Dolphin, and D. Lexa, Perchlorinated and highly chlorinated meso-tetraphenylporphyrins, *Angew. Chem. Int. Ed. Engl., 29*: 1028 (1990).

50. J.F. Bartoli, O. Brigaud, P. Battioni, and D. Mansuy, Hydroxylation of linear alkanes catalyzed by iron porphyrins: particular efficacy and regioselectivity of perhalogenated porphyrins, *J. Chem. Soc. Chem. Commun.,* 440 (1991).
51. S. Tsuchiya, M. Seno, and M. Kawai, Fluorine-substituted porphyrin complexes, *Jpn. Kokai* 90-250883, Oct. 8 (1990).
52. S. Tsuchiya and M. Seno, Novel synthetic method of phenol from benzene catalyzed by perfluorinated hemin, *Chem. Lett.,* 263 (1989).
53. P.E. Ellis, Jr., and J.E. Lyons, Halogen substituent effects on the catalytic activity of iron porphyrin complexes for selective air-oxidation of alkanes in the liquid phase, *Cat. Lett., 3*: 389 (1989).
54. J.E. Lyons and P.E. Ellis, Jr., Selective low-temperature hydroxylation of isobutane by molecular oxygen catalyzed by an iron perhaloporphyrin complex, *Cat. Lett., 8*: 45 (1991).
55. J.E. Lyons, P.E. Ellis, Jr., and V.A. Durante, Active iron oxo centers for the selective catalytic oxidation of alkanes, in *Structure–Activity Relationships in Heterogeneous Catalysis* (R.E. Grasselli and A.W. Sleight, eds.), Elsevier, Amsterdam, 1991, pp. 99–116.
56. N. Ono, H. Kawamura, and K. Maruyama, A convenient synthesis of trifluoromethylated pyrroles and porphyrins, *Bull. Chem. Soc. Jpn., 62*: 3386 (1989).
57. P.E. Ellis, Jr., and J.E. Lyons, Chromium halogenated coordination complexes for the oxidation of butane to methylethylketone, U.S. Patent 4,970,348, Nov. 13, 1990.
58. H.J. Callot, A. Giraudeau, and M. Gross, Cyanoporphyrins. Co-ordinating and electrochemical properties, *J. Chem. Soc. Perkin Trans. 2*, 1321 (1975).
59. P.E. Ellis, Jr., and J.E. Lyons, Cyano- and polycyanometalloporphyrins as catalysts for alkane oxidation, U.S. Patent 5,118,886, June 2, 1992.
60. E. Watanabe, S. Nishimura, H. Ogoshi, and Z. Yoshida, Orientation of electrophilic meso-substitution in metallooctaethylporphyrins, *Tetrahedron, 31*: 1385 (1975).
61. P.E. Ellis, Jr., and J.E. Lyons, Nitrated metalloporphyrins as catalysts for alkane oxidation, U.S. Patent 5,120,882, June 9, 1992.
62. P.E. Ellis, Jr., and J.E. Lyons, Nitrated metallooctaethylporphines as catalysts for the oxidation of alkanes with molecular oxygen, *Cat. Lett.,* in press.
63. C.K. Chang, M.S. Kuo, and C.B. Wang, Stacked double-macrocyclic ligands II. Synthesis of cofacial diporphyrins, *J. Heterocyclic Chem., 14*: 943 (1977).
64. J.P. Collman, J.E. Hutchinson, M.A. Lopez, A. Tabard, R. Guilard, W.K. Seok, J.A. Ibers, and M. L'Her, Synthesis and characterization of a superoxo complex of the dicobalt cofacial diporphyrin $[(\mu\text{-}O_2)Co_2(DPB)(1,5\text{-diphenylimidazole})_2][PF_6]$, the structure of the parent dicobalt diporphyrin $Co_2(DPB)$, and a new synthesis of the free-base cofacial diporphyrin $H_4(DPB)$, *J. Am. Chem. Soc., 114*: 9869 (1992).
65. N. Herron, G.D. Stucky, and C.A. Tolman, Shape selectivity in hydrocarbon oxidations using zeolite encapsulated iron phthalocyanine catalysts, *J. Chem. Soc. Chem. Commun.,* 1521 (1986).

66. H. Kameyama, H. Suzuki, and A. Amano, Intercalation of Co(II) meso-tetrakis (1-methyl-4-pyridyl)porphyrin into montmorillonite, *Chem. Lett.,* 1117 (1988).
67. M. Chibwe and T.J. Pinnavaia, Stabilization of a cobalt(II)phthalocyanine oxidation catalyst by intercalaction in a layered double hydroxide host, *J. Chem. Soc. Chem. Common.,* 278 (1993).
68. H. Fischer, G. Schultz-Ekloff, T. Buck, D. Wohrle, M. Vassileva, and A. Andreev, Mercaptan adsorption capacity and catalytic oxidation activity of silica-supported phthalocyanines, *Langmuir, 8*: 2720 (1992).
69. J.E. Lyons, P.E. Ellis, R.W. Wagner, R.E. Thompson, M.E. Hughes, J.A. Hodge, and H.B. Gray, Symposium, *Am. Chem. Soc. Div. Petroleum Chem.,* April, 1992.

11
Supported Metalloporphyrin Models for Heme-Catalyzed Oxidations

John R. Lindsay Smith
University of York, York, England

I. BACKGROUND

Enzymes catalyze a wide range of reactions under mild conditions and often with high selectivity. Understanding their mechanisms and attempting to mimic their reactions with chemical systems has provided chemists with important challenges and research goals. In this respect the monooxygenases and, in particular, the cytochrome P450 monooxygenases have attracted much interest in recent years [1–4].

The challenge has been to discover how these monooxygenases activate dioxygen to bring about the controlled monooxygenation of organic compounds (Reaction 1). Great advances have been made in our understanding of these processes through studies on the biological systems and also on inherently simpler chemical models. This in turn has led chemists to consider whether these enzymes might provide the key to the development of a new generation of catalysts for organic oxidations.

$$S + O_2 + 2H^+ + 2e^- \xrightarrow{\text{monooxygenase}} SO + H_2O \tag{1}$$

Groves et al. [5] were the first to show that an iron(III)porphyrin and a chemical oxidant, such as iodosylbenzene, would mimic many of the oxidations of cytochrome P450 monooxygenases. Since this initial report, there has been a plethora of studies exploring variants of this simple system involving changes in the metal ion, porphyrin, oxidant, and solvent. Almost all of these reactions have been carried out in homogeneous solution on a very small scale.

A significant difference between these models and the enzymes is the presence of the protein matrix in the latter systems. The protein binds and site-isolates the iron(III)protoporphyrin IX prosthetic group, controls the reactivity of the active oxidant, and prevents inactivation of the enzyme through aggregation or bimolecular self-oxidation of the iron(III)porphyrin. The protein also controls access of the substrate to the active oxidant and hence the selectivity of the oxidation, and it provides the hydrophobic environment for substrate binding. The deficiencies of the model systems have been partly overcome through structural modification of the porphyrin rings. Thus, bulky and electron-withdrawing substituents help to prevent deactivation of the catalyst from aggregation and self-oxidation [6–8]. Further, picket-fenced, strapped, and capped porphyrins have been developed to control the access of the substrate to the active oxidant and thereby to introduce selectivity into the oxidation [7,9–11]. Indeed, the use of chiral superstructures on the porphyrins can lead to enantioselective oxidations [12–15].

If the full diversity of the metalloporphyrin-catalyzed oxidations is to be realized in synthetic procedures for laboratory, medium, or even large-scale reactions, two important difficulties will need to be overcome, namely, catalyst recovery and reuse. The cost of these compounds is such that methods will have to be developed to ensure maximum product output per gram of catalyst. One approach to solving these problems has been to anchor the metalloporphyrin on a solid surface, which should then allow the easy separation of catalyst from reactants and products. In this way the aim has been to combine the advantages of a heterogeneous catalyst with the activity of the metalloporphyrin for catalyzing oxidations in homogeneous solutions. Such a supported catalyst might then be used in large-scale continuous flow or batch oxidations of organic compounds.

Supporting the metalloporphyrin could also have added benefits arising from the structure of the support since the support provides the local environment of the catalyst. In heme enzymes this is the role of the protein matrix which surrounds the prosthetic group and defines the types of oxidation that will occur. Thus the same prosthetic group, iron(III)protoporphyrin IX, brings about oxygen transfer in cytochrome P450 monooxygenases, hydrogen atom abstraction in peroxidases, and electron transfer in cytochrome C oxidase. In this way the surface structure, rigidity, and polarity of the support would all be expected to influence the behavior and selectivity of the catalysts. A further advantage of supporting the catalyst is that it can be used to ensure that, like the enzymes, there is site isolation of the metalloporphyrin molecules. This in turn should prevent loss of activity through catalyst aggregation or intermolecular self-oxidation. However, in this respect it should be noted that if the metalloporphyrin molecules are not fixed but are able to move on the surface, the support can, by concentrating the metalloporphyrin on the surface, lead to increased aggregation and decreased catalyst efficiency [16].

It is worth stressing that, as noted above, there can be disadvantages in using supported catalysts. These include interference from the support, e.g., the backbone of an organic polymer support can act as a competing substrate in oxidations and lead to low overall yields of the desired product, and low reaction rates arising from poor diffusion of the reactants and products to and from the catalyst. Another serious problem is the difficulty in characterizing the catalyst on the surface of the support. Thus, although the surface of the solid can be studied by standard techniques, the methods available for determining detailed information about the structure of the bound metalloporphyrin and of intermediates formed in the catalytic process are not as well developed for solid samples as they are for homogeneous systems. In general, without information to the contrary, it is assumed that the behavior of the bound catalyst in the oxidation is the same or similar to that of the homogeneous analog, which can be more readily defined.

The versatility of supported metalloporphyrins as oxidation catalysts is determined by the range of available metalloporphyrins and supports, the nature of the binding between support and catalyst, and the conditions under which the catalyst is used. The aim of this chapter is to provide an overview of the progress and achievements to date and to look ahead to likely future developments.

The origins of the work described in this chapter go back to research in the 1960s and 1970s, and this has been extensively covered in several reviews on polymer-supported metal complexes [17–19]. The aim of these early studies was to prepare artificial mimics of heme proteins. These involved the use of metalloporphyrins and other metal complexes and were directed mainly to reversible dioxygen binding to model hemoglobin and myoglobin. As noted in a recent review by Sherrington [19], relatively little attention has been paid to oxidations catalyzed by supported metal complexes. This was attributed to the general lack of understanding and/or absence of effective homogeneous catalysts suitable for attachment to solid supports and also the general assumption that since organic polymers would be prone to oxidation, they would be unsuitable supports [19]. However, in the current climate of "clean technology" catalytic oxidations, and particularly those using heterogeneous catalysts, have become important and attractive targets to replace the more traditional stoichiometric processes and hence to help minimize the problem of industrial waste treatment and disposal.

II. METHODS USED TO ANCHOR METALLOPORPHYRINS

The attachment of metalloporphyrins to supports can be classified as either chemical or physical and within each of these groups the methods employed can be subdivided further by type of binding or type of support. In this chapter the former subdivision will be adopted so that for each method organic polymers and inorganic supports will be considered together. The chemical attachment of the metalloporphyrin to the support uses covalent and coordinative binding, whereas the physical approach involves adsorption (including electrostatic interaction), intercalation, and entrapment. The methods have general applicability and have indeed been used with many metal complexes, but this chapter will focus on metalloporphyrin catalysts.

A. Chemical Binding of Metalloporphyrins to Supports

1. Covalent Binding

Bonding the Porphyrins to a Preformed Support. The most common approach used to prepare metalloporphyrins covalently bound to supports is to react one or more substituents on the preformed porphyrin ring with a functional group on the surface of the support. Amine [20–25] and amide

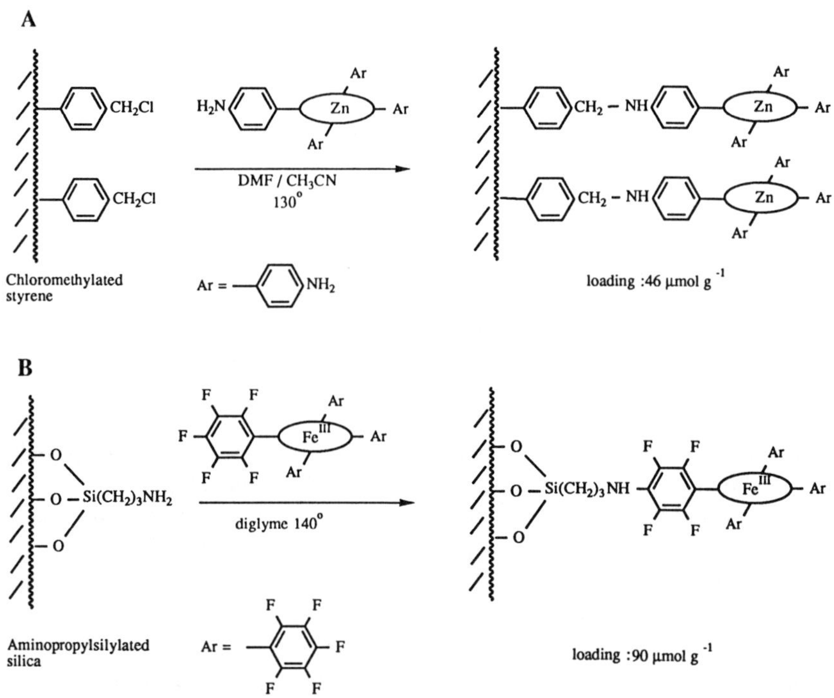

Figure 1 Covalent binding of metalloporphyrins to supports by nucleophilic amination [23,25].

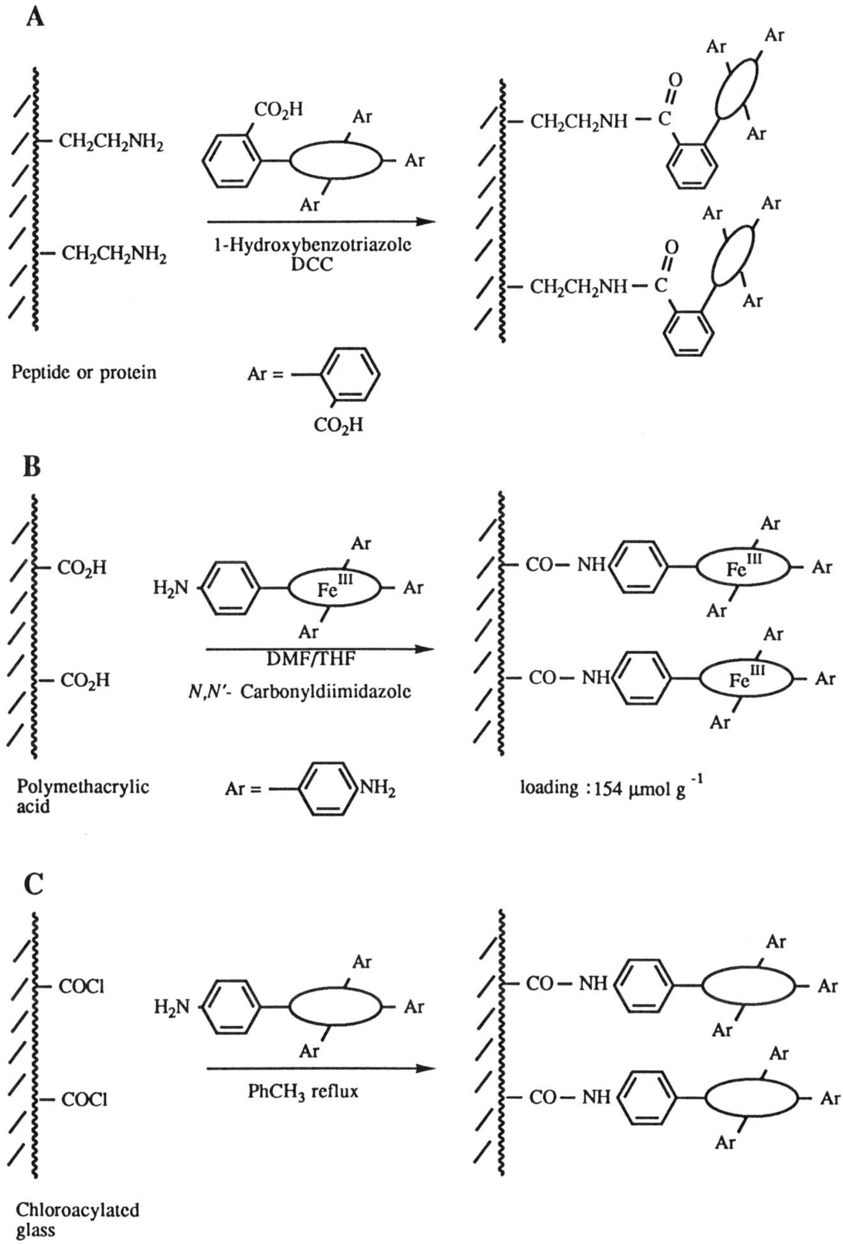

Figure 2 Covalent binding of porphyrins and metalloporphyrins to supports by amidation [26,29,31,37].

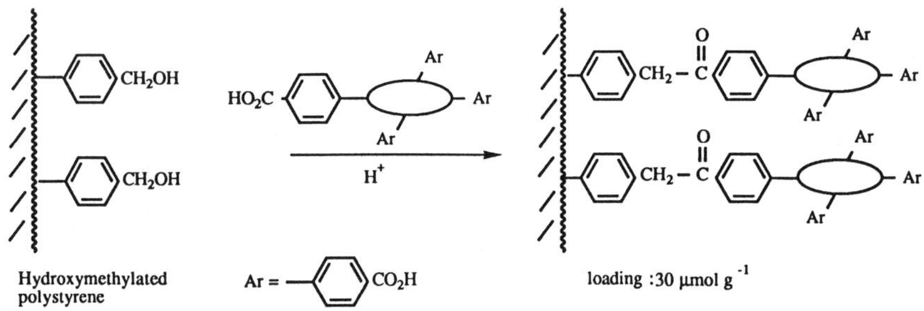

Figure 3 Covalent binding of a tetraarylporphyrin to a polymer support by esterification [20,21].

[21,24,26–31] linkages have been used most frequently, although other methods involving the formation of esters [20,21,26,32], ethers [33,34], ketones [20, 21], and quaternary ammonium ions [35,36] have been employed. This method has advantages over others described below for it allows greater choice in the nature of the linker unit, metal, and support employed. Furthermore, the metalloporphyrin can be attached to a support of well-defined structure. This in turn allows the properties of the support to be tailored to the needs of the reaction under study. Thus the support can be a flexible organic polymer such as a polymethacrylate [24,26,29], polystyrene [22–24,29], poly(vinylpyridine) [34], poly(mono-*O*-methylethyleneglycol) [32], polypeptide [27,28,30,31], and their copolymers [24,29,34], or a more rigid material such as a polymerized isocyanide [33,34] or crosslinked resin [20,21,25]. Alternatively, the hydroxylated surfaces of inorganic solids such as silica [25,35,36], clays [16], and platinum [37] or glassy carbon [38] electrodes have been used.

Some representative examples of the range of reactions that have been used to bind metalloporphyrins covalently to solid supports are illustrated in Figures 1–7.

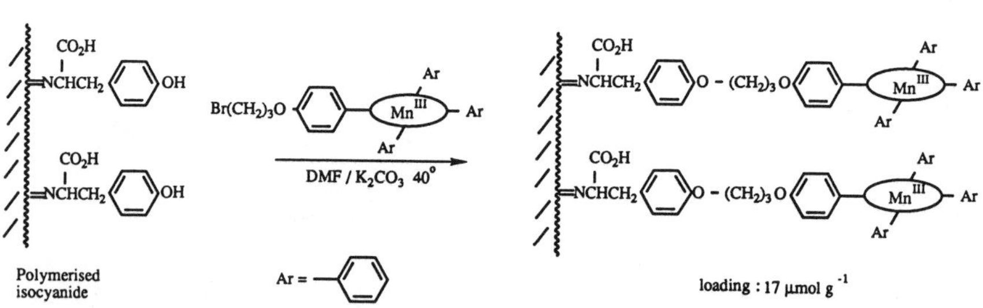

Figure 4 Covalent binding of a metalloporphyrin to a polymerized isocyanide by etherification [33,34].

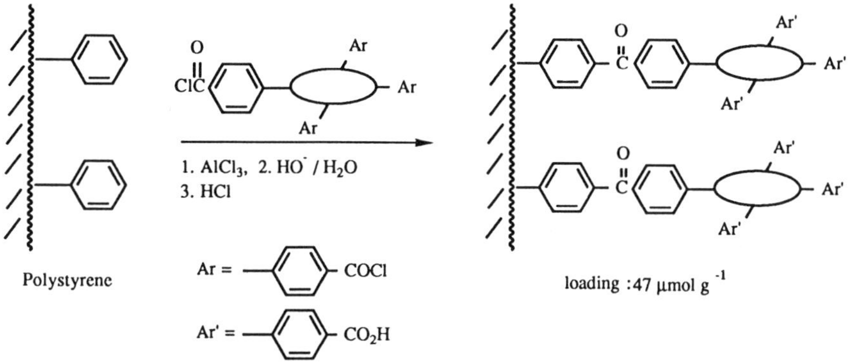

Figure 5 Covalent binding of porphyrin to polystyrene by acylation [20].

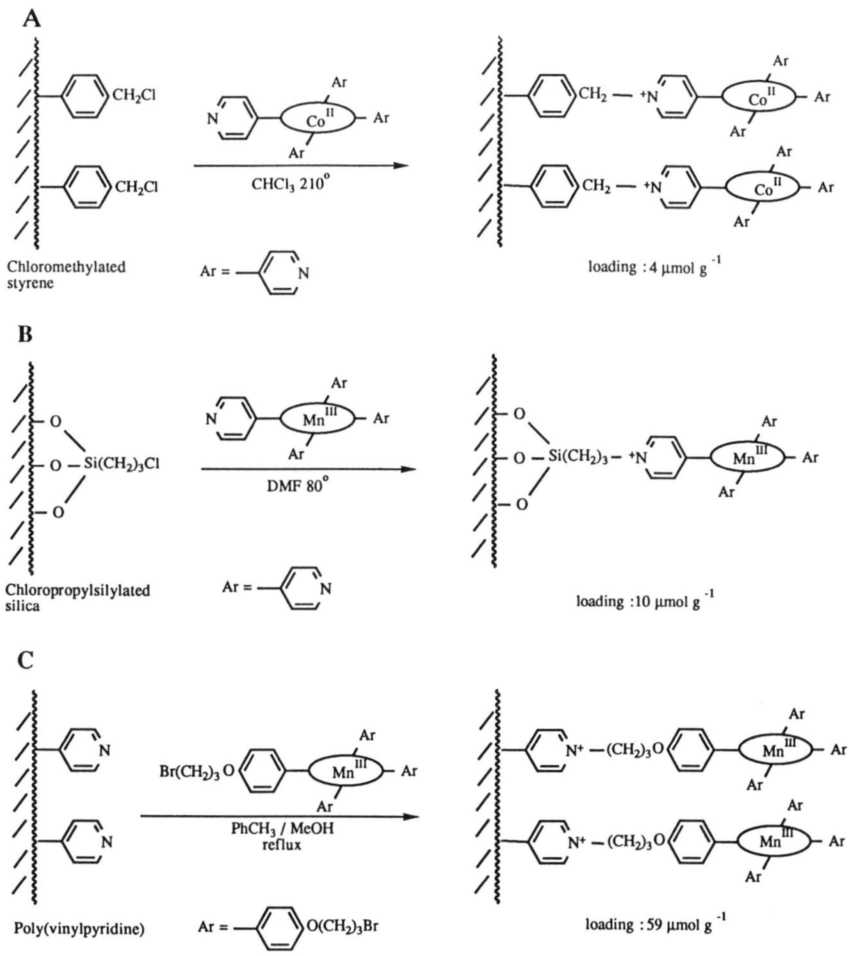

Figure 6 Covalent binding of metalloporphyrins to supports by formation of an ammonium salt [21,34–36].

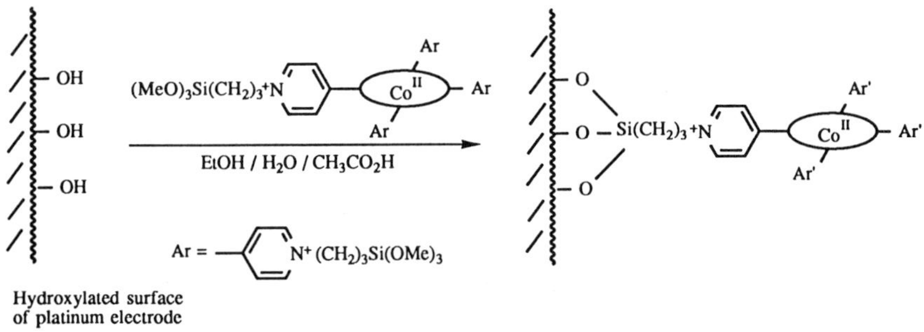

Figure 7 Covalent binding of a metalloporphyrin to a platinum electrode by silylation [37].

Radical-Initiated Copolymerization of Alkenes with Vinyl-Substituted Porphyrins. Copolymers of alkene monomers with vinyl-substituted porphyrins or metalloporphyrins have been much less extensively studied than the materials from the covalent bonding of the metal complex to a preformed support. There are no obvious advantages of the copolymerization method, which is restricted to the preparation of metalloporphyrins anchored to organic polymers. Furthermore, the resulting materials are more difficult to purify and their physical characteristics are less well defined than the metalloporphyrins covalently bound to a preformed support.

Styrene has been most commonly employed in a radical polymerization with a variety vinyl-substituted porphyrins (Figure 8) and the metal ion can either be present in the porphyrin before polymerization (e.g., Fe^{III}, Fe^{II}, Co^{II}, Mn^{III}, Ni^{II}, Cu^{II}, Zn^{II}, Mg^{II}) [39] or inserted into the polymerized free base [40,41]. Copolymers with styrene and *N*-vinylimidazoles have also been synthesized with the intention of modifying the reactivity of the metal complex by providing an imidazole group for axial ligation [39,41]. Copolymers of methacrylate esters and an acrylonitrile-substituted manganese(III)tetraphenylporphyrin with or without *N*-vinylimidazole have also been reported [42].

Copolymerization by Nucleophilic Substitution. Kellet and Spiro [38] examined a number of methods for preparing cobalt porphyrin films on glassy carbon electrodes, one of which involved the copolymerization of cobalt(II)tetra(4-pyridyl)porphyrin with 1,4-di(bromomethyl)benzene by nucleophilic substitution. This provided a crosslinked polyporphyrin polycation (Figure 9A). More recently, Traylor and Mansuy and their coworkers [43] prepared polyporphyrins by nucleophilic substitution of the *para*-fluorines in iron(III)tetra(pentafluorophenyl)porphyrin by sodium sulfide (Figure 9B). This method is more versatile than the one above since, by the

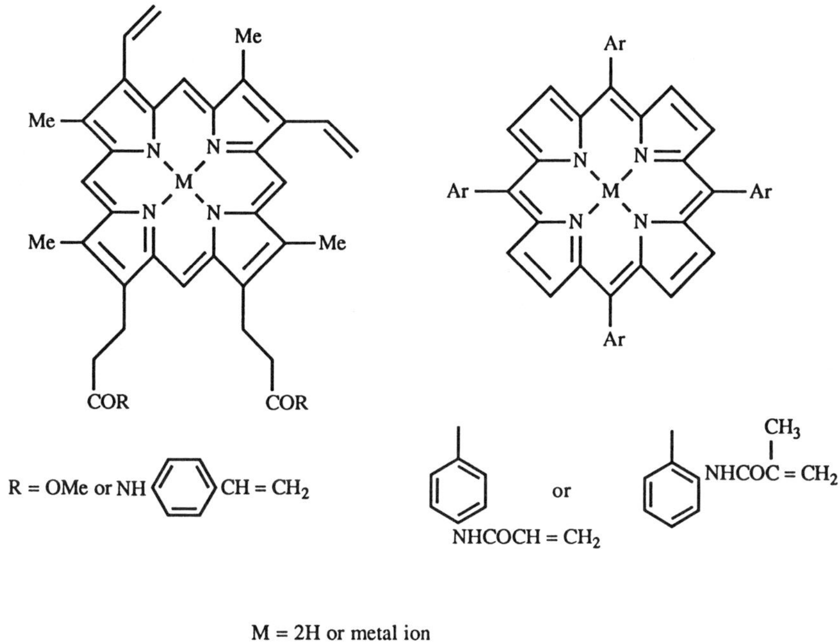

Figure 8 Vinyl-substituted porphyrins that have been used to prepare polymer porphyrins by radical copolymerization [39–42].

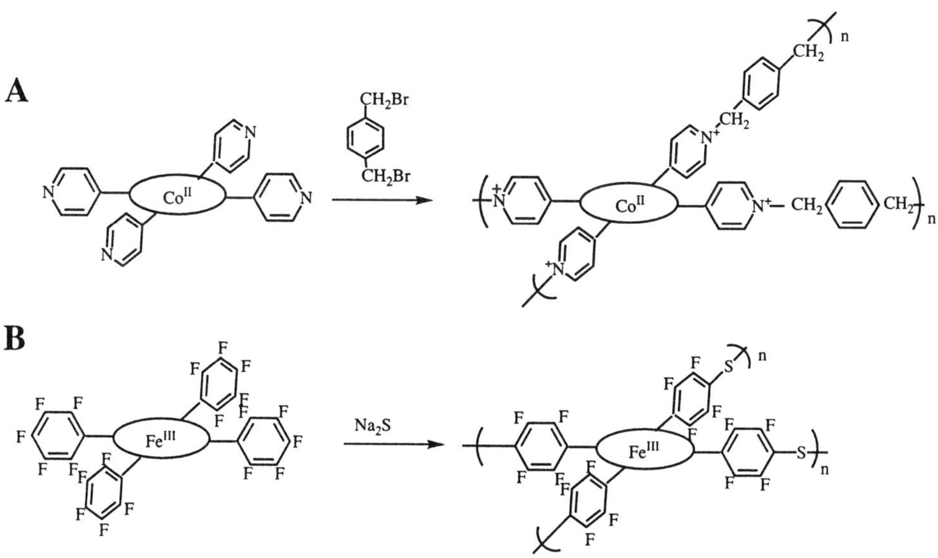

Figure 9 Metalloporphyrin copolymerization by nucleophilic substitution [38,43].

addition of perfluorobiphenyl, copolymers with varying porphyrin concentrations can be prepared.

Electropolymerization of Substituted Porphyrins. Substituted tetraphenylporphyrins and their metallo derivatives [44] and cobalt(II)protoporphyrin IX [38] have been electropolymerized by anodic oxidation. This provides a simple and controlled method for the modification of electrode surfaces (Figure 10).

2. Coordinative Bonding

Since the role of iron protoporphyrin IX in biological systems is strongly dependent on the axial ligand to the iron center, a logical approach to the design of models for heme proteins is to attach the metalloporphyrin by coordination to ligands on the surface of the support. The coordinative bond then serves the dual role of anchor and moderator of the metal ion's activity. This approach to preparing supported metalloporphyrins is simpler and, in principle, more versatile than covalent binding; however, it has received relatively little attention. This probably arises from the perceived problems of reversible binding and loss of metalloporphyrin from the support. Labat and Meunier [45] recently devised a novel solution to these difficulties by binding the catalyst through coordination and by electrostatic interactions between ionic groups on the porphyrin and counterionic groups on the support (Figure 11).

The early work in this field was directed to mimicking the chemistry of hemoglobin and myoglobin [39,46–49]. Subsequently the method has

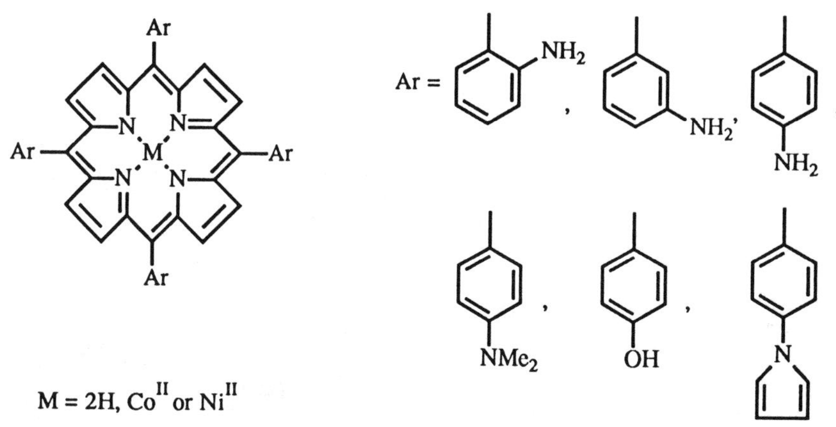

Figure 10 Substituted tetraphenylporphyrins used in electropolymerization [38, 44].

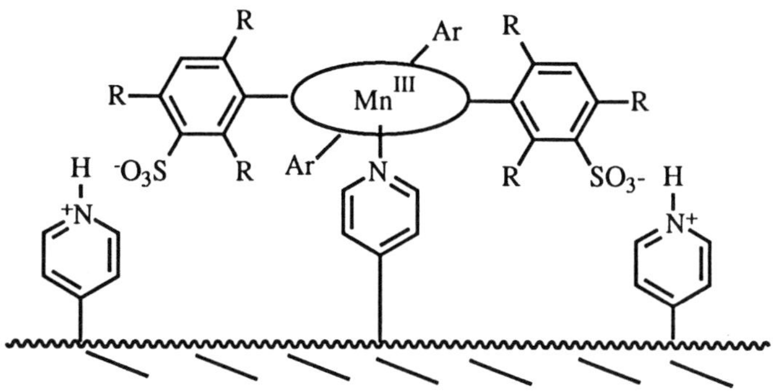

Figure 11 Anionic metalloporphyrin anchored to a partially protonated poly(vinylpyridine) by coordinative and electrostatic bonding [45].

been extended to models for peroxidases [45] and cytochrome P450 monooxygenases [50–52].

Although most of the research into coordinatively bound metalloporphyrins has involved iron porphyrins, a variety of other metals, including Mn^{III}, Co^{II}, Au^{III}, Ru^{IV}, and Rh^{III}, have also been investigated. The initial studies used protoporphyrin IX [47,48] and tetraphenylporphyrin [46,49]. However, the range of porphyrin macrocycles was recently extended to sterically hindered and polyhalogenated compounds, e.g., metallotetra(2,6-dichlorophenyl)porphyrins [52] and metallo-β-pyrrole–halogenated porphyrins [53].

Organic Polymer Supports. A variety of linear and crosslinked organic polymer ligands have been employed to support metalloporphyrins. These include poly(4-vinylpyridine) (PVP) [47,48,52,53] and its *N*-oxide [52], poly-L-lysine [47,48], polyethyleneimine [48], poly(2-methyl-*N*-vinylimidazole) [47], copolymers of styrene with PVP [54] and with 2-substituted *N*-vinylimidazole [39], and polystyrene functionalized with (*N*-imidazolyl)methyl groups [46,52]. The extent of crosslinking of these materials affects the flexibility of the support which in turn affects the activity of the catalyst. Thus with linear and lightly crosslinked polymers, bis ligation to the metalloporphyrin [47,48,52] (Figure 12) and intermolecular reactions between bound porphyrins [46] are possible, whereas with rigid resins these processes are prevented.

Silica Supports. There have been a few reports of using nonpolar metallotetraphenylporphyrins adsorbed on silica as oxidation catalysts [51, 55,56]. The nature of the bonding between the porphyrin and the support is

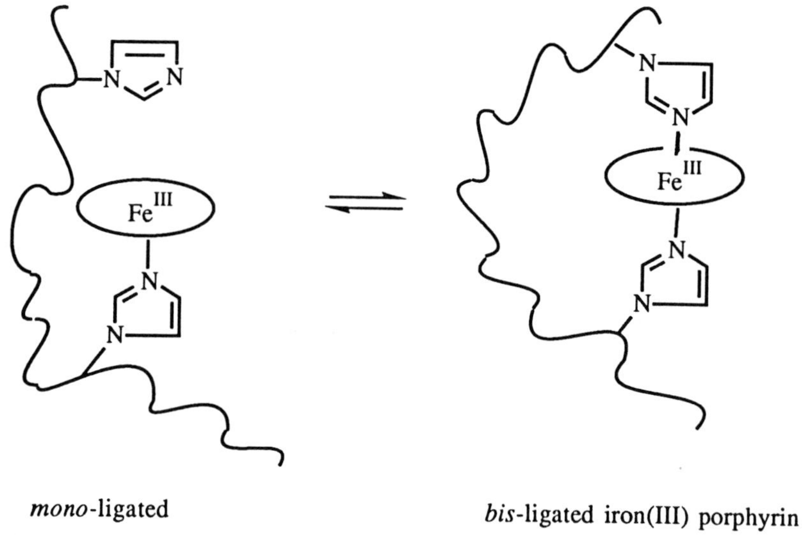

Figure 12 Bis ligation of an iron(III)porphyrin coordinately bound to a flexible polymer.

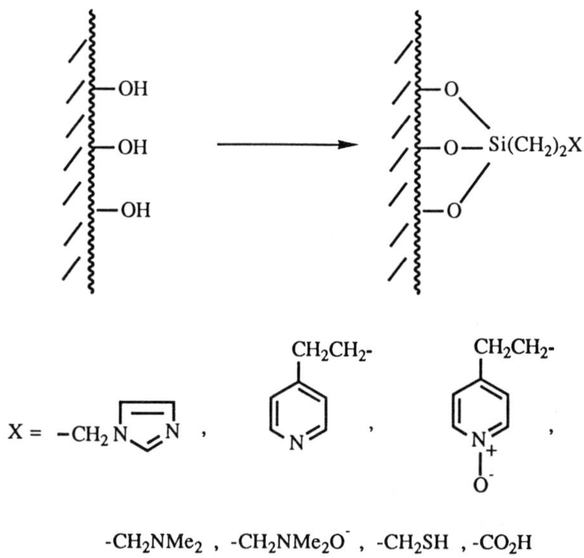

Figure 13 Surface modification of silica with silylpropyl ligands.

not clear although it has been suggested that this involves direct coordination of the surface silanol groups to the metal center. The more common approach using silica as a coordinative support requires converting the surface silanol groups into silylpropyl derivatives (Figure 13) which act as ligands for the metalloporphyrin [49]. The large majority of the studies have employed bound imidazole [49–52]; others have used pyridine and its *N*-oxide [57], trialkylamines and trialkylamine *N*-oxides, thiol and carboxylate groups [58]. The rigid support and short linker units ensure that with these materials few if any of the metalloporphyrins are bound by bis ligation.

B. Physical Binding of Metalloporphyrins to Supports

1. *Electrostatic Binding*

The principle behind this approach to preparing supported metalloporphyrins is to make use of electrostatic interactions between ionic metalloporphyrins and counterionic groups on the surface of the support (Figure 14). Although the method is limited to ionic metalloporphyrins, this has not prevented it from being widely used (Table 1). The main advantages are the strong interactions between porphyrin and support and the simplicity of the preparations. The latter usually only require stirring a solution of the ionic metalloporphyrin with a suspension or solution of the counterionic support.

The large majority of the materials have been prepared from ionic organic polymers. These include lightly and extensively crosslinked polystyrene-based anion [61–64,67,69–71,76,77] and cation (16,37,67,69,78] exchangers,

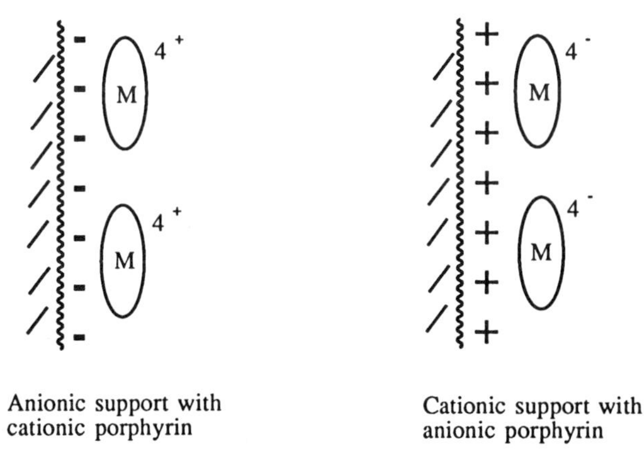

Figure 14 Schematic diagram of electrostatic binding of ionic metalloporphyrins to counterionic supports.

Table 1 Ionic Metallotetraarylporphyrin (MTArP) Catalysts Bound to Counterionic Supports

MTArP			
M	Ar	Abbreviation	References
Fe^{III}, Mn^{III}, Co^{III}, Cu^{II}, Zn^{II}	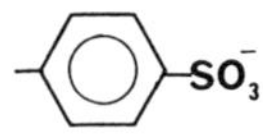	MTSPP	45,60–69
Fe^{III}, Mn^{III}, Co^{II}, Co^{III}, Cu^{II}	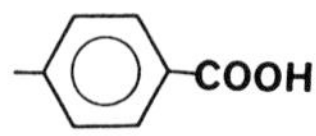	MTCPP	21,61–63,66,69,70
Fe^{III}, Mn^{III}	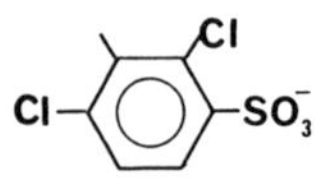	MTDCSPP	45,53,67,69,71–73
Fe^{III}, Mn^{III}	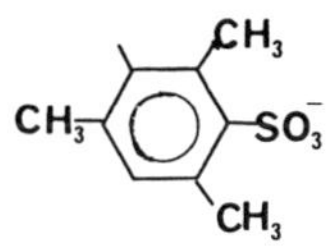	MTSMP	45
Fe^{III}, Mn^{III}	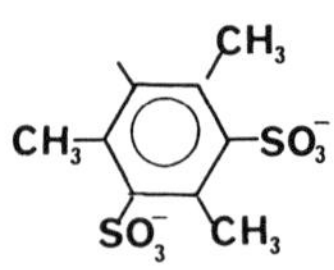	MTDSMP	53
Co^{II}	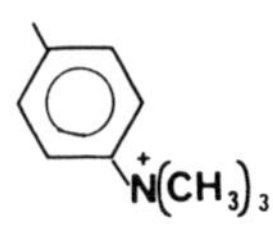	MTTAPP	21
Fe^{III}, Mn^{III}, Co^{II}, Co^{III}, Au^{III}, Cu^{II}, Zn^{II}	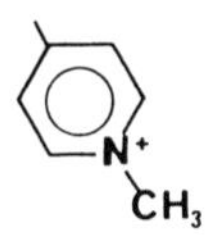	MT4MPyP	16,21,59,60,69,74,75
Fe^{III}	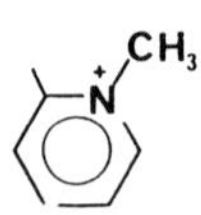	MT2MPyP	16

quaternized crosslinked PVP [45,53,73,77], colloidal cationic latexes [72, 73], and linear flexible water-soluble polyelectrolytes such as poly(vinylsulfonate) (PVS) [59,79], polyacrylic acid (PAA) [79], and poly(styrenesulfonate) (PSS) [45,79]. Ionic metalloporphyrins have also been successfully anchored to glassy carbon electrodes [74,80], the zeolite faujasite [60], and onto the oxides silica, alumina, and magnesia [66,69].

2. *Intercalation*

The layered structures of smectite clays, such as montmorillonite, provide large interlamellar anionic surfaces for the intercalation of cations. When coupled with the clay's ability to expand the interlayer spacing, this allows access to the internal surfaces of both large and small cations and has led to their use as cation exchangers. The extent of the interlayer swelling depends on the cation and the nature of the internal surface.

These unique properties of smectite clays has led to their use as supports for cationic metal complex catalysts [81]. In this respect protonated tetraphenyl- and tetra(4-pyridyl)porphyrin [82] and cationic tetra(4-*N*-methylpyridyl)- and tetra(4-trimethylammoniumylphenyl)porphyrin [83] and their metallated derivatives [69,83–86] are strongly adsorbed onto the internal surfaces of montmorillonite and saponite. X-ray diffraction and UV-Vis spectroscopy show that the tetraarylporphyrins form an interlamellar monolayer in which the aryl groups are flattened more toward the porphyrin plane than they are in free solution [82,84,87].

The clay-supported materials are easily prepared by stirring aqueous or polar organic solutions of the metalloporphyrin with the clay. The clay serves the dual role of immobilizing the cationic catalyst on a stable, unreactive solid surface and controlling the local environment of and access to the catalyst. However, despite their obvious potential, there has been relatively little published research on the use of clay-supported metalloporphyrins as oxidation catalysts [69,85,86].

In a recent study Mansuy and his coworkers [69] extended the range of layered supports to include phosphatoantimonic acids, $HSbP_7O_8$ and $H_3Sb_3P_2O_{14}$, and the layered dihydroxide $Mg_6Al_2(OH)_6CO_3$. The former have been used with Mn^{III}T4MPyP while the charge on the latter enables anionic manganese(III)porphyrin catalysts such as Mn^{III}TSPP* and Mn^{III}TDCSPP to be employed.

*For definitions, see list of abbreviations preceding the references.

3. *Entrapment of Metalloporphyrins in Polymers and Inorganic Materials*

The principle behind this method of immobilizing metalloporphyrins is to entrap the macrocycle in the support matrix. This "ship-in-a-bottle" approach mimics certain heme proteins where the iron porphyrin prosthetic group is buried in a protein matrix. Thus, provided the matrix support is inert to oxidation, regio- and substrate-selective oxidations might be possible by controlling access to the metal center. The difficulty in preparing such entrapped metalloporphyrins has, however, meant that there are very few examples of this type of catalyst.

The early work used entrapment to site-isolate iron(II)porphyrins in polystyrene [88,89] or, more recently, cobalt(II)porphyrins in poly(butylmethacrylate) for reversible dioxygen binding studies [90]. Oxidation catalysts have been prepared by Herron et al. by in situ synthesis of iron phthalocyanines inside zeolites [91,92] and by Bowers and Dutta [93] using manganese(III)salen in zeolites. However, the limited size of the zeolite supercages has prevented these materials from being used for tetraarylporphyrins. There is one report of the in situ preparation of iron(III)tetramethylporphyrin in a zeolite and its use as an oxidation catalyst [94].

III. SUMMARY OF SUPPORTED METALLOPORPHYRINS AND THE PROPERTIES OF AN IDEAL OXIDATION CATALYST

From the examples of supported metalloporphyrins described above it is clear that a large range of catalysts is available. Thus the support can be soluble, insoluble, or colloidal; flexible or rigid; polar or nonpolar; and have a structured or irregular surface. Furthermore, these materials can be used with different metal ions and porphyrin ligands. These controllable parameters have the potential of providing a wide choice of oxidation catalysts.

Although many methods have been used, the ideal preparation of supported catalysts should be simple to carry out and adaptable for use with a large number of supports, metal ions, and porphyrins, thereby providing easy access to a range of different catalysts. It should also be reproducible and suitable for controlled support loading.

For general application the supported metalloporphyrin catalyst should be

1. Oxidatively stable
2. Tough and resistant to physical abrasion
3. Reusable

4. Resistant to metalloporphyrin leaching or removal
5. Suitable for batch or continuous flow systems
6. Suitable for use in a wide range of solvents and conditions, and
7. Capable of being "tailor-made" for selective oxidations

In the following sections the oxidations catalyzed by supported metalloporphyrins and how well the catalysts approach these ideal properties will be discussed. The subject is divided into two sections. The first describes the reactions catalyzed by metalloporphyrins on polymers and this is followed by reactions with catalysts supported on inorganic solids. Each of these sections is subdivided by reaction type.

IV. OXIDATIONS CATALYZED BY POLYMER-SUPPORTED METALLOPORPHYRINS

A. Polymer-Supported Metalloporphyrin Models for Cytochrome P450 Monooxygenases

1. *Catalyzed Oxidations with Hypochlorite*

One of the first successful polymer–porphyrin models for cytochrome P450 monooxygenases was reported by Nolte and his coworkers [33,34]. The catalyst in this system was manganese(III)tetraarylporphyrin anchored by one or more covalent bonds to a polymerized isocyanide

$$(R-N{=}C\langle)_n \text{ where } R = \text{4-HOC}_6\text{H}_4\text{CH}_2\overset{|}{\text{C}}\text{HCO}_2\text{H}$$

The support provided a rigid helical backbone from which the Mn^{III} porphyrin groups protruded into the solution allowing easy access for the reagents. Epoxidation of cyclohexene by sodium hypochlorite in an aqueous/dichloromethane system catalyzed by the Mn^{III} porphyrin attached by a single covalent bond to this support and by the same catalyst bound to a flexible PVP showed that the supported catalysts were more active than the homogeneous analog despite the fact that the former involved a three-phase system (catalyst, water, dichloromethane) with consequential reactant diffusion problems, whereas the latter was biphase. The enhanced catalytic activity was attributed to site isolation of the metalloporphyrin molecules on the support preventing the formation of less reactive μ-oxomanganese(IV) dimers (Figure 15). Interestingly, a catalyst with multiple bonds to the polyisocyanide was less reactive, probably because this resulted in extensive crosslinking which restricted access of reactants to the manganese(III)porphyrin. The use of these supported catalysts in other oxidations and in high-catalyst-turnover reactions was not reported.

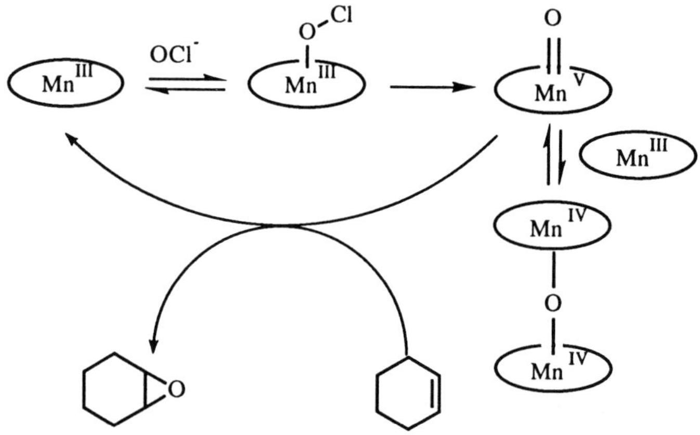

Figure 15 Generation of oxomanganese(V)porphyrin active oxidant using hypochlorite and its competitive reactions with cyclohexene and manganese(III)porphyrin.

Wöhrle et al. [24,29] also investigated alkene epoxidation by hypochlorite catalyzed by Mn^{III}, Fe^{III}, and Cr^{III} porphyrins covalently bound to neutral and charged supports. However, by using a water-soluble substrate, 2,5-dihydrofuran, polymer-supported catalyst, and oxidant they were able to carry out the reactions in homogeneous aqueous solution, thereby avoiding the complications of phase transfer of reactants. They found, rather surprisingly, that the ionic charge on the support had little influence on the epoxidation and that supported and unsupported catalysts had comparable catalytic activity. The stability of the supported catalysts toward oxidative destruction using large amounts of oxidant relative to catalyst was not reported.

Ford and his coworkers [72,73] prepared an anionic manganese(III)porphyrin electrostatically bound to colloidal anion exchange particles (44–71 nm) and used them to catalyze the epoxidation of styrene in aqueous solution and of other alkenes in aqueous dichloromethane systems by hypochlorite. Like Nolte and his coworkers [33], they found that the support-anchored manganese(III)porphyrins were more active catalysts than their unsupported analogs. However, they attributed this not to suppression of μ-oxodimer formation but to the support's ability to concentrate the anionic oxidant and the nonpolar substrate into the polymer phase, thereby generating a high local concentration of reactants close to the catalyst.

A major advantage in using either soluble or colloidal supported catalysts is the ability to study the reactions by transmission UV-Vis spectroscopy. This not only enables the characterization of the catalyst before and after reaction, and hence determination of catalyst stability, but also allows the detection of high-valent metalloporphyrin intermediates, such as polymer-bound oxomanganese(V) species [72]. However, although this is important for mechanistic studies of these oxidations, simple catalyst reuse and recovery requires that the supported metalloporphyrins be insoluble and hence easily separated from the other reagents. Ultrafiltration techniques required to recover colloidal catalysts limit the potential applications for such systems.

An exciting recent development of the polymer-bound metalloporphyrins in hypochlorite-induced epoxidation of alkenes was the use of a protein support as a chiral auxiliary [30]. When manganese(III)tetra(4-carboxyphenyl)porphyrin was adsorbed onto or covalently bound to bovine serum albumin the resulting materials catalyzed the enantioselective (albeit small, 4–11.7% ee) epoxidation of styrene by hypochlorite in aqueous dichloromethane or chloroform. Interestingly, the catalytic behavior of the covalently bound and physically adsorbed Mn^{III} porphyrins were very comparable. Enantioselective epoxidation of simple alkenes that are not amenable to the Sharpless procedure is an important goal for chemists, and further studies will no doubt be directed to improving this method.

2. Catalyzed Oxidations with Iodosylarenes

In the study of homogeneous metalloporphyrin-based models for cytochrome P450 monooxygenases, iodosylarenes and, in particular, iodosylbenzene have been most commonly used as the oxidant. This has arisen despite the fact that iodosylarenes are not commercially available, have poor solubilities in most solvents, and contain relatively little active oxygen per given mass of oxidant. The main advantages are that they are relatively inert toward most organic compounds without catalysis and they react unambiguously with metalloporphyrins by two-electron processes (Figure 16).

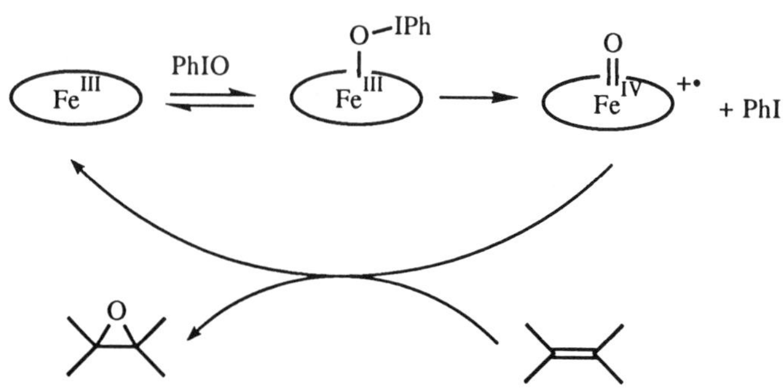

Figure 16 Alkene epoxidation by iodosylbenzene catalyzed by an iron(III)porphyrin.

With supported metalloporphyrins the potential problem of reacting an insoluble catalyst with a poorly soluble oxidant has not proved to be a limitation. Thus, dichloromethane [25] and dichloromethane mixed with trifluoroethanol and water [43] are effective solvents, or methanol, which is one of the few solvents for iodosylbenzene, can be employed [16,52,67].

Leanord and Lindsay Smith [16,67] used ionic iron(III)tetraarylporphyrin catalysts electrostatically bound to counterionic exchange resins to catalyze the epoxidation of alkenes by iodosylbenzene in methanol. The results show that increased crosslinking of the resin support and ortho substituents on the aryl groups attached to the porphyrin ring favor epoxidation by iodosylbenzene. However, with even the best systems substantial oxidant remained unaccounted for at the end of the reaction and the investigators concluded that the missing oxidant was consumed in the competitive oxidation of the polystyrene support. Thus, the active oxidant was diverted from epoxidation of the alkene to oxidation of the benzylic C-H groups and any vinyl groups present on the support (Figure 17). This is a major problem with the development of organic polymer supports and probably also accounts for the low oxidation yields noted by Mansuy and Meunier and their coworkers when using ion exchange supports [53,69]. Although it can be minimized by increasing the rigidity of the polymer by crosslinking, this limits the use of these materials in oxidation.

A further feature of the reactions catalyzed by these supported metalloporphyrins is their slowness compared with the corresponding homogeneous oxidations, indicating that for the solid catalysts either access to the active oxidant is more restricted or the catalyst is bound as less active aggregates. The latter explanation was suggested since catalyst reuse experiments, with iron(III)tetra(4-*N*-methylpyridyl)porphyrin on Dowex MSC-1, revealed that the supported catalysts, far from being more stable toward oxidative bleaching, were more readily deactivated than their homogeneous analogs [16]. This implies that the metalloporphyrins are not site-isolated on the ion exchange resin but rather the support increases the local concentration of the catalyst and encourages aggregation. Thus, adsorption on the resin favors, rather than prevents, intermolecular self-destruction of the catalyst. Interestingly, iron(III)tetra(2,6-dichloro-3-sulfonatophenyl)porphyrin, which is sterically protected from aggregation and intermolecular self-oxidation, gave a much more stable catalyst on ion exchange resins than supported unhindered iron(III)tetra(4-sulfonatophenyl)porphyrin or Fe^{III}T4MPyP [67]. Very recent work with the cationic Fe^{III}T4MPyP bound to water-soluble anionic polymers in water confirms that the polymer, rather than site-isolating the metalloporphyrin molecules, leads to extensive cofacial aggregation [79].

Although metalloporphyrins bound to ion exchange resins are easily prepared and aggregation can be prevented using sterically hindered porphy-

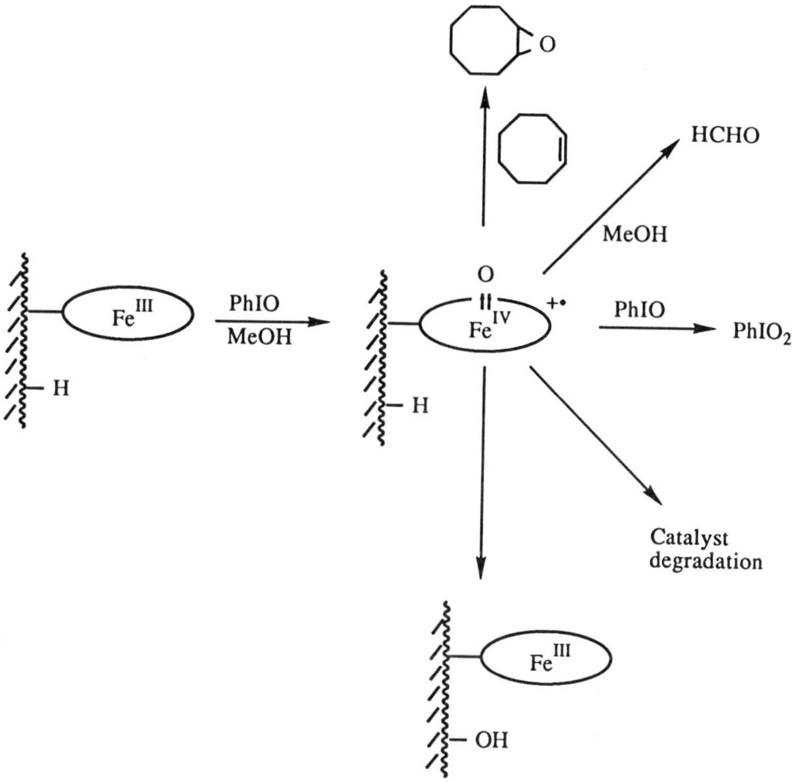

Figure 17 Competitive reactions of polymer-supported oxoiron(IV)porphyrin π-radical cations formed by reaction of iodosylbenzene with iron(III)porphyrins.

rins, the problem of support oxidation restricts their use to oxidatively inert organic polymers such as Nafion.

In an alternative approach, Cooke and Lindsay Smith [52] employed (1-imidazolyl)methylated polystyrene (PS-Im) (1% crosslinked) and PVP (6% crosslinked) to bind the sterically hindered iron(III)tetra(2,6-dichlorophenyl)porphyrin (Fe^{III}TDCPP) by coordination. Again, although these catalysts gave similar oxidation yields to Fe^{III}TDCPP in free solution, their reactions were slower. It is likely that, with these relatively flexible polymers, the majority of the catalyst was bound as unreactive bis-ligated complexes [96] and this, rather than aggregation, accounts for the low rates of epoxidation (Figure 12).

Repeat use experiments revealed that although the catalyst activity decreased with each addition of oxidant, the iron(III)porphyrin remained visibly unbleached, unlike the ion exchange–supported catalysts suggesting that with these materials loss of activity was not associated with oxidative bleaching of all the catalyst but only the small number of catalytically active species near the polymer surface. Bis-ligated Fe^{III}TDCPP in the interstices of the polymer remained intact but unavailable for catalysis. The problem associated with bis ligation can be avoided by using a rigid support such as surface-modified silica (see Section V.A.1) [52]. Interestingly, the polyhalogenated porphyrins $Fe^{III}TDCPCl_8P$ and $Fe^{III}TPFPBr_8P$, bind very readily to PS-Im and are almost inactive as catalysts for alkene epoxidation by iodosylbenzene [98]. This can be attributed to the very strong tendency of these porphyrins to form bis-ligated complexes.

Campestrini and Meunier [53] employed both the coordinative ligation and electrostatic binding in the same supported catalyst by using anionic metalloporphyrins on partially protonated or quaternized PVP. In this way they were able to combine the known benefits of a proximal pyridine ligand in metalloporphyrin-catalyzed oxidations [97,99–101] with a strong ionic interaction between catalyst and support (Figure 11) and investigate the catalytic activity of several robust sulfonated manganese(III)- and iron(III)porphyrins in epoxidations and aliphatic hydroxylations (Figure 18). Comparison of these reactions with epoxidation catalyzed by some anionic metalloporphyrins on anion exchange resins, where no coordinative ligation to the support was present, showed the expected advantages of the proximal pyridine groups.

The manganese(III)porphyrins were found to be more active than their iron(III) analogs and the best catalysts were halogenated on the porphyrin β-pyrrole positions. However, in catalyst recycling experiments even the most stable catalyst ($Mn^{III}TCSMCl_8P$) lost significant activity in the epoxidation of cyclooctene after 1500 turnovers. Interestingly, the proportion of iodosylbenzene converted to epoxycyclooctane with the supported catalysts was lower than that from the corresponding homogeneous oxidations, suggesting that either some of the oxidant was consumed in oxidizing the support, as noted above [16,67], or the disproportionation of iodosylbenzene to iodo- and iodoxybenzene was more important with the supported catalysts.

The supported $Mn^{III}TCSMCl_8P$ was also active in the hydroxylation of adamantane, giving a high ratio of adamantanols to adamantanone (44:1) and a low ratio of C_{tert} to C_{sec} oxidation (3.2–4.2:1). The selectivities for alcohol and secondary C-H oxidation were slightly higher with the supported catalyst than the homogeneous analog, suggesting that the support induced some extra shape selectivity to the oxidation. A similar effect was observed in the oxidation of octane.

Traylor et al. [43] recently reported very encouraging results from oxidations by perfluoroiodosylbenzene catalyzed by crosslinked polymeric poly-

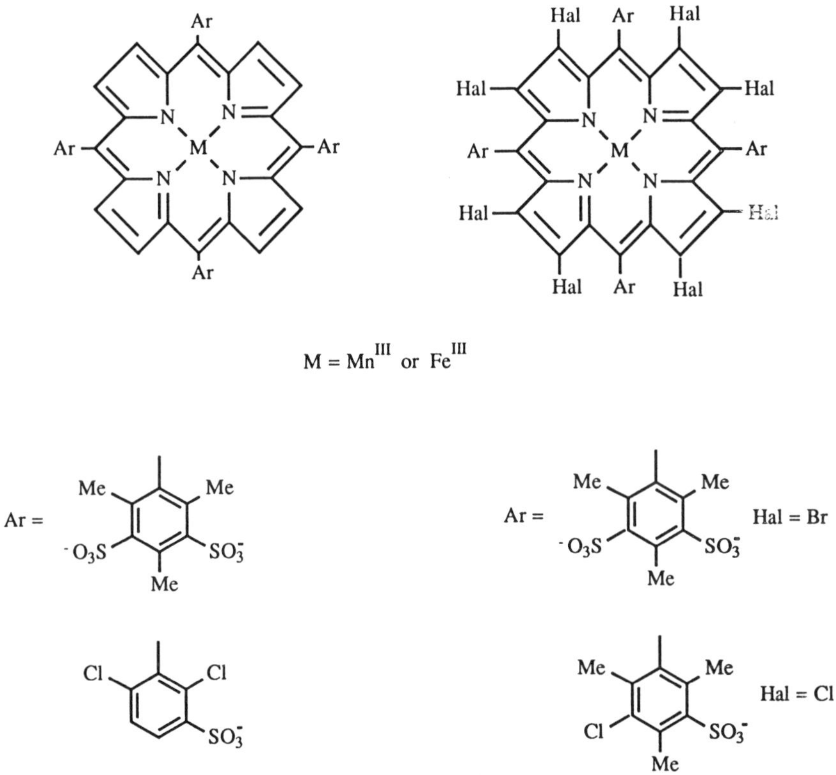

Figure 18 Robust sulfonated metalloporphyrins [53].

halogenated iron(III)porphyrins. These catalysts were prepared by making use of the facile nucleophilic substitution of fluoride from perfluoroaromatics to link and crosslink Fe^{III}TPFPP [102] with the bifunctional nucleophile sulfide. Copolymers of Fe^{III}TPFPP and other perfluoroaromatics were also prepared. These rigid polymeric porphyrins were very effective catalysts for norbornene epoxidation and cyclohexane hydroxylation and, interestingly, they showed a very high selectivity for alkane hydroxylation with relatively little ketone formation (cyclohexanol/cyclohexanone, 45–100:1). Repeat use experiments revealed that the catalysts carried out ~50,000 turnovers per iron porphyrin before oxidative bleaching reduced their activity. The feasibility of carrying out the oxidations in a continuous flow reactor was also established.

3. *Catalyzed Oxidations with Hydrogen Peroxide and Hydroperoxides*

There have been few reports of oxidations with hydrogen peroxide or alkylhydroperoxides catalyzed by supported metalloporphyrins. The main reason for this is possibly the difficulty in preparing supported catalysts with the good donor ligand necessary to generate and stabilize the high-valent oxomanganese(V)- or oxoiron(IV)porphyrin π-radical cation species [103].

Campestrini and Meunier, with their combined ligand and electrostatically bound Mn^{III} catalysts, investigated the oxidation of cyclooctene with H_2O_2 in acetonitrile but reported that no reaction occurred [53]. However, Hirobe and his coworkers [27,28] reported that iron(III)porphyrins covalently bound to poly-L-glutamate catalyzed the H_2O_2 oxidation of aniline to 4-hydroxyaniline and of styrene to styrene oxide, although the yields were low (<10 catalyst turnovers or $<5\%$ based on oxidant). These workers also reported results using a polypeptide–iron(III)porphyrin with a thiolate tail [28]. This mimicked cytochrome P450 monooxygenases in that it showed an absorption at 450 nm for the CO complex of the reduced catalyst and it catalyzed epoxidation of styrene and cyclohexene by cumene hydroperoxide in low yield. However, there were no reports on whether or not the epoxidation of styrene showed any enantioselectivity.

4. *Catalyzed Oxidations with Dioxygen*

Metalloporphyrins are known to catalyze oxidations with dioxygen in the presence [104–106] or absence [107–109] of a sacrificial reductant. The former reactions can be considered as mimics of the cytochrome P450 oxidation cycle, whereas the latter are autoxidations and as such are not within the scope of this chapter. With polymer-supported metalloporphyrins there have been a few brief reports of the use of dioxygen in the presence of sodium borohydride to oxidize alkenes [27,28] and cholesterol [42], hydroxylate and oxidatively demethylate *N*-methylaniline [28], and one where the reducing agent employed was NADPH [110]. However, a thorough study giving full details of these systems has not been published.

B. Polymer-Supported Metalloporphyrin Models for Peroxidases

Although the majority of the investigations of homogeneous metalloporphyrin-catalyzed oxidations have involved models for cytochrome P450 monooxygenases, there has also been a strong interest in models for oxidations catalyzed by other heme enzymes, particularly peroxidases. The mechanism of the peroxidatic cycle is generally agreed to involve a two-electron heterolytic cleavage of the peroxide bond to give an intermediate (compound 1) which is reduced in two one-electron steps via compound 2 to the resting enzyme (Reactions 2–5). The protein is believed to encourage the heterolytic cleavage by acid–base catalysis and by proximal ligation to the heme group. The structure of compound I for individual peroxidase depends on the nature of the protein matrix. One oxidizing equivalent is provided by the iron of the heme group and the other by the porphyrin macrocycle or a neighboring amino acid group [112,113].

$$\text{Peroxidase} + RO_2H \rightarrow \text{compound I} + ROH \quad (2)$$

$$\text{Compound I} + AH_2 \rightarrow \text{compound II} + AH\cdot \quad (3)$$

$$\text{Compound II} + AH_2 \rightarrow \text{peroxidase} + AH\cdot \quad (4)$$

$$2AH\cdot \rightarrow \text{products} \quad (5)$$

The mechanisms of metalloporphyrin models, without the protein matrix, are less certain, with extensive research producing conflicting evidence for both one- and two-electron cleavage of the peroxides [113,114]. To date there have been no reports of attempts to reproduce the important functions of the protein in peroxidases using a synthetic polymer. Instead, the published work with polymer-supported metalloporphyrins has largely concentrated on mimicking the overall oxidations of peroxidases with a heterogeneous catalyst, irrespective of whether they are one- or two-electron processes, with a view to developing practical applications for these systems.

The most extensive studies have been carried out by Saito and his coworkers [61–64,70] who, in a series of papers, explored the potential of a selection of Mn^{III}, Fe^{III}, and Co^{III} porphyrins electrostatically bound to the anion exchange resin Amberlite IRA 900, as peroxidatic catalysts for oxidation of phenol [61,63,70], *N,N*-diethylaniline [63,64], and methanol [64] by H_2O_2 in aqueous solution. The oxidations were monitored by trapping the oxidized substrate as a colored product (Figure 19).

These investigations showed that although all three metal ions catalyzed the oxidations, manganese(III) was the best, with Mn(III)TSPP being the preferred catalyst. *N,N*-Diethylaniline was a better substrate than phenol because the product, being cationic, was not prone to adsorption on the positively charged ion exchange resin, which was found to be a problem with the uncharged quinoneimine product from phenol. The best of these synthetic systems showed 85% of the activity of horseradish peroxidase (HRP) as assessed by the yield of the colored products. In repeat use oxidations the catalysts gave the same overall activity after 10 cycles, although the reported data suggest that the catalysts were subjected to less than two turnovers. Since these supported catalysts are easily stored, are stable, and require no special precautions in handling, they have the potential of being used in place of HRP to estimate H_2O_2 and alkylhydroperoxides.

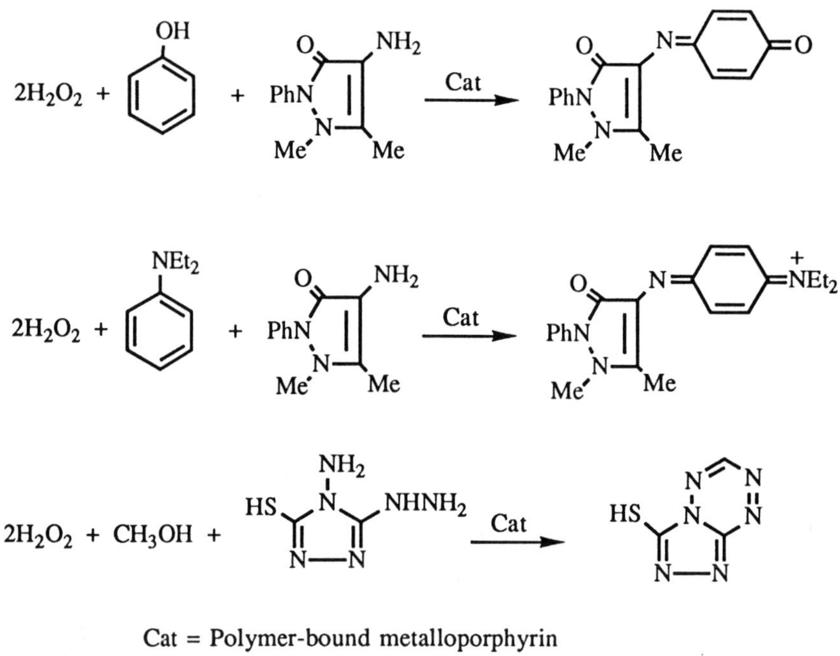

Figure 19 Peroxidative oxidations by hydrogen peroxide catalyzed by polymer-supported metalloporphyrins [61,62,64].

Saito and his coworkers have also investigated the peroxidase activity of hemin covalently bound to a poly(mono-*O*-methylethyleneglycol) [32]. This material, which was readily soluble in water and organic solvents such as chloroform and 1,1,1-trichloroethane, was shown to contain unaggregated heme groups. The rates of reaction of this material with H_2O_2 and peroxidized linoleic acid were measured using *ortho*-phenylenediamine as an oxidizable trap. Interestingly, the second-order rate constant for reaction with H_2O_2 at pH 7 is significantly lower than the values for comparable reactions of H_2O_2 and *t*-BuO_2H with ionic iron(III)tetraarylporphyrins in free solution [114,115]. A similar rate retardation induced by a polymer support was observed by Lindsay Smith and Lower, who measured the second-order rate constant of *t*-BuO_2H with the cationic Fe^{III}T4MPyP in the presence [79] and absence [115] of the water-soluble anionic polymer poly(styrenesulfonate). In these studies evidence was presented to suggest that there is extensive porphyrin aggregation on the polymer and the rate retardation was then attributed to the known low reactivity of porphyrin aggregates. Clearly, further work is needed to study the influence of the flexible polymer supports on the rates of these peroxidase models.

Labat and Meunier investigated sulfonated iron(III)- and manganese(III)-porphyrins supported on a cation exchange resin and on partially protonated PVP (see Section II.A.2) as models for chloroperoxidase in the oxidative chlorination of dimedone by H_2O_2 in aqueous sodium chloride solution [77]. These studies revealed that the iron porphyrins behaved like a peroxidase and gave hydroxydimedone; however, the manganese analogs also produced significant amounts of chlorodimedone (Figure 20). The best catalysts for both oxidations were those supported on PVP and it was concluded that their activity was in part due to the pyridine groups attached as proximal ligands to the metal center.

In a series of recent papers, Labat and Meunier also describe their work in developing efficient chemical models for ligninase (a peroxidase that catalyzes the oxidation of lignin in the presence of H_2O_2) [45,65,71,70]. In these studies they used sulfonated iron(III)- and manganese(III)tetraarylporphyrins in homogeneous solution and immobilized on cationic organic polymers to catalyze the oxidation of two lignin model compounds by potassium monopersulfate and by H_2O_2 (Figure 21). On the anion exchange resin, Amberlite IRA 900, both Mn^{III} and Fe^{III} catalysts brought about the same reactions as in free solution, although at a slower rate. Furthermore, the manganese(III) catalysts, unlike their iron(III) analogs, required added pyridine, which served as a proximal ligand, for maximum efficiency. Interestingly, when the support was changed to partially protonated PVP the manganese(III)porphyrins were active in the absence of additional pyridine, showing that the support was able to provide the necessary pyridine ligand for the catalyst [45]. In recycling experiments the best manganese(III)porphyrin systems retained 95% of their activity on the third use although their iron(III) ana-

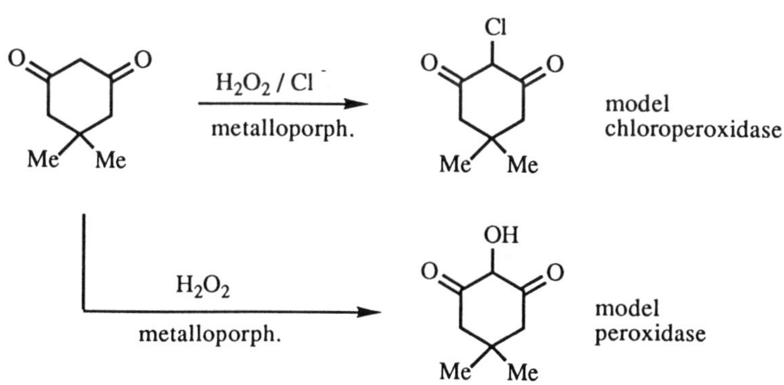

Figure 20 Peroxidase and chloroperoxidase model oxidations of dimedone catalyzed by metalloporphyrins [77].

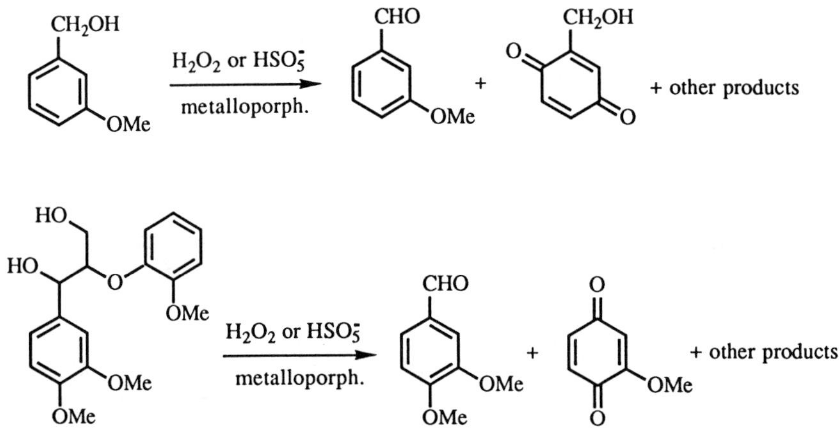

Figure 21 Oxidations of lignin models catalyzed by polymer-supported metalloporphyrins.

logs were less stable [45]. More extensive repeat use experiments were not described.

V. OXIDATIONS CATALYZED BY METALLOPORPHYRINS ON INORGANIC SUPPORTS

Inorganic solids, being rigid and oxidatively stable, have distinct advantages over organic polymers as supports for metalloporphyrin oxidation catalysts. Furthermore, the inorganic solid can have an amorphous (such as silica or alumina) or structured (such as zeolites and clays) surface, thereby allowing control of the local environment of the catalyst and access of substrates to the active oxidant. It is therefore somewhat surprising that despite these obvious potential advantages, there have been relatively few studies on oxidations catalyzed by metalloporphyrins anchored to inorganic supports. It seems likely, in view of the interesting results that have recently been obtained, that this area of supported catalysis will receive more attention in the near future.

A. Metalloporphyrins on Inorganic Supports as Models for Cytochrome P450 Monooxygenases

1. Catalyzed Oxidation by Iodosylbenzene

The large majority of the oxidation studies with metalloporphyrins on inorganic supports have been concerned with modeling the chemistry of cytochrome P450 monooxygenases and the most successful of these have employed iodosylbenzene as the oxidant. Amongst the earliest reports were those of Herron and his coworkers [91,92] who described the use of zeolites to control the selectivity of alkane hydroxylation. Although the catalyst was an iron(III)phthalocyanine rather than a porphyrin, the principles behind their investigations are important and future research is likely to explore this approach with porphyrins entrapped in other inorganic solids.

The aim of the research was to encapsulate the catalyst molecules within the supercages of a zeolite and to use the interlinking pores within the rigid framework of the support to select and direct the substrate molecules to the active oxidant. Using this so-called ship-in-a-bottle approach it was argued that if the active oxidant was sufficiently reactive it should not only oxidize the first molecule entering the supercage but also the first C-H bond of the molecule it encounters. Thus control of access to the catalyst should lead to selective oxidation.

Comparison of the encapsulated and homogeneous iron(III)phthalocyanine in oxidations showed the former to be more stable, since site isolation of the catalyst molecules prevented them from reacting with each other in a bimolecular self-destruction process. Other differences included a distinct, albeit small, shape selectivity for small alkane substrates in competitive oxidations and increased ω-1 regioselectivity in linear alkane hydroxylation, endo selectivity in norbornane hydroxylation, and trans-4-hydroxylation of methylcyclohexane by the encapsulated catalyst (Figure 22).

However, the zeolite catalysts have two major drawbacks. First, the rates of reaction were very slow because of the need to diffuse reactants and products in and out of the narrow pores and, second, catalyst turnovers were low as a result of pore blockage by products of oxidation.

In a more recent but related study of alkene oxidations by iodosylbenzene catalyzed by Mn^{III} salen encapsulated in a zeolite, Bowers and Dutta came to conclusions very similar to those of Herron and his coworkers [93].

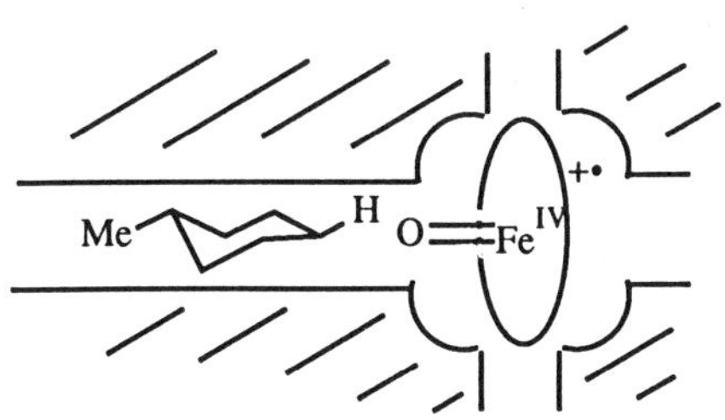

Figure 22 Schematic diagram of the trans-4-hydroxylation of methylcyclohexane by a zeolite-encapsulated oxoiron phthalocyanine-active oxidant [92].

Mansuy and his coworkers have shown that cationic and anionic manganese(III)tetraarylporphyrins adsorbed onto silica or alumina, respectively, are very effective catalysts for alkene epoxidation and alkane hydroxylation by iodosylbenzene [66,69]. The best of these was Mn^{III}T4MPyP on silica, which gave higher yields of oxidation products than the Mn^{III} porphyrin in free solution and gave a higher alcohol to ketone product ratio (Table 2). It was argued that the increased yield of alcohol may arise from better control of the alkyl radical in the oxygen rebound mechanism (Figure 23) favoring hydroxylation over other reactions of the radical. Surprisingly, Mn^{III}T4MPyP showed the same regioselectivity adsorbed on silica as it did in free solution.

In recycling experiments with cyclooctene epoxidation, the supported Mn^{III}T4MPyP was found to be fully active after 30,000 turnovers, showing it to be a very robust catalyst and certainly far more oxidatively stable than in free solution. A further useful feature of this catalyst was its ability to be used for alkane hydroxylation in neat alkane in the absence of a solvent, opening up the possibility of carrying out solvent-free oxidations.

Pautet and Daudon [56] recently used uncharged Fe^{III}TPP adsorbed onto silica which they reported to be a more selective catalyst for the mono-S-oxidation of sulfides by iodosylbenzene than the porphyrin in free solution. The former system did not suffer from overoxidation of sulfoxide to sulfone, which they attributed to the decreased reactivity of sulfoxide adsorbed onto silica.

Table 2 Yields of Alcohols and Ketones from the Oxidations of Alkanes by Iodosylbenzene in Acetonitrile/Dichloromethane Catalyzed by Mn^{III}T4MPyP in Solution and Supported on Silica or Montmorillonite

Alkane	Catalyst	Product yield (%)		Alcohol/ketone yield ratio
		Alcohol	Ketone	
Cyclohexane	Homogeneous	4	2	2
	On silica	26	4	6.5
	On clay	35	4	8.8
Cyclooctane	Homogeneous	5	6	0.8
	On silica	34	6.5	5.2
	On clay	53	6	8.8
Pentane	Homogeneous	1	—	—
	On silica	15	4	3.7
	On clay	21	4	5.2
Adamantane	Homogeneous	7	0.5	14
	On silica	37	1	37
	On clay	57	2	28.5

Source: Data from Ref. 69.

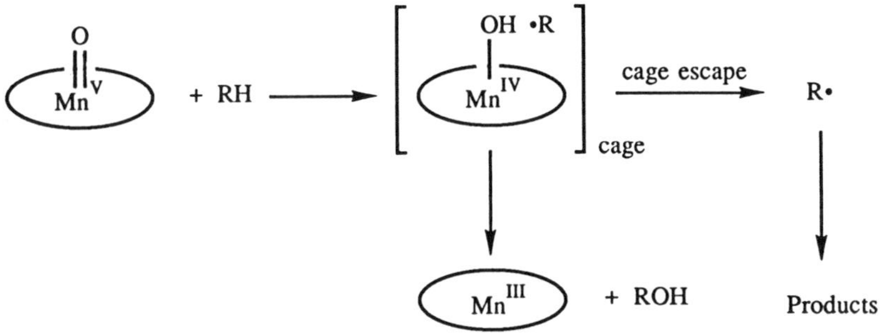

Figure 23 Alkane hydroxylation by oxomanganese(V) and a competitive pathway of alkyl radical arising from cage escape.

Mansuy's group [69,86] also explored the catalytic potential of ionic manganese(III)tetraarylporphyrins, adsorbed in the interlamellar space of montmorillonite, phosphatoantimonic acids, and layered dihydroxides, in alkene epoxidation and alkane hydroxylation. The best overall catalyst was Mn^{III}T4MPyP intercalated in montmorillonite, which is easy to prepare, and the porphyrin is held by strong ionic interactions with the support allowing it to be used, without leaching, in a wide range of solvents (CH_2Cl_2, CH_3OH or H_2O). Comparison of cyclooctene epoxidation catalyzed by this material and by Mn^{III}T4MPyP on silica showed that both catalysts gave very similar overall epoxide yields but that the reactions of the latter catalyst were faster, probably because access to the metalloporphyrin by iodosylbenzene and substrate was easier. However, no competitive epoxidation experiments designed to investigate the shape selectivity of the reactions were reported. By contrast, the yield of alcohols from alkanes was significantly higher with Mn^{III}T4MPyP in montmorillonite than on silica (Table 2) and the former catalyst in competitive reactions showed a preference for small unhindered over more branched alkanes. Combined with the higher overall yield of alcohols, there was also a higher ratio of alcohol to ketone products in oxidations with the clay-bound Mn^{III}T4MPyP (Table 2), making this material the most efficient catalyst for alkane hydroxylation by iodosylbenzene reported to date.

For linear alkanes, the regioselectivity of hydroxylation by iodosylbenzene was surprisingly independent of whether the manganese(III)porphyrin catalyst was in solution, supported on silica or alumina, or intercalated in montmorillonite, phosphatoantimonic acids, or layered dihydroxides. However, with more hindered substrates, such as adamantane, there was a small shift in favor of the less hindered secondary C-H bonds with Mn^{III}T4MPyP in montmorillonite compared with the porphyrin free solution.

In repeat use, Mn^{III}T4MPyP in montomorillonite was found to retain all its catalytic activity for alkane hydroxylation after four additions of oxidant. However, this amounts to a total of only 80 turnovers, so that further work is needed to compare the stability of the intercalated catalysts with the very robust Mn^{III}T4MPyP on silica [66].

Mansuy and Traylor and their coworkers also recently developed a simple one-step method of covalently binding polyhalogenated iron(III)- and manganese(III)porphyrins to supports [25]. As described above (Section II.A.1) this involved nucleophilic displacement by support nucleophiles of halide from haloaromatic groups on the porphyrin. In this way they prepared a range of iron(III) and manganese(III) catalysts bound to polystyrene, montmorillonite, and silica. All these supported metalloporphyrins catalyzed alkene epoxidation and alkane hydroxylation by iodosylbenzene; however, the best were Fe^{III}TPFPP and Mn^{III}TPFPP on silica. Further studies will reveal the full potential of these novel catalysts.

In another recent communication, Cooke and Lindsay Smith [52] reported the use of Fe^{III}TDCPP and Mn^{III}TDCPP coordinatively bound to silica modified with 3-(1-imidazolyl)propyl groups (Si-Im) as catalysts for epoxidation of cyclooctene by iodosylbenzene. These catalysts gave comparable yields of epoxycyclooctane to those from equivalent homogeneous oxidations but their reactions were four to five times slower. The catalysts were also shown to be very robust, giving large catalyst turnovers without loss of catalytic activity, e.g., for Si-Im-Fe^{III}TDCPP and Si-Im-Mn^{III}TDCPP 7900 and 24,700, respectively. The latter reaction, which was not run to complete consumption of substrate, gave an epoxide yield of 86% based on oxidant and when based on substrate, which was in excess, >70%.

2. Catalyzed Oxidations with Hydrogen Peroxide and Hydroperoxides

Extensive work by Mansuy and his coworkers [103] and others [116] revealed the key mechanistic features of cytochrome P450 monooxygenase models using H_2O_2 and manganese(III)porphyrin catalysts. This shows that it is important to encourage the heterolytic cleavage of the peroxide bond and generate the oxomanganese(V) intermediate through acid–base catalysis and a good donor proximal ligand (Figure 24), in a manner analogous to that of the generation of compound I in HRP. Imidazole and *N*-methylimidazole can fulfill both roles and as such promote the manganese(III)porphyrin–catalyzed epoxidation of alkenes.

Attempts to bring about these catalyzed oxidations with supported manganese(III)porphyrins has met with only limited success. Tatsumi and his coworkers [50,51] employed Mn^{III}TPP coordinated to silica, surface-modified with 3-(1-imidazolyl)propyl groups, which they showed to be a significantly more effective catalyst for cyclohexane hydroxylation by H_2O_2 than the same

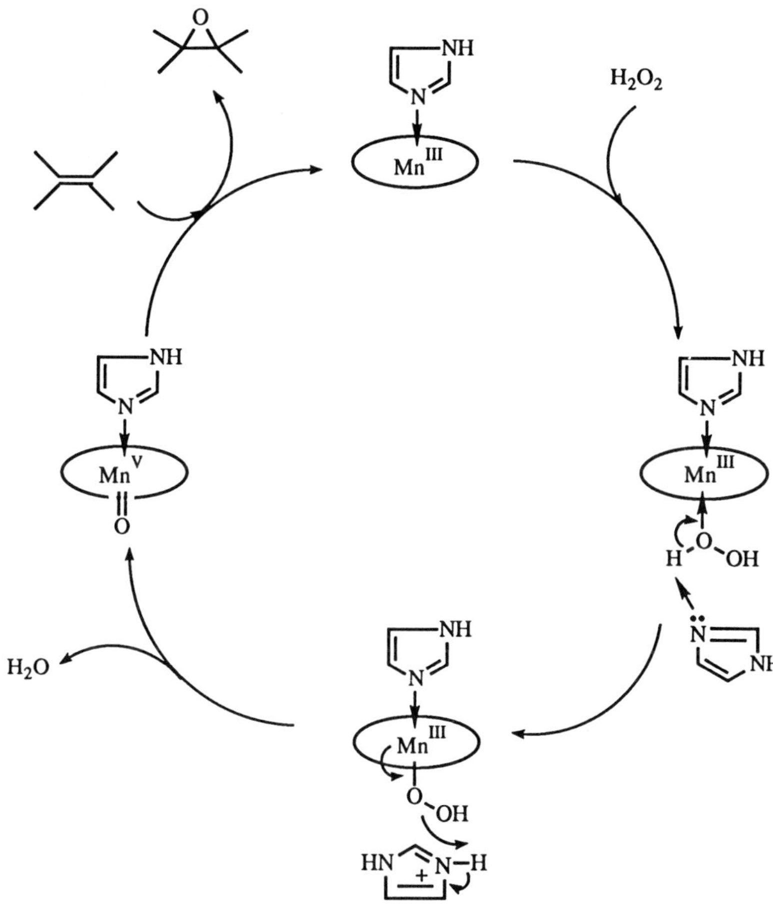

Figure 24 Role of imidazole in the heterolytic cleavage of hydrogen peroxide by a manganese(III)porphyrin [103].

metalloporphyrin in solution or adsorbed onto silica, or onto silica which had been pretreated with *N*-methylimidazole. However, even the best catalyst gave only a modest 40 turnovers. Cooke and Lindsay Smith [52] with the same approach used Mn^{III}TDCPP as catalyst to epoxidize cyclooctene and obtained catalyst turnovers of ~100. However, with *t*-BuO_2H as the oxidant epoxidation yields were very low [117]. It is likely that the efficiency of these systems is low since the support only supplies a proximal axial ligand and is unable to provide the acid–base catalysis needed to promote the heterolytic cleavage of the peroxide bond. As a consequence the catalyst is degraded by competitive radical reactions.

Using another approach Tatsumi's group prepared iron(III)- and manganese(III)tetramethylporphyrin catalysts encapsulated in a zeolite matrix (the more normally used tetraarylporphyrins are too large to fit in the zeolite supercages) [94]. These materials were, however, found to be poor catalysts for hydroxylation of cyclohexane by H_2O_2 and it was argued that the majority of the catalyst was located inside the zeolite and unavailable for reaction.

It is clear that further work is needed to improve the efficiency of these supported metalloporphyrin models for cytochrome P450 monooxygenases. This should include the incorporation of an acid–base function into the catalyst in order to promote the heterolysis of H_2O_2.

3. Catalyzed Oxidation with Dioxygen

An interesting study by Tanaka and his coworkers has compared the catalytic behavior of Mn^{III}TPP and Fe^{III}TPP in homogeneous solution and ligated to silanol groups on silica or silica-alumina in the oxidation of cyclohexene in the presence of $NaBH_4$ [55]. With the Mn^{III} catalysts the products obtained were independent of whether the reactions were hetero- or homogeneous. However, the homogeneous and supported Fe^{III}TPP systems were found to show distinct differences. Kinetic and product studies suggested that in solution Fe^{III}TPP catalyzes the radical chain autoxidation of cyclohexene leading to predominantly cyclohexenone [107] whereas the supported catalyst behaves like a cytochrome P450 monooxygenase model giving epoxycyclohexane and cyclohexenol. The former catalyst was converted to μ-oxoiron(III)porphyrin dimers, whereas the latter remained largely unchanged. The differences were attributed to the support which site-isolated the iron porphyrin molecules, thus preventing the formation of μ-oxo- and peroxo dimers (Reaction 6) and encouraged the formation of high-valent oxoiron species. A further role of the support may have been to bind the substrate. In this respect the authors likened the support to the protein matrix in cytochrome P450 monooxygenases (Figure 25). Both the homogeneous and heterogeneous manganese(III) catalysts, however, favored the oxomanganese intermediates and cytochrome P450-type oxygenations.

$$\mathrm{Fe^{II}TPP} + \mathrm{O_2} \longrightarrow \overset{\mathrm{O_2^{\bullet}}}{\overset{|}{\mathrm{Fe^{III}TPP}}} \xrightarrow{\mathrm{Fe^{II}TPP}} (\mathrm{Fe^{III}TPP})_2\mathrm{O_2} \xrightarrow{\mathrm{Fe^{II}TPP}} (\mathrm{Fe^{III}TPP})_2\mathrm{O} \qquad (6)$$

VI. FUTURE DEVELOPMENTS OF SUPPORTED METALLOPORPHYRIN-CATALYZED OXIDATIONS

An important goal in recent studies has been the development of oxidatively stable and catalytically active supported metalloporphyrins. This work suggests

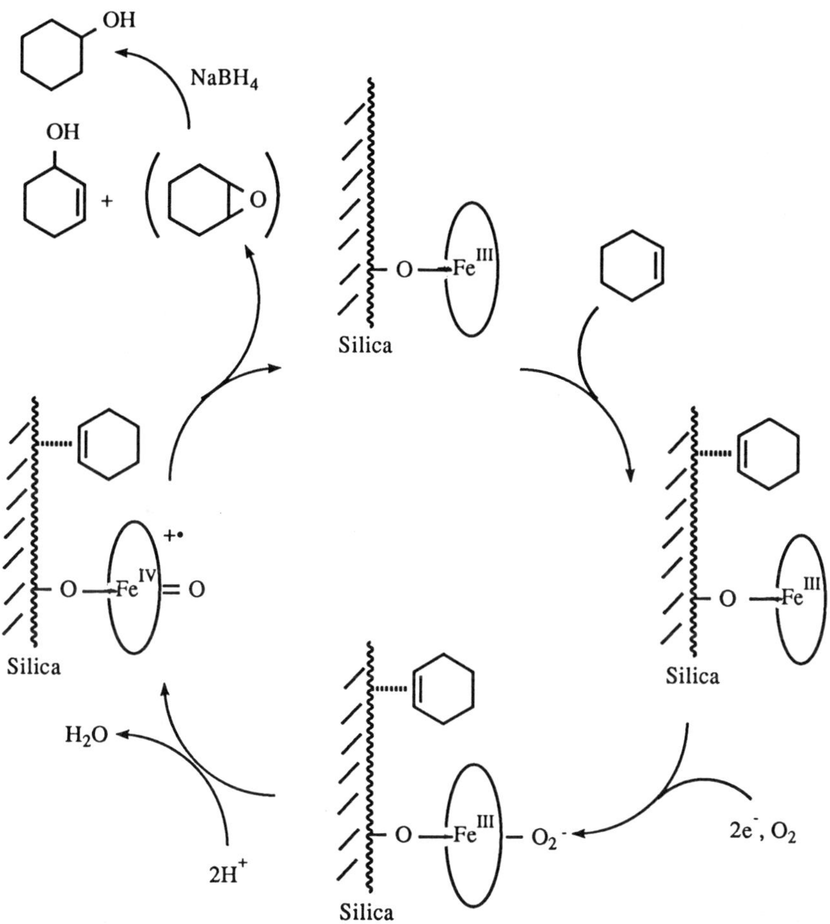

Figure 25 Oxidation of cyclohexene by dioxygen and sodium borohydride catalyzed by an iron(III)porphyrin supported on silica [55].

that the best catalysts in future will be supported on inorganic solids or on rigid inert polymers to minimize porphyrin–porphyrin self-destruction processes and the wasteful consumption of oxidant in support oxidation. Catalyst stability should be improved still further through the use of sterically hindered robust metalloporphyrins.

Combined with the search for stable supported metalloporphyrins has been the challenge of developing selective catalysts. In this respect the clay-intercalated and silica-supported cationic metalloporphyrins show promise in alkane hydroxylation [69]. More extensive studies are required to explore the potential of surface structured supports, including, clays, pillared clays, templated silica [118], and shape-selective polymers [119], in promoting chemo- and regioselective oxidations.

The increasing demand for enantioselective oxidants will require the development of supported chiral oxidation catalysts. This may be achieved through the use of chiral supports, by immobilizing chiral metalloporphyrins, or a combination of both methods. An alternative approach is the development of catalytic antibodies (abzymes) with metalloporphyrin cofactors [120] to provide a route into enantioselective oxidations. In principle these catalytic systems, either supported or in solution, should be capable of being tailored to very specific oxidations.

If oxidations catalyzed by supported metalloporphyrins are to be synthetically useful, they must be applicable to at least laboratory and preferably larger scale reactions. Future research must examine the problems in scaling up the reactions for batch or continuous flow reactors. The need for cheap, environmentally friendly catalytic oxidations also imposes important restrictions on the future developments of metalloporphyrin systems. The research emphasis will need to be directed more to the use of oxidants such as hydrogen peroxide and dioxygen than iodosylarenes, hypochlorite, or peracids. This will build on the optimum features of the homogeneous systems and will require designing supports that will favor the formation of high-valent oxometalloporphyrin intermediates. With dioxygen systems, the necessary reducing equivalents could be provided electrochemically, using a porphyrin supported on an electrode, rather than by a sacrificial reductant.

Other areas for future studies should include oxidations in solvent-free or aqueous systems, photo-assisted oxidations [121,122], and controlled autoxidations [107–109].

ABBREVIATIONS

Tetraarylporphyrin Anions (TArP)	Aryl Group
TCPP	4-carboxyphenyl
TDCPP	2,6-dichlorophenyl
TDCSPP	2,6-dichloro-3-sulfonatophenyl
TDSMPP	3,5-disulfonato-2,4,6-trimethylphenyl
T2MPyP	2-*N*-methylpyridyl
T4MPyP	4-*N*-methylpyridyl
TPFPP	pentafluorophenyl
TPP	phenyl
TSPP	4-sulfonatophenyl
TSMP	3-sulfonato-2,4,6-trimethylphenyl
TTAPP	4-trimethylammoniumylphenyl

β-pyrrole–halogenated porphyrins ($TArHal_8P$)	Aryl group
$TCSMCl_8P$	3-chloro-5-sulfonato-2,4,6-trimethylphenyl
$TDSMBr_8P$	3,5-disulfonato-2,4,6-trimethylphenyl
$TDCPCl_8P$	2,6-dichlorophenyl
$TPFPBr_8P$	pentafluorophenyl

Supports

PAA	polyacrylic acid
PS-Im	imidazole-modified polystyrene
PSS	poly(styrene-4-sulfonate)
PVP	poly(4-vinylpyridine)
PVS	poly(vinylsulfonate)

REFERENCES

1. B. Meunier, Metalloporphyrin-catalysed oxygenation of hydrocarbons, *Bull. Soc. Chim. Fr.*, 578 (1986).
2. D. Mansuy, P. Battioni, and J.P. Battioni, Chemical model systems for drug metabolising cytochrome P450 dependent monooxygenases, *Eur. J. Biochem., 184*: 267 (1989).
3. D. Mansuy, Biomimetic catalysts for selective oxidation in organic chemistry, *Pure Appl. Chem., 62*: 741 (1990).
4. B. Meunier, Metalloporphyrins as versatile catalysts for oxidation reactions and oxidative DNA cleavage, *Chem. Rev., 92*: 1411 (1992).
5. J.T. Groves, T.E. Nemo, and R.S. Myers, Hydroxylation and epoxidation catalysed by iron-porphine complexes. Oxygen transfer from iodosylbenzene, *J. Am. Chem. Soc., 101*: 1032 (1979).
6. P.S. Traylor, D. Dolphin, and T.G. Traylor, Sterically protected hemins with electronegative substituents: efficient catalysts for hydroxylation and epoxidation, *J. Chem. Soc. Chem. Commun.*, 279 (1984).
7. M.J. Nappa and C.A. Tolman, Steric and electronic control of iron porphyrin catalysed hydrocarbon oxidations, *J. Am. Chem. Soc., 24*: 4711 (1985).
8. T.G. Traylor and S. Tsuchiya, Perhalogenated tetraphenylhemins: stable catalysts of high turnover catalytic hydroxylations, *Inorg. Chem., 26*: 1338 (1987).
9. K. Suslick, B. Cook, and M. Fox, Shape selective alkane hydroxylation, *J. Chem. Soc. Chem. Commun.*: 580 (1985).
10. K. Suslick and B.R. Cook, Regioselective epoxidations of dienes with manganese(III) porphyrin catalysts, *J. Chem. Soc. Chem. Commun.*, 200 (1987).
11. J.P. Collman, X. Zhang, R.T. Hembre, and J.I. Brauman, Shape-selective olefin epoxidation catalysed by manganese picnic basket porphyrins, *J. Am. Chem. Soc., 112*: 5356 (1990).
12. J.T. Groves and R.S. Myers, Catalytic asymmetric epoxidation with chiral iron porphyrins, *J. Am. Chem. Soc., 105*: 5791 (1983).

13. D. Mansuy, P. Battioni, J.-P. Renaud, and P. Guerin, Asymmetric epoxidation of alkenes catalyzed by a "basket-handle" iron-porphyrin bearing amino acids, *J. Chem. Soc. Chem. Commun.*, 155 (1985).
14. R.L. Halterman and S.-T. Jan, Catalytic asymmetric epoxidation of unfunctionalised alkenes using the first D_4-symmetric metallotetraphenylporphyrin, *J. Org. Chem., 56*: 5253 (1991).
15. J.P. Collman, X. Zhang, V.J. Lee, and J.I. Brauman, Asymmetric epoxidation catalysed by an iron "binap capped" porphyrin, *J. Chem. Soc. Chem. Commun.*, 1647 (1992).
16. D.R. Leanord and J.R. Lindsay Smith, Model systems for cytochrome P450 dependent monooxygenases. 8. A study of the epoxidation of (Z)-cyclooctene by iodosylbenzene catalysed by cationic iron(III) tetra(N-methylpyridyl)porphyrins on Dowex MSC1, *J. Chem. Soc. Perkin Trans., 2*: 25 (1991).
17. E. Tsuchida and H. Nishide, Polymer–metal complexes and their catalytic activity, *Adv. Polym. Sci., 24*: 1 (1977).
18. D. Wöhrle, Polymer square planar metal chelates for science and industry, *Adv. Polym. Sci., 50*: 45 (1983).
19. D.C. Sherrington, Polymer-supported metal complex oxidation catalysts, *Pure Appl. Chem., 60*: 401 (1988).
20. L.D. Rollman, Porous, polymer-bound metalloporphyrins, *J. Am. Chem. Soc., 97*: 2132 (1975).
21. K. Maruyama, H. Tamiaki, and S. Kawabata, Exothermic isomerisation of water-soluble quadricyclanes to norbornadienes by soluble and insoluble catalysts, *J. Chem. Soc. Perkin Trans., 2*: 543 (1986).
22. J. Gitzel, H. Ohno, E. Tsuchida, and D. Woehrle, Electron transfer reaction between metalloporphyrins covalently bound to polystyrene in N,N-dimethylformamide, *Polymer, 27*: 1781 (1986).
23. D. Wöhrle and G. Krawczyk, Polymeric bound porphyrins and their precursors. 3. Photoredox properties of combined moieties of porphyrin and phthalocyanine, covalently bound to polystyrene, *Makromol. Chem., 187*: 2535 (1986).
24. D. Wöhrle, J. Gitzel, G. Krawczyk, E. Tsuchida, H. Ohno, I. Okura, and T. Nishisaka, Synthesis, redox behaviour, sensitiser activity and oxygen transfer of covalently bound polymeric porphyrins, *J. Macromol. Sci-Chem., A25*: 1227 (1988).
25. P. Battioni, J.F. Bartolli, D. Mansuy, Y.S. Byun, and T.G. Traylor, An easy access to polyhalogenated metalloporphyrins covalently bound to polymeric supports as efficient catalysts for hydrocarbon oxidation, *J. Chem. Soc. Chem. Commun.*, 1051 (1991).
26. A. Bettelheim, R.J.H. Chan, and T. Kuwana, Electroanalysis of oxygen reduction. 3. Selective reduction to hydrogen peroxide or water using polymeric attachment of metalloporphyrins, *J. Electroanal. Chem., 110*: 93 (1980).
27. T. Mori, T. Santa, and M. Hirobe, Synthesis and cytochrome P450-like reactivity of polypeptide-bound porphyrinatoiron(III), *Tetrahedron Lett., 26*: 5555 (1985).
28. T. Higuchi, T. Mori, S. Uzu, and M. Hirobe, Polypeptide-bound porphyrinatoirons as cytochrome P450 model compounds, in *The Role of Oxygen in Chemistry and Biochemistry* (W. Ando and Y. Moro-oka, eds.), Elsevier, Amsterdam, 1988, p. 455.

29. D. Wöhrle and J. Gitzel, Polymeric bound porphyrins and their precursors. 9. Synthesis of water-soluble polymers with covalently bound tetraphenylporphyrins and their use in the epoxidation of 2,5-dihydrofuran, *Makromol. Chem. Rapid Commun., 9*: 229 (1988).
30. K. Yoshinaga, N. Itoh, and T. Kito, Enantioselective oxidation catalysed by manganese(III) tetraphenylporphyrin-bovine serum albumin systems modified or unmodified by poly(ethylene glycol), *Polymer J., 23*: 65 (1991).
31. T. Fujimoto, H. Umekawa, and N. Nishino, A new picket-fence porphyrin precursor, synthesis and atropisomerism of free base and zinc complex of meso-tetra(o-methoxycarbonylphenyl)porphyrin, *Chem. Lett.*, 37 (1992).
32. K. Takahashi, A. Matsushima, Y. Saito, and Y. Inada, Polyethylene glycol-modified hemin having peroxidase activity in organic solvents, *Biochem. Biophys. Res. Commun., 138*: 283 (1986).
33. A.W. van der Made, J.W.H. Smeets, R.J.M. Nolte, and W. Drenth, Olefin epoxidation by a mono-oxygenase model. Effect of site isolation, *J. Chem. Soc. Chem. Commun.*, 1204 (1983).
34. J.A.S.J. Razenberg, A.W. van der Made, J.W.H. Smeets, and R.J.M. Nolte, Cyclohexene epoxidation by the mono-oxygenase model (tetraphenylporphyrinato)-manganese(III) acetate-sodium hypochlorite, *J. Mol. Catal., 31*: 271 (1985).
35. H.S. Hilal, M.L. Sito, and A.F. Schreiner, Preparation and characterisation of poly(siloxane)-supported metalloporphyrins: the effect of the support on the electronic absorption spectra, *Inorg. Chim. Acta, 189*: 141 (1991).
36. H.S. Hilal, C. Kim, M.L. Sito, and A.F. Schreiner, The preparation and characterisation of tetra(4-pyridyl)porphyrinatomanganese(III) cation supported covalently on poly(siloxane), *J. Mol. Catal., 64*: 133 (1991).
37. C.A. Marrese, E.A. Blubaugh, and R.A. Durst, Chemical modification of Pt electrodes by tetra(4-pyridyl)porphyrin through siloxane polymer formation, *J. Electroanal. Chem., 243*: 193 (1988).
38. R.M. Kellett and T. Spiro, Cobalt porphyrin electrode films as H_2 evolution catalysts, *Inorg. Chem., 24*: 2378 (1985).
39. J.-H. Fuhrop, S. Besecke, W. Vogt, J. Ernst, and J. Subramanian, Metalloporphyrins in polymeric matrices and in micelles. 2. Iron porphyrins and cobalt porphyrins in copolymers of styrene and 1-vinylimidazole, *Macromol. Chem., 178*: 1621 (1977).
40. E. Hasegawa, T. Kanayama, and E. Tsuchida, Synthesis of polystyrene-covalently bonded metalloporphyrin complexes and their reaction with molecular oxygen, *J. Polym. Sci., 15*: 3039 (1977).
41. E. Tsuchida, E. Hasegawa, and T. Kanayama, Oxygenation of polymer-covalently bound metalloporphyrins, *Macromolecules, 11*: 947 (1978).
42. N.S. Enikolopyan and A.B. Solovéva, Catalysis of non-chain olefin oxidation reactions by metal porphyrins, *Zhur. Fiz. Khim., 62*: 2289 (1988).
43. T.G. Traylor, Y.S. Byun, P.S. Traylor, P. Battioni, and D. Mansuy, Polymeric polyhalogenated metalloporphyrin catalysts for hydroxylation of alkanes and epoxidation of alkenes, *J. Am. Chem. Soc., 113*: 7821 (1991).
44. A. Bettelheim, B.A. White, S.A. Raybuck, and R.W. Murray, Electrochemical polymerisation of amino-, pyrrole-, and hydroxy-substituted tetraphenylporphyrins, *Inorg. Chem., 26*: 1007 (1987).

45. G. Labat and B. Meunier, Easy preparation of supported metalloporphyrin catalysts with a proximal ligand arising from the support itself, *C. R. Acad. Sci. Paris, 311*: 625 (1990).
46. J.P. Collman and C.A. Reed, Synthesis of ferrous-porphyrin complexes. A hypothetical model for deoxymyoglobin, *J. Am. Chem. Soc., 95*: 2048 (1973).
47. E. Tsuchida, K. Honda, and E. Hasegawa, Complexes between synthetic polymer ligands and ferri- and ferro-protoporphyrin IX, *Biochim. Biophys. Acta, 393*: 483 (1975).
48. E. Tsuchida, E. Hasegawa, and K. Honda, Reaction of polymer-heme complexes with carbon monoxide and molecular oxygen, *Biochem. Biophys. Res. Commun., 67*: 864 (1975).
49. O. Leal, D.L. Anderson, R.C. Bowman, F. Basolo, and R.L. Burwell, Reversible adsorption of oxygen on silica gel modified by imadazole-attached iron tetraphenylporphyrin, *J. Am. Chem. Soc., 97*: 5125 (1975).
50. T. Tatsumi, M. Nakamura, and H. Tominaga, Hydroxylation of alkanes catalysed by manganese tetraphenylporphyrin immobilised on imidazole-modified silica gel, *Chem. Lett.*, 419 (1989).
51. T. Tatsumi, M. Nakamura, and H. Tominaga, Hydroxylation of alkanes catalysed by manganese tetraphenylporphyrin immobilised on imidazole-modified silica gel, *Catal. Today, 6*: 163 (1989).
52. P.R. Cooke and J.R. Lindsay Smith, Alkene epoxidation catalysed by ligand-bound supported metalloporphyrins, *Tetrahedron Lett., 33*: 2737 (1992).
53. S. Campestrini and B. Meunier, Olefin epoxidation and alkane hydroxylation catalysed by robust sulfonated manganese and iron porphyrins on cationic ion-exchange resins, *Inorg. Chem., 31*: 1999 (1992).
54. W. Bushong, S. Shupack, E. Blubaugh, and R. Durst, Polymer film electrodes: crosslinking to enhance stability, *Polym. Material Sci. Eng., 53*: 123 (1985).
55. S. Fukuzumi, S. Mochizuki, and T. Tanaka, P450 Type activation of dioxygen by heterogenised metal porphyrins: comparison with the corresponding homogeneous systems, *Isr. J. Chem., 28*: 29 (1987).
56. F. Pautet and M. Daudon, Reactions on solid supports: oxidation of sulphides to sulphoxides catalysed by metalloporphyrins supported on silica gel, *Tetrahedron Lett., 32*: 1457 (1991).
57. C. Gilmartin and J.R. Lindsay Smith, unpublished results.
58. J.R. Lindsay Smith and J. Nener, unpublished results.
59. T. Shimidzu and K. Fukui, Photochemical decomposition of water with polymer electrolyte complex catalysts, *Kyoto Daigaku Nippon Kagaku Saní Kenkyusho, 38*: 93 (1981).
60. B. de Vismes, F. Bedioui, and J. Derynck, Catalytic oxidation of di-tertbutylphenol by molecular oxygen electroassisted by manganese-porphyrin fixed on zeolite, *New J. Chem., 10*: 81 (1986).
61. Y. Saito, M. Mifune, T. Kawaguchi, J. Odo, Y. Tanaka, M. Chikuma, and H. Tanaka, Catalase-like catalytic activity of ion-exchange resins modified with metalloporphyrins, *Chem. Pharm. Bull., 34*: 2885 (1986).
62. Y. Saito, M. Mifune, S. Nakashima, H. Nakayama, J. Odo, Y. Tanaka, M. Chikuma, and H. Tanaka, Peroxidase-like catalytic activity of anion-exchange

resins modified with metalloporphyrins in the dye formation reaction of N,N-diethylaniline with 4-aminoantipyrine by hydrogen peroxide, *Chem. Pharm. Bull., 35*: 869 (1987).
63. Y. Saito, M. Mifune, S. Nakashima, J. Odo, Y. Tanaka, M. Chikuma, and H. Tanaka, Determination of hydrogen peroxide with N,N-diethylaniline and 4-aminoantipyrine by use of an ion-exchange resin modified with manganese-tetrakis(sulphonatophenyl)porphine as a substitute for peroxidase, *Talanta, 34*: 667 (1987).
64. Y. Saito, M. Satouchi, M. Mifune, T. Tai, J. Odo, Y. Tanaka, M. Chikuma, and H. Tanaka, Catalase-like activity of anion-exchange resin modified with metalloporphyrin in the oxidation of methyl alcohols and its application to the deterination of hydrogen peroxide, *Bull. Chem. Soc. Jpn., 60*: 2227 (1987).
65. G. Labat and B. Meunier, Efficient ligninase models based on water-soluble manganese and iron porphyrin complexes associated with potassium monopersulphate, *New J. Chem., 13*: 801 (1989).
66. P. Battioni, J.-P. Lallier, L. Barloy, and D. Mansuy, Monooxygenase-like oxidation of hydrocarbons using supported manganese-porphyrin catalysts: beneficial effects of a silica support for alkane hydroxylation, *J. Chem. Soc. Chem. Commun.,* 1149 (1989).
67. D.R. Leanord and J.R. Lindsay Smith, Model systems for cytochrome P450 dependent mono-oxygenases. 7. Alkene epoxidation by iodosylbenzene catalysed by ionic iron(III) tetraarylporphyrins supported on ion-exchange resins, *J. Chem. Soc. Perkin Trans., 2*: 1917 (1990).
68. K.T. Yue, M. Lin, T.A. Gray, and L.G. Marzilli, Nickel(II) porphyrin binding to anionic biopolymers investigated by resonance Raman and optical spectroscopy, *Inorg. Chem., 30*: 3214 (1991).
69. L. Barloy, J.-P. Lallier, P. Battioni, and D. Mansuy, Manganese-porphyrins absorbed and intercalated in different mineral matrices: preparation and compared properties as catalysts for alkene and alkane oxidation, *New J. Chem., 16*: 71 (1992).
70. Y. Saito, M. Mifune, S. Nakashima, Y. Tanaka, M. Chikuma, and H. Tanaka, Peroxidase-like accelerative activity of ion-exchange resins modified with metalloporphyrins with the dye formation reaction of phenol with 4-aminoantipyrine by hydrogen peroxide, *Chem. Pharm. Bull., 34*: 5016 (1986).
71. G. Labat, J.-L. Seris, and B. Meunier, Oxidative degradation of aromatic pollutants by chemical models of ligninase based on porphyrin complexes, *Angew. Chem. Int., 29*: 1471 (1990).
72. H. Turk and W.T. Ford, Epoxidation of styrene with aqueous hypochlorite catalysed by a manganese(III) porphyrin bound to colloidal anion-exchange particles, *J. Org. Chem., 56*: 1253 (1991).
73. W. Zhu and W.T. Ford, Oxidations catalysed by a manganese(III) porphyrin and cationic polymer latexes, *J. Polym. Sci: Pt. A Polym. Chem., 30*: 1305 (1992).
74. Y.O. Su and T. Kuwana, The deactivation/reactivation of cobalt(III) tetrakis(4-pyridyl)porphyrin electrocatalyst, *Chem. Lett.,* 459 (1985).
75. H. Segawa, T. Shimidzu, and K. Honda, Control of π-radical anion state of porphyrin with a polymer matrix, *Polym. J., 20*: 441 (1988).

76. G. Labat and B. Meunier, Factors controlling the reactivity of a ligninase model based on the association of potassium monopersulfate to manganese and iron porphyrin complexes, *J. Org. Chem., 54*: 5008 (1989).
77. G. Labat and B. Meunier, First example of a chloroperoxidase-type chlorination of dimedone using a supported manganese porphyrin catalyst, *J. Chem. Soc. Chem. Commun.,* 1414 (1990).
78. R.C. Smierciak and P.J. Giordano, Selective catalysis of exoergic isomerisations using ionically polymer anchored metalloporphyrins, *Appl. Catal., 18*: 353 (1985).
79. J.R. Lindsay Smith and R.J. Lower, The structure of iron(III) tetra(4-N-methylpyridyl)porphyrin species bound to synthetic water-soluble anionic polymers and their reactions with tert-butyl hydroperoxide in aqueous solution, *J. Chem. Soc. Perkin Trans., 2*: 2187 (1992).
80. R.J.H. Chan, S.O. Su, and T. Kuwana, Electrocatalysis of oxygen reduction. 5. Oxygen to hydrogen peroxide conversion by cobalt(II) tetrakis(N-methyl-4-pyridyl)porphyrin, *Inorg. Chem., 24*: 3777 (1985).
81. T.J. Pinnavaia, Intercalated clay catalysts, *Science, 220*: 365 (1983).
82. H. Van Damme, M. Crespin, F. Obrecht, M.I. Cruz, and J.J. Fripiat, Acid-base and complexation behaviour of porphyrins on the intracrystal surface of swelling clays: meso-tetraphenylporphyrin and meso-tetra(4-pyridyl)porphyrin on montmorillonite, *J. Colloid Interf. Sci., 60*: 43 (1978).
83. K.A. Carrado and R.E. Winans, Intercalation of water-soluble porphyrins and metalloporphyrins with smectite clay surfaces, *Chem. Mater., 2*: 328 (1990).
84. H. Kameyama, H. Susuki, and A. Amano, Intercalation of cobalt(II) meso-tetrakis(1-methyl-4-pyridyl)porphyrin into montmorillonite, *Chem. Lett.,* 1117 (1988).
85. P. Battioni, L. Barloy, D. Mansuy, P. Palvadeau, M. Tournoux, Y. Piffard, and J. Rouxel, Selective oxidation of hydrocarbons catalysed by metalloporphyrins intercalated in layered inorganic structure, *Curr. Trend. Appl. [Proc. Workshop],* 195 (1989).
86. L. Barloy, P. Battioni, and D. Mansuy, Manganese porphyrins supported on montmorillonite as hydrocarbon mono-oxygenation catalysts: particular efficiency for linear alkane hydroxylation, *J. Chem. Soc. Chem. Commun.,* 1365 (1990).
87. S.S. Cady and T.J. Pinnavaia, Porphyrin intercalation in mica-type silicates, *Inorg. Chem., 17*: 1501 (1978).
88. J.H. Wang, Synthetic biochemical models, *Acc. Chem. Res., 3*: 90 (1970).
89. C.K. Chang and T.G. Traylor, Synthesis of the myoglobin active site, *Proc. Natl. Acad. Sci. USA, 70*: 2647 (1973).
90. H. Nishide, M. Ohyanagi, O. Okada, and E. Tsuchida, Highly selective transport of molecular oxygen in a polymer containing cobalt porphyrin complex as a fixed carrier, *Macromolecules, 19*: 494 (1986).
91. N. Herron, G.D. Stucky, and C.A. Tolman, Shape selectivity in hydrocarbon oxidations using zeolite encapsulated iron phthalocyanine catalysts, *J. Chem. Soc. Chem. Commun.,* 1521 (1986).
92. N. Herron, The selective partial oxidation of alkanes using zeolite based catalysts. Phthalocyanine (PC) "ship-in-a-bottle" species, *J. Coord. Chem., 19*: 25 (1988).

93. C. Bowers and P.K. Dutta, Olefin oxidation by zeolite-encapsulated manganese (salen)$^+$ complexes under ambient conditions, *J. Catal., 122*: 271 (1990).
94. M. Nakamura, T. Tatsumi, and H. Tominaga, Synthesis and catalytic capability of zeolite-encapsulated iron and manganese tetramethylporphyrin complexes, *Bull. Chem. Soc. Jpn., 63*: 3334 (1990).
95. L.-C. Yuan and T.C. Bruice, Influence of nitrogen base ligation and hydrogen bonds on the rate constants of oxygen transfer from percarboxylic acids and alkyl hydroperoxides to (meso-tetraphenylporphinato)manganese(III) chloride, *J. Am. Chem. Soc., 108*: 1643 (1986).
96. W.-H. Wong, D. Ostović, and T.C. Bruice, Kinetics and mechanism of oxygen transfer in the reaction of p-cyano-N,N-dimethylaniline N-oxide with metalloporphyrin salts. 5. The influence of imidazole ligation of (meso-tetrakis(2,6-dimethylphenyl)porphinato)manganese(III) chloride on the rates of oxygen transfer from N-oxide to metalloporphyrin salt and the efficiency of oxidations by the intermediate higher valent manganese-oxo species, *J. Am. Chem. Soc., 109*: 3428 (1987).
97. B. Meunier, M.E. de Carvalho, O. Bortolini, and M. Momenteau, Proximal effect of the nitrogen ligands in the catalytic epoxidation of olefins by the NaOCl/manganese(III) porphyrin system, *Inorg. Chem., 27*: 161 (1988).
98. P. Battioni, P.R. Cooke, J.R. Lindsay Smith, and D. Mansuy, unpublished results.
99. F. Montanari, S. Banfi, and S. Quici, Oxidations catalysed by stable metallotetraarylporphyrins under aqueous-organic two-phase conditions, *Pure Appl. Chem., 61*: 1631 (1989).
100. S. Banfi, F. Montanari, and S. Quici, Investigation on factors ruling catalytic efficiency and chemical stability of Mn(III) porphyrins in HOCl olefin epoxidation: conditions for practical application, *J. Org. Chem., 54*: 1850 (1989).
101. A. Robert, B. Loock, M. Momenteau, and B. Meunier, Catalase modeling with metalloporphyrin complexes having an oxygen ligand in a proximal position. Comparison with complexes containing a proximal nitrogen, *Inorg. Chem., 30*: 706 (1991).
102. K.M. Kadish, C. Araullo-McAdams, B.C. Han, and M.M. Franzen, Synthesis and spectroscopic characterisation of (T(p-Me_2N)F_4PP)H_2 and (T(p-Me_2N)F_4PP)M where T(p-Me_2N)F_4PP is the dianion of meso-tetrakis(o,o,m,m-tetrafluoro-p-(dimethylamino)phenyl)porphyrin and M = Co(II), Cu(II) or Ni(II). Structures of T(p-Me_2N)F_4PP)Co and (meso-tetrakis)pentafluorophenyl)porphinato)cobalt(II), (TF_5PP)Co, *J. Am. Chem. Soc., 112*: 8364 (1990).
103. P. Battioni, J.-P. Renaud, M. Reina-Artilles, M. Fort, and D. Mansuy, Monooxygenase-like oxidation of hydrocarbons by H_2O_2 catalysed by manganese porphyrins and imidazole: selection of the best catalytic system and nature of the active oxidant species, *J. Am. Chem. Soc., 110*: 8462 (1988).
104. I. Tabushi and N. Koga, P-450 type oxygen activation by porphine-manganese complex, *J. Am. Chem. Soc., 101*: 6456 (1979).
105. I. Tabushi and A. Yazaki, P-450 type dioxygen activation using H_2/colloidal Pt as an effective electron donor, *J. Am. Chem. Soc., 103*: 7371 (1981).
106. M. Fontecave and D. Mansuy, Monooxygenase-like oxidations of olefins and alkanes catalysed by manganese porphyrins: comparison of systems in-

volving either O_2 and ascorbate or iodosylbenzene, *Tetrahedron, 40*: 4297 (1984).

107. D.R. Paulson, R. Ullman, and R.B. Sloane, Catalysis of autoxidation by metalloporphyrins, *J. Chem. Soc. Chem. Commun.,* 186 (1974).
108. P.E. Ellis and J.E. Lyons, Effect of fluorination of the meso-phenyl groups on selective tetraphenylporphyrinatoiron(III)-catalysed reactions of propane with molecular oxygen, *J. Chem. Soc. Chem. Commun.,* 1315 (1989).
109. J.E. Lyons and P.E. Ellis, Selective low temperature hydroxylation of isobutane by molecular oxygen catalysed by iron perhaloporphyrin complex, *Catal. Lett., 8*: 45 (1991).
110. M. Kuhn and P. Möhr, Metalloporphyrin complexes: polymer diimidazoliron porphyrin-complex as a functional model of cytochrome P450, *Pharmazie, 36*: 383 (1981).
111. H.B. Dunford, Horseradish peroxidase: structure and kinetic properties, in *Peroxidases in Chemistry and Biology* (J. Everse, K.E. Everse, and M.B. Grisham, eds.), CRC Press, Boca Raton, 1991, p. 1.
112. P.R. Ortiz de Montellano, Catalytic sites of hemoprotein peroxidases, *Annu. Rev. Pharmacol. Toxicol., 32*: 89 (1992).
113. T.G. Traylor and F. Xu, Mechanisms of reactions of iron(III) porphyrins with hydrogen peroxide and hydroperoxides: solvent and solvent isotope effects, *J. Am. Chem. Soc., 112*: 178 (1991).
114. T.C. Bruice, Reactions of hydroperoxides with metalloporphyrins in aqueous solution, *Acc. Chem. Res., 24*: 243 (1991).
115. J.R. Lindsay Smith and R.J. Lower, The mechanisms of the reaction between t-butyl hydroperoxide and 5,10,15,20-tetra(N-methyl-4-pyridyl)porphyrinatoiron(III) pentachloride in aqueous solution, *J. Chem. Soc. Perkin Trans., 2*: 31 (1991).
116. A.M. d'A.R. Gonsalves, R.A.W. Johnstone, M.P. Pereira, and J. Shaw, Metal-assisted reactions. 21. Epoxidation of alkenes catalysed by manganese-porphyrins: the effects of various oxidatively-stable ligands, *J. Chem. Soc. Perkin Trans., 1*: 645 (1991).
117. P.R. Cooke and J.R. Lindsay Smith, unpublished results.
118. K. Morihara, M. Kurokawa, Y. Kamata, and T. Shimada, Enzyme-like enantioselective catalysis over chiral "molecular footprint" cavities on a silica (alumina) gel surface, *J. Chem. Soc. Chem. Commun.,* 358 (1992).
119. K.J. Shea and D.Y. Sasaki, On the control of microenvironment shape of functionalised network polymers prepared by template polymerisation, *J. Am. Chem. Soc., 111*: 3442 (1989).
120. E. Keinan, S.C. Sinha, A. Sinha-Bagchi, E. Benory, M.C. Ghozi, Z. Eshbar, and B.S. Green, Towards antibody-mediated metalloporphyrin chemistry, *Pure Appl. Chem., 62*: 2013 (1990).
121. K.S. Suslick, F.V. Acholla, and B.R. Cook, Photocatalytic oxidation of hydrocarbons by (5,10,15,20-tetraphenylporphyrinato)manganese(III) perchlorate and periodate, *J. Am. Chem. Soc., 109*: 2818 (1987).
122. A. Maldotti, C. Bartocci, R. Amadelli, E. Polo, P. Battioni, and D. Mansuy, Oxidation of alkanes by dioxygen catalysed by photoactivated iron porphyrins, *J. Chem. Soc. Chem. Commun.,* 1487 (1991).

Epilogue
Future Prospects

The use of metalloporphyrins as selective oxidation catalysts continues to attract attention and is focused on two different aspects. The first one involves the rational design of biomimetic models, of increasing sophistication and complexity, that more closely resemble the active site of heme-dependent enzymes. The second area pertains to the modification of the porphyrin ring, by the replacement of the labile C-H bonds with halogen and/or strongly attracting groups in order to produce highly active and/or stable suprabiotic catalysts. The ultimate goal of this research is the design of catalysts capable of mediating the selective oxidation of simple hydrocarbons without the need for a sacrificial reductant.

The design of biomimetic models is focused on providing insights into the intricate details of the oxidation mechanism in vivo. For example, the heterolytic or homolytic nature of O-O bond homolysis in an intermediate peroxoheme complex in cytochrome P450–mediated oxidations, and the role that the axial ligand plays in this process, are important questions. Recently, Hirobe and coworkers [1] synthesized the model iron(III) porphyrin (I) containing a built-in axial thiolate ligand.

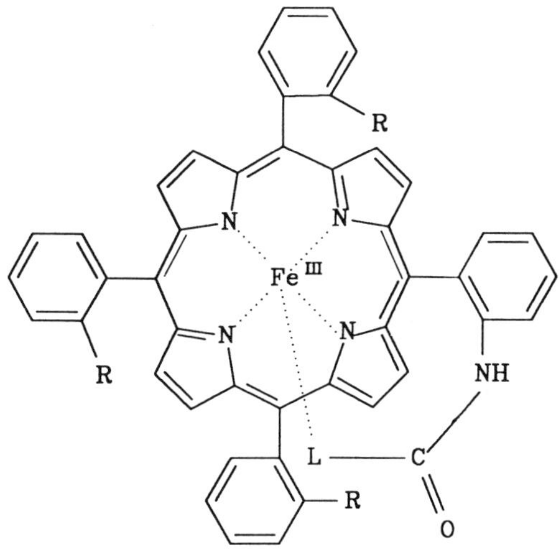

I

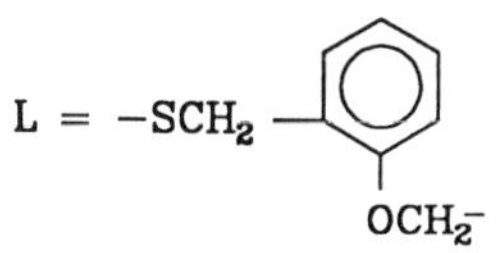

$R = NHCOCMe_3$

The oxidation of 2,4,6-tri-*tert*-butylphenol (TBPH) with peroxyphenylacetic acid was used as a mechanistic probe to distinguish between O-O bond heterolysis and homolysis. The former affords only phenylacetic acid, whereas the latter gives mainly toluene and benzyl alcohol via rapid decarboxylation of phenylacetoxy radicals to benzyl radicals (see Figure 1).

The oxidation of TBPH with peroxyphenylacetic acid catalyzed by (I) and Fe(TPP)Cl were compared in benzene as solvent, which was presumed to mimic the hydrophobic environment of the active site of cytochrome P450. Complex (I) afforded exclusively phenylacetic acid, while Fe(TPP)Cl gave mainly toluene and benzyl alcohol. These results are consistent with the axial thiolate ligand promoting heterolytic O-O bond cleavage, even in a highly hydrophobic medium, without the need for an acid or a base. Furthermore, it was shown that complex (I) is a very effective catalyst for alkane hydroxylation with peroxy carboxylic acids. Adamantane, for example, afforded a 10:1 mixture of 2-adamantol and 1-adamantol in an 88% yield. The enhancement of the formation of the putative high-valent oxoiron intermediate via heterolytic O-O bond cleavage provides a rational explanation for the role of axial thiolate ligation in cytochrome P450 chemistry.

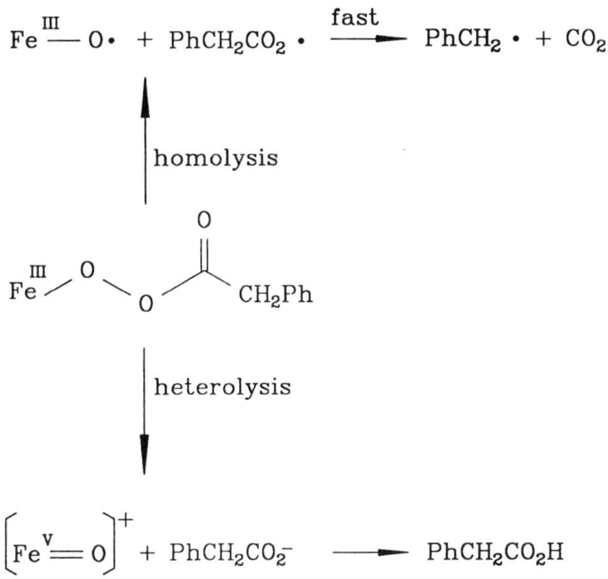

Figure 1 Heterolytic versus homolytic O-O bond homolysis in a peroxoheme.

Collman and coworkers [2] have recently reported a powerful general method for synthesizing metalloporphyrins with superstructures that model the function of the protein backbones of various hemoproteins. Michael acceptors oriented in fixed geometries around the porphyrin plane are first introduced by preparing acrylamides from atropisomers of tetra(*o*-aminophenyl)porphyrin. Macrocyclic superstructures are subsequently introduced by multiple congruent Michael additions of primary or secondary amine groups to these fixed acrylamide sites. When the amine has a geometry complimentary to that of the Michael acceptor, high yields are obtained in a single step.

For example, a novel crown ether porphyrin (II) was synthesized by using 1,10-diaza-18-crown-6 as the amine component in the Michael addition. The host–guest chemistry of the crown ether structure was then employed as a novel axial ligand delivery system. When 2-aminoethane thiol is used as the guest molecule, for example, the high binding affinity of the 1,10-diaza-18-crown-6 structure for primary alkylammonium ions promotes complexation of the zwitterion, $H_3\overset{+}{N}CH_2CH_2S^-$, which then coordinates via the thiolate group to the ironporphyrin center. This fascinating concept should also be applicable to the modeling of other hemoproteins, e.g., phenolate ligands to mimic catalase and imidazole to mimic peroxidase and hemoglobin. In this context we also note that Bruice and coworkers [3] have designed phenol pendant-capped porphyrins as biomimetic models of catalase.

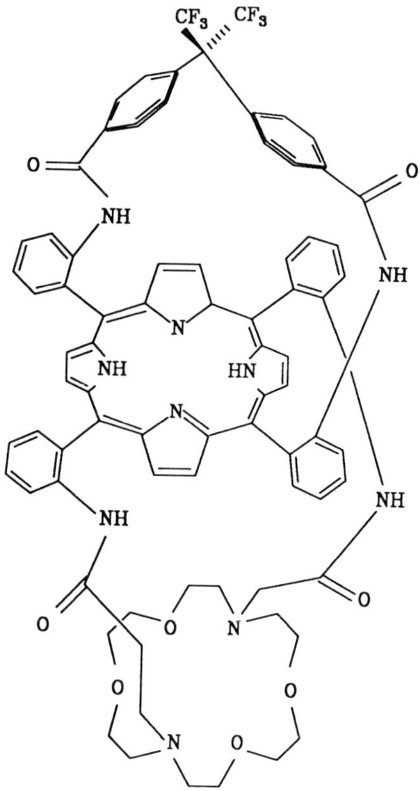

II

Some heme-dependent enzymes contain a second redox metal in the active site, e.g., cytochrome c oxidase contains copper ion in addition to the heme. Collman and coworkers [2] employed the congruent Michael addition of 1,4,7-triazacyclononane to produce the porphyrin (III) capped with a tetradentate nitrogen macrocycle, which could be a useful model for the binuclear active site of cytochrome c oxidase.

In a similar vein, Maruyama and coworkers [4] have described the synthesis of salen-capped porphyrins, such as (IV), by condensation of bis-(salicylaldehyde)-linked porphyrins with ethylene diamine. These complexes can also be used to mimic the active site of cytochrome c oxidase by simultaneous complexation of two metal ions.

The design of effective chiral metalloporphyrins for use as enantioselective oxidation catalysts remains an important and elusive goal. As Collman noted [5] in a recent review of metalloporphyrin-catalyzed enantioselective epoxidations: elaborate and expensive syntheses coupled with low turnover

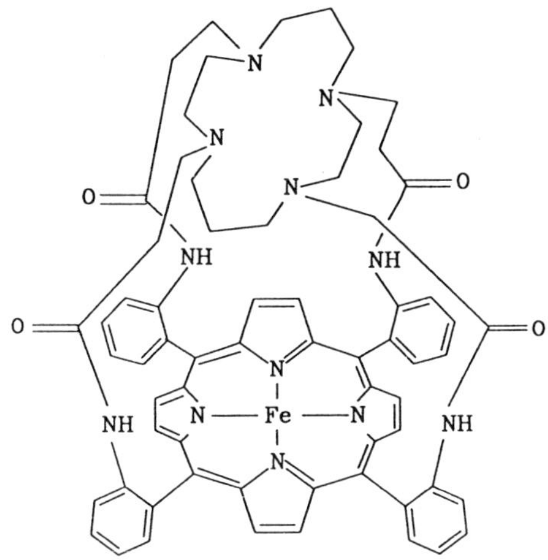

III

numbers, resulting from low oxidative stability of the porphyrin ligand, severely limited their practical utility. Moreover, the development of systems that can effectively utilize "clean" oxidants, such as H_2O_2, RO_2H, and O_2 remains an important challenge.

In the area of suprabiotic catalysis, the search for more active, more stable metalloporphyrins capable of selective hydrocarbon oxidation continues. In this context, Lyons has recently shown [6] that the oxidation of light alkanes mediated by highly electrophilic, perhalogenated metallopor-

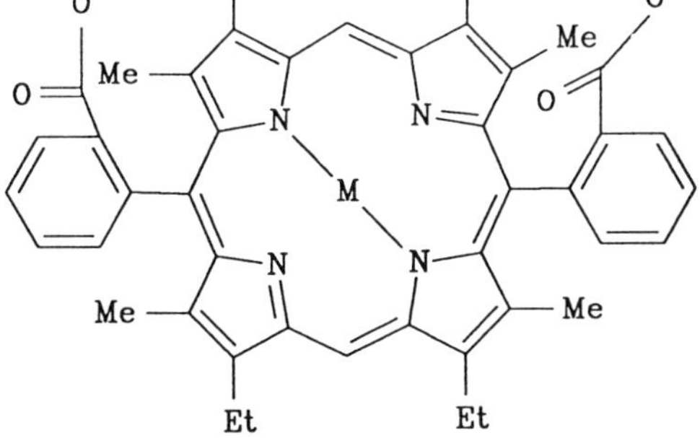

IV

phyrins proceeds via a classical autoxidation mechanism. Apparently these "suprabiotic" metalloporphyrins are highly efficient catalysts for both the formation and the decomposition of an intermediate hydroperoxide. This also explains why these systems are singularly ineffective for olefin epoxidation.

In conclusion, the quest for a better understanding of in vivo oxidative transformations and for effective, economically viable systems for in vitro oxidations continues unabated.

REFERENCES

1. T. Higuchi, K. Shimada, N. Maruyama, and M. Hirobe, Heterolytic oxygen–oxygen bond cleavage of peroxy acid and effective alkane hydroxylation in hydrophobic solvent mediated by an iron porphyrin coordinated by thiolate anion as a model for cytochrome P450, *J. Am. Chem. Soc., 115*: 7551–7552 (1993).
2. J. P. Collman, X. Zhang, P. C. Herrmann, E. S. Uffelman, B. Boitrel, A. Straumanis, and J. I. Brauman, Congruent multiple Michael addition for the synthesis of biomimetic heme analogues, *J. Am. Chem. Soc.,* in press.
3. A. Blasko, B. Garcia, and T. C. Bruice, Pendant-capped porphyrins. 2. Structural analysis and dynamics of the biphenyl pendant-capped porphyrin model of catalase and its Fe(III) complex by one and two dimensional ^{1}H NMR spectroscopy and distance geometry/molecular modeling refinement, *J. Org. Chem., 58*: 5738–5747 (1993).
4. K. Maruyama, F. Kobayashi, and A. Osuka, Salen-capped porphyrins as an active site model of metalloenzymes: synthesis and their intramolecular interactions between the metal complexes, *Bull. Chem. Soc. Japan, 63*: 2672–2681 (1990).
5. J. P. Collman, X. Zhang, V. J. Lee, E. S. Uffelman, and J. I. Brauman, Regioselective and enantioselective epoxidation catalyzed by metaloporphyrins, *Science, 261*: 1404–1411 (1993).
6. J. E. Lyons, Selective catalytic oxidation of isobutane to tert-butyl alcohol, lecture presented at the Catalytica ACT 1993 seminar in San Francisco, October, 1993.

Glossary

Achiral A molecule is achiral when it is superimposable on its mirror image (i.e., when it exhibits reflectional symmetry).

Asymmetric Lacking all elements of symmetry.

Asymmetric induction (a) Preferential formation of one enantiomer from a prochiral substrate induced by, for example, a chiral catalyst. (b) Preferential formation of one diastereomer by the creation of a new stereogenic center in a chiral molecule.

Asymmetric synthesis There are two types of asymmetric synthesis: (a) **Enantioselective syntheses** involve the preferential formation of one enantiomer from a prochiral substrate; (b) **Diastereoselective syntheses** involve the selective formation of one diastereomer by creating a new stereogenic center in a chiral molecule. (See also **Asymmetric induction**.)

Autoxidation Oxidation of an organic compound with dioxygen (molecular oxygen) usually via a free radical chain process.

Biocatalyst An enzyme in cell-free or whole-cell form that catalyzes metabolic reactions in living organisms and/or substrate conversions in various chemical reactions.

Biomimetic catalysis Catalysis in vitro by simple model systems, e.g., metalloporphyrins, that mimic transformations mediated by enzymes in vivo. (See also **Suprabiotic catalyst.**)

Chiral An object or molecule is chiral when it is nonsuperimposable on its mirror image, that is, when it lacks reflectional symmetry. Chiral is not synonymous with *optically active* or *enantiomerically pure.*

Coenzymes See **Cofactors.**

Cofactors Low-molecular-weight organic compounds or metal ions that are required by an enzyme for its activity. They may act as cosubstrates (e.g., NADH, FADH, ascorbic acid) or as a cocatalyst (e.g., biotin, pyridoxal phosphate); in the latter case, they are sometimes referred to as *coenzymes*. The coenzyme may be loosely or tightly bound to the enzyme. When it is tightly bound it is generally referred to as a *prosthetic group*. A prosthetic group, e.g., a metalloporphyrin, binds with its associated protein (*apoenzyme*), which is functionally inactive, to form the active enzyme (*holoenzyme*).

Conversion Amount of reactant consumed in a chemical process, expressed as a percentage relative to the original charge.

Cooxidation Simultaneous oxidation of two or more organic substrates, often resulting in a synergistic effect, compared with oxidation of the separate substrates.

Cytochrome A hemoprotein, the characteristic mode of action of which is electron and/or proton transport by virtue of a reversible valency change of the iron. Cytochrome P450 is not, strictly speaking, a cytochrome as it catalyzes oxidation reactions.

Cytochrome P450 An extensive group of isozymes, containing iron protoporphyrin IX as the prosthetic group, that catalyze the oxidation of a wide variety of substrates with dioxygen.

Diastereomers Stereoisomers that are not mirror images of each other (i.e., that are not enantiomers).

Diastereoselective synthesis Preferential formation of one diastereomer as a result of the creation of a new stereogenic center in a chiral molecule.

Distal ligand The sixth (axial) coordination site in the iron porphyrin complex of a hemoprotein, on the opposite side of the porphyrin ring to the *proximal site*, is occupied by the distal ligand.

Enantiomeric excess (ee) Mole fraction denoting the ratio of enantiomers in a mixture:

$$\%\ \text{ee} = \frac{[R] - [S]}{[R] + [S]} \times 100$$

where [R] and [S] are the concentrations of the (R)- and (S)-enantiomers.

Enantiomers Stereoisomers that are not superimposable on their mirror image.

Enantioselective synthesis Preferential formation of one enantiomer from a prochiral substrate (i.e., an asymmetric synthesis).

Enzyme A protein that functions as a catalyst for one or more reactions.

Gene An ordered sequence of nucleotide bases that encodes for a particular protein.

Heme An iron-porphyrin complex (in most instances iron protoporphyrin IX) which in its ferrous form is called protoporphyrin IX iron(II), protoheme, or simply heme. Hemes are the prosthetic groups of *hemoproteins*. Sometimes the term is applied more loosely to include other metalloporphyrins, e.g., manganese.

Hemin The iron(III) complex of protoporphyrin IX is called protohemin or simply hemin. When chloride is the axial ligand, for example, the complex is called chlorohemin or chloroproporphyrin IX iron(III).

Hemoprotein A protein containing an iron porphyrin complex as the prosthetic group.

Isozymes (isoenzymes) A group of structurally related enzymes that catalyze the same (type of) reaction but possess different amino acid sequences, e.g., the extensive group of cytochrome P450–dependent enzymes.

Metabolism The entire physical and chemical processes involved in the maintenance and reproduction of life in which nutrients are broken down to generate energy and simpler molecules (catabolism), which may be subsequently converted to more complex molecules (anabolism).

Metabolite Any intermediate or product resulting from metabolism. *Primary metabolites* are the primary building blocks (e.g., amino acids) formed from metabolism of carbohydrates and nutrients; *secondary metabolites* (e.g., vitamins and hormones) are formed via further biosynthetic reactions of the primary metabolites.

Microorganisms (microbes) Microscopic living entities some of which, e.g., bacteria, molds and yeasts, are important sources of industrial enzymes.

Metalloporphyrins Metal complexes of porphyrins. *First-generation metalloporphyrins* refer to complexes of synthetic porphyrins containing simple alkyl or aryl substituents, e.g., *meso*-tetraphenyl porphyrin(TPP). *Second-generation metalloporphyrins* refer to those containing halogen groups in the *meso* aryl substituents. *Third-generation metalloporphyrins* refer to the robust complexes containing perhalogenated aryl groups at the *meso* positions and halogen or other electron withdrawing groups at the β (pyrrolic) positions.

Oxidoreductase Enzymes that catalyze oxidations and reductions, classified into several groups according to their respective electron or oxygen donors and acceptors.

Oxygen activation Enhancement of the reactivity of dioxygen in autoxidation, usually by coordination to a metal center.

Oxygenases Enzymes that catalyze oxidations of organic substrates with dioxygen, whereby one (monooxygenases) or both (dioxygenases) oxygen atoms of the dioxygen molecule are incorporated in the substrate molecule. In monooxygenase-catalyzed oxidations the second oxygen atom is reduced to water.

Oxygen atom donor An oxidant that can transfer one oxygen atom to a substrate, e.g., H_2O_2, RO_2H, NaOCl, R_3NO, usually in the presence of a metal catalyst. Also referred to as a *single oxygen donor.*

Oxygen transfer An oxidation reaction in which an oxygen atom is formally transferred from the oxidant (an *oxygen atom donor*) to the substrate, usually mediated by a metal catalyst.

Peroxidases Enzymes that catalyze oxidations of organic substrates with hydrogen peroxide (or alkyl hydroperoxides).

Peroxide shunt The pathway whereby the need for a coreductant (cofactor) in an oxygenase-mediated oxidation is circumvented by using an oxygen atom donor such as H_2O_2, RO_2H, or NaOCl.

Prochiral A molecule is said to be prochiral when addition to an unsaturated group, e.g., a double bond or divalent sulfur, or replacement of one of two equivalent groups at a particular atom leads to the creation of a new stereogenic center in the molecule, e.g., a prochiral olefin.

Proximal ligand The proximal ligand in a hemoprotein is defined as the axial ligand in the iron porphyrin complex that is provided by an amino acid residue in the polypeptide chain, e.g., in hemoglobin it is an imidazole group of a histidine residue and is referred to as the proximal histidine. *Heme thiolate* enzymes (e.g., cytochrome P450 dependent) contain cysteine as the proximal ligand.

Porphyrin A tetrapyrrolic macrocycle consisting of a 20-carbon skeleton, the four pyrrole rings being linked by single-carbon atom bridges. The macrocycle contains a total of 22-π electrons, 18 of which are in direct conjugation, conforming to Huckel's $(4n + 2)$ rule for aromaticity. The pyrrolic positions are referred to as *beta*, and the four bridging positions as *meso*. In abbreviations for porphyrin structures, the *meso* substituents are generally given preference, e.g., *meso*-tetrakis(pentafluorophenyl)-β-octachloro porphyrin is denoted as F_{20}TPP-β-Cl_8.

Porphin Refers to the unsubstituted porphyrin ring.

R or **S** Absolute configuration of a molecule about a particular stereogenic (asymmetric) center. Assigned by application of the Cahn-Ingold-Prelog sequence rules. The R or S refers to whether a circle passes through certain atoms surrounding the stereogenic center, clockwise to the right (R, *rectus*) or counterclockwise to the left (S, *sinister*).

Racemate or **racemic mixture** A mixture composed of equal amounts of a pair of enantiomers; also denoted as dl, $(\pm)$, or R/S.

Resting state The resting state of an enzyme refers to the (oxidation) state in which the enzyme normally exists in vivo prior to reaction, e.g., the iron(III) state in cytochrome P450–dependent monooxygenases.

Selectivity The amount of a particular product formed (yield) divided by the amount of substrate consumed (conversion), expressed as a percentage. *Chemoselectivity* refers to competition between different functional groups. *Regioselectivity* refers to the formation of regioisomers via competing modes of attack on a functional group, e.g., an olefinic double bond. *Stereoselectivity* pertains to the selective formation of one *stereoisomer.*

Stereoisomers Isomers that differ *only* in the way that the atoms are oriented in three-dimensional space.

Substrate selectivity (specificity) Refers to the selectivity that an enzyme, for example, exhibits with regard to different substrates, e.g., S-specific.

Suprabiotic catalyst A simple abiotic catalyst that exhibits capabilities that surpass those of *biomimetic catalysts*, e.g., in mimicking reactions of oxygenases without the need for a coreductant.

Terminal oxidant In a catalytic process, the oxidizing agent that is consumed and expressed in the overall stoichiometry.

Turnover (number) The amount (in moles) of substrate converted per gram (or sometimes per mole) of catalyst. The *turnover frequency* refers to the amount of substrate converted per gram (or mole) of catalyst per hour.

Index